# Lecture Notes in Physics

Springer
*Berlin*
*Heidelberg*
*New York*
*Barcelona*
*Hong Kong*
*London*
*Milan*
*Paris*
*Singapore*
*Tokyo*

**Physics and Astronomy**  ONLINE LIBRARY

http://www.springer.de/phys/

## Editorial Policy

The series *Lecture Notes in Physics* (LNP), founded in 1969, reports new developments in physics research and teaching -- quickly, informally but with a high quality. Manuscripts to be considered for publication are topical volumes consisting of a limited number of contributions, carefully edited and closely related to each other. Each contribution should contain at least partly original and previously unpublished material, be written in a clear, pedagogical style and aimed at a broader readership, especially graduate students and nonspecialist researchers wishing to familiarize themselves with the topic concerned. For this reason, traditional proceedings cannot be considered for this series though volumes to appear in this series are often based on material presented at conferences, workshops and schools (in exceptional cases the original papers and/or those not included in the printed book may be added on an accompanying CD ROM, together with the abstracts of posters and other material suitable for publication, e.g. large tables, colour pictures, program codes, etc.).

## Acceptance

A project can only be accepted tentatively for publication, by both the editorial board and the publisher, following thorough examination of the material submitted. The book proposal sent to the publisher should consist at least of a preliminary table of contents outlining the structure of the book together with abstracts of all contributions to be included.
Final acceptance is issued by the series editor in charge, in consultation with the publisher, only after receiving the complete manuscript. Final acceptance, possibly requiring minor corrections, usually follows the tentative acceptance unless the final manuscript differs significantly from expectations (project outline). In particular, the series editors are entitled to reject individual contributions if they do not meet the high quality standards of this series. The final manuscript must be camera-ready, and should include both an informative introduction and a sufficiently detailed subject index.

## Contractual Aspects

Publication in LNP is free of charge. There is no formal contract, no royalties are paid, and no bulk orders are required, although special discounts are offered in this case. The volume editors receive jointly 30 free copies for their personal use and are entitled, as are the contributing authors, to purchase Springer books at a reduced rate. The publisher secures the copyright for each volume. As a rule, no reprints of individual contributions can be supplied.

## Manuscript Submission

The manuscript in its final and approved version must be submitted in camera-ready form. The corresponding electronic source files are also required for the production process, in particular the online version. Technical assistance in compiling the final manuscript can be provided by the publisher's production editor(s), especially with regard to the publisher's own Latex macro package which has been specially designed for this series.

## Online Version/ LNP Homepage

LNP homepage (list of available titles, aims and scope, editorial contacts etc.):
http://www.springer.de/phys/books/lnpp/
LNP online (abstracts, full-texts, subscriptions etc.):
http://link.springer.de/series/lnpp/

Michael Ziese   Martin J. Thornton   (Eds.)

# Spin Electronics

 Springer

**Editors**

Michael Ziese
Dept. of Superconductivity and Magnetism
University of Leipzig
Linnestrasse 5
04103 Leipzig, Germany

Martin J. Thornton
Clarendon Laboratory
Oxford University
Parks Road
Oxford 3PU OX1, UK

---

*Cover picture*: Schematic illustration of the passage of an electron through a spin field. The field was calculated using the OOMMF micromagnetic solver developed by Mike Donahue and Don Porter.

---

Library of Congress Cataloging-in-Publication Data applied for.

Die Deutsche Bibliothek - CIP-Einheitsaufnahme

Spin electronics / Michael Ziese ; Martin J. Thornton (ed.). - Berlin ;
Heidelberg ; New York ; Barcelona ; Hong Kong ; London ; Milan ;
Paris ; Singapore ; Tokyo : Springer, 2001
  (Lecture notes in physics ; 569)
  (Physics and astronomy online library)
  ISBN 3-540-41804-0

ISSN 0075-8450
ISBN 3-540-41804-0 Springer-Verlag Berlin Heidelberg New York

Springer-Verlag Berlin Heidelberg New York
a member of BertelsmannSpringer Science+Business Media GmbH

http://www.springer.de

© Springer-Verlag Berlin Heidelberg 2001
Printed in Germany

Typesetting: Camera-ready by the authors/editor
Cover design: *design & production*, Heidelberg

Printed on acid-free paper
SPIN: 10783210      57/3141/du - 5 4 3 2 1 0

OXide Spin Electronics Network

# Foreword

Electrons are tiny magnets as well as elementary charged particles, yet for 50 years conventional electronics has ignored the spin on the electron. Distinguishing the spin up ($\uparrow$) and spin down ($\downarrow$) currents, and manipulating the spin as well as the charge information adds a new dimension to the practice of electronics. The first applications are now beginning to make an impact.

'Spin Electronics' provides an introduction for newcomers to the field, whether condensed-matter physicists whose notions of electronics may be hazy, or electronic engineers who know little of magnetism. The book is based on lectures delivered at a Summer School organized by John Gregg in Magdalen College, Oxford during September 1998 in the framework of the Oxide Spin Electronics Network. OXSEN was a group of eight laboratories that formed a network for training and mobility of researchers, funded by the European Commission from 1996–2000. Two of the OXSEN researchers have edited the contributions, many of which have been substantially updated. The material has been divided into three sections. First is an overview of the whole field with an introduction to electronic devices and the way in which magnetism can impact them. The second section presents the basic concepts including transport, theory, metallic and oxide magnetism, semiconductors, giant and colossal magnetoresistance, spin tunnelling, micromagnetism and noise. The third section focuses on materials, methods and spin electronic circuits and devices.

As we have not yet found a magnetic semiconductor that is usable at room temperature, and attempts at spin injection into conventional semiconductors are problematic, the first wave of spin electronics has been concerned with simple magnetoresistors and switches. Exploration of the magnetoresistance in magnetic multilayers, spin valves and tunnel junctions has culminated in the development of ultra-high density magnetic recording and magnetic random-access memory. The direction leading to a new wave of active spin electronic devices and eventually to single-spin devices is signposted. It is expected that the route will lead to closer integration of magnetic with conventional semiconductor technology.

This book is intended to encourage more scientists to make their way into an exciting new field. We are grateful to the European Commission for their support.

Dublin                                                                 *J. M. D. Coey*
December 2000

# Preface

Recent years have seen a rapid development of spin electronics (also called magnetoelectronics or spintronics). This new field of research combines two traditional branches of physics: magnetism and electronics. The aim is to find ways to manipulate the electron spin in transport processes. The approach to spin electronics is very broad and includes the investigation of spin dependent processes in various systems ranging from metallic multilayers via oxide magnets to semiconductors and tunneling junctions. Precise control of the electron spin adds a new degree of freedom for the engineering of electronic devices. In the last decade we have seen the introduction of spin electronic ideas into the first generation of practical devices.

Among the major achievements of spin electronics is the understanding of spin dependent transport processes in various physical systems. These include metallic multilayers showing giant magnetoresistance, ferromagnetic tunnel junctions exhibiting tunneling magnetoresistance and certain ferromagnetic oxides showing colossal magnetoresistance near metal-insulator transitions. The transport of spin polarized currents in semiconductors is barely understood to date, but interesting first results have been achieved.

Evidently there is a need for spin electronic researchers coming from various areas of physics to efficiently grasp the basic concepts of the other fields involved as well as to communicate their results. Moreover, since the understanding of transport processes in magnetic systems has evolved so rapidly in recent years, there exists a large gulf between advanced textbooks and the research front. The aim of this book, the first of its kind on spin electronics, is to bridge this gap as it attempts to describe all the topical themes essential for new researchers entering the field. Most of the chapters include exercises and solutions to help the student become familiar with the material.

The book consists of eighteen chapters, written with a uniform notation, with cross references in each chapter to related subjects in other chapters. The choice of material is intended to provide the basic concepts of the various fields of physics involved in spin electronic research as well as to cover recent developments in spintronics. The book is divided into three parts: the first part consists of a general survey of the field, in the second part basic concepts from magnetism, transport theory and semiconductor physics are introduced and the third part is devoted to an overview of materials, thin film characterization techniques and spin electronic devices.

In the first chapter, J. F. Gregg reviews the field of spin electronics detailing landmark, recent and current experimental devices. Two and three terminal, hybrid and current novel devices are reviewed. Spin injection in semiconductors is discussed and the current difficulties in spin electronics are outlined. He concludes by suggesting possible applications for future spin electronics devices.

The second chapter, written by G. A. Gehring, presents a basic account of magnetic phenomena and transport processes in metals. Starting from Fermi liquid theory, band magnetism and the Stoner theory are discussed. The chapter ends with a brief description of strong coupling theories, including the formation of localized moments, the Kondo–effect and heavy fermion compounds.

Chapter 3 by B. J. Hickey, G. J. Morgan and M. A. Howson is devoted to both theoretical methods for the calculation of metallic conductivity and experimental techniques for resistivity measurements. In the first part the Boltzmann equation and the Kubo–Greenwood formula are introduced, the Fuchs–Sondheimer model for thin metallic films is discussed and a brief account of Lorentz magnetoresistance and quantum interference effects is given. In the second part experimental techniques for the measurement of resistivity, Hall resistivity and thermopower are explained.

In Chap. 4, G. Mathon gives an introduction to the phenomenon of giant magnetoresistance in metallic multilayers such as Fe–Cr or Co–Cu. This chapter presents experimental data on giant magnetoresistance, as well as an analysis of spin dependent scattering processes and quantitative network resistor theory.

Chapter 5 by D. Khomskii gives an overview on the electronic structure, exchange mechanisms and magnetic states in oxides. This theoretical account starts with general properties of transition metal ions in crystals and discusses orbital effects, Jahn–Teller effect, exchange interactions in and classification of insulators, Goodenough–Kanamori–Anderson rules and the double exchange mechanism.

In Chap. 6, M. Viret reviews recent results on the properties of mixed-valence manganites showing colossal magnetoresistance. Starting from an analysis of the electronic structure within an ionic and a band model, the resistivity, magneto-transport and magnetic properties of this class of oxides is characterized from an experimental as well as theoretical viewpoint. Some applications relevant for spin electronics are discussed.

The physics of spin dependent tunneling processes is briefly reviewed by F. Guinea, M. J. Calderón and L. Brey in Chap. 7. Simple models for the tunneling current and the influence of ferromagnetic electrodes are discussed. The significance of magnetic impurities, magnetic excitations and charging effects is studied.

Chapters 8 and 9 cover traditional semiconductor physics. There H. Jenniches gives an overview of the basic properties of semiconductors including band structure, carrier concentration, mobility and p-n junctions and D. I. Pugh briefly reviews metal-semiconductor contacts.

Chapter 10 by R. Skomski is on micromagnetic properties. The intrinsic properties of ferromagnetic materials are defined and extrinsic properties such as domain formation, domain walls, hysteresis and coercivity are discussed within simple models. The chapter concludes with a new model of grain-boundary magnetism that is especially relevant for spin dependent scattering processes in polycrystalline materials.

The second part of the book ends with Chap. 11 which is an extensive review by B. Raquet on electronic noise in magnetic materials and devices. The author starts with a theoretical treatment of the problem and discusses the different types of noise such as thermal, shot, 1/f and telegraph noise. In the second part of this chapter noise measurements on a variety of magnetic and spin electronic systems are reviewed.

The third part of the book starts with Chap. 12 by J. M. D. Coey presenting an overview on materials for spin electronics. Here the properties of Fe-, Ni- and Co- based alloys are reviewed and an account of antiferromagnets, half-metals and ferromagnetic semiconductors is given, as well as an introduction to thin film device structures.

Chapter 13 by E. Steinbeiss covers thin film deposition methods such as thermal evaporation, MBE, pulsed laser deposition and various forms of sputtering. Thin film growth mechanisms are also discussed.

In Chap. 14 A. K. Petford–Long gives an overview on magnetic imaging methods taking into account Bitter patterns, Lorentz microscopy, scanning force microscopy and polarized light microscopy.

In Chapter 15 by K. Ounadjela, I. L. Prejbeanu, L. D. Buda, U. Ebels and M. Hehn the magnetic force microscopy thechnique is introduced and recent results on the magnetic states of nanosized dots, rings and wires are reviewed; furthermore, domain wall scattering in Co wires is discussed.

In Chap. 16 micro– and nanofabrication techniques are described by C. Fermon. This chapter contains a description of basic patterning processes, deposition techniques, lithography processes and etching techniques. It concludes by presenting a couple of novel spin electronic devices.

Chapter 17, written by M. Ziese, is a review of recent developments in the field of spin dependent transport processes in semiconductors. Starting from the basics of spin polarized transport in semiconductors, the author reviews recent experimental results on spin coherent electron transport in semiconductors as well as spin injection and spin detection. The chapter concludes with a brief account of spin electronic semiconductor devices.

In Chap. 18 J. F. Gregg and M. J. Thornton describe the basic principles of circuit theory. The introduction of Norton–Thevenin transforms and transfer functions in ac circuit theory gives the spin electronician a powerful tool for the analysis and design of electronic devices. In the chapter small signal analysis, equivalent circuits, load lines, the Miller effect, Nyquist amplifier stability theory, noise and dc motors are discussed.

Finally, Chap. 19 by P. P. Freitas provides an overview of recent developments of spin valve and spin tunnelling devices. This chapter covers topics such as magnetic data storage, design and fabrication of spin valve sensors, magnetic random access memories and general sensor applications.

We wish to thank all the authors for their cooperation. It is our hope that the book can serve as a textbook for graduate students, for lecturers at universities for preparing course material, for professionals in the electronics industry who need to obtain information on physical concepts and for researchers joining the new field of spin electronics.

Oxford                                                              *Martin Thornton*
Leipzig                                                              *Michael Ziese*
December 2000

# Contents

## Part III    Materials, Techniques and Devices

# List of Contributors

**Luis Brey**
Instituto de Ciencia de Materiales
Consejo Superior de Investigaciones
Científicas, Cantoblanco,
E-28049, Madrid, Spain
brey@naomi.icmm.csic.es

**Liliana D. Buda**
Institut de Physique et Chimie
des Matériaux de Strasbourg,
Groupe d'Etude des
Matériaux Metalliques,
23, rue du Loess,
67037 Strasbourg Cedex, France
Liliana.Buda@ipcms.u-strasbg.fr

**Maria Jose Calderón**
Instituto de Ciencia de Materiales
Consejo Superior de Investigaciones
Científicas, Cantoblanco,
E-28049, Madrid, Spain
maryjoe@eclipse.icmm.csic.es

**J. M. D. Coey**
Physics Department,
Trinity College,
Dublin 2, Ireland
jcoey@tcd.ie

**Ursula Ebels**
Institut de Physique et Chimie
des Matériaux de Strasbourg,
Groupe d'Etude des
Matériaux Metalliques,
23, rue du Loess,
67037 Strasbourg Cedex, France
Ursula.Ebels@ipcms.u-strasbg.fr

**Claude Fermon**
Service de Physique
de l'État Condensé,
CEA-Saclay,
91191 Gif/Yvette, France
fermon@spec.saclay.cea.fr

**Paulo P. Freitas**
Instituto de Engenharia
de Sistemas e Computadores,
R.Alves Redol, 9, Lisbon,
Portugal
and
Instituto Superior Tecnico,
Departamento de Fisica,
Av. Rovisco Pais, P-1000 Lisbon,
Portugal
ppf@eniac.inesc.pt

**Gillian A. Gehring**
Department of Physics
and Astronomy,
University of Sheffield,
Hicks Building, Hounsfield Road,
Sheffield S3 7RH,
United Kingdom
g.gehring@sheffield.ac.uk

**John F. Gregg**
Clarendon Laboratory,
Oxford University,
Parks Road,
Oxford OX1 3PU,
United Kingdom
john.gregg@magdalen.oxford.ac.uk

**Francisco Guinea**
Instituto de Ciencia de Materiales
Consejo Superior de
Investigaciones Científicas,
Cantoblanco, E-28049, Madrid,
Spain
paco.guinea@uam.es

**M. Hehn**
Laboratoire de Physique
des Matériaux,
Université Henri Poincaré,
54506 Vandoeuvre lès Nancy,
France
Michel.Hehn@lpm.u-nancy.fr

**Bryan Hickey**
Department of Physics,
University of Leeds,
Leeds LS2 9JT,
United Kingdom
b.j.hickey@leeds.ac.uk

**Mark A. Howson**
2 Duchy Grove,
Harrogate HG2 0ND,
United Kingdom
mark.howson@physics.org

**Hartmut Jenniches**
Department of Physics,
University of Leeds,
Leeds LS2 9JT,
United Kingdom
phyhj@phys-irc.leeds.ac.uk

**Daniel Khomskii**
Laboratory of Solid State Physics,
Groningen University,
Nijenborgh 4, 9722 AG Groningen,
The Netherlands
khomskii@phys.rug.nl

**George Mathon**
Department of Mathematics,
City University,
London EC1V 0HB,
United Kingdom
j.mathon@city.ac.uk

**Gwynne James Morgan**
Department of Physics,
University of Leeds,
Leeds LS2 9JT,
United Kingdom
g.j.morgan@leeds.ac.uk

**Kamel Ounadjela**
Institut de Physique et Chimie
des Matériaux de Strasbourg,
Groupe d'Etude des
Matériaux Metalliques,
23, rue du Loess,
67037 Strasbourg Cedex,
France
Kamel.Ounadjela
@ipcms.u-strasbg.fr

**Amanda K. Petford-Long**
Department of Materials,
Oxford University,
United Kingdom
amanda.petford-long@
materials.ox.ac.uk

**I. L. Prejbeanu**
Institut de Physique et Chimie
des Matériaux de Strasbourg,
Groupe d'Etude des
Matériaux Metalliques,
23, rue du Loess,
67037 Strasbourg Cedex,
France
Prejbeanu@ipcms.u-strasbg.fr

**David I. Pugh**
Department of Physics,
University of York,
York Y01 5DD,
United Kingdom
dip101@york.ac.uk

**Bertrand Raquet**
Laboratoire de Physique
de la Matière Condensée
de Toulouse,
LPMCT-LNCMP-INSA,
France
raquet@insa-tlse.fr

**Ralph Skomski**
Department of Physics
and Astronomy
and Center for
Materials Research and Analysis,
University of Nebraska,
Lincoln NE 68588, USA
rskomski@unlserve.unl.edu

**Erwin Steinbeiss**
Institut für Physikalische
Hochtechnologie Jena e.V.,
Winzerlaer Strasse 10,
07745 Jena,
Germany
steinbeiss@ipht-jena.de

**Martin J. Thornton**
Clarendon Laboratory,
Oxford University,
Parks Road, Oxford OX1 3PU,
United Kingdom
martin.thornton
@physics.oxford.ac.uk

**Michel Viret**
Service de Physique de
l'État Condensé,
CEA-Saclay, 91191 Gif/Yvette,
France
viret@drecam.saclay.cea.fr

**Michael Ziese**
Department of Superconductivity
and Magnetism,
University of Leipzig,
Linnéstrasse 5,
04103 Leipzig,
Germany
ziese@physik.uni-leipzig.de

# Part I

# Introduction

# 1 Introduction to Spin Electronics

J. F. Gregg

Clarendon Laboratory, Oxford University, Parks Road, Oxford OX1 3PU, U.K.

## 1.1 Coey's Lemma

The driving force behind Spin Electronics is neatly summarized in J. M. D. Coey's incisive observation [1] that "Conventional Electronics has ignored the spin of the electron". In every hi-fi and radio set, 50% of the conducting electrons tend to be spin-up and the remainder are spin down (where up and down relate to some locally induced quantisation axis in the relevant wires and devices). Yet, although electron spin was known about for most of the 20th Century, no technical use is made of this fact.

## 1.2 The Two Spin Channel Model

The mechanistic basis for Spin Electronics is almost as old as the concept of electron spin itself. In the mid-thirties, Mott postulated [2] that certain electrical transport anomalies in the behaviour of metallic ferromagnets arose from the ability to consider the spin-up and spin-down conduction electrons as two independent families of charge carriers, each with its own distinct transport properties. Mott's hypothesis essentially is that spin-flip scattering is sufficiently rare on the timescale of all the other scattering processes canonical to the problem that defections from one spin channel to the other may be ignored, hence the relative independence of the two channels [3–5].

### 1.2.1 Spin Asymmetry

The other necessary ingredient of this model is that the two spin families contribute very differently to the electrical transport processes. This may be because the number densities of each carrier type are different, or it may because they have different mobilities – in other words that the same momentum or energy scattering mechanisms treat them very differently. In either case, the asymmetry which makes spin-up electrons behave differently to spin-down electrons arises because the ferromagnetic exchange field splits the spin-up and spin-down conduction bands, leaving different bandstructures evident at the Fermi surface. If the densities of electron states differs at the Fermi surface, then clearly the number of electrons participating in the conduction process is different for each spin channel. However, more subtly, different densities of states for spin-up and spin-down implies that the susceptibility to scattering of the two spin types is different, and this in turn leads to their having different mobilities.

## 1.2.2   Spin Accumulation

Let us consider two spin channels of different mobility (Fig. 1.1). When an electric field is applied to the metal, there is a shift, $\Delta \boldsymbol{k}$, in momentum space of the spin-up and spin-down Fermi surfaces in accordance with the equation:

$$\boldsymbol{F} = e\boldsymbol{E} = \hbar \frac{d\boldsymbol{k}}{dt} = \hbar \frac{\Delta \boldsymbol{k}}{\tau} \tag{1.1}$$

where $\boldsymbol{F}$ is force on carrier, $\boldsymbol{E}$ is electric field, e is electronic charge, $\tau$ is electron scattering time given by $\mu = e\tau/m^*$, $\mu$ being the electron mobility and $m^*$ the electron effective mass. Since the channels have different mobilities, this shift is different for the spin-up and spin-down Fermi surfaces as illustrated.

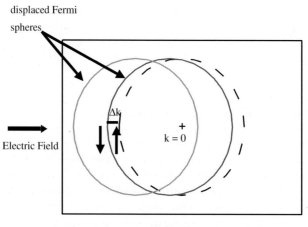

**Fig. 1.1.** The shift of the Fermi surface when an electric field is applied to a ferromagnet is shown. The solid circles represents the Fermi sphere of up and down spin electrons in a field, the dashed circle represents the Fermi sphere in zero external field.

From Fig. 1.1, it is evident that the spin-up electrons are performing the lion's share of the electrical conducting, and, moreover, that if a current is passed from such a spin-asymmetric material – for example cobalt – into a paramagnet like silver (where there is no asymmetry between spin channels [6]), there is a net influx into the silver of up-spins over down-spins. Thus a surplus of up-spins appears in the silver and with it a small associated magnetic moment per volume. This surplus is known as a "spin accumulation". Evidently, for constant current flow, the spin accumulation cannot increase indefinitely; this is because as fast as the spins are injected into the silver across the cobalt-silver interface, they are converted into down-spins by the slow spin-flip processes which we have hitherto ignored. This spin-flipping goes on throughout all parts of the silver which have been invaded by the spin accumulation. So now we have a dynamic equilibrium between influx of up-spins and their death by spin-flipping. This in turn defines a characteristic lengthscale which describes how far the spin accumulation extends into the silver.

Incidentally, to establish the concept of spin accumulation, we have assumed that both spin-up and spin-down electrons were present in the ferromagnet in equal numbers but that their mobilities are different. The same result could have been achieved by assuming a half-metallic ferromagnet in which one spin channel is entirely absent and no assumption need be made about the mobility of its spins. In other words, we can produce a spin accumulation as a direct consequence of an asymmetric density of states or as an indirect consequence via asymmetry in electron mobility.

### 1.2.3   Spin Diffusion Length

It follows from the above discussion that the spin accumulation decays exponentially away from the interface on a lengthscale called the "spin diffusion length". It is instructive to do a rough "back of the envelope" calculation to see how large is this spin diffusion length, $\lambda_{sd}$, and on what parameters it depends. The estimate proceeds as follows. Consider a newly injected up-spin arriving across the interface into the nonmagnetic material. It undergoes a number $N$ of momentum-changing collisions before being flipped (on average after time $\tau_{\uparrow\downarrow}$). The average distance between momentum scattering collisions is $\lambda$, the mean free path. We can now make two relations. By analogy with the progress of a drunken sailor leaving a bar and executing a random walk up and down the street, we can say (remembering to include a factor of 3 since, unlike the sailor, our spin can move in 3 dimensions) that the average distance which the spin penetrates into the nonmagnetic material (perpendicular to the interface) is $\lambda\sqrt{N/3}$. This distance is $\lambda_{sd}$, the spin diffusion length which we wish to estimate. Moreover, the total distance walked by the spin is $N\lambda$ which in turn equals its velocity (the Fermi velocity, $v_F$) times the spin-flip time $\tau_{\uparrow\downarrow}$. Eliminating the number $N$ of collisions gives

$$\lambda_{sd} = \sqrt{\frac{\lambda v_F \tau_{\uparrow\downarrow}}{3}} \qquad (1.2)$$

### 1.2.4   The Role of Impurities in Spin Electronics

This relation is interesting because it highlights the critical importance of impurity concentration in determining spin diffusion length. If the impurity levels are increased in the silver, not only does the spin diffusion length drop because of the shortened mean free path, it also drops because the impurities reduce the spin-flip time by introducing more spin-orbit scattering [7].

### 1.2.5   How Long is the Spin Diffusion Length?

The relation also allows us to estimate values for the size of the spin diffusion length. Again taking silver as an example, the spin diffusion length can vary between microns for very pure silver to of order 10 nm for silver with 1% gold impurity. Yang etc. [8–11] have made elegant measurements of this parameter in other materials. For a mathematically rigorous analysis of the spin-accumulation in terms of the respective electrochemical potential of the spin channels, the reader is referred to Valet and Fert [12] from which it can be seen, numerical factors apart, that the crude "drunken sailor" model gives a remarkably accurate insight into the physics of this problem.

### 1.2.6   How Large is a Typical Spin Accumulation?

It is also of interest to estimate how large is the spin accumulation for typical current densities. The calculation is done by balancing the net spin injection across the interface:

$$\frac{dn}{dt} = \frac{A\alpha j}{e} \tag{1.3}$$

with the total decay rate of spins due to spin flipping in the entire volume influenced by the spin accumulation:

$$\frac{A}{\tau_{\uparrow\downarrow}} \int_0^\infty n dx = \frac{n_0 A}{\tau_{\uparrow\downarrow}} \int_0^\infty \exp\left(\frac{-x}{\lambda_{sd}}\right) dx = \frac{A n_0 \lambda_{sd}}{\tau_{\uparrow\downarrow}} \tag{1.4}$$

$A$ is sectional area, $j$ is current density, $n$ is number density of excess spins, $x$ is distance from the interface, $\alpha$ is ferromagnet spin polarization. This in turn gives a spin accumulation just inside the interface of

$$n_0 = \frac{\alpha j \tau_{\uparrow\downarrow}}{e\lambda_{sd}} = \frac{3\alpha j \lambda_{sd}}{e v_F \lambda} \tag{1.5}$$

Putting in typical numbers of $j = 1000$ Amps/cm$^2$, $\alpha = 1$, $v_F = 10^6$ m/s, $\lambda = 5$ nm, $\lambda_{sd} = 100$ nm, gives $n_0 = 4 \times 10^{22}$ m$^{-3}$. Thus, given an electron density of $3 \times 10^{28}$ m$^{-3}$, it is seen that only one part in $10^6$ of the electrons are spin polarized. The significance of this will be discussed below. Incidentally the magnetic field $B$ associated with this spin accumulation is:

$$B = \mu_0 M = \mu_0 \mu_B n_0 \tag{1.6}$$
$$= 10^{-6} \times 10^{-24} \times 10^{22} = 10\,\text{nTesla!!} \tag{1.7}$$

This is experimentally very hard to detect, especially considering the magnetic fields caused by the current which generates the spin accumulation in the first place.

## 1.3   Two Terminal Spin Electronics

The next step in the Spin Electronic story is to make a simple device and this is realized by making a sandwich in which the "bread" is two thin film layers of ferromagnet and the "filling" is a thin film layer of paramagnetic metal (Fig. 1.2).

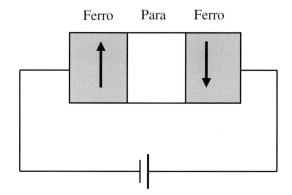

**Fig. 1.2.** Passive two terminal spin electronic device.

This is the simplest Spin Electronic device possible. It is a two-terminal passive device which in some realizations is known as a "spin valve" and it passes muster in the world of commerce as a Giant Magnetoresistive hard-disk read-head.

Empirically, the device functions as follows [13]: The electrical resistance is measured between the two terminals and an externally applied magnetic field (supplied for example by the magnetic information bit on the hard disk whose orientation it is required to read) is used to switch the relative magnetic orientations of the ferromagnetic layers from parallel to antiparallel. It is observed that the parallel magnetic moment configuration corresponds to a low electrical resistance and the antiparallel state to a high resistance. Changes in electrical resistance of order 100% are possible in quality devices, hence the term **giant magnetoresistance**, since by comparison with, for example, anisotropic magnetoresistance in ferromagnets, the observed effects are about 2 orders of magnitude bigger.

### 1.3.1   The Analogy with Polarized Light

There are a variety of different ways – of varying rigour – to consider the operation of this spin valve structure. To keep things simple, let us analyse it by analogy with the phenomenon of polarized light. In the limit in which the ferromagnets are half-metallic, the left hand magnetic element supplies a current consisting of spin-up electrons only which produce a spin accumulation in the central layer. If the physical thickness of the silver layer is comparable with or smaller than the spin diffusion length, this spin accumulation reaches across to the right hand magnetic layer which, on account of its being half-metallic, acts as a spin filter, just as a piece of Polaroid spectacle lens acts as a polarized light filter. The spin accumulation presents different densities of up and down electrons to this spin filter which thus lets through different currents depending on whether its magnetic orientation is parallel or antiparallel to the orientation of the polariser (i.e. the first magnetic layer). The only difference with the case of crossed optical polarisers is that in optics the extinction angle is 90 degrees. In the spin electronic case it is 180 degrees [14].

### 1.3.2    Spin Tunneling Processes [15–18]

If two metallic electrodes are separated by a thin layer of insulating material and a voltage applied between them, a current may pass the insulator by quantum mechanical tunneling of the current carriers. The tunnel current depends on the bias applied, but also on the energy height and physical width of the barrier. Insulators may be regarded as semiconductors in which the electronic bandgap between the full valence and the next (empty) conduction band is so large that population of the conduction band is laughably unlikely at the operating temperature. The effective quantum mechanical barrier height (for small bias) is thus the difference between the Fermi level of the metal and the bottom of the insulator conduction band.

Moreover, it is established theoretically and experimentally that the spin of the tunneling carriers is preserved in transit. An analogous structure to the spin valve described above may be made by making the two metallic electrodes of half-metallic ferromagnet (HMF) and separating them with a thin layer of insulator. Now, if the magnetizations of the electrodes are opposite, no current may flow across the junction since the electrons which might tunnel have no density of final states on the far side to receive them. However if the electrode magnetizations are parallel, tunneling current may flow as usual. We thus have a spin electronic switch whose operation again mirrors that of a pair of crossed optical polarisers and which may be switched on and off by application of external magnetic fields. If the electrodes are not ideal HMFs, then the on/off conductance ratio is finite and reflects the majority and minority density of states for the ferromagnet concerned. Spin tunnel junctions as described depend for their operation only on density of states and do not invoke carrier mobility. Moreover, unlike all-metallic systems they have lower conductances per unit area of device and hence larger signal voltages (of order millivolts or more) are realizable for practical values of operating current. Moreover, the device characteristics such as the size of the "on" resistance, current densities, operating voltages and total current may be tuned by playing with the device cross-section, the barrier height and the barrier width. As we shall see below, this is just one reason why they are very promising candidates for the spin-injector stages of future Spin Electronic devices. They are also the basis of the next generation of Tunnel MRAM, as illustrated in Figs. 1.3 & 1.4.

### 1.3.3    The Dominance of the Fermi Surface

Following the estimate above of the size of a typical spin accumulation, it might be asked how an effect which involves changes of order 100% in electrical transport could derive from a phenomenon in which only one part in a million of the spins are polarized. The answer is that it is yet another demonstration of how the properties of metallic systems are controlled exclusively by the mafia of

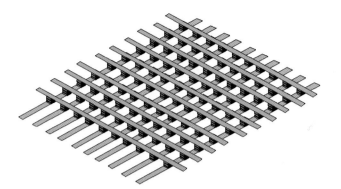

**Fig. 1.3.** A 10×10 matrix with the memory elements 0.1 microns in size. One of the project goal of the European funded framework 5 network NANOMEM (courtesy of M. Hehn, Université Henri Poincaré, Nancy, France).

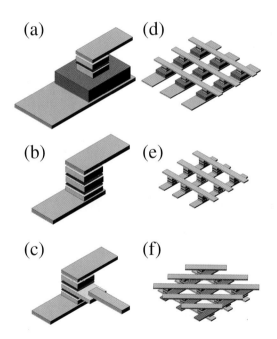

**Fig. 1.4.** Currently state of the art MRAMs use: (a) semiconductor diodes to prevent current shortcuts. Shown in (b) MIM diodes and (c) TTRAMs with selective polarisation are being developed to replace the semiconductor diodes and prevent current shortcuts. With (d), (e) & (f) the respective MRAMs in array form (courtesy of M. Hehn, Université Henri Poincaré, Nancy, France).

electrons at or very near the Fermi surface whose bandstructure properties the metal reflects. The spin polarized electrons may be few in number but they are injected at the point in the bandstructure which counts – and with devastating results. There is a useful lesson here for later design work: always make sure your spin polarization is injected at the right part of the energy bandstructure.

### 1.3.4   CIP and CPP GMR [19]

In fact there are two configurations in which our simple two terminal device can work – they are respectively described as current in plane (CIP) and current perpendicular to plane (CPP). Above, we have discussed only the latter in which the critical lengthscale for the magnetic phenomena is the spin diffusion length. The physics involved in CIP operation is rather different and the critical lengthscale here is the mean free path. However we shall leave the discussion of this case since it is not central to the theme of this chapter. The reader is referred to G. Mathon's chapter for further details.

## 1.4   Three Terminal Spin Electronics

Electronically, the natural progression is from this two terminal device to a three terminal one, and this step was made by Mark Johnson [20–22] who achieved it simply by introducing a third contact to the intermediate paramagnetic base layer to create the Johnson Transistor (Fig. 1.5). Now in the language of bipolar transistors, we can speak of a base, an emitter, and a collector, the last two being

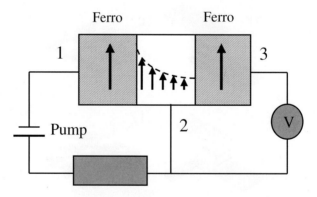

**Fig. 1.5.** Johnson transistor.

the ferromagnetic layers. Just like its bipolar counterpart, the Johnson transistor may be used in various configurations; the one we discuss here is chosen because it gives insight into yet another way to analyse spin filtering and spin accumulation. We leave the collector floating and monitor the potential at which it floats using a high impedance voltmeter. Meanwhile a current is pumped round the emitter-base circuit and this causes a spin accumulation in the base layer as before. The potential at which the collector floats now depends on whether its magnetic moment is parallel or antiparallel to the magnetization of the polarizing emitter electrode which causes the spin accumulation. Evidently this potential may be altered by using an external magnetic field to switch the relative orientation of the emitter and collector magnetic moments. To analyse this behaviour, consider again the limiting case of a half-metallic ferromagnet as the collector electrode. It floats in equilibrium with the base electrode – in other words in the steady state, no net current flows. But because it is half metallic it can only trade electrons with the base whose spin is (say) parallel to its magnetization and the "no current" condition then means that its electrochemical potential is equal to the electrochemical potential in the base layer for the same electron spin type. In other words, the collector is sampling the electrochemical potential of the appropriate spin type (spin-up) in the base. Reversing the collector magnetization means it now samples the spin-down electrochemical potential in the base. Since there is a spin accumulation in the base, these spin-up and spin-down electrochemical potentials are different (see [12]) and the collector potential is thus dependent on the orientation of its magnetic moment. Thus we have a three terminal Spin Electronic device for which the conditions at terminal 3 may be set by suitable adjustment of the conditions at terminals 1 and 2, as for a traditional electronic three terminal device. However, in addition, these conditions are also switchable by applying an external magnetic field. This encapsulates the essence of Spin Electronic device behaviour.

## 1.5   Mesomagnetism

Evidently in the above discussion, it is essential that the spin accumulation penetrates right across the thickness of the base layer in order that the collector may sample it. Likewise, in the two terminal device, it was important that the base layer thickness was small on the lengthscale of the spin diffusion length. This provides us with an interesting new way to view spin electronic devices. We can regard their behaviour as a write-read process in which an encoder writes spin information onto the itinerant electrons in one part of the device and this information is then conveyed to a physically different part of the device where it is read off by a decoder. The encoder and decoder elements are nanoscale ferromagnets and the spin information decays in transit on the lengthscale of the spin diffusion length. The message then is that for successful Spin Electronic device operation, the device **must be physically engineered on this length scale or smaller**.

This is just one particular manifestation of the general phenomenon of Mesomagnetism which concerns itself with the appearance of novel physical phenomena when magnetic systems are reduced to the nanoscale. The underlying tenet of Mesomagnetism is that magnetic processes are characterized by a variety of lengthscales and that when the physical dimensions of a magnetic system are engineered to dimensions comparable with or smaller than these characteristic lengths, new and unusual magnetic phenomena appear – such as Giant Magnetoresistance, Superparamagnetism, perpendicular recording media. These characteristic lengthscales have various origins. Many of them – domain size, domain wall width, exchange length, thin film perpendicular anisotropy threshold – are governed by a balance of energy terms. Others are the result of diffusion processes for energy, momentum, magnetization.

### 1.5.1   Giant Thermal Magnetoresistance

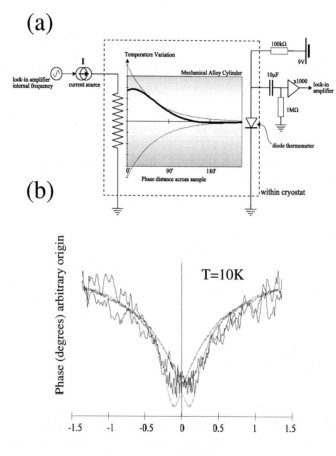

**Fig. 1.6.** Schematic set-up for measurement of the giant thermal magnetoresistance in a GMR mechanical alloy shown in (a). With the thermal GMR effect in a mechanical alloy shown in (b). For comparision the electrical GMR is also shown inverted and superimposed on the lower trace, with the axes arbitary.

As an interesting aside, the Wiedemann Franz Law (WFL) tells us that there is a close relationship between electrical transport and heat transport in most materials. Thermal and electrical conductivities are limited in most regimes by the same scattering processes and the WFL tells us that in these circumstances their quotient is a constant times absolute temperature. Moreover, this close relationship extends to magnetotransport in mesomagnetic systems. Figure 1.6 shows measurement of the Giant Thermal Magnetoresistance in a giant magnetoresistive mechanical alloy. The analysis is identical to the electrical case. Spin information is encoded onto a thermal current in one part of the device and read off again in a different part of the device: the result is a thermal resistance which varies with applied magnetic field by many percent [23].

### 1.5.2   The Domain Wall in Spin Electronics

Another example of the intrigue of Mesomagnetism may be seen by considering the geometrical similarity between a spin-valve structure and a ferromagnetic domain wall as illustrated in Fig. 1.7. In both cases, regions of differential mag-

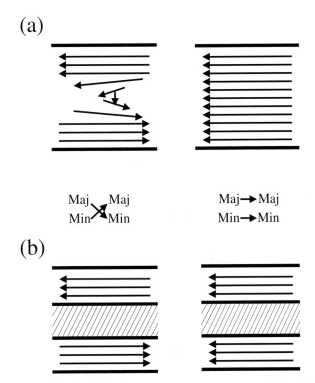

**Fig. 1.7.** Geometric similarities of (a) FM domain wall and (b) a GMR trilayer.

netization are separated by an intermediate zone which takes the form of a thin film of nonmagnetic metal and a region of twisted magnetization in the respective cases. The spin valve functions provided that spin conservation occurs across the intermediate zone. This suggests a model of domain wall resistance [24–26]

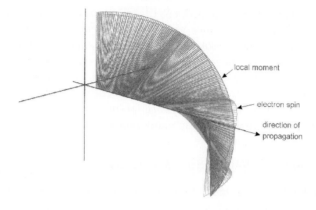

local moment

electron spin

direction of
propagation

**Fig. 1.8.** The spin trajectory is shown for the electrical carriers in transit through domain walls in Co (typically Co wall thickness ∼ 15 nm).

in which the value of the resistance is determined by the degree of spin depolarization in the twisted magnetic structure which forms the heart of the domain wall. The model invokes magnetic resonance in the ferromagnetic exchange field to determine the degree of electron spin mistracking on passing through the domain wall. This mistracking of, say, an up-spin leads to its making an average angle $\theta$ with the local magnetization direction in the domain wall and this corresponds to its wavefunction being contaminated by a fraction $\sin(\theta/2)$ of the down-spin wavefunction. It is then susceptible to additional scattering by an amount equivalent to $\langle \sin^2(\theta/2) \rangle$ multiplied by the down-spin scattering rate. This model leads to (1.8), an expression for the spin-dependent contribution to domain wall resistivity (shown in Fig. 1.8):

$$\frac{\delta\rho_w}{\rho_0} = \left( \frac{\lambda^*}{\lambda} + \frac{\lambda}{\lambda^*} - 2 \right) \langle \sin^2(\theta/2) \rangle \qquad (1.8)$$

where $\lambda$ and $\lambda^*$ are the majority and minority spin mean free paths, $\rho_0$ and $\delta\rho_w$ are respectively the bulk ferromagnetic resistivity and the resistivity increase for domain wall material.

This spin-dependent contribution differs from the contributions from the many possible mechanisms for domain wall resistance in that it predicts not a fixed value of resistance for the wall but rather a ratio increase based on the bulk value for the material. In principle therefore the validity of the model may be tested by measuring domain walls in increasingly impure samples of the same ferromagnet and observing if the ratio $\delta\rho_w/\rho_0$ stays fixed. The model has been re-analyzed [27] by replacing this simple rotating frame approach with a sophisticated quantum mechanical analysis: to within a simple numerical factor, identical results are obtained.

## 1.6    Hybrid Spin Electronics

The Johnson transistor is a useful and versatile demonstrator device but it has practical limitations. The voltage changes measured are small and it has no power gain without the addition of two extra electrodes and a transformer structure. The underlying design problem with the device is that it is entirely Ohmic in operation since all its constituent parts are metals.

Clearly another technology progression is needed and this is the introduction of Hybrid Spin Electronics – the combination of conventional semiconductors with spin-asymmetric conducting materials. At a stroke, this releases to the Spin Electronic designer all the armoury of semiconductor physics such as exploiting diffusion currents, depletion zones and the tunnel effect to create new high-performance spin-devices.

### 1.6.1    The Monsma Transistor

The first Hybrid Spin Electronic device was the Monsma transistor [28–30] produced by the university of Twente which was fabricated by sandwiching a traditional spin valve device between two layers of silicon. Three electrical contacts are made to the spin-valve base layer and to the respective silicon layers. The spin valve is more sophisticated than that illustrated in Fig. 1.9a and comprises multiple magnetic/nonmagnetic bilayers, but its operating principle is the same. Schottky barriers form at the interfaces between the silicon and the metal structure and these absorb the bias voltages applied between pairs of terminals.

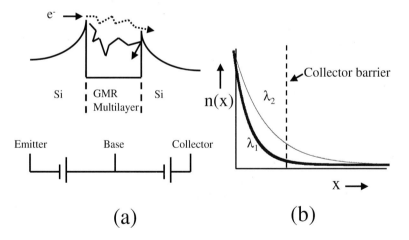

(a)                          (b)

**Fig. 1.9.** Monsma transistor: first attempt to integrate ferromagnetic metals with silicon shown in (a). In (b) the average energy of both spin types plotted as a function of distance. The thick line denotes scattering for both spin types in an antiferromagnetically aligned mutilayer (both species experiences strong scattering) and the thin line denotes the scattering when the layers are ferromagnetically aligned (only one species will experiences strong scattering).

The collector Schottky barrier is back biased and the emitter Schottky is forward biased. This has the effect of injecting (unpolarised) hot electrons from the semiconductor emitter into the metallic base high above its Fermi energy. The question now is whether the hot electrons can travel across the thickness of the base and retain enough energy to surmount the collector Schottky barrier. If not, they remain in the base and get swept out the base connection.

By varying the magnetic configuration of the base magnetic multilayer the operator can determine how much energy the hot electrons lose in their passage across the base. If the magnetic layers are antiferromagnetically aligned in the multilayer then both spin types experience heavy scattering in one or other magnetic layer orientation, so the average energy of both spin types as a function of distance into the base follows the thick line exponential decay curve ($\lambda_1$) of Fig. 1.9b. On the other hand, if the magnetic multilayer is in applied field and its layers are all aligned, one spin class gets scattered heavily in every magnetic layer, whereas the other class has a passport to travel through the structure relatively unscathed and the average energy vs distance of this privileged class follows the thin curve ($\lambda_2$). It may thus be seen that for parallel magnetic alignment, spins with higher average energy impinge on the collector barrier and the collected current is correspondingly higher. Once again we have a transistor whose electrical characteristics are magnetically tunable. This time, however, the current gain and the magnetic sensitivity are sufficiently large that, with help from some conventional electronics, this is a candidate for a practical working device.

It may be seen from comparison of the two traces of Fig. 1.9b that there is a trade-off to be made in determining the optimum base thickness. A thin base allows a large collector current harvest but affords little magnetic discrimination. A thick base on the other hand means a large factor between the collector currents corresponding to the two magnetic states of the multilayer but an abysmally small current gain. (The low current gain has always been the Achilles Heel of metal base transistors, and is probably the main reason for their fall from grace as practical devices despite their good high frequency performance owing to the absence of base charge storage.)

An interesting feature of the Monsma transistor is that the transmission selection at the collector barrier is done on the basis of energy. Thus the scattering processes in the base which determine collected current are the inelastic ones. Elastic collisions which change momentum but not energy are of less significance. This contrasts with the functioning of a spin valve type system in which all momentum changing collision processes have the same status in determining device performance [31].

### 1.6.2  Spin Transport in Semiconductors

The Monsma transistor represents a very important step in the evolution of Spin Electronics. It is the first combination of spin-selective materials with a semiconductor. However, as yet, the semiconductor is used only to generate barriers and shield the spin-dependent part of the device from electric fields. To release the full potential of Hybrid Spin Electronics we need to make devices which exploit spin-dependent transport in the semiconductor itself.

### 1.6.3   The SPICE Transistor [32,33]

The current gain of a conventional bipolar transistor is in part due to the screening action of the junctions either side of the base which absorb the bias voltages and leave the base region relatively free of electric fields. The current which diffuses across the base is primarily driven by a carrier concentration gradient and to a rather lesser extent by electric field and the randomness associated with concentration driven current flow helps to improve the current gain. The carriers injected by the emitter are forced to wander towards the base along the top of an extended cliff in voltage, at the bottom of which lies the collector. Of order, say, 99% of the carriers stumble over the cliff and are swept out the collector and the remaining 1% make it to the base connection; this gives a very satisfactory current gain $\beta = \mathrm{I}_C/\mathrm{I}_B$ of 99.

Implementing spin polarized current transport in a semiconductor enables a new concept in Spin Transistor design – the **S**pin **P**olarised **I**njection **C**urrent **E**mitter device (SPICE) in which the emitter launches a spin polarized current into the electric field screened region and a spin-selective guard-rail along the top of the cliff determines if these polarized carriers are allowed to fall into the collector or not. Thus we have a device with a respectable current gain from which power-gain may easily be derived, but whose characteristics may again be switched by manipulating the magnetic guard rail via an externally applied magnetic field. A wide variety of designs are possible which answer to this general principle. For example the emitter and collector interfaces may be realized by p-n junctions, Schottky barriers or spin tunnel junctions and the geometry of the device may be adjusted to allow a greater or lesser degree of electric field driving component to the diffusion current in the base depending on the application.

### 1.6.4   Measuring Spin Decoherence in Semiconductors

The crucial question which needs to be answered in order to realize this kind of Spin Transistor is whether spin transport is possible at all in semiconductors, and, if so, whether it is possible over the sort of physical dimensions on which a typical transistor is built. In other words, we need an estimate of the spin diffusion length in a typical semiconductor. A subsidiary question concerns the role of dopants in the semiconductor and whether they introduce spin-orbit scattering which militates against the spin transport by reducing the spin flip times.

An immediate way to address this question is to directly spin-inject into a semiconductor [34,35] and observe the polarization of the current which emerges on the other side. Figure 1.10 shows an experiment in which this was performed. Doped channels of silicon with various dopant types and concentrations and of different lengths (from 1 to 64 microns) were contacted at each end with differentially magnetisable cobalt pads of well defined magnetizing behaviour. The

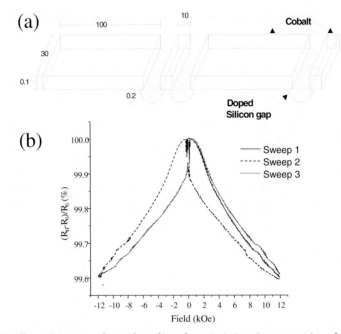

**Fig. 1.10.** Experiment performed to directly spin-inject into a semiconductor and observe the polarization of the current which emerges on the other side is shown in (a). The resulting transport measurements (b) suggests that the spin diffusion length in silicon is at least many tens of microns, but the spin injection process at the metal/silicon interface is highly inefficient.

transport results shown in Fig. 1.10b are insensitive to magnetic field direction, have even symmetry (thereby eliminating AMR and Hall effect as a possible cause) and they are compatible with the observed domain magnetization processes for the cobalt pads. They appear to correspond to spin transport through the semiconductor, and as such they correlate well with earlier experiments using nickel injectors [34]. Interestingly however the spin transport effects are of order a few percent at best, yet the effect decays only very slightly with silicon channel length and was still well observable for 64 micron channels. The message would seem to be that the spin diffusion length in silicon is many tens of microns at the least, but that the spin injection process at the metal/silicon interface is highly inefficient. This direct injection inefficiency is being widely observed and its cause is still hotly debated. It may arise from spin depolarization by surface states [36], or it may be explainable by the Valet/Fert model in which spin injection is less efficient for materials of very different conductivities [37]. It may also be because the spin injection is not being implemented at the optimum point in the semiconductor bandstructure. From the latter point of view, spin tunnel injection into semiconductors is a more versatile technique, since, for a given injected tunnel-current density, the necessary bias (and hence the point in the band-structure where injection occurs) may be tuned by varying the thickness and/or the tunnel barrier height.

A very beautiful direct measurement of semiconductor spin diffusion length has been made by decoupling from the spin injection problem [38,39] and generating the spin polarized carriers in the semiconductor itself (see Fig. 1.11). Gallium Arsenide [40,41] was used as the host which has the property that,

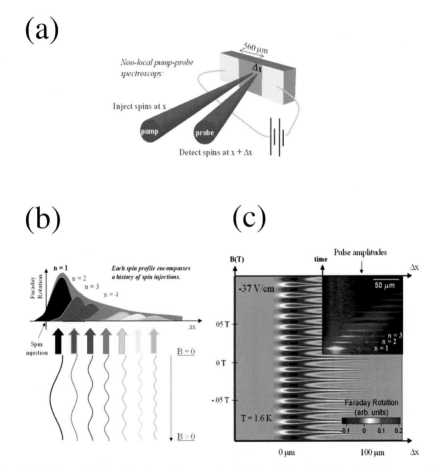

**Fig. 1.11.** Lateral drag of spin coherence in Gallium Arsenide has been measured by Faraday rotation as shown in (a). A new spin population is created every time a pump pulse hits the sample as shown in (b). The electrons in each new population then drift along the electric field. When observed at some time after injection, each population will have drifted an amount proportional to its age as well as experienced an exponential decay in its Faraday signal. A number of field scans can be taken over a range of displacements in order to identify the spatial extent of each spin population and track its movement in time, as shown in (c). Spin transport can be observed at distances exceeding 100 microns (after [39]).

when pumped with circularly polarized light, the selection rules are such as to populate the conduction band with predominantly one spin type. These spins can be made to precess by application of a small magnetic field. The resulting precessing magnetization is then detected using optical Faraday rotation using a probe beam from the same optics as provides the pump. The magnetization drifts under the application of a driving electric field and the spatial decay in precession signal gives a measure of the spin diffusion length. The results are of order many tens of microns, in accordance with the silicon measurements of the direct injection experiment discussed above. See Chap. 17 for further details.

Thus it would seem beyond doubt that the spin diffusion length in semiconductors is adequate for the design and realization of SPICE type transistor structures – provided that means are provided for efficient delivery of the initial spin polarized current.

### 1.6.5  How to Improve Direct Spin-Injection Efficiency

With this problem in mind it is interesting to examine the results of an experiment which injects spin polarized carriers from a **magnetic** semiconductor into a normal semiconductor light emitting diode structure [42–46] (see Fig. 1.12). The polarization of the injected carriers is dependent on the magnetization direction of the magnetic semiconductor which supplies them. This is reflected in the polarization of the light emitted by the LED – its polarisation is related to the spin of the electrons which cause it via the same selection rules as discussed above in the Awschalom experiment. The polarization of the light emitted correlates well with the hysteresis loop for the magnetic semiconductor and decays with temperature as the magnetic moment of the magnetic semiconductor, leaving little doubt that spin injection has been achieved. The percentage injection realised here is more favourable than has been possible by direct injection from metals and it may be that magnetic semiconductors have an important role to play in future Spin Electronics development, notwithstanding the non-negligible material problems which they pose.

Otherwise, experiments suggest that spin-tunnel injection into semiconductors is a promising technique which offers higher injection efficiency than direct spin-injection. Further results in this area are imminent.

### 1.6.6  Novel Spin Transistor Geometries – Materials and Construction Challenges

The various Spin Transistors designed along the SPICE principle all require ferromagnetic polariser and analyzer stages each side of the semiconductor assemblies. For contamination reasons the magnetic fabrication must be performed only after the semiconductor processing is complete. The materials must be compatible, the process must allow high quality tunnel junctions to be implemented, the nanomagnetic elements must be differentially magnetisable, the physical dimensions must satisfy spin diffusion length requirements and the fabrication

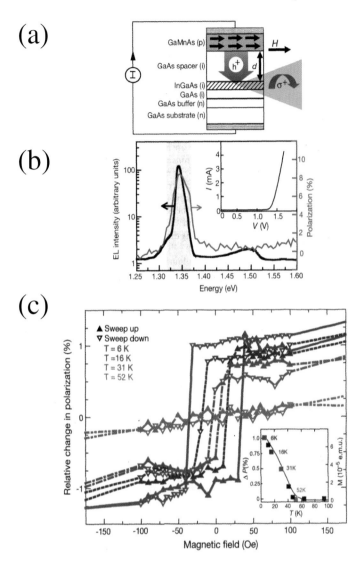

**Fig. 1.12.** Electrical spin injection into an epitaxially grown ferromagnetic semiconductor shown in (a). In (b) the total photoluminescence intensity of the device. In (c) the presence of hysteretic polarization observed in magnetic samples with $d = 20$–$220$ nm, and its absence in the control samples, indicates that hole spins can be injected and transported over 200 nm (after [44]).

must comprise a lithographic stage which defines the three distinct electrodes, all with a minimum of process steps.

Faced with these challenges, the author and his colleagues in York, Strasbourg and Southampton have found the configuration illustrated in Figs. 1.13 & 1.14

most satisfactory for making this type of device. The basis of the structure is a silicon-on-insulator (SOI) wafer into the base of which is etched a micron sized pit with relieved sides. The spin polarized injection emitter is built into the pit and the base and collector structures are deposited and etched on the device quality silicon side.

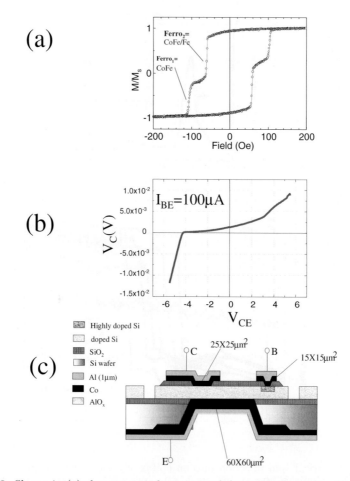

**Fig. 1.13.** Shown in (a) the magnetic hysteresis of the transistor before the base and collector were defined. Note that magnetization of the top and bottom magnetic layers switch at different fields. In (b) the electrical characteristics of the transistor and (c) a schematic diagram of the device with the spin injector emitter built into the pits and the base and collector structures fabricated on the quality silicon side (courtesy of C. Tiusan, IPCMS, Strasbourg, France).

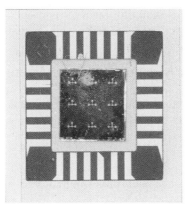

**Fig. 1.14.** Spin polarised injector emitter transistor.

## 1.7  The Rashba Effect and the Spin FET [47,48]

The Lorentz transform applied to electromagnetism shows that to a relativistic traveller, a stationary electric field looks partially magnetic. Since charge carrier velocities in devices are of order $10^6$ m/s or larger, relativistic considerations apply, and electrons in the channel of a field effect transistor see the gate-imposed electric field as having a magnetic component. Depending on orientation this field may be diagonal or off-diagonal and accordingly it causes either band splitting or precession. This is known as the Rashba effect. It follows that if the channel current in the FET is spin polarized, the spins will interact differently with the electrically imposed gate signal depending on whether they are spin-up or spin-down. This is the principle of the Spin FET and although the device has not yet escaped from the drawing board, some of the essential building blocks have been established [49,50].

## 1.8  Refinements in the Understanding of Spin Tunneling

An outline of the principles of spin tunneling was given earlier in this chapter. In practice this simple analysis of the physics of spin tunneling is unable to explain the experimental details observed. The simple theory predicts that a particular ferromagnet will always exhibit the same polarization (i.e. that the ratio of majority to minority density of states is always the same). In practice the polarization of some ferromagnets not only varies in magnitude when different tunnel barrier material are used but they are even known to change sign [51]! The explanation of this riddle is thought to be due to the fact that the tunnel current emerges from the thin layer of metallic electrode right next to the barrier and this material has a bandstructure unlike the bulk metal owing to hybridisation with the insulating material. It follows that, for spin tunneling processes, it is inappropriate to attempt to assign a given spin polarization to a particular spin-asymmetric electrode material: rather it is proper to assign polarizations to combinations of metal ferromagnets and insulator materials [52].

## 1.9    Methods for Measuring Spin Asymmetry

With the caveat, particularly for spin tunneling, that the concept of degree of spin polarisation is more appropriate to combinations of materials, it is interesting to establish the expected polarization which a particular material might offer in a device. Several methods exist and include spin-polarised photoemission spectroscopy [53] and Andreev reflection [54] in which the transport properties are examined of an interface between a superconductor and point-contact of the spin-asymmetric material. Another technique involves characterisation of tunneling currents from an electrode of the material under investigation to a known electrode/insulator combination [51].

A fourth technique [55] is to analyse the magnetic variation in Schottky characteristics of a barrier formed between the ferromagnetic conductor under analysis and a semiconductor. The Schottky current varies as:

$$I = I_0 \exp\left(\mu_B B \left[\rho_\uparrow - \rho_\downarrow / \rho_\uparrow + \rho_\downarrow\right] / k_B T\right) \left(\exp\left[eV/k_B T\right] - 1\right) \qquad (1.9)$$

where $V$ is the bias voltage, $B$ is applied magnetic field and $\left[\rho_\uparrow - \rho_\downarrow / \rho_\uparrow + \rho_\downarrow\right]$ is the required spin asymmetry which may therefore be extracted by observing the modifications to the Schottky characteristic in a magnetic field.

## 1.10    FSETs

The electrostatic energy of a charged capacitor is $\frac{1}{2}Q^2/C$. If $C$ is sufficiently small, this energy can compete with thermal quanta of size $k_B T$, even for $Q = e$, the electronic charge. Small metallic spheres or pads with physical dimensions in the nanometer range have capacitances in the right ballpark for this condition to obtain [56]. If such a metallic island is sandwiched between two physically close metallic electrodes (the source and the drain), we have a Single Electron Transistor (SET) [57–59] through which current may be made to pass one electron at a time (or in bunches of electrons depending on biasing conditions). A third electrode (the gate) which is capacitatively coupled to the metallic island is pulsed in order to trigger the passage of each charge packet (see Fig. 1.15). The physics involved is a competition between three energy terms; the electrostatic energy, $E_i$, of the island due to the presence on it of just one electron, the thermal quantum $k_B T$, and the energy $eV_b$ gained by an electron in falling through the bias voltage $V_b$. The first electron which arrives on the island from the source electrode charges it to a potential $e/C$ which, since it is larger than $V_b$, is sufficient to prevent any further electrons hopping to the island until the first electron has left via the drain electrode. The charges are encouraged to jump from the island to the drain (and hence make room for more charges to arrive from the source) by negative-going pulses on the gate electrode. If the thermal

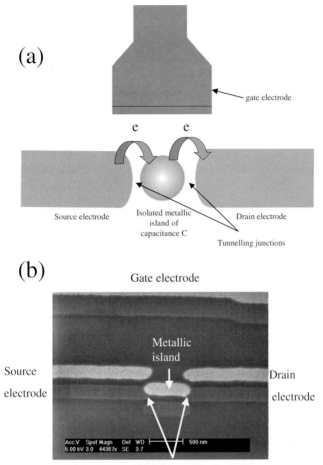

**Fig. 1.15.** Shown in (a) is a schematic digram of a FSET and (b) a micrograph of a FSET (courtesy of I. Petej, Clarendon Laboratory, Oxford).

quantum size is arranged to be small compared with the electrostatic energies in play, random thermal interference with the current control is reduced.

There is a fourth energy term which we can now introduce into the problem, namely the electrochemical potential difference for spin-up and spin-down electrons associated with a spin accumulation. In practice this is achieved by making the electrodes and/or the island from ferromagnetic material [60–63]. A ferromagnetic source electrode will in principle produce a spin accumulation on a nonmagnetic island and, under certain bias conditions, the associated electrochemical potential holds the balance of power between the main energy terms and hence has a large degree of control over the current flow to the ferromagnetic drain. Other configurations are possible in which the island also is magnetic. Fert and Barnas have made extensive calculations for various temperature regimes of the various possible modes of behaviour of such devices which are called Ferromagnetic Single Electron Transistors (FSETs) or Spin SETs. They are of particular interest to the experimental development of quantum computing since they offer a nice opportunity for the manipulation of spatially localized qubits as discussed later.

### 1.10.1   Spin Blockade

Another interesting possibility which arises also if the magnetic island is itself a ferromagnet is that of a spin-blockaded system in which electrical transport across the device is switched by magnetizing the island [64]. An example of a Schottky barrier at low temperature which has been spin blockaded in this fashion is shown in Fig. 1.16 [65]. The MR effect is as large as 25% at 20 K which is unprecedented in a silicon device (shown in Fig. 1.17). The bandstructure consists of the Schottky barrier on the edge of which have been placed a series of magnetic islands which are antiferromagnetically coupled (and hence blockaded) in zero applied magnetic field. Applying a field orients these superparamagnetic particles and the resistance of the structure decreases owing to a tunnel-hopping current between adjacent islands. Exposure to light increases the resistance of the structure owing to photon-promotion of electrons from the islands to the large density of adjacent surface states. The geometry of this system is not unlike that of a High Electron Mobility Transistor (HEMT) in which the performance of the main current channel is controlled by localized states in a physically distinct but nearby region of the device.

## 1.11   Unusual Ventures in Spin Electronics.

Just as conventional electronics insinuates itself into all walks of life, so Spin Electronics shows the same invasive tendency. Even the carbon nanotube has not escaped [66]. Figure 1.18 shows the spin-valve effect observed from a cobalt contacted nanotube, from which it is deduced that the spin diffusion length in such nanotubes is a surprisingly large 130 nm. This would seem to promise well for future device applications of such materials.

## 1.12   The Future of Spin Electronics.

Outside the realms of Politics and Economics it is most foolhardy to predict the future of anything. Who would have thought that, after a mere decade of existence (starting for real in 1988), Spin Electronics would underpin a major industry like hard-disk read technology. It seems clear that its next conquest is likely to be to carve itself a large niche in the MRAM industry using existing tunnel junction technology and perhaps eventually refinements of the spin-tunnel transistors discussed above. Ultimately it may spawn a new philosophy in computer memory in which the distinction between storage memory and active memory becomes less defined.

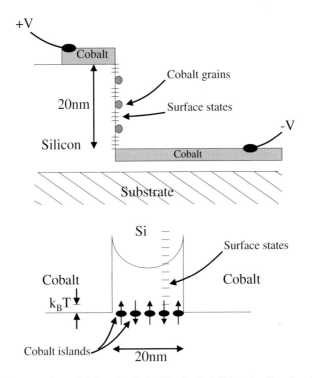

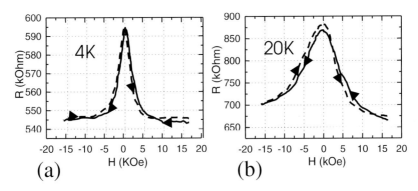

**Fig. 1.16.** Proposed model for the Spin Blockaded Schottky Barrier device [65].

**Fig. 1.17.** Spin Blockaded Schottky Barrier magnetoresistance at (a) 4 K and (b) 20 K. The MR effect is as large as 25% at 20 K which is unprecedented in a silicon device.

On an equally speculative note, it would seem that Spin Electronics has a bright potential future in the world of Quantum Information Technology [69].

The simple Spin Electronic devices which have been demonstrated to date – such as GMR devices and the various spin transistors – function by coding spin information onto the electrical carriers in one part of the device and reading it back in another remote region of the device. In short, contemporary Spin

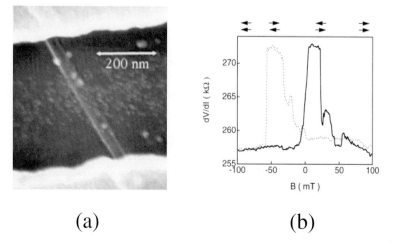

(a)                                    (b)

**Fig. 1.18.** Are carbon nanotubes the future of spin electronics? In (a) Micrograph of a Co-contacted 40 nm diameter carbon nanotube and (b) A nanotube has a spin-flip scattering length of at least 130 nm (after [66]).

Electronics functions by transfer of streams of single qubits from one part of the Spin Electronic circuit to another. Viewed thus, this is just the simplest possible type of quantum information transfer in which no entanglement is involved. The next stage in Spin Electronics is to implement devices which function by displacing spin information by means of entangled qubit pairs. So for example, multi-terminal spin devices of the future might be envisaged in which streams of entangled qubits enable communication between different device terminal, each of which receives one qubit component of the entangled ensemble. The practical realization of such a device might be attempted by employing combinations of Spin SETs.

The FSET (or Spin SET) is a particularly important stepping-stone on the path to quantum information processing. Its distinguishing feature is that it is a unique example of a quantum processor in which the qubits (i.e. the spins) may be physically displaced, allowing the gates and their implementation hardware to be spatially localized as in conventional computing. Competing quantum processor hardware, such as nuclear magnetic resonance processors have fixed qubits and peripatetic gates. Coupled with this configurational advantage, the Spin SET is also endowed with an automatic electrical facility for measuring and collapsing the qubit. These two attributes alone position it in the forefront of potential candidates for future quantum information processing hardware.

While the realization of a full-blown quantum computer is a long way into the future, owing to the monumental problems of overcoming uncontrolled quantum decoherence and parasitic interactions of qubits, nonetheless, the more modest aim of implementing demonstrators of basic quantum information processing hardware is feasible in the medium term. Particularly intriguing would be to explore their use in quantum dense coding, in which fractions of entangled qubits are used to carry increased information capacity compared to classical bit-streams. This might be achieved by using pairs of Spin SETs, each of which is fed entangled qubit spins by a central generator, and each of which is equipped with gate hardware capable of executing the basic single qubit operators X ,Y, Z, H and P($\theta$), which are used to decode the entangled dual spin states. In the simplest case, the gates might consist simply of ferromagnetic layer sandwich structures with differing anisotropy axes in combination with ultra-fast switching microwave pulsing.

A rather simpler task, which could be investigated to gain insight into the functioning of this hardware is the matter of transmitting quantum encrypted data. This has been achieved experimentally using polarized light (see for example [67,68]) but never with localized qubits. The problem is one of transmitting single qubits with one of two orthogonal quantisation axes and projecting them on arrival onto similar axes. Interception of the data may then be detected by monitoring the bitstream error rate which must remain lower than 25% for guaranteed secure transmission. This is a configuration which lends itself to implementation by assemblies of three connected Spin SETs.

The main obstacles in Quantum Information Processing are unsolicited interaction, quantum decoherence and data corruption by noise. A key element in any successful programme will be to reduce these effects to a working minimum necessary to demonstrate functioning of such primitive quantum hardware as has been outlined above. In particular, ways need to be developed to introduce Quantum Error Correction and spin regeneration by methods which do not seek to violate the "no-qubit cloning rule".

## 1.13   Acknowledgements

The author wishes to thank Martin Thornton for his invaluable help with the text, Ivan Petej for his FSET and Mike Coey for allowing him to drive the lawnmower.

## References

1. J. M. D. Coey private communication.
2. N. F. Mott, Proc. R. Soc. **153**, 699 (1936).
3. A. Fert and I. A. Campbell, Phys. Rev. Lett. **21**, 1190 (1968).
4. A. Fert and I. A. Campbell, J. Phys. F **6**, 849 (1976).
5. A. Barthelemy and A. Fert, Phys. Rev. B **43**, 13124 (1991).
6. M. Johnson and R. H. Silsbee, Phys. Rev. Lett. **55**, 1790 (1985).
7. A. Fert, J.-L. Duvail, and T. Valet, Phys. Rev. B **52**, 6513 (1995).
8. Q. Yang, S. F. Lee, L. L. Henry, R. Lololee, P. A. Schroeder, W. P. Pratt and J. Bass, Phys. Rev. Lett. **72**, 3274 (1994).
9. W. P. Pratt, S. F. Lee, J. M. Slaughter, R. Lololee, P. A. Schroeder and J. Bass, Phys. Rev. Lett., **66**, 3060 (1991).
10. S. Y. Hsu, P. Holody, R. Lololee, J. M. Rittner, W. P. Pratt and P. A. Schroeder, Phys. Rev. B **54**, 9027 (1996).

11. H. J. M. Swagten, G. J. Strijkers, M. M. H. Willekens and W. J. M. de Jonge, Phys. Rev. B **54**, 9365 (1996).
12. T. Valet and A. Fert, Phys. Rev. B **48**, 7099 (1993).
13. S. F. Lee, W. P. Pratt, Q. Yang, P. Holody, R. Lololee, P. A. Schroeder and J. Bass, J. Magn. Magn. Mater. **126**, 406 (1993).
14. B. Dieny, V. S. Speriosu, S. S. P. Parkin, B. A. Gurney, D. R. Whilhoit and D. Mauri, Phys. Rev. B **43**, 1297 (1991).
15. J. G. Simmons and G. J. Unterkofler, J. Appl. Phys. **34**, 1793 (1963).
16. P. M. Tedrov and and R. Merservey, Phys. Rev. Lett. **26**, 192, (1971).
17. M. Julliere, Phys. Lett. A **54**, 225 (1975).
18. R. Merservey and P. M. Tedrov, Phys. Rep. **238**, 173 (1994).
19. M. N. Baibich, J. M. Broto, A. Fert, F. Nguyen Van Dau, F. Petroff, P. Eitenne, G. Creuzet, A. Friederich and J. Chazelas, Phys. Rev. Lett. **61**, 2472 (1988).
20. M. Johnson, Science, **260**, 320 (1993).
21. M. Johnson, Mat. Sci. and Eng. B **31**, 199 (1995).
22. M. Johnson, Appl. Phys. Lett. **63**, 1435 (1993).
23. D. S. Daniel, J. F. Gregg, S. M. Thompson, J. M. D. Coey, A. Fagan, K. Ounadjela, C. Fermon and G. Saux, J. Magn. Magn. Mater. **140**, 493 (1995).
24. J. F. Gregg, W. Allen, K. Ounadjela, M. Viret, M. Hehn, S. M. Thompson and J. M. D. Coey, Phys. Rev. Lett., **77**, 1580 (1996).
25. M. Viret, D. Vignoles, D. Cole, J. M. D. Coey, W. Allen, D. S. Daniel and J. F. Gregg, Phys. Rev. B **53**, 8464 (1996).
26. J. E. Wegrowe, A. Comment, Y. Jaccard, J. Ph. Ansermet, N. M. Dempsey and J. P. Nozieres, Phys. Rev. B **61**, 12216 (2000).
27. S. F. Zhang and P. M. Levy, Phys. Rev. Lett. **79**, 5110 (1997).
28. D. J. Monsma, J. C. Lodder, J. A. Popma and B. Dieny, Phys. Rev. Lett. **74**, 5260 (1995).
29. D. J. Monsma, R. Vlutters and J. C. Lodder, Science **281**, 407 (1998).
30. J. C. Lodder, D. J. Monsma, R. Vlutters and T. Shimatsu, J. Magn. Magn. Mater. **198**, 119 (1999).
31. R. Jansen, P. S. Anil Kumar, O. M. J. van't Erve, R. Vlutters, P. de Haan, and J. C. Lodder, Phys. Rev. Lett. **85**, 3277 (2000).
32. J. F. Gregg and P. D. Sparks, British Patent 9608716.8 (1997).
33. J. F. Gregg, British Patent 0006142.4 (2000).
34. Y. Q. Jia, R. C. Shi and S. Y. Chou, IEEE Trans. Magn. **32**, 4707 (1996).
35. J. F. Gregg, W. D. Allen, N. Viart, R. Kirschman, C. Sirisathitkul, J P. Schille, M. Gester, S. M. Thompson, P. Sparks, V. da Costa, K. Ounadjela, M. Skvarla, J. Magn. Magn. Mater. **175,** 1 (1997).
36. F. G. Monzon and M. L. Roukes, J. Magn. Magn. Mater. **198**, 632 (1999).
37. G. Schmidt, L. W. Molenkamp, A. T. Filip and B. J. van Wees, Phys. Rev. B **62**, 4790 (2000).
38. D. Hägele, M. Oestreich, W. W. Rühle, N. Nestle and K. Ebert, Appl. Phys. Lett. **73**, 1580 (1998).
39. J. M. Kikkawa and D. D. Awschalom, Nature **397**, 139 (1999).
40. J. M. Kikkawa, I. P. Smorchkova, N. Samarth and D. D. Awschalom, Science **277**, 1284 (1997).
41. D. D. Awschalom and N. Samarth, J. Supercon. **13**, 201 (2000).
42. M. Oestreich, J. Hübner, D. Hägele, P. J. Klar, W. Heimbrodt, W. W. Ruhle, D. E. Ashenford and B. Lunn, Appl. Phys. Lett. **74**, 1251 (1999).
43. R. Fiederling, M. Keim, G. Reuscher, W. Ossau, G. Schmidt, A. Waag and L. W. Molenkamp, Nature **402**, 787 (1999).

44. Y. Ohno, D. K. Young, B. Beschoten, F. Matsukura, H. Ohno and D. D. Awschalom, Nature **402**, 790 (1999).
45. B. T. Jonker, Y. D. Park, B. R. Bennett, H. D. Cheong, G. Kioseoglou and A. Petrou, Phys. Rev. B **62**, 8180 (2000).
46. H. Ohno, Science **281**, 951 (1998).
47. S. Datta and B. Das, Appl. Phys. Lett. **56**, 665 (1990).
48. M. Johnson, Phys. Rev. B, **58**, 9635 (1998).
49. P. R. Hammar, B. R. Bennett, M. J. Yang and M. Johnson, Phys. Rev. Lett. **83**, 203 (1999).
50. S. Gardelis, C. G. Smith, C. H. W. Barnes, E. H. Linfield and D. A. Ritchie, Phys. Rev. B **60**, 7764 (1999).
51. J. M. de Teresa, A. Barthelemy, A. Fert, J. P. Contour, F. Montaigne and P. Seneor, Science, **286**, 507, (1999).
52. P. LeClair, H. J. M. Swagten, J. T. Kohlhepp, R. J. M. van de Veerdonk and W. J. M. de Jonge, Phys. Rev. Lett., **84**, 2933 (2000).
53. J.-H. Park, E. Vescovo, H.-J. Kim, C. Kwon, R. Ramesh and T. Venkatesan, Nature **392**, 794 (1998).
54. R. J. Soulen, J. M. Byers, M. S. Osofsky, B. Nadgorny, T. Ambrose, S. F. Cheng, P. R. Broussard, C. T. Tanaka, J. Nowak, J. S. Moodera, A. Barry and J. M. D. Coey, Science **282**, 85 (1998).
55. C. Sirisathitkul, W. D. Allen, J. F. Gregg, P. D. Sparks, J. M. D. Coey, R. Kirschman and S. M. Thompson, "Measuring Spin Asymmetry via the Chemical Potential Zeeman shift", in preparation.
56. J. B. Barner and S. T. Ruggiero, Phys. Rev. Lett. **59**, 807 (1987).
57. M. Amman, K. Mullen, and E. Ben-Jacob, J. Appl. Phys. **65**, 339 (1989).
58. M. H. Devoret and H. Grabert: *Single Charge Tunneling*, (Plenum Press, New York, 1992).
59. S. Altimeyer, B. Spangenberg and H. Kurz, Appl. Phys. Lett. **67**, 569 (1995).
60. J. Barnas and A. Fert, Europhys. Lett. **80**, 1058 (1998).
61. K. Majumdar and S. Hershfield, Phys. Rev. B **57**, 11521 (1998).
62. K. Ono, H. Shimada and Y. Ootuka, J. Phys. Soc. Jap. **67**, 2852 (1998).
63. J. Barnas and A. Fert, Europhys. Lett. **44**, 85 (1998).
64. L. F. Schelp, A. Fert, F. Fettar, P. Holody, S. F. Lee, J. L. Maurice, F. Petroff and A. Vaures, Phys. Rev. B **56**, 5747 (1997).
65. C. Sirisathitkul, W. D. Allen, J. F. Gregg, P. D. Sparks, J. M. D. Coey, R. Kirschman and S. M. Thompson, "Spin blockaded silicon Schottky barrier", in preparation.
66. B.W. Alphenaar, K, Tsukagoshi and H Ago, Physica E **6**, 848 (2000).
67. S. J. D. Phoenix and P.D. Townsend, *Quantum cryptography: Protecting our future networks with quantum mechanics* in: *Cryptography and Coding. 5th IMA Conference.*, (Springer-Verlag, Berlin, Germany; 1995) pp112-31.
68. H. Zbinden, J. D Gautier, N. Gisin, B. Huttner, A. Muller and W. Tittel, Elect. Lett. **33**, 586 (1997).
69. A. Steane, Rep. Prog. Phys. **61**, 117 (1998).

# Part II

# Basic Concepts

# 2 An Introduction to the Theory of Normal and Ferromagnetic Metals

G. A. Gehring

Department of Physics and Astronomy, University of Sheffield, Sheffield S3 7RH,
United Kingdom

## 2.1 Introduction

The proper understanding of spin electronics in metallic systems relies on an understanding of some basic metal physics. The material presented here is very brief and is covered in more detail in several excellent textbooks [1,2]. The discussion starts with the definition of a metal and the justification for the frequently used independent particle model even when the interactions between the electrons are included. The role of impurities is also discussed and localized and extended states are defined. Because we are concerned with magnetic properties the most relevant perturbation will be that of a periodic magnetic field. This is characterized by the generalized susceptibility, which is a very useful concept as it allows us to consider the instability of a metal to ferromagnetism or to spin-density waves as well as the response of a paramagnetic metal to a magnetic impurity. Finally we consider strong coupling theory. This includes a discussion of the formation of local moments in a metal. Such a moment will be coupled to the conduction electrons and may be screened out by the Kondo effect. A dense array of Kondo type impurities will form a *heavy Fermion* compound, which is also described briefly. Alternatively, a dense array of such moments can interact with each other to form a spin glass or a ferromagnet.

## 2.2 What is a Metal ?

In this introduction we summarize the important ingredients for a material to behave like a metal.

### 2.2.1 Definition of the Fermi Energy

Many of the effects which are important depend on the metallic Fermi surface. Let us review the theory, which allows a Fermi surface to be defined. A Fermi *energy* exists for any system with a large number of electrons. The chemical potential, $\mu$, is defined by the condition that the total occupation of all available states is equal to the number of electrons, $N$. Here the density of states is written as a function of the energy $\epsilon$ as $D(\epsilon)$ and the temperature $\beta = 1/kT$:

$$N = \int_0^\infty d\epsilon\, D(\epsilon)\, \frac{1}{\exp[\beta(\epsilon - \mu)] + 1} \tag{2.1}$$

The Fermi energy is given as the limit of the chemical potential as the temperature goes to zero and is equal to the energy of the highest energy state which can be occupied.

$$N = \int_0^{\epsilon_F} d\epsilon \, D(\epsilon) \tag{2.2}$$

In a metal $D(\epsilon_F) \neq 0$. But there is more than that: the states at $\epsilon_F$ must be *extended*. As we discuss later this means that the electrons may not be interacting too strongly with each other, or lattice vibrations or defects.

We start from a model in which the electrons move independently in a perfectly periodic potential. Such theories form a good starting point for simple metals. This is known as the independent particle model. Consider the electrons, which are in the last incomplete shell in a perfect crystal – each unit cell contains the same ion core and electron wave functions. In an eigenstate, the electron density must be the same in all unit cells. This symmetry requires that the wave function of the 'extra' electrons will be of the following form

$$\Psi_k^n(r,t) = u_k^n(r) \exp(i k \cdot r - i\epsilon t/\hbar) \quad \text{where} \quad u_k^n(r + R_i) = u_k^n(r) \tag{2.3}$$

for each Bravais lattice vector $R_i$. The energy, $\epsilon$, will depend on $k$ and $n$, which is the band index.

## 2.2.2   Electron Energy Bands in Metals

Calculated electron bands are shown in Fig. 2.1 for copper. In these calculations, the individual electron's energy includes the kinetic energy, the attractive potential energy of the band electrons with the ion cores and an approximation to the electron-electron interaction energy. This latter term is treated within the Linear Density Functional Theory, which works very well for a metal like copper such that the results of experimental measurements such as photoemission or de Haas–van Alphen effect may be modelled accurately.

It is useful to also consider an approximation to Cu, which is known as tight binding theory. In this approach one constructs bands that are linear combinations of the five atomic $d$ orbitals. As the orbitals have a rather small radius the overlap of a $d$ orbital on one site with another $d$ orbital on one of its neighbours will be relatively small and the bands will be narrow. The energies fall in two groups at the $\Gamma$ point because of the cubic crystal-field splitting of the states into $t_{2g}$ and $e_g$ states and then disperse over the zone. There are bonding and anti-bonding states at lower and higher energies, respectively. The calculated bands for copper show clearly the five $d$ bands hybridizing with the very wide band which starts at an energy of $\sim -0.064$ Ry at the $\Gamma$ point and then re-emerges above the $d$ electrons at $\sim 0.5$ Ry close to the edge of the Brillioun zone near the $X$ and $L$ points. This pattern of the narrow hybridized $d$ bands crossing the conduction band is typical of a transition metal. Copper is special because the Fermi energy is above the region where the hybridization occurs. The electrons at the Fermi level are essentially pure conduction electrons with very little $d$ character.

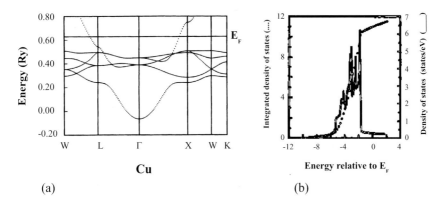

**Fig. 2.1.** (a) The energy bands plotted as a function of wave vector for Cu [3]. The point labelled $\Gamma$ is at the zone center and the other points are at high symmetry points on the zone boundary. $\Gamma - X$ corresponds to $q \parallel$ [100] and $\Gamma - L$ corresponds to $q \parallel$ [111]. (b) The density of states and the integrated density of states for Cu – the integrated density of states is equal to 11 at the Fermi energy which is defined to be at $\epsilon = 0$. (This corresponds to the ten $3d$ and one $4s$ electron, which are present on the free atom.)

Two electrons with opposite spins may occupy each $\boldsymbol{k}$ state in the Brillioun zone. From the bands we may evaluate the density of states. This is done by evaluating the energy bands over the entire Brillioun zone, which is covered by a dense, regular mesh of $\boldsymbol{k}$ points each of which may be occupied by one electron of each spin. The total number of states for each energy, the density of states $D(\epsilon)$, is found by summing up over all $\boldsymbol{k}$ states. These are shown in Fig. 2.1b.

The five $d$ bands contain a total of ten states, which are all occupied. The last, $11th$, electron is in the conduction band, which itself could hold a maximum of two electrons. From the band energies we actually expect the density of states at the Fermi surface to be low but there is a large density of states associated with the $d$ bands, which is some 2 eV below the Fermi level. This is confirmed in Fig. 2.1b. Thus there will be strong optical absorption associated with removing an electron from the $d$ bands to just above the Fermi level – it is this which gives copper its characteristic pink colour.

We see that although the figures for the full band structures look complicated at first it is straightforward to see that they are formed from the hybridization of the much wider band arising from the atomic $4s$ with the $3d$ bands. At many points in the Brillioun zone it is possible to identify electron states as being either predominantly $d$ or $s$ like. This is very useful when transport is discussed in later chapters.

The difference between Cu and the transition metals such as Ni, Co and Fe is that the Fermi level lies well above the $d$ bands in Cu but crosses it for the transition metals. This means that the density of states at the Fermi level in the transition metals would be much higher – but we already know that these metals are also ferromagnetic, so the assumption that both spin bands are equally occupied needs to be dropped. This is discussed in Sect. 2.3.

The Fermi energy was defined by (2.2). We define a Fermi temperature, $T_F$ by $T_F = \epsilon_F/k_B$. Since the value of $T_F$ is normally very high ($\sim 50,000$ K), the assumption of an absolutely sharp Fermi energy is still useful at finite temperatures. The condition that the band energies equal the Fermi energy defines surfaces in $\boldsymbol{k}$ space:

$$\epsilon_{\boldsymbol{k}}^n = \epsilon_F . \tag{2.4}$$

There is a surface in $\boldsymbol{k}$ space for each value of $n$, which separates the occupied from the unoccupied states and plays a very important role in the development of the theory. The electrons at, or above, the Fermi energy are free to move to other states of higher energy or to be scattered elastically to states of equal energy but different $\boldsymbol{k}$ value.

Strictly speaking the Fermi surface is only absolutely sharp at $T = 0$ but as the characteristic temperature, the Fermi temperature $T_F$ is very big the result has general validity for metals at all accessible temperatures.

We should ask why the electrons do not scatter off each other and so give a broadening to the Fermi surface – this is described below.

### 2.2.3   Justification of the Independent Particle Model

It was shown by Landau many years ago that the electrons cannot scatter off each other, since the Fermi statistics implies that there is no available phase space. An excellent treatment is given in Nozières [4]; here we give a simplified discussion.

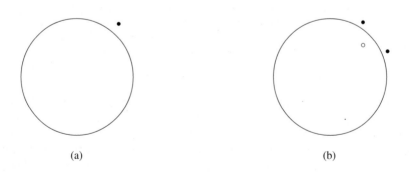

(a)                                        (b)

**Fig. 2.2.** (a) One electron is excited outside the Fermi sea with total energy $\epsilon$. (b) The extra electron has scattered off an electron in the Fermi sea leaving a hole and both electrons lie outside the Fermi sea. The energy and momentum of the configuration shown in (b) must be equal to that in (a).

Consider an electron with an energy $\epsilon$ which is a little bit higher than $\epsilon_F$ at $T = 0$. Figure 2.2 shows a section through the Fermi surface where the energy of a particle depends on its distance from the origin. Figure 2.2b shows an allowed scattering state, which is accessible to the electron shown in Fig. 2.2a. In any electron-electron scattering event the total energy and momentum must be conserved. In this case the Pauli principle is putting very severe extra constraints on the accessible final states.

When the electron scatters off one of the electrons inside the Fermi surface the final two electrons must both end up in states which were previously empty. This ensures that the density of final states is squeezed into a vanishingly thin shell around $\epsilon_F$ and hence that the lifetime varies as $\tau(\epsilon)^{-1} \sim (\epsilon - \epsilon_F)^2 \to 0$ [1,4]. This is very important because the broadening of a quantum state, $\delta\epsilon$, is given by $\hbar\tau^{-1}$ where $\hbar$ denotes Planck's constant divided by $2\pi$ and hence the Fermi energy remains perfectly defined. In a full version of Landau theory [4] it is shown that the individual electron states considered here map on to quasi-particles that have renormalized masses but also have no scattering at the Fermi energy. Thus the scattering that needs to be considered at very low temperatures comes from the defects and not from the electron-electron interactions.

### 2.2.4   Imperfect Crystals

Any defect causes scattering and hence gives a breakdown in the rule that the energy states are characterized by the crystal momentum. Elastic scattering will take an electron from one $\mathbf{k}$ value outside the Fermi surface to another state of equal energy. It is useful to define a mean free path, $\lambda$, that is the distance that an electron will travel in one momentum state before scattering.

For very dense arrays of strong scatterers the states at the Fermi energy may be localized [5]. A characteristic of this change is that the mean free path appears to become comparable with the lattice spacing. (This is more likely to occur in low dimensions.) However, we shall consider here only the case that the scattering is weak and so the states are extended, which is necessary for the material to be metallic. In a metal, the only scattering that is allowed at $T = 0$ is elastic scattering from defects or the sample boundaries. As the temperature is raised various inelastic scattering mechanisms become allowed for electrons; this includes the electron-electron scattering mentioned above, phonon scattering and, in a ferromagnet, scattering from spin waves or spin disorder. All of these effects lead to a resistance, which rises as a function of temperature. In poor conductors, which show a transition from metallic to semiconducting behaviour as a function of doping (for example the manganites), the experimentally applied criterion for a metallic state is that the resistance rises as a function of temperature.

## 2.3   Band Magnetism

We consider the properties of magnetic materials which are best described within a band picture. This is a perturbation scheme where the interactions between the electrons can be treated as an effective field. The theory is developed using the generalized susceptibility which is introduced first.

## 2.3.1   Magnetic Susceptibility

We should consider the response of the metal to a sinusoidally varying magnetic field. This is appropriate since the electron wave functions are given by Bloch waves.

$$\boldsymbol{B}(\boldsymbol{r},t) = \boldsymbol{B}_0 \exp(\mathrm{i}\boldsymbol{q} \cdot \boldsymbol{r} - \mathrm{i}\omega t) \qquad (2.5)$$

The response to this will be a magnetization $\boldsymbol{M}(\boldsymbol{r},t) = \boldsymbol{m}\exp(\mathrm{i}\boldsymbol{q} \cdot \boldsymbol{r} - \mathrm{i}\omega t)$ where $m_{\boldsymbol{q}}^{\alpha} = \chi^{\alpha\beta}(\boldsymbol{q},\omega)B_0^{\beta}$. Symmetry requires that the induced magnetization must have the same dependence on the frequency, $\omega$, and wave vector, $\boldsymbol{q}$, as the driving field.

We evaluate the susceptibility, $\chi^{\alpha\beta}$, using perturbation theory [1,2]. The result below allows for a ferromagnet in which the energy of the electron states depends on spin – we shall use it first to discuss paramagnetic metals for which the electron energies are independent of spin.

$$\chi^{+-}(\boldsymbol{q},\omega) = -4\mu_{\mathrm{B}}^2 \sum_{k} \frac{n_{\boldsymbol{k}\uparrow} - n_{\boldsymbol{k}+\boldsymbol{q}\downarrow}}{\epsilon_{\boldsymbol{k}\uparrow} - \epsilon_{\boldsymbol{k}+\boldsymbol{q}\downarrow} - \omega + \mathrm{i}\epsilon} \, . \qquad (2.6)$$

We comment on this result. In second order perturbation theory, an electron makes a virtual transition between an occupied state with low energy to an unoccupied state with higher energy. This is why we have the two Fermi factors, $n$, on the top – the result is zero unless they differ. At $T = 0$ the factors $n$ are either zero or unity if the energy lies below or above the Fermi energy. We see that we get a contribution if and only if *one* of the factors $n$ in the numerator is unity. If we have $\epsilon_{\boldsymbol{k}\uparrow} \leq \epsilon_{\mathrm{F}}$ and $\epsilon_{\boldsymbol{k}+\boldsymbol{q}\downarrow} \geq \epsilon_{\mathrm{F}}$ then this necessarily implies that $\epsilon_{\boldsymbol{k}\uparrow} < \epsilon_{\boldsymbol{k}+\boldsymbol{q}\downarrow}$ so that the denominator is negative and the whole function is positive for $\omega \to 0$. The analogous expression for the dielectric susceptibility is known as the Lindhard function.

We can show that (2.6) reduces to the well-known expression for the Pauli susceptibility

$$\lim_{q \to 0} \chi(\boldsymbol{q},0) = 2\mu_0\mu_{\mathrm{B}}^2 D(\epsilon_{\mathrm{F}}) \, . \qquad (2.7)$$

In a ferromagnetic metal, the electron-electron repulsion is reduced if the electron spins are parallel. This is because the Pauli exclusion principle keeps them apart. The simplest description is in terms of a self-consistent field. We assume that the $d$ electrons interact when they come on to the same atomic site, so the interaction between the electrons may be taken as short range in real space (the Hubbard model). A point interaction in real space transforms into a constant in momentum space and so the interaction constant $I$ may be assumed to be independent of $\boldsymbol{q}$. In this approximation the magnetization is related to an external field, $B_0$, by

$$m(\boldsymbol{q}) = \chi(\boldsymbol{q})\left[B_0 + Im(\boldsymbol{q})\right] \, . \qquad (2.8)$$

This may be solved to give the magnetization and the interacting susceptibility.

$$m(\boldsymbol{q}) = \frac{\chi(\boldsymbol{q})}{1 - I\chi(\boldsymbol{q})} B_0 \, , \qquad \chi^{int}(\boldsymbol{q}) = \frac{\chi(\boldsymbol{q})}{1 - I\chi(\boldsymbol{q})} \, . \qquad (2.9)$$

In some non-magnetic metals, particularly Pd, the interacting susceptibility may be very considerably enhanced, however we shall only be concerned with the situation in which the enhancement is large enough to cause a transition to an ordered phase.

## 2.3.2  Ordered Phases

The ordered phases arise because the value of $I$ exceeds the value required to make the interacting susceptibility diverge.

A metal is unstable with respect to ferromagnetism if $I\chi(q = 0) > 1$. Since (2.7) showed that the Pauli susceptibility depends on the density of states at the Fermi level we see that ferromagnetism is favoured if the density of states is high. From Fig. 2.1 it is seen that a very high density of states will occur as the Fermi level drops into the region where the $d$ bands lie, as indeed occurs for Fe, Co and Ni.

If an instability occurs at finite $\boldsymbol{q}_0$ this means that the material develops a spin-density wave. In order for this to happen, a maximum must occur in $\chi(\boldsymbol{q})$ at the value of $\boldsymbol{q}_0$. From (2.6) we see that if there are parallel sheets of the Fermi surface such that $\epsilon_{\boldsymbol{k}+\boldsymbol{q}_0} \simeq \epsilon_{\boldsymbol{k}}$ for a range of $\boldsymbol{k}$ then the value of $\chi(\boldsymbol{q}_0)$ is particularly large. Fermi surfaces where one part can be folded on to another by a uniform translation through a wave vector $\boldsymbol{q}_0$ are said to be *nested*. The magnetic order has periodicity given by $2\pi/q_0$.

Since $q_0$ is determined by the geometry of the Fermi surface there is no reason why the periodicity should be an exact number of lattice spacings. A situation in which the ratio of the periodicity to the lattice constant is irrational (or at least not a simple multiple!) is known as incommensurate. The best-known example of a spin-density wave material is metallic chromium, which does have an incommensurate order; however, this is actually rather close to simple antiferromagnetism. This led people to expect that in a multilayer such as Cr/Fe the magnetic coupling between the Fe layers would have a period of approximately one lattice spacing – as is explained in later chapters the period may be considerably longer than this – in agreement with observation [6].

## 2.3.3  Stoner Theory

In the ferromagnetic phase, the densities of states for the electrons with up and down spins are shifted relative to each other by a constant splitting, $\Delta$, where $\Delta = \epsilon_{\boldsymbol{k}\downarrow} - \epsilon_{\boldsymbol{k}\uparrow}$. Stoner postulated that this occurred with no change to the shape of the bands. The first principles calculations shown in Fig. 2.3 indicate that this is an excellent approximation. From the figures we can estimate the exchange splitting to be approximately 1.4 eV in Fe and 1.7 eV in Co. Cobalt has two crystallographic phases, which are very close to each other in energy. The hcp phase is stable at room temperature and an fcc phase is stable above room temperature. However, as it is convenient to compare the two metals when they are both in the cubic phases, we show the density of states for fcc cobalt. It is also the relevant phase for most spin-electronics applications because Co usually adopts its cubic phase when it is grown in a layer structure with other cubic metals.

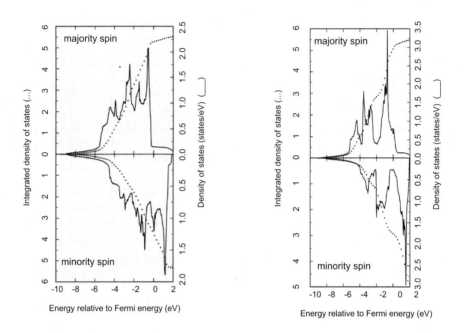

(a) Co (fcc)            (b) Fe (bcc)

**Fig. 2.3.** (a) The density of states for Co (fcc). The density of states is much higher for the minority spin band as the large density of states associated with the majority $d$ band electrons lies entirely below the Fermi energy. (b) The density of states for bcc Fe. The Fermi energy is at $\epsilon = 0$ and lies in a region of high density of states for both the minority and the majority density of states.

Figure 2.3 shows a first principle calculation of the density of states for Fe and Co (fcc) [3]. The Fermi energy is defined to be at zero. It is clear that the density of states for the majority spin has moved down relative to that for the minority spin. The density of states for Co also shows very clearly the separation of the density of states into a low, broad contribution which comes from the $s$ electrons and the narrower, very large, density of states which comes from the $d$ bands. In Co the width of the $d$ bands is approximately 6 eV. The Fermi momenta for the two spin bands are quite different: $k_{F\uparrow} \neq k_{F\downarrow}$.

### 2.3.4   Strong and Weak Ferromagnets

The densities of states for Fe and Co as shown in Fig. 2.3 differ in one very important respect. The density of states in the majority-spin band for Fe is much higher than for Co. The reason for this is that in Co the majority spin $d$ band has dropped to just below the Fermi level and so is full whereas in Fe there is partial occupation in both $d$ bands. Metals in which the majority $d$ band is entirely full are known as *strong ferromagnets* and those, which have both $d$ bands partially filled are known as *weak*. Thus Co is a strong ferromagnet and Fe is a weak ferromagnet.

The low density of states in the majority $d$ band as occurs for strong ferromagnets and shown in the figure for Co has important consequences for spin dependent electronic transport and will be discussed in detail in other chapters of the book.

There is another important distinction between strong and weak ferromagnets that may be seen from the Slater–Pauling curve, which is shown in Fig. 2.4. Consider the elements Fe, Co, Ni and Cu and their alloys. An alloy of $Cu_{40}Ni_{60}$

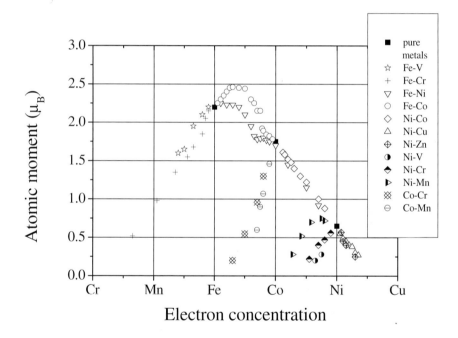

**Fig. 2.4.** Slater–Pauling curve. The magnetic moment per site is given in units of Bohr magnetons ($\mu_B$) for elements and alloys, which are characterized by the average number of electrons per site. After [7].

has a vanishing magnetic moment. As the electron concentration is reduced by increasing the fraction of Ni the magnetic moment increases linearly as each hole is taken from the minority-spin band. This continues until an alloy of approximately 50% FeCo is reached, when the curve bends over. Deviation from linear behaviour corresponds to the holes first appearing in the majority-spin band (there is also a break in the curve between Fe and Co which corresponds to the point where the crystal structure changes from bcc (Fe rich) to fcc (Co rich)).

The early period elements V, Cr, Mn have a tendency towards antiferromagnetism and so do not lie on the straight line for $z > 8$.

### 2.3.5    Excitations in Ferromagnets

For low momenta there are two types of spin excitations we should consider. Stoner excitations in which a single electron is excited from within the Fermi sea and changes both its spin and its momentum and so moves to an unoccupied region of reciprocal space. This energy is given by

$$E_q = \epsilon_{k+q\downarrow} - \epsilon_{k\uparrow} \quad \text{where} \quad E_0 = \Delta\,(q = 0)\,. \tag{2.10}$$

Another form of excitation is a spin wave in which there is a coherent rotation of the magnetization over a wavelength $\lambda = 2\pi/q$ – the energies of these excitations vary as $q^2$ for small values of $q$. Both types of excitation are shown schematically in Fig. 2.5.

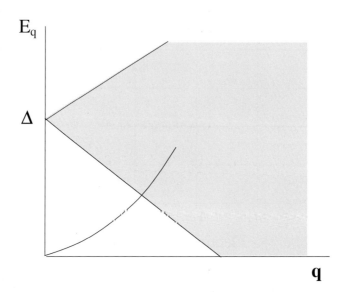

**Fig. 2.5.** Schematic diagram of excitation energies in a ferromagnet. The Stoner excitations are allowed for all values of energy between the two straight lines. The spin-wave excitations broaden and weaken as they cross into the Stoner band.

The Stoner excitations fill the whole area between the two straight lines. At $q = 0$, the energy is $\Delta$ but for larger values of $q$, the energy of the Stoner excitation which is given by (2.10) takes a range of different values depending on the magnitude of $k$ and the relative orientation of $k$ and $q$. A Stoner excitation can have zero energy, when $q$ is equal to the difference between the two Fermi momenta, $q = k_{F\uparrow} - k_{F\downarrow}$.

The spin-wave branch broadens as it crosses into the region where there are Stoner excitations as the spin wave can then decay more readily. The spin waves in ferromagnets may be observed by neutron scattering.

## 2.3.6   The Phase Transition

The free energy, $F = U - TS$, is a minimum at all temperatures. This occurs because of a balance between the energy $U$ (a minimum in the ground state) and the entropy $S$ which is a maximum in the disordered phase. The phase transition occurs at $T_c$ as a result of the benefit of increasing entropy, $S$ (disorder), at the expense of losing benefit of large negative energy, $U$, due to order. In a metal the spin-wave excitations are more readily excited than the Stoner excitations and it is these excitations which are responsible for the reduction of magnetism at high temperatures. Another consequence of this is that above $T_c$ the magnitude of the magnetization at a given site $\langle m_n^2 \rangle$ is not reduced much below its low temperature value and the paramagnetic susceptibility is given by a Curie–Weiss law rather than the result of applying the Stoner theory to high temperatures. Individual sites are likely to have a net magnetic moment albeit randomly oriented. In the original, Stoner, picture, which is valid for weak itinerant ferromagnets all the magnetic moments also vanish above $T_c$.

## 2.3.7   Impurities in Nonmagnetic Metals

Impurities may have *charge* – the electric field must be screened out. It may have magnetic moment – this will also produce a magnetic moment distribution around it.

For a local magnetic field in the $\beta$ direction, $h^\beta$ at $r'$, the response is written in terms of the Fourier transform of the susceptibility that we derived earlier.

$$m^\alpha(r) = \int d^3r' \, \chi^{\alpha\beta}(r - r') \, h^\beta(r') \,. \tag{2.11}$$

Similarly we find the polarization, $p(r)$, due to an electric field $E$ in the $\mu$ direction

$$p^\lambda(r) = \int d^3r' \, \alpha^{\lambda\mu}(r - r') \, E^\mu(r') \,, \tag{2.12}$$

where $\alpha^{\lambda\mu}$ denotes the electric polarizability. The magnetic susceptibility in real space is given below.

$$\chi^{\alpha\beta}(r - r') = \frac{1}{\sqrt{N}} \int d^3q \, \exp[-iq \cdot (r - r')] \, \chi^{\alpha\beta}(q)$$

$$= \frac{-4\mu_B^2}{\sqrt{N}} \sum_{kq} \frac{\exp\left[-iq \cdot (r - r')\right](n_k - n_{k+q})}{\epsilon_k - \epsilon_{k+q} + i\epsilon} \tag{2.13}$$

The existence of a sharp Fermi surface means that $\chi^{\alpha\beta}(\boldsymbol{q})$ has a *weak singularity* (change in slope) at $q = 2k_F$. This is shown schematically for a spherical Fermi surface in Fig. 2.6. We remember that if $\epsilon_{\boldsymbol{k}}$ lies inside the Fermi surface then $\epsilon_{\boldsymbol{k}+\boldsymbol{q}}$ must lie outside. If $q$ is smaller than $2k_F$ then this will be satisfied for some but not all $\boldsymbol{k}$. However if $q$ is greater than $2k_F$ then *all* states at $\boldsymbol{k} + \boldsymbol{q}$ will lie outside. This is shown schematically below.

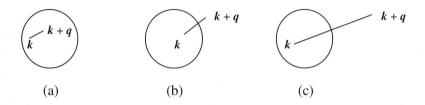

(a)                    (b)                    (c)

**Fig. 2.6.** (a) This term cannot contribute to $\chi$ because both states corresponding to $\boldsymbol{k}$ and $\boldsymbol{k} + \boldsymbol{q}$ lie inside the Fermi sphere. (b) This term does contribute to $\chi$, although $q$ is small. (c) This term has a value of $\boldsymbol{q}$, which is so large that it will contribute to $\chi$ whatever value of $\boldsymbol{k}$ is chosen.

In the double sum over $\boldsymbol{k}$ and $\boldsymbol{q}$ there is a weak discontinuity (a change in slope ) at $q = 2k_F$. This can be understood from the diagram. Imagine that we fix $\boldsymbol{q}$ and sum over $\boldsymbol{k}$. If $q < 2k_F$ some values of $\boldsymbol{k}$ will contribute to the sum as shown in Figs. 2.6a and b. For values of $\boldsymbol{q}$ such that $q > 2k_F$ all values of $\boldsymbol{k}$ can contribute. This is a very weak singularity of $\chi$ in momentum space and as usual this causes the Fourier transform in real space to show oscillations. The simple, free electron, Fermi surface sketched here has only one spanning wave vector. For a more complicated Fermi surface there may be several different extremal spanning wavevectors; this is shown in Fig. 2.7 for copper [8].

**Friedel Oscillations**

These are the charge oscillations which occur around a charged impurity at $r = 0$. A simplified derivation of this result is obtained from the susceptibility that we have given in (2.13). The impurity charge generates a local electric potential, which then gives rise to a spatially distributed response as determined by (2.13). At large distances, the dominant term is the oscillation arising from the sharp Fermi surface. The screening charge decays like $r^{-3}$ at large distances and shows the characteristic oscillation with $2k_F$; there is a phase shift $\phi$ (this is actually important in fixing the total screening charge equal to the impurity charge so that the impurity charge is exactly compensated by the oscillating screening charge).

$$\Delta\rho(r) \sim \frac{\cos(2k_F r + \phi)}{r^3} \tag{2.14}$$

**Ruderman–Kittel Oscillations**

In a similar way a nonmagnetic metal responds to an impurity that has a magnetic moment and an oscillating magnetic density is set up. The local impurity spin $S$ has an exchange interaction with the conduction electron spins $\sigma$,

$$\Delta E = J\boldsymbol{S} \cdot \boldsymbol{\sigma}\,. \tag{2.15}$$

The conduction electrons see a local magnetic field, $JS$, and using the result (2.13) the response at a distance is given by

$$\Delta m(r) \sim \frac{\cos(2k_\mathrm{F}r + \varphi)}{r^3}\,. \tag{2.16}$$

This has a very important consequence if there is a small concentration of magnetic ions in a paramagnetic host. Each ion produces the oscillating magnetization around itself that in turn generates a magnetic field which acts on the other ions. Hence there is a long range and oscillating interaction between the magnetic ions with a strength, which is characterized by $(JS)^2$. This is known as the RKKY interaction (Ruderman–Kittel–Kasuya–Yoshida) [1].

The theory presented here was for a spherical Fermi surface for which there is one value for the spanning wave vector given by $2k_\mathrm{F}$. In a material with a more complicated Fermi surface there may be several 'spanning' Fermi momenta. These give rise to different periods of oscillation. Early theories of the coupling between magnetic layers assumed that the oscillations arose from the RKKY interaction described here. More recently it is seen that it is more correct to consider the effects of confinement in the spacer layer [8] (see also the chapter by J. Mathon in this volume). However, the spanning $\boldsymbol{q}$ vectors still determine the periods. These are shown in Fig. 2.7 for Cu [8].

## 2.4   Strong Coupling Theories

In this section we consider the interaction between electrons more carefully because perturbation theory breaks down. We consider first what criterion must be met in order for a single metal ion to carry a local moment when it is dissolved in a non-magnetic host.

### 2.4.1   Formation of Local Moments

When a transition element A is dissolved into metal B it may carry a local moment. If it does then the magnetic susceptibility has a Curie–Weiss form $\chi \simeq C/T$ at high temperatures and if the element B were superconducting, the value of the superconducting transition temperature, $T_c$, drops dramatically for even small concentrations of A, because the local moment causes pair breaking.

We need to consider an impurity with single $d$ orbital $\varphi_d$ which can hold two electrons with opposite spins. The possible energies are 0, $E_0$, $E_0$, $2E_0 + U$

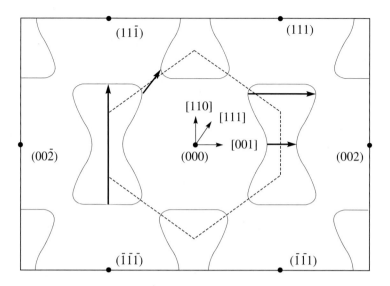

**Fig. 2.7.** A cross-section through the Fermi surface of Cu showing the spanning $q$ vectors. The section is along the $(1\bar{1}0)$ plane passing through the origin. The solid arrows indicate the $q$ vectors giving the oscillation periods for the (001), (111) and (110) orientations. Reproduced from [8] with kind permission of the author.

corresponding to an occupation of zero, one or two. The states are written as $|0\rangle$, $|\uparrow\rangle$, $|\downarrow\rangle$, $|\uparrow\downarrow\rangle$. This is the Anderson model.

Qualitatively a local moment will occur if the state with *one d* electron lies below the Fermi surface, i.e. $E_0 < \epsilon_F$ but the state with two $d$ electrons lies above the Fermi energy $2E_0 + U > \epsilon_F$. However, we need to be more careful as a localized state will *hybridize* with the conduction electrons; the strength of this hybridization is designated as $V$. This allows the state with one $d$ electron to decay into the state where the occupation is zero and there is one extra conduction electron. This means that the localized state acquires a width, $\Delta$, where

$$\Delta \sim \pi V^2 D(\epsilon_F). \tag{2.17}$$

We should solve for the occupation of the $d$ level self-consistently [1,9]. The density of the $d$ electron states is given in terms of a Lorenzian with a width given by $\Delta$.

$$\varphi_\sigma(\epsilon) = +\frac{1}{\pi} \frac{\Delta}{(E - E_\sigma)^2 + \Delta^2}, \tag{2.18}$$

with $\sigma = \uparrow, \downarrow$. This allows us to calculate the occupation of the spin up state.

$$\langle n_{d\uparrow}\rangle = \int_0^{\epsilon_F} d\epsilon\, D_\uparrow(\epsilon) = \frac{1}{\pi} \cot^{-1}\left[\frac{E_\uparrow - \epsilon_F}{\Delta}\right]. \tag{2.19}$$

The energy of the state with an electron with spin $\uparrow$ depends on the occupation of the state with spin $\downarrow$ because the electrons are interacting. In this theory we use a simple Hartee–Fock approximation to estimate the energy and obtain

$$E_\uparrow = E_0 + U\langle n_{d\downarrow}\rangle .$$  (2.20)

This leads to two self-consistent equations for $\langle n_{d\uparrow}\rangle$ and $\langle n_{d\downarrow}\rangle$.

$$\langle n_{d\uparrow}\rangle = \frac{1}{\pi}\cot^{-1}\left[\frac{E_0 - \epsilon_F + U\langle n_{d\downarrow}\rangle}{\Delta}\right]$$  (2.21)

$$\langle n_{d\downarrow}\rangle = \frac{1}{\pi}\cot^{-1}\left[\frac{E_0 - \epsilon_F + U\langle n_{d\uparrow}\rangle}{\Delta}\right] .$$  (2.22)

These must be solved self-consistently. There is a non-magnetic solution in which $\langle n_{d\uparrow}\rangle = \langle n_{d\downarrow}\rangle$ and a magnetic solution $\langle n_{d\uparrow}\rangle \neq \langle n_{d\downarrow}\rangle$. The condition for local moments depends on $E_0 - \epsilon_F$, $\varphi(\epsilon_F)$, $V$ and $U$ [10]. The criterion depends on the particular combination of elements involved, however a large $U$ is certainly necessary. Mn which has a very large $U$ carries a moment in most hosts whereas Fe where $U$ is more modest has a local moment in Pd but not in Ru [1].

## 2.4.2  Ordered Arrays of Moments

The susceptibility of a single spin moment in a metallic host behaving classically follows the Curie law, $\chi \sim 1/T$. This gives a finite entropy change (for field or temperature changes) at very low temperatures in conflict with the third law of thermodynamics. The third law is satisfied when the system settles into its quantum mechanical ground state. As the individual susceptibilities are becoming very large at low temperatures only a very small interaction between the moments is enough to give an ordered state.

In this section we look at the way in which a ground state may be reached by interaction with other magnetic particles.

We have already seen that the local spin interacts with the conduction electrons by an exchange, $JS$, and that this generates an exchange interaction, $J_{RKKY}$, between localized spins which varies as, $J_{RKKY} \sim (JS)^2 \cos(2k_F r + \varphi)/r^3$. Each spin will interact with several of its neighbours. The interaction is oscillating between ferromagnetic and antiferromagnetic, so the fields from the neighbours may add or subtract such that each spin is finally acted on by a field, which is given by a random distribution. This leads to frustration as there is no easy way for a system to find the true ground state (even if one exists!). The spins 'freeze out' at a certain temperature below which they are locked into their particular local configuration. One experimental manifestation of this is that the alloy exhibits non-reversible behaviour – in particular a sample, which was cooled in a magnetic field, will have a magnetic moment until it is warmed above the temperature where the freezing occurred. This occurs for example for small quantities of Mn dissolved in Cu. As the RKKY interactions which couple the Mn atoms are long range the spin glass properties are seen down to very low concentration even for parts per million. A different scenario holds for alloys of Pd because this element is very close to ferromagnetism itself. The interacting susceptibility given by (2.9) is strongly enhanced because $[1 - I\chi(q = 0)]^{-1} \sim 10$. This enhancement is peaked strongly near $q = 0$. The susceptibility in real space given by (2.13) is the Fourier transform of $\chi^{int}(q)$. A sharply peaked function in $q$ space will give rise to a slowly varying function in real space. Hence when an element carrying a local moment such as Fe or Co is dissolved into Pd it induces a very large island of positive polarization around itself. The effective number of Bohr magnetons which can be associated with a single atom of Fe, for example, may be as high as 13; of these $\sim 2\mu_B$ will be on the Fe ion and the rest spread over the large polarized island of Pd.

Since one magnetic atom produces such a change it is not surprising to learn that only a very small concentration is required to produce ferromagnetism.

### 2.4.3   The Kondo Effect

So far we introduced the exchange, $J$, between the spin of the conduction electron and the localized spin on the impurity, $S$, as a phenomenological constant. In fact this was not necessary and it may be derived from the Anderson model for local moments discussed in Sect. 2.4.1. A necessary condition for a local moment to form is that $U$ (the on-site electron repulsion) is large. We consider putting a second conduction electron on a site where there is already a local moment. The two electrons must have opposite spins. The overlap integral is $V$ and the extra energy penalty is $U$. In second order perturbation theory, the energy may be lowered by an amount $V^2/U$ provided the spins were antiparallel. The Pauli exclusion principle allows no such lowering for the parallel spin configuration. Thus the antiparallel configuration is lowered and this can be written as an antiferromagnetic exchange energy as introduced in (2.15), $J\boldsymbol{S} \cdot \boldsymbol{\sigma}$, where we now know that $J \sim V^2/U$. This was a very simple derivation, a more rigorous discussions will be found in Hewson [9].

We discussed earlier that an isolated local moment will have a susceptibility, which obeys the Curie law and hence diverges at absolute zero in contradiction to the third law of thermodynamics. For a high concentration of such impurities we argued that one would get some spin ordering. Here we discuss what occurs if we have an isolated impurity.

The interaction between the conduction electrons and the impurity, $J\boldsymbol{S} \cdot \boldsymbol{\sigma}$, has been shown to give rise to the Ruderman–Kittel oscillations when treated in lowest order perturbation theory. If we go beyond that we see that the isolated spin can scatter the conduction electrons and also cause a spin flip. This scattering becomes very large indeed at low temperatures causing a dramatic rise in the resistivity. This is what was originally called the Kondo effect. At even lower temperatures a coherent singlet ground state is formed between the localized spin and the sea of conduction electrons – this is known as the Kondo singlet [9]. The resistivity then becomes independent of temperature. The singlet ground state is, of course, perfectly allowed by the third law.

### 2.4.4   Heavy Fermion Compounds

The last topic we consider is what happens if we have many Kondo centers. We have seen that for many localized spins we can have an ordered magnetic ground state or a spin glass for a random alloy. For a single isolated spin we see that we can have a Kondo singlet ground state. In some rare earth compounds where every unit cell contains a spin it is possible to have a Kondo singlet form before the spins can order. In this case a coherent Kondo state is formed where the Kondo singlets themselves become coherent. Such materials are known as Heavy Fermion compounds because the electron effective mass is usually strongly enhanced (by a factor of up to $10^3$!). In rare earth compounds there is a competition between spin ordering via the RKKY interaction and formation of the heavy Fermion ground state. The two scenarios are not compatible – in any compound the ordering is of one type or the other [9]. The phase diagram was first given by Doniach [11]. In compounds where the spin is associated with a transition metal ion the RKKY interaction is always dominant and there is either spin ordering at high concentrations or the single site Kondo effect at very low concentrations.

## 2.5   Problems

1. A multilayer has 5 layers of Cu and one of Co grown perpendicular to (100). Assuming that both metals are fcc with the same lattice constant $a_0$ evaluate the size of the unit cell. How large is the Brillioun zone ? Make a rough sketch of a possible band structure.
2. List ways in which (a) the density of states and (b) the Fermi surface may be measured for a non-magnetic metal. What measurements may be made if it is ferromagnetic ?
3. Assume that copper may be described in free electron theory as a one electron fcc metal. Estimate the period of the RKKY oscillations.

## References

1. C. Kittel, *Quantum Theory of Solids*, Wiley 1963.
2. W. A. Harrison, *Electronic Structure and the Properties of Solids*, W H Freeman 1980.
3. V. L. Moruzzi, J. F. Janak, and A. R. Williams *Calculated Properties of metals*, Pergamon 1978.
4. P. Nozières, *Interacting Fermi Systems*, Benjamin 1964.
5. P. W. Anderson, *Phys. Rev.* **109**, 1492 (1958).
6. S. S. P. Parkin, N. More, and K. P. Roche, *Phys. Rev. Lett.* **64**, 2304 (1990).
7. R. M. Bozorth, *Ferromagnetism*, Van Nostrand (1961).
8. P. Bruno, *J. Phys.: Cond. Matter* **11**, 9403 (1999).
9. A. C. Hewson, *The Kondo Problem to Heavy Fermions*, Cambridge University Press (1993).
10. P. W. Anderson, *Phys. Rev.* **124**, 41 (1961).
11. S. Doniach, *Physica* **B&C 91**, 231 (1977).

# 3 Basic Electron Transport

B. J. Hickey, G. J. Morgan and M. A. Howson

Department of Physics, University of Leeds, Leeds LS2 9JT, United Kingdom

## 3.1 Introduction

This chapter will take you through a simple introduction to transport theory covering the Boltzmann equation, the Fuchs–Sondheimer model for thin films, the normal magnetoresistance and quantum interference effects in metals with strong electron scattering. At the end of the chapter we will also introduce you to a number of the basic techniques involved in electron transport measurements. All of this is by way of introduction to basic transport properties common to all metals. In later chapters these ideas will be developed and applied to systems in which spin dependent transport is important.

## 3.2 The Boltzmann Equation

The Boltzmann equation is a semiclassical approach to the calculation of the electrical conductivity which assumes the electrical field ($\mathcal{E}$) is sufficiently small that the response to the current, $j$, is linear, so that $j = \sigma\mathcal{E}$ where $\sigma$ is the electrical conductivity. It also assumes that the momentum is well defined and so $k$ is a good quantum number. This is true if the wavelength of the electron is small compared to the mean free path – the mean distance between scatterers. Figure 3.1a shows the case where the distance between scatterers is large compared to the wavelength so the electron appears as a plane wave before each scattering event. In these circumstances the condition $k_F\lambda \gg 1$ is satisfied where $k_F$ is the Fermi wavevector and $\lambda$ is the mean free path of the electrons. On the other hand, Fig. 3.1b shows the interference effects which can arise when the separation between impurities is small compared to the electron wavelength. We will consider the significance of this later in the chapter. In the meantime it is $k_F\lambda \gg 1$ which allows us to think of the electrons as semiclassical particles.

We describe the electrons by their distribution function $f(k)$ which at equilibrium is simply the Fermi–Dirac distribution

$$f_0(k) = \frac{1}{\exp\left[(E(k) - E_F)/k_B T\right] + 1}. \tag{3.1}$$

In the presence of an electric field the distribution deviates from the equilibrium such that $f(k) = f_0(k) + g(k)$. In a steady state condition the current, and hence

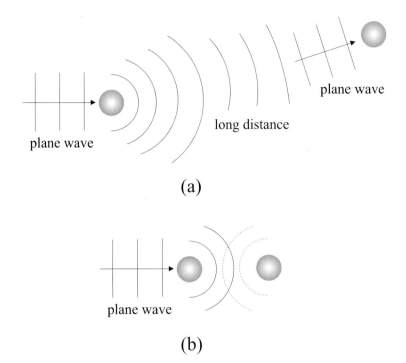

(a)

(b)

**Fig. 3.1.** Schematic drawing of electron waves scattered by impurities in the limits (a) $k_{\mathrm{F}}\lambda \gg 1$ and (b) $k_{\mathrm{F}}\lambda \ll 1$.

the conductivity, can be found from a simple average of the electron velocity using this distribution function.

$$j_x = \sigma\mathcal{E} = \frac{e\int \nu_x f(\nu)d\nu}{\int f(\nu)d\nu} = \frac{e\int \nu_x g(\nu)d\nu}{\int f(\nu)d\nu} \tag{3.2}$$

where $\nu = \hbar k/m$ and note that the integral involving $f_0$ is zero, since this distribution function describes the equilibrium state where there is no net flow of electrons. The way in which the distribution function changes in an electric field is shown schematically in Fig. 3.2. Part (a) of the figure shows the change (shaded) in the Fermi function, wheras part (b) shows the shift in the projected Fermi sphere in response to an applied field.

The problem of calculating the conductivity is now reduced to calculating the distribution function $f$, in fact since we know $f_0$, we have to calculate $g$; this is done using the Boltzmann equation.

In its simplest form the Boltzmann equation is a simple expression of the steady state condition that the distribution function is not changing with time:

$$\frac{df}{dt} = \frac{df}{dt}\bigg|_{\text{field}} + \frac{df}{dt}\bigg|_{\text{scattering}} = 0 \tag{3.3}$$

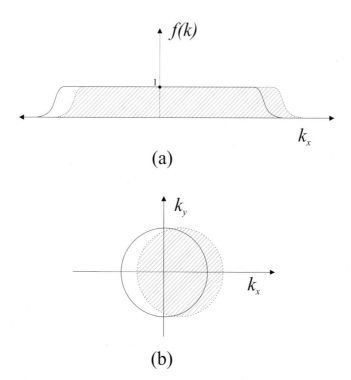

(a)

(b)

**Fig. 3.2.** Schematic drawing of the distribution function in an applied electric field: (a) change (shaded) in the Fermi function and (b) shift of the projected Fermi sphere.

It is made up of two terms, the first describing the fact that $f$ is being driven away from equilibrium as the electrons are accelerated by the electric field and the second describing the relaxation back to equilibrium due to scattering processes. Below we shall show how this simple expression can be used to calculate the conductivity in a simple free electron model, in more complex models and in thin films.

Let's start by looking at the simple free electron case since this will help us to understand some basic ideas. Thus

$$\frac{df}{dt}\bigg|_{\text{field}} = \frac{df}{dk_x}\frac{dk_x}{dt} = \frac{df_0}{dk_x}\frac{e\mathcal{E}}{\hbar} ; \qquad (3.4)$$

the last step arises as $dk/dt$ is simply related to the rate of change of momentum and so to the accelerating force on the electrons due to the electric field, $\mathcal{E}$. Here we have allowed the electron charge e to include its sign. In the last step we have only kept the derivative of $f_0$ in order to keep terms linear in $\mathcal{E}$. The next step is to change the derivative with respect to $k$ to one with respect to energy using the free electron relationship $E = \hbar^2 k^2/2m$:

$$\frac{df}{dt}\bigg|_{\text{field}} = -\frac{df_0}{dE}v_x e\mathcal{E}. \qquad (3.5)$$

The scattering term is usually treated within the relaxation time approximation which assumes that $g$ will relax back to zero exponentially with a relaxation time, $\tau$, so that

$$\left.\frac{df}{dt}\right|_{\text{scattering}} = \left.\frac{dg}{dt}\right|_{\text{scattering}} = -\frac{g}{\tau}\,,\tag{3.6}$$

so

$$g = -\frac{df_0}{dE}\,v_x e \mathcal{E} \tau\,,\tag{3.7}$$

and substituting this into (3.2) we obtain:

$$\sigma = -\frac{e^2}{4\pi^3}\int v\tau\,\frac{df_0}{dE}\,d^3k\,.\tag{3.8}$$

This is an integral over the vector $\mathbf{k}$, the constants simply reflect the normalisation from the integral of $f$ in the denominator of (3.2). In the free electron model we can think of this as an integral over the magnitude of $k$ which can be converted to an integral over $k^2$ and thus energy (which is proportional to $k^2$). Then we can integrate over the spherical energy surface in reciprocal space. This is made easy by the presence of the $df_0/dE$ factor. At absolute zero $f_0$ is step function equal to 1 below $E_F$ and 0 above. The derivative of this is the Dirac delta function and so the integral over energy simply picks out the Fermi surface leaving us with a simple integral over the Fermi sphere. We now find:

$$\sigma = \frac{e^2}{12\pi^3\hbar}\oint v(k)\tau(k)dS\,.\tag{3.9}$$

Here we have explicitly shown a dependence of the velocity and relaxation time on $k$ and the integral is over the Fermi surface so that $\oint dS = 4\pi k_F^2$ within the free electron model. We can further simplify this expression to:

$$\sigma = \frac{e^2}{12\pi^3\hbar}\langle\lambda\rangle S_F\,.\tag{3.10}$$

This form is useful to consider when the Fermi surface is not a simple sphere – as in Cu for example. $\langle\lambda\rangle$ is an average of the mean free path over the Fermi sphere and $S_F$ is the area of the Fermi surface. It is straightforward to show that this equation is equivalent to the Drude formula for the conductivity.

## 3.3   The Relationship Between the Boltzmann Equation and the Kubo–Greenwood Formula

In order to go beyond the simple nearly free electron model we have to go back to the von Neumann equation which is the quantum mechanical version of the classical Liouville equation describing a steady state for the number density of particles:

$$\frac{df(\mathbf{r},\mathbf{p},t)}{dt} = 0\tag{3.11}$$

and which we can write as

$$\frac{\partial f}{\partial t} + \frac{d\boldsymbol{r}}{dt} \cdot \frac{\partial f}{\partial \boldsymbol{r}} + \frac{d\boldsymbol{p}}{dt} \cdot \frac{\partial f}{\partial \boldsymbol{p}} = 0 \tag{3.12}$$

and using $\partial H/\partial \boldsymbol{p} = d\boldsymbol{r}/dt$ and $\partial H/\partial \boldsymbol{r} = -d\boldsymbol{p}/dt$ where $H$ is the Hamiltonian this can be written as:

$$\frac{\partial f}{\partial t} + \{f, H\} = 0. \tag{3.13}$$

The quantum mechanical version of this simply replaces the Poisson bracket with a commutator and the number density of particles with the electron distribution function operator, thus:

$$\frac{\partial f}{\partial t} + \frac{i}{\hbar}[H, f] = 0 \qquad \text{and} \qquad H = H_0 + H' \tag{3.14}$$

with $H_0 = p^2/2m + V(\boldsymbol{r})$ and $H' = e\mathcal{E}x$ is the perturbation due to the electric field. Here $\boldsymbol{p}$ and $\boldsymbol{r}$ are operators.

If we substitute this Hamiltonian into the von Neumann equation, keeping only terms which are linear in the electric field and assuming that the system is homogeneous so $f$ has no spatial dependence, then we find:

$$\frac{i}{\hbar}[H_0, g] + \frac{ie\mathcal{E}}{\hbar}[x, f_0] = 0. \tag{3.15}$$

The first term is linear in $\mathcal{E}$ because of $g$, the deviation of $f_0$ from equilibrium. We can replace the second term with the derivative of the Fermi function and the velocity operator but it is beyond the scope of this introduction to show this derivation. Its origin is the identity

$$\frac{i}{\hbar}[x, F(H)] = \nu_x \frac{dF(H)}{dH} \tag{3.16}$$

and so we find

$$\frac{i}{\hbar}[H_0, g] - e\mathcal{E}\nu_x \frac{df_0}{dE}. \tag{3.17}$$

We can see the resemblance with the Boltzmann equation from the earlier section by comparing this with (3.5) and (3.6). The second term is the term driving the electrons from equilibrium and the first term is the relaxation.

This is the equation we now have to solve to find $g$ and to calculate the conductivity. In this case, however, instead of a simple integral we now evaluate the trace of $\nu_x$ with $g$, i.e.

$$j = 2eTr(\nu_x g), \tag{3.18}$$

where $\nu_x$ is the velocity operator.

We can now follow two routes. If we use a $k$-representation to evaluate the commutator and trace and use the Born approximation, we end up with what many refer to as the Boltzmann equation. If we use exact energy eigenfunctions to evaluate the commutator and trace, we end up with the Kubo–Greenwood formula. The details involve some tedious algebra and so we refer the reader to the bibliography, here we simply give a quick overview and discuss the results.

Using the energy-eigenfunction representation with eigenfunctions labelled by $n$ whose eigenvalues are $E_n$, using (3.16) and (3.17) we find

$$g_{nn'} = \frac{\mathrm{i} e \mathcal{E} \left[ x, f_0(H) \right]_{nn'} / \hbar}{\delta + \dfrac{\mathrm{i}}{\hbar} (E_n - E_{n'})} , \qquad (3.19)$$

which then is substituted into (3.18) and using the identity (3.17), taking the real part gives us

$$\sigma = \frac{2e^2 \pi}{\Omega} \sum_{nn'} | \nu_{nn'} |^2 \; \delta(E_n - E_{n'}) \frac{df(E_n)}{dE} , \qquad (3.20)$$

which may be written in the form

$$\sigma = \frac{2e^2 \pi}{\Omega} \sum_{nn'} | \nu_{nn'} |^2 \; \delta(E_n - E_{n'}) \, \delta(E_{n'} - E_{\mathrm{F}}) . \qquad (3.21)$$

$\Omega$ denotes a $k$-space volume. If we introduce $G''$, the imaginary part of the Green function operator

$$G(E) = \frac{1}{H - E + \mathrm{i}\delta} , \qquad (3.22)$$

then this expression for the conductivity is equivalent to

$$\sigma = \frac{2e^2}{\pi \Omega} Tr \left( G'' \nu G'' \nu \right) . \qquad (3.23)$$

This is known as the Kubo–Greenwood formula for the conductivity. It has the advantage that it is an exact expression for the electrical conductivity. Of course, it presupposes one can derive the exact Green's functions.

On the other hand, using the $k$-representation we find

$$\left. \frac{df}{dt} \right|_{\mathrm{collisions}} = \frac{\mathrm{i}}{\hbar} \, [H_0, g]_{k,k} = \sum_{k'} P_{kk'} \left[ g(k) - g(k') \right] \qquad (3.24)$$

from which we derive a slightly more 'refined' Boltzmann equation:

$$\sum_{k'} P_{kk'} \left[ g(k) - g(k') \right] = e \mathcal{E} \nu_x \frac{df_0}{dE} \qquad (3.25)$$

which can be compared with (3.7).

Here $P_{kk'}$ is the transition rate for scattering from state $k$ to $k'$ on the Fermi sphere. This is derived in the Born approximation and is proportional to $|\langle k|V|k'\rangle|^2$ where $V$ is the scattering potential which appears in $H_0$.

## 3.4 How the Energy and Momentum Relaxation Rates are Related

This latter derivation of the Boltzmann equation allows us to demonstrate an important difference between the energy and the transport relaxation time. We can use (3.9) to obtain the Drude formula $\sigma = ne^2\tau/m$, but the $\tau$ here is explicitly the energy relaxation time – the time for $g$ to relax back to zero or the Fermi sphere to relax back to the origin. If we use the more exact derivation, (3.25), we find $\sigma = ne^2\tau_{tr}/m$ where $\tau_{tr}$ is the transport relaxation time which takes account of the effectiveness of large angle scattering over small angle scattering in destroying the electric current. The relationship between $\tau_{tr}$ and $\tau$ can be shown by the following argument.

$$\frac{df}{dt}\bigg|_{\text{collisions}} = \sum_{k'} P_{kk'}\left[g(k) - g(k')\right] = \frac{g(k)}{\tau_{tr}} \tag{3.26}$$

and so

$$\frac{1}{\tau_{tr}} = \sum_{k'} P_{kk'}\left[1 - \frac{g(k')}{g(k)}\right]. \tag{3.27}$$

Now from (3.7) we see that $g$ will be proportional to $\mathcal{E} \cdot k$ and so

$$\frac{1}{\tau_{tr}} = \sum_{k'} P_{kk'}\left[1 - \frac{k' \cdot k}{k \cdot k}\right] \tag{3.28}$$

$$= \sum_{k'} P_{kk'}\left[1 - \cos(\Theta_{kk'})\right], \tag{3.29}$$

where $\Theta_{kk'}$ is the angle between the initial wavevector $k$ and the scattered wavevector $k'$ which are both on the Fermi sphere. Here scattering events are weighted by a $(1 - \cos(\Theta))$ factor, so large angle scattering is more effective at reducing the conductivity.

But the energy relaxation time in the original simple derivation is simply $\sum_{k'} P_{kk'}$ and represents the relaxation of the electron distribution function back to $f_0$ through all scattering events equally weighted.

The transport relaxation time takes account of the fact that small angle scattering does not diminish the current as effectively as large angle scattering. This is important for example at low temperature where normal phonon scattering involves phonons with a small momentum which can only produce small angle scattering and so become ineffective at scattering electrons.

## 3.5 Thin Films and the Fuchs–Sondheimer Model

In the above derivation of the Boltzmann equation we assumed that the system was homogeneous and so the distribution function has no spatial dependence. If we consider thin films or multilayers then the system is clearly inhomogeneous in one dimension so that for a thin film $f$ goes to zero at the surfaces of the film.

The Fuchs–Sondheimer model [1] re-derives the Boltzmann equation and determines $g$, but includes the spatial dependence of $g$. If we refer back to the Liouville equation we need to include the term involving $(d\boldsymbol{r}/dt) \cdot (\partial f/\partial \boldsymbol{r})$ and if we consider the electric field to be in the $x$-direction and the $z$-axis to be perpendicular to the plane of the film, the Boltzmann equation in the same form as (3.7) becomes:

$$\frac{g}{\tau} = -\frac{df_0}{dE} \nu_x e\mathcal{E} + \frac{\hbar \boldsymbol{k}}{m} \cdot \frac{\partial g}{\partial \boldsymbol{r}} \qquad (3.30)$$

which becomes

$$\frac{g}{\tau} = -\frac{df_0}{dE} \nu_x e\mathcal{E} + \frac{\hbar k_z}{m} \frac{\partial g}{\partial z} \qquad (3.31)$$

which has the solution

$$g(z,k) = \nu_x e\mathcal{E} \frac{df_0}{dE} \left[ 1 + \exp\left( -\frac{mz}{\hbar \tau k_z} \right) \right] \qquad (3.32)$$

and to calculate the conductivity (3.9) becomes

$$\sigma = \frac{e^2}{4\pi^3 \hbar} \oint \nu(k)\tau(k) \frac{df_0}{dE} \left[ 1 + \exp\left( -\frac{mz}{\hbar \tau k_z} \right) \right] dr\,dk . \qquad (3.33)$$

The details of how to do this integral can be found in Sondheimer's Advances in Physics paper [1]. The result is very simple for a thin film for which the boundary conditions are $g(z) = 0$ at $z = 0$ and $z = t$, and where $t$, the thickness of the film, is greater than $\lambda$, the mean free path:

$$\frac{\rho}{\rho_{\mathrm{B}}} = \left[ 1 + \frac{3\lambda}{8t} \right] , \qquad (3.34)$$

where $\rho$ is the resistivity and $\rho_{\mathrm{B}}$ is the bulk resistivity in the limit of infinite thickness. In Fig. 3.3 we show how the resistivity varies with thickness for a thin film.

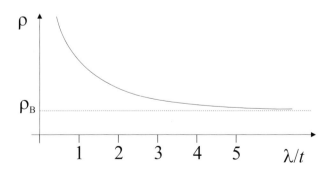

**Fig. 3.3.** Schematic drawing of the resistivity variation of a thin film as a function of thickness.

Sondheimer's initial motivation for this calculation was as a means to directly measure the mean free path. A simple resistivity measurement in a bulk sample depends on many complex factors. But the dependence of $\rho/\rho_B$ on thickness in a film whose thickness is comparable with the electron mean free path is dependent only on the mean free path. A number of workers have used this to determine the mean free path in thin films and multilayers and for example, find typical values of about 50 nm in a Co/Cu multilayer.

Camley and Barnas [2], and Valet and Fert [3] have used this Boltzmann approach to calculate the conductivity in multilayers with boundary conditions at the interfaces of the metal layers rather than just the surfaces and have included modifications to the equations to treat the spin up and spin down electrons separately. This is dealt with in more detail in a later chapter.

## 3.6   The Normal Magnetoresistance

The magnetoresistance of magnetic thin films and multilayers is one of their most interesting transport properties and results from the different scattering properties of spin up and spin down electrons. This will be discussed at great length in the following chapters but here we give a simple account of the normal magnetoresistance which is present in all metals. In Fig. 3.4 we see the magnetoresistance of a relatively thick Co film with a thickness of about 100 nm, a little thicker than the mean free path and, with respect to the Sondheimer model, it would be considered bulk-like. The data is shown with the field both perpendicular and transverse to the current direction. Below the saturation field we see the behaviour typical of the anisotropic magnetoresistance found in all magnetic materials. But for fields greater than the saturation magnetoresistance we see the normal magnetoresistance which is positive and varies as $B^2$.

In Fig. 3.5 we show data for a very thin Co film of thickness 3 nm. Below saturation we see the anisotropic magnetoresistance and this time we see a small negative magnetoresistance above saturation. Broto *et al.* [4] showed that this magnetoresistance extends out beyond fields of 40 T. This is also a normal magnetoresistance but in the limit of the thickness being smaller than or comparable with the mean free path.

We can see how the normal magnetoresistance arises by a simple semi-classical argument. The force on an electron is due to the Lorentz force and the electric field:

$$\boldsymbol{F} = m\frac{d\boldsymbol{\nu}}{dt} = e\boldsymbol{\mathcal{E}} + e\boldsymbol{\nu} \times \boldsymbol{B} \tag{3.35}$$

and if we write the current density as $j = (e/V)\sum_{i=1}^{N} \nu_i$ where $V$ is the volume of the sample and $N$ the number of electrons then within the relaxation time approximation we can write:

$$\frac{d\boldsymbol{j}}{dt} = \frac{ne^2}{m}\boldsymbol{\mathcal{E}} + \frac{e^2}{m}\boldsymbol{\nu} \times \boldsymbol{B} = \frac{\boldsymbol{j}}{\tau} \tag{3.36}$$

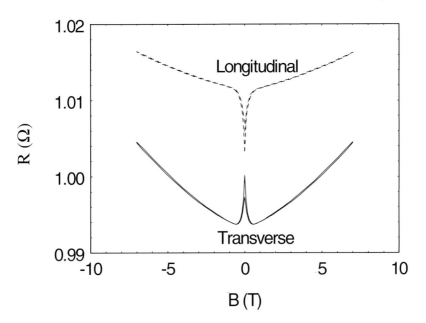

**Fig. 3.4.** Resistance of a 100 nm thick Co film with the magnetic field parallel (longitudinal) and transverse to the applied current.

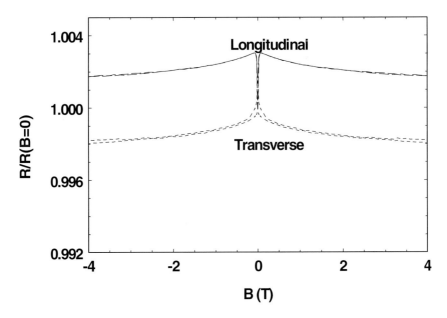

**Fig. 3.5.** Resistance of a 3 nm thick Co film normalized to the zero field resistance, measured with the magnetic field parallel (longitudinal) and transverse to the applied current.

and so

$$j = \frac{ne^2\tau}{m}\, \mathcal{E} + \frac{e^2\tau}{m}\, \nu \times B\,. \tag{3.37}$$

This can be represented vectorially as in Fig. 3.6 where $\Phi$ is the Hall angle. If we consider the situation where $j$ is along the x-axis and the $B$-field is perpendicular to $j$, then the x-component of the electric field is $\mathcal{E}_x = \mathcal{E}\cos(\phi)$ and from Fig. 3.6 we find $\cos(\phi) = j/\sigma\mathcal{E}$ so that $j = \sigma\mathcal{E}_x$. The interpretation of this is that in the presence of a $B$–field the component of the electric field along the direction of $j$ is unchanged. The $B$–field simply generates a Hall field perpendicular to $j$. In other words the conductivity is independent of the magnetic field and the normal magnetoresistance is zero.

The only way to get a normal magnetoresistance within a free electron model is to consider a system with two types of electrons. This is often referred to as a two-band model because of its application to semiconductors where the magnetoresistance results from the presence of electrons and holes. In metals like copper though, the presence of electrons from the neck region of the Fermi surface and others from the belly region are enough to produce a magnetoresistance. Now we have to treat the two types of electrons, labelled 1 and 2, separately and combine the vector diagram for $j_1$ and $j_2$. More details can be found in Ziman's textbook on Electrons and Phonons listed in the bibliography. The net result is that to first order the normal magnetoresistance is given by

$$\frac{\Delta\rho}{\rho} = (\omega_c\tau)^2 = \left(\frac{eB}{m}\,\tau\right)^2 = \left(\frac{ne^2\tau}{m}\frac{1}{ne}\,B\right)^2 \tag{3.38}$$

and so

$$\frac{\Delta\rho}{\rho} = \left(\frac{R_H}{\rho}\right)^2 B^2\,. \tag{3.39}$$

The normal magnetoresistance is always positive and can vary in magnitude widely depending on the resistivity of the sample. The magnetoresistance can be very substantial in very pure metals at low temperatures with resistivities as low as $10^{-4}$ $\mu\Omega$cm when the normal magnetoresistance can be 10000 in a few Tesla.

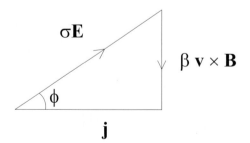

**Fig. 3.6.** Vector diagram showing the current density, electric field and the Lorentz force. $\phi$ denotes the Hall angle.

In general most metals follow Kohler's rule that the magnetoresistance is a function of $B/\rho$ .

The 'normal' magnetoresistance in very thin films in the Fuchs–Sondheimer regime can in fact be negative as the curvature of the electron trajectory takes the electrons away from the surface of the sample. This was discussed by Chambers [5] and explains the small negative magnetoresistance often observed above the saturation field in magnetic multilayers.

## 3.7   Beyond the Boltzmann Theory: Quantum Interference Effects

At the beginning of this chapter we pointed out that the Boltzmann equation is correct in the limit that the mean free path was long compared to the electron wavelength, i.e. $k_F \lambda \gg 1$. We can see under these conditions that when a plane wave is scattered from an impurity atom, it sends out a scattered spherical wave but by the time this wave reaches the next impurity atom it is so large that in the region around the second scatterer it appears to be a plane wave. When the scattering is very strong, or the impurities are very close together as in disordered metals, or if the wavelength of the electrons is very long as in impurity band-conduction in semiconductors, then the spherical wave from the second scattering event can interfere with the incoming spherical wave. This interference leads to substantial corrections to the Boltzmann conductivity, eventually leading to localisation in the extreme case. In Fig. 3.1 we showed a schematic diagram to illustrate the two conditions $k_F \lambda \gg 1$ and $k_F \lambda \sim 1$.

One manifestation of these interference effects is a gradual reduction in the temperature coefficient of the resistivity so that for resitivities greater than around 150 μΩcm the resistivity actually decreases as the temperature rises. Such a decrease in the resistivity is usually considered a feature of semiconductor behaviour, but in fact small negative temperature coefficients are typical in strong scattering disordered metals. This correlation between the magnitude of the resistivity and the temperature coefficient is often referred to as the Mooij correlation.

The crossover from positive slope to negative slope occurs when $k_F \lambda \sim 1$. If we take (3.10) above and use free electron values for the area of the Fermi sphere we find

$$\sigma = \frac{1}{3\pi^2} \frac{e^2}{\hbar} \frac{1}{\lambda} (k_F \lambda)^2 \qquad (3.40)$$

and so for a typical strong scattering metal with $k_F \lambda \sim 1$, $\lambda$ is of the order of the Fermi wavelength and $\sigma$ about 500000 $(\Omega m)^{-1}$, i.e. a resistivity of about 200 μΩcm.

The quantum interference effects arise from the coherent interference of electron waves which have followed different scattering paths. Most of the paths interfere incoherently and make an additional contribution to the resistivity. But if we look at the special case of two waves following the same scattering

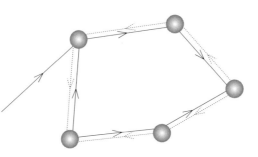

**Fig. 3.7.** Schematic drawing of the interference between two identical electron trajectories traversed in opposite directions.

path but in opposite directions they return to the starting point in phase and interfere constructively. This is shown schematically in Fig. 3.7.

In the regime where $k_F\lambda \sim 1$ we often think of the electron motion as diffusive and executing a sort of random walk. In Fig. 3.8 we show the probability distribution of distance diffused from the origin after a time $t$. It follows the usual Gaussian distribution of a random walk. The constructive interference enhances the amplitude of the wavefunction at the origin and so the effect of the interference is to increase the probability that the electron returns to the starting point. This appears as a peak at $r = 0$ on top of the Gaussian distribution. The effect is to increase the electrical resistivity or reduce the conductivity:

$$\sigma = \sigma_B + \sigma_{QIE} = \frac{ne^2\tau}{m} - \frac{1}{2\pi^2}\frac{e^2}{\hbar}\left[\frac{1}{\sqrt{D\tau_o}} - \frac{1}{\sqrt{D\tau_i}}\right]. \qquad (3.41)$$

Here $D$ is the diffusion constant $(D = (1/3)v_F\tau_o)$ and $\tau_o$ is the elastic relaxation time.

The negative temperature coefficient is the result of a reduction in the interference as the temperature rises. In the above equation the first term is the Boltzmann contribution to the conductivity which dominates the resistivity. However, when $k_F\lambda \sim 1$ the conductivity has a very small positive temperature coefficient. The second term is the decrease in conductivity due to interference, the full effect of which is experienced at $T = 0$. The third term describes the reduction in the interference effect as the temperature increases through the inelastic relaxation time, $\tau_i$, from electron-phonon scattering or electron-electron scattering. This is because the constructive interference of the two waves must be in phase when they return to the starting point but if one has scattered from a phonon, energy is lost by one electron and the two waves are no longer coherent.

A negative magnetoresistance also results from the reduction in the interference in a magnetic field. In Fig. 3.7 we can see that the electrons of interest follow a closed path which will enclose a magnetic flux. This shifts the phase of the electron wavefunction, and the two electrons following opposite paths have their phase shifted in opposite directions. This of course destroys the coherence between the waves and so the constructive interference is reduced. The interference

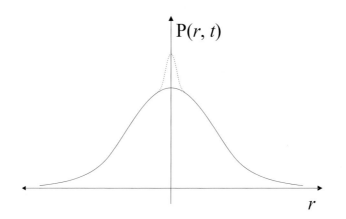

**Fig. 3.8.** Probability of distance $r$ diffused by an electron after time $t$. In the case of weak localisation the probability that the electron returns to the origin is increased. This is manifested by the peak superimposed on the Gaussian distribution for the normal diffusion process.

contribution to the resistivity is reduced resulting in a negative magnetoresistance.

$$\frac{\Delta\rho}{\rho} = -\frac{1}{2\pi^2}\frac{e^2}{\hbar}\sqrt{\frac{eB}{\hbar}}. \tag{3.42}$$

## 3.8   Experimental Methods

Having given an overview of the theoretical aspects of the conventional spin independent electrical transport we now take a brief look at some of the experimental techniques involved in transport measurements. The resistivity itself is a relatively straightforward measurement to perform: after all we all learnt how to use a voltmeter and amperemeter at school! But there are a few points that are worth making.

### 3.8.1   Resistivity

It is usual to make a four probe measurement of the resistivity in order to eliminate contributions from the contacts and wires. For a long thin conductor or rod shaped conductor you can use the approximation that $\rho = Rwt/L_e$ where $w$ is the width of the rod, $t$ the thickness and $L_e$ the distance between the voltage probes. For other shapes of sample it is not so straightforward. For a 'rectangular' sample the current lines are not parallel and so we use four contacts which are evenly spaced and the resistance per square $R_\square = C(V/I)$ and the resistivity $\rho = R_\square t$. The 'correction' factors $C$ are listed in table 3.1 and the geometry is shown in Fig. 3.9.

**Table 3.1.** Correction factors for calculating the resistance per square for a circle and a rectangular sample

|  | | Circle | Square | Rectangle | Rectangle | Rectangle |
|---|---|---|---|---|---|---|
| $W/S$ | | | $L/D = 1$ | $L/D = 2$ | $L/D = 3$ | $L/D = 4$ |
| 1 | | | | | 0.9988 | 0.9994 |
| 1.5 | | | | 1.4788 | 1.4893 | 1.4893 |
| 2.0 | | | | 1.9475 | 1.9475 | 1.9475 |
| 2.5 | | | | 2.3541 | 2.3541 | 2.3541 |
| 3.0 | | 2.2662 | 2.4575 | 2.7000 | 2.7005 | 2.7005 |
| 4.0 | | 2.9289 | 3.1137 | 3.2246 | 3.2248 | 3.2248 |
| 5.0 | | 3.3625 | 3.5098 | 3.5749 | 3.5750 | 3.5750 |
| 10.0 | | 4.1716 | 4.2209 | 4.2357 | 4.2357 | 4.2357 |
| 20.0 | | 4.4364 | 4.4516 | 4.5553 | 4.5553 | 4.5553 |
| 40.0 | | 4.5076 | 4.5120 | 4.5129 | 4.5129 | 4.5129 |
| infinite | | 4.5324 | 4.5324 | 4.5325 | 4.5325 | 4.5325 |

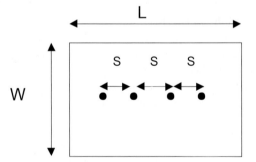

**Fig. 3.9.** Geometry of a four-point resistivity measurement on a rectangular sample.

If the sample is of a completely irregular shape then you can use the van der Pauw method in which four contacts are placed on the edge of the sample and several measurements are made with the current and voltage leads interchanged. This is complicated to show but is discussed in several papers [6,7].

Contacts are a major issue. If your contact resistance is an order of magnitude higher than the sample resistance, you start to run into noise problems and here what you do depends on the sample. You can use pressure contacts, solder, silver or gold paint, ultrasonic bonding or spot welding. With contacts you also need to worry about the superconducting or magnetic properties which could influence your measurements either by changing the stray fields around your sample or the contact resistance. Workers also use a variety of contact wires on delicate samples including thin Cu (just because it is to hand), Ni, Au and Pt are good because they are easy to keep from oxidising, and for superconducting contacts Nb or NbTi. On magnetic multilayers we have used two main contact methods: 1) pressure using Coda system contacts (Coda systems Ltd, Braintree, UK) and 2) solder using indium – here we have found it useful sometimes to use an ultrasonic soldering iron which helps to break through any oxide coating. Once In pads are placed on the sample it is possible to cold weld wires to the In to produce contacts with a resistance less than 1 mΩ.

With dc measurements and a good nanovoltmeter you can measure quickly to 1 part in $10^5$ for samples between 0.1 Ω and 100 Ω. The method suffers from thermal emfs which are typically 100 μV in rough and ready systems but if you take time to reduce the number of contacts in the system between the sample and the voltmeter and temperature stabilise the sample they can be 500 nV or less. They can be cancelled out by making measurements with the current in two different directions but you need to watch for sample heating and Peltier effects at the contacts if you want to make careful measurements.

Ac measurements are often faster and can be more accurate if looking for small changes. Using a lock-in amplifier in place of a dc voltmeter will get you about 1 part in $10^4$ and allows you to see changes of factors of 2 easily. You can achieve 1 part in $10^6$ but only with a bridge. A typical bridge for measuring samples between 0.1 Ω and a few hundred Ohms is described by Anderson [8]. There are some quite accurate conductance bridges on the market now, but they are expensive if you want good precision.

Common sources of error are: thermal emfs, Hall voltages from contacts that are not in line (best to place them on one side of the sample only), self heating (just check for non-ohmic behaviour) and bad contacts (you see noise and non-ohmic behaviour).

### 3.8.2   Hall Effect and Thermopower

The two other main transport properties which are investigated are the Hall effect and thermopower. It is worth just mentioning these in passing and point people in the right direction for more details.

The Hall effect is usually measured using a 5 probe technique on a crucifix shaped sample when investigating metals where the Hall voltage can be as low as 100 nV. Apart from the current leads there are three voltage leads, one on one side of the sample and the other two close to each other on the other side. It is practically impossible to align the Hall voltage leads and so the two leads on one side are used to null off any misalignment voltage in zero magnetic field. The Hall voltage is large in magnetic samples below the saturation field as the magnetisation contributes to the field experienced by the electrons and it is possible, but not advisable, to get away with only one Hall lead on each side of the sample. There are two ways of determining the Hall coefficient. One is to vary the field from negative to positive values and any residual misalignment voltage can be removed since the misalignment voltage will be an even function of field, but the Hall voltage is an odd function of the field. The other is to hold the field steady and rotate the sample in the field, although you can hold the sample steady and rotate the electromagnet around the sample, which reduces any mechanical disturbance of the sample. For odd shaped samples the van der Pauw method can also be used but it is not very precise. Hurd's text listed in the bibliography is a good reference for the Hall effect in metals.

The thermopower is probably the most difficult of the transport properties discussed here to measure correctly. There are two approaches. The differential method in which a small temperature gradient is established across the sample and the thermal voltage is measured so that $S = \Delta V / \Delta T$, or the integral method in which one end of the sample is held at a fixed temperature and the other is continually heated. The temperature of the hot end and the thermal emf are measured continuously and the thermopower is the derivative of the thermal emf versus temperature. These methods are outlined in Barnard's book listed in the bibliography. An important problem with the thermopower of a material is that it can only be measured as part of a thermocouple with reference to another material. In which case a standard is required, the thermopower of which is known. This standard is Pb. It was chosen because Pb is in some sense very low in dislocation density. Any small dislocations in the sample produced by handling tend to self anneal because Pb is so soft. This makes the thermopower of pure Pb very reproducible. Roberts [9] in 1977 performed a very careful experiment measuring the Thomson effect which is related to the thermopower but doesn't require a reference. Before 1977 Cu was used as a reference using data from the early 60's but this proved to be significantly in error particularly below 100 K. At low temperature, superconductors can be used as a reference since they have zero thermopower below $T_c$.

## 3.9    Problems

1. Show how the Drude formula can be obtained from (3.9).
2. Derive (3.17).
3. Estimate the normal magnetoresistance in Cu and in Co at room temperature.
4. Estimate the magnetoresistance due to quantum interference effects in a disordered metal of resistivity 100 μΩcm. (Typical high quality multilayers of Co/Cu have a resistivity of less than 10 μΩcm but some sputtered Fe/Cr multilayers can have resistivities of up to 100 μΩcm.)

## 3.10    Solutions

1.

$$\sigma = \frac{e^2}{12\pi^3\hbar} \oint v\tau dS,$$

with $\nu$ and $\tau$ being constants over the Fermi surface for free electrons. Thus

$$\sigma = \frac{e^2}{12\pi^3\hbar}\,\nu\tau \oint dS = \frac{e^2}{12\pi^3\hbar}\,\nu\tau\,4\pi k_F^2$$

and substituting $\nu = \hbar k_F/m$ and $n = k_F^3/3\pi^2$ leads to $\sigma = ne^2\tau/m$.

2. Switch on $f$ adiabatically: $f \to f\exp(\delta t)$ for $t < 0$ and $f \to f$ for $t > 0$.

$$\frac{\partial f}{\partial t} + \frac{i}{\hbar}[H_0 + H', f_0 + g] = 0$$

$$\Rightarrow f\delta + \frac{i}{\hbar}[H_0, f_0 + g] + \frac{i}{\hbar}[H', f_0 + g] = 0$$

$$f\delta + \frac{i}{\hbar}[H_0, f_0] + \frac{i}{\hbar}[H', f_0] + \frac{i}{\hbar}[H_0, g] + \frac{i}{\hbar}[H', g] = 0$$

$H' \propto \mathcal{E}$ and $g \propto \mathcal{E}$ in the linear response approximation. $[H_0, f_0] = 0$ as $f_0$ is a function of $H_0$. Let $\delta \to 0$, then:

$$\frac{i}{\hbar}[H_0, g] + \frac{i}{\hbar}[H', f_0] = 0$$

$$\frac{i}{\hbar}[H_0, g] + \frac{i}{\hbar}\,e\mathcal{E}\,[x, f_0] = 0\,.$$

3.

$$\frac{\Delta R}{R} = (\omega_c\tau)^2 = \left[\frac{eB}{m}\frac{\sigma m}{ne^2}\right]^2 = \left[\frac{\sigma}{ne}\right]^2 B^2 = \left[\frac{R_H}{\rho}\right]^2 B^2$$

At room temperature typical values are: $\rho(\text{Co}) = 10\ \mu\Omega\text{cm}$, $\rho(\text{Cu}) = 1\ \mu\Omega\text{cm}$ and $R_H(\text{Cu}) = -5 \times 10^{-10}\ \text{C}^{-1}\text{m}^{-3}$ and the ordinary Hall effect in Co is $R_H(\text{Co}) = +1 \times 10^{-9}\ \text{C}^{-1}\text{m}^{-3}$. Therefore the magnetoresistance of both Cu and Co is of the order of $10^{-3}$ and $10^{-4}$ at 1 T and room temperature, respectively. At low temperature it can be much higher if the residual resistivity ratio is very high, as in very pure samples.

4.

$$\Delta\sigma \simeq -\frac{1}{2\pi^2}\frac{e^2}{\hbar}\sqrt{\frac{eB}{m}} \Rightarrow \frac{\Delta\rho}{\rho} = -\frac{1}{2\pi^2}\frac{e^2}{\hbar}\sqrt{\frac{eB}{m}}\,\rho$$

At a field of 1 T the magnetoresistance is about $1 \times 10^{-4}$. This is at a temperature below 10 K. By about 60 K the effect is unmeasurable as inelastic scattering processes destroy the quantum interference.

## Bibliography

J. S. Dugdale, *The Electrical Properties of Metals and Alloys*, Taylor and Francis 1977.

R. D. Barnard, *Thermoelectricity in Metals and Alloys*, Taylor and Francis

1972.

C. M. Hurd, *The Hall Effect in Metals and Alloys*, Plenum 1972.

L. Berger in *The Hall Effect and its Applications*, edited by Chien and Westgate, Plenum 1979.

*Low Level Measurements* published by Keithley Instruments.

A. C. Rose–Innes, *Low Temperature Techniques*, The English Universities Press Ltd 1973.

I. A. Campbell and A. Fert in *Ferromagnetic Metals* Vol.3, edited by P Wohlfarth (1982).

M. A. Howson and B. L. Gallagher, Phys. Rep. **170**, 267 (1988).

J. S. Dugdale, *The Electrical Properties of Disordered Metals*, Cambridge (1995).

# References

[*]  Present address: M. A. Howson, 2 Duchy Grove, Harrogate, HG2 0ND, United Kingdom, e-mail: mark.howson@physics.org.

1.  E. H. Sondheimer, Adv. Phys. **1**, 1 (1952).

2.  R. E. Camley and J. Barnas, Phys. Rev. Lett. **53**, 664 (1989).

3.  T. Valet and A. Fert, Phys. Rev. B **48**, 7099 (1993).

4.  A. Sdaq, J. M. Broto, H. Rakoto, J. C. Ousset, B. Raquet, B. Vidal, Z. Jiang, J. F. Bobo, M. Piecuch, and B. Baylac, J. Magn. Magn. Mater. **121**, 409 (1993).

5.  R. G. Chambers, W. A. Harrison, and M. B. Webb (Eds.), *The Fermi Surface*, Wiley, New York, 1960.

6.  L. J. van der Pauw, Philips Reports **113**, 1 (1958).

7.  H. C. Montgomery, J. Appl. Phys. **42**, 2971 (1971).

8.  A. C. Anderson, Rev. Sci. Instruments **41**, 1446 (1970).

9.  R. B. Roberts, Phil. Mag. B **36**, 91 (1977).

# 4  Phenomenological Theory of Giant Magnetoresistance

J. Mathon

Department of Mathematics, City University, London EC1V 0HB, United Kingdom

## 4.1  Introduction

The era of spin electronics began almost exactly ten years ago with the discovery [1], [2] that the electric current in a magnetic multilayer consisting of a sequence of thin magnetic layers separated by equally thin non-magnetic metallic layers is strongly influenced by the relative orientation of the magnetizations of the magnetic layers. In fact, it is found that the resistance of the magnetic multilayer is low when the magnetizations of all the magnetic layers are parallel (Fig. 4.1a) but it becomes much higher when the magnetizations of the neighbouring magnetic layers are ordered antiparallel (Fig. 4.1b). This implies that the internal

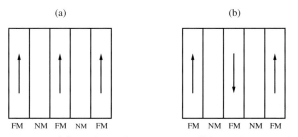

**Fig. 4.1.** Ferromagnetic (a) and antiferromagnetic (b) configurations of a magnetic multilayer.

magnetic moment of electrons associated with their spin plays an important role in transport of electric charge. Hence the term spin electronics. The most commonly used combinations of magnetic and non-magnetic layers are cobalt–copper and iron–chromium but multilayers based on permalloy as the magnetic component are also frequently used.

The second key ingredient was the discovery by Stuart Parkin [3] that the relative orientation of the magnetic moments of two neighbouring magnetic layers depends on the thickness of the intervening non-magnetic layer. In fact, he found that the orientation of the magnetic moments of the magnetic layers oscillates between parallel (ferromagnetic) and antiparallel (antiferromagnetic) as a function of the non-magnetic layer thickness. This phenomenon is referred to as an oscillatory exchange coupling. The oscillatory exchange coupling is very interesting in its own right since it is one of the rare manifestations of quantum interference effects in metals but the underlying physics is not essential for understanding of the transport of current in magnetic multilayers. The interested reader is referred to specialist articles such as [4].

Assuming that the thickness of the non-magnetic spacer layer is chosen so that the spontaneous orientation of the adjacent magnetic layers is antiferromagnetic, a change of the magnetic configuration from antiferromagnetic to ferromagnetic, and hence a change of the resistance, can be effected by an applied magnetic field. The relative change of the resistance can be larger than 200%, and that is the reason why the effect is called giant magnetoresistance (GMR). The giant magnetoresistance should not be confused with the anisotropic magnetoresistance (AMR) or the ordinary positive magnetoresistance discussed elsewhere in this book. These additional magnetoresistance effects are the intrinsic properties of elemental solids (or alloys), whereas the giant magnetoresistance is always the property of a device consisting of alternating layers of different materials.

The 'optimistic' magnetoresistance ratio, most commonly used, is defined by

$$\frac{\Delta R}{R} = \frac{R^{\uparrow\downarrow} - R^{\uparrow\uparrow}}{R^{\uparrow\uparrow}}, \tag{4.1}$$

where $R^{\uparrow\downarrow}$ and $R^{\uparrow\uparrow}$ are the resistances of the magnetic multilayer in its antiparallel (zero field) and parallel (saturation field) magnetic configurations. The optimistic GMR ratio is unbounded but the 'pessimistic' ratio $(\Delta R/R) = (R^{\uparrow\downarrow} - R^{\uparrow\uparrow})/R^{\uparrow\downarrow}$, which is also in use, is never greater than 1. The dependence

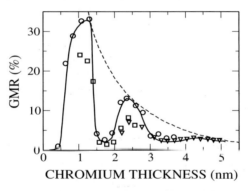

**Fig. 4.2.** Dependence of the GMR ratio of an Fe/Cr multilayer on Cr thickness. After [3].

of the GMR ratio of a Fe/Cr multilayer on the thickness of the non-magnetic chromium layer, observed by Parkin in his original experiment [3], is reproduced in Fig. 4.2. Oscillations of the GMR as a function of chromium thickness occur because the magnetoresistance effect is measurable only for those thicknesses of chromium for which the interlayer exchange coupling aligns the magnetic moments of all the iron layers antiparallel.

A typical magnetoresistance curve [5] for an Fe/Cr multilayer of fifty repeats of an iron layer 0.45nm thick and a chromium layer 1.2nm thick is shown in

Fig. 4.3 for two temperatures $T = 1.5K$ and $300K$. The gradual decrease of the resistance with increasing magnetic field, seen in Fig. 4.3, occurs because the

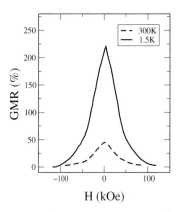

**Fig. 4.3.** Magnetoresistance curve of an Fe/Cr multilayer.

magnetic field which tends to align the moments of the magnetic layers parallel has to overcome the oscillatory exchange coupling which favours the antiparallel arrangement (for this particular thickness of chromium). Complete alignment is achieved only in a saturating field equal in magnitude to the exchange field.

The discovery of the GMR has created great excitement since the effect has important applications, particularly in magnetic information storage technology. Information is stored on a magnetic disc in the form of small magnetised regions (domains) arranged in concentric tracks. A conventional reading head used to be an induction coil which senses the rate of change of the magnetic field as the disc rotates. The signal and hence the density of magnetised bits is thus limited by the speed of rotation of the disc. Magnetoresistive sensors do not suffer from this defect since they sense the strength of the field rather than its rate of change. They are, therefore, capable of reading discs with a much higher density of magnetic bits. Magnetoresistive reading heads are commercially available and will be the leading technology beyond the year 2000 [6].

## 4.2   Physical Origin of GMR

To clarify the origin of the GMR we begin with a simple overview. There are two principal geometries of the GMR effect. They are shown schematically in Fig. 4.4. In the first case (Fig.4a), the current flows  perpendicular to the layers (CPP geometry). Figure 4.4b illustrates the more usual geometry when the current flows in plane of the layers (CIP). As we shall see, the CPP geometry is easier to treat theoretically but much more difficult to realize experimentally. This is because the transverse dimensions of typical multilayers are of the order of $cm^2$ whereas their thickness is only of the order of a few nanometers. It follows

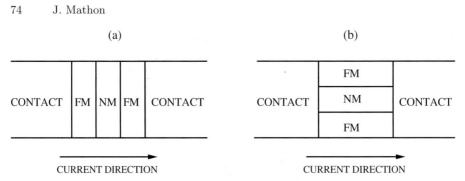

Fig. 4.4. Current perpendicular to plane (a) and current in plane (b) GMR geometries.

that the resistance of the multilayer in the CPP geometry is extremely low and, therefore, very sophisticated experimental techniques [7] are required to measure accurately the very small voltage drop across the sample. However, since the underlying physical mechanism is the same in the CPP and CIP geometries, there is no need in such an introductory account to distinguish between them.

Consider a trilayer with two magnetic layers separated by a non-magnetic metallic spacer layer. The GMR effect relies on the experimentally established fact that electron spin is conserved over distances of up to several tens of nanometers, which is greater than the thickness of a typical multilayer. We can thus assume that electric current in the trilayer flows in two channels, one corresponding to electrons with spin projection ↑ and the other to electrons with spin projection ↓ [8]. Since the ↑ and ↓ spin channels are independent (spin is conserved) they can be regarded as two wires connected in parallel.

The second essential ingredient is that electrons with spin projections parallel and antiparallel to the magnetization of the ferromagnetic layer are scattered at different rates when they enter the ferromagnet. This is called spin-dependent scattering. Let us assume that electrons with spin antiparallel to the magnetization are scattered more strongly. We shall see later that this is the case for the Co/Cu combination but the opposite is true for the Fe/Cr system. The GMR effect in a trilayer can be now explained qualitatively using a simple resistor model shown in Fig. 4.5. In the ferromagnetic configuration of the trilayer, electrons with ↑ spin are weakly scattered both in the first and second ferromagnet whereas the ↓ spin electrons are strongly scattered in both ferromagnetic layers. This is modelled by two small resistors in the ↑ spin channel and by two large resistors in the ↓ spin channel in the equivalent resistor network shown in Fig. 4.5a. Since the ↓ and ↑ spin channels are connected in parallel, the total resistance of the trilayer in its ferromagnetic configuration is determined by the low-resistance ↑ spin channel which shorts the high-resistance ↓ spin channel. It follows that the total resistance of the trilayer in its ferromagnetic configuration is low. On the other hand, ↓ spin electrons in the antiferromagnetic configuration are strongly scattered in the first ferromagnetic layer but weakly scattered in the second ferromagnetic layer. The ↑ spin electrons are weakly scattered in the first ferromagnetic layer and strongly scattered in the second. This is modelled in

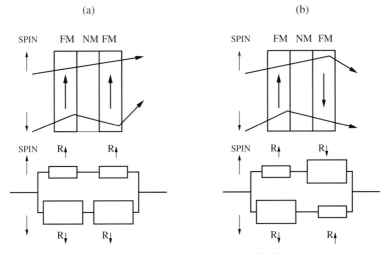

**Fig. 4.5.** Resistor model of GMR.

Fig. 4.5b by one large and one small resistor in each spin channel. There is no shorting now and the total resistance in the antiferromagnetic configuration is, therefore, much higher that in the ferromagnetic configuration.

This simple physical model of the GMR effect is believed to be correct but needs to be converted into a quantitative theory that can explain the differences between the CIP and CPP geometries, the observed dependence of the GMR on the layer thicknesses and also the material dependence of the effect. Moreover, we need to understand the microscopic origin of the spin-dependent scattering and clarify under what conditions the ↑ and ↓ spin channels in magnetic multilayers can be treated as independent.

In spite of great efforts made over the last ten years a fully predictive theory of the GMR is still not available. The first theories were based on the Boltzmann equation [9]. A more microscopic description used the quantum Kubo formula [10]. In either approach the electronic band structure of the magnetic and non-magnetic layers was approximated by a simple parabolic band common to the whole multilayer and spin dependent scattering was introduced phenomenologically. When these two assumptions are made, the results obtained from the Boltzmann and Kubo formulations are essentially equivalent. A more recent refinement is to incorporate in the Boltzmann equation a fully realistic band structure [11]. The main advantage of this approach is that the spin dependent scattering is introduced from first principles and the dependence of the GMR on different magnet/non-magnet combinations can be thus discussed. Finally, *ab initio* quantum calculations based on the Kubo formula and realistic band structure have recently been made but are possible only for the CPP geometry under some simplifying assumptions about the disorder in the system. The reader is referred to an excellent recent review [12] of the status of the theory of the CPP GMR. In this introductory chapter we shall describe a quantitative resistor net-

work theory of the GMR [13]. The resistor network theory is equivalent to the
Boltzmann equation approach but has the advantage that it allows us to derive
simple analytic formulae both for the CIP and CPP GMR.

## 4.3   Spin Dependent Scattering of Electrons
in Magnetic Multilayers

We begin our discussion with the different types of scattering electrons may experience in magnetic multilayers. In the Boltzmann equation approach we are
mainly concerned with elastic (energy conserving) scattering. In each scattering act only the direction of propagation of electrons changes. It is essential to
distinguish between spin dependent scattering which causes the GMR and spin
flip scattering which is detrimental to the GMR. The two types of scattering are
illustrated in Fig. 4.6. In the case of spin dependent scattering the orientation
of the electron spin is conserved in each scattering event but the probabilities
of scattering for electrons with $\uparrow$ and $\downarrow$ spin projections are different. On the
other hand, when an electron undergoes a spin-flip scattering, its spin orienta-
tion changes from $\uparrow$ $(s_z = \hbar/2)$ to $\downarrow$ $(s_z = -\hbar/2)$ or vice versa and, at the same
time, the spin of the scattering centre changes by $\Delta = \hbar$ so that the total spin
is conserved.

There are several sources of spin flip scattering. When magnetic multilayers
are prepared, some of the magnetic atoms may enter the non-magnetic spacer
layer to form magnetic impurities. When an electron is scattered off a magnetic
impurity the spins of the electron and that of the impurity can interchange
provided the impurity spin is free to rotate. This is the case when the impurity
spin is not strongly coupled to the spins of the ferromagnetic layers, i.e. when
the impurity is not near the ferromagnet/spacer interface.

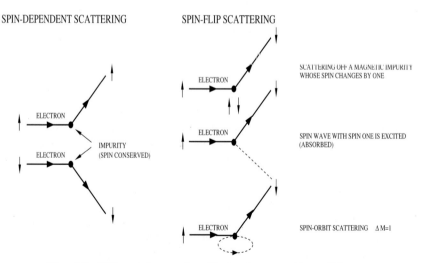

**Fig. 4.6.** Different types of scattering in magnetic multilayers.

Electrons can also be scattered from spin waves in the ferromagnetic layers. Spin waves are quasiparticles with spin one and, therefore, creation (annihilation) of a spin wave in a collision with an electron leads to a flip of the electron spin. Since creation (annihilation) of spin waves involves the spin-wave energy, this is an inelastic process which is only important at elevated temperatures.

Finally, when impurities with a strong spin-orbit interaction, such as gold, are present in the multilayer, the spin of an electron incident on such an impurity may be reversed due to the spin-orbit interaction.

Since all these processes mix $\uparrow$ and $\downarrow$ spin channels, they are detrimental to the GMR. In what follows, we shall assume that spin-flip scattering is weak so that no mixing of the $\uparrow$ and $\downarrow$ spin channels takes place. This assumption may break down for relatively thick multilayers and the implications of spin flip scattering for GMR are discussed in detail in Ref. [14].

We now turn to the spin dependent scattering which conserves the electron spin. The key feature here is that electrons with different spin orientations ($\uparrow, \downarrow$) are scattered at different rates when they enter the ferromagnetic layers. Given that electrons obey the Pauli exclusion principle, an electron can be scattered from an impurity only to quantum states that are not occupied by other electrons. At zero (low) temperatures, all the states with energies $E$ below the Fermi energy $E_F$ are occupied and those with $E > E_F$ are empty. Since scattering from impurities is elastic, electrons at the Fermi level (which carry the current) can be scattered only to states in the immediate vicinity of the Fermi level. It follows that the scattering probability is proportional to the number of states available for scattering at $E_F$, i.e. to the density of states $D(E_F)$. The densities of states of copper, cobalt and iron for $\uparrow$ (upper panel) and $\downarrow$ (lower panel) spin orientations are shown in Fig. 4.7. The Fermi level in copper (and other noble metals) intersects only the conduction band whose density of states $D(E_F)$ is low. It follows that the scattering probability in copper is also low, which explains why copper is a very good conductor. On the other hand, the $d$ band in transition metals is only partially occupied and, therefore, the Fermi level in these metals intersects not only the conduction but also the $d$ bands. Moreover, since the atomic wave functions of $d$ levels are more localized than those of the outer s levels, they overlap much less, which means that the $d$ band is narrow and the corresponding density of states is high. This opens up a new very effective channel for scattering of conduction electrons into the $d$ band. This new scattering mechanism (Mott scattering [15]) explains why all transition metals are poor conductors compared with noble metals.

In the case of magnetic transition metals, we need to consider an additional crucial factor, namely that $d$ bands for $\uparrow$ and $\downarrow$ spin electrons are split by the exchange interaction. This amounts to an almost rigid relative shift of the $\uparrow$ and $\downarrow$ spin $d$ bands which is clearly seen for cobalt and iron in Fig. 4.7. The $\uparrow$ spin $d$ band in cobalt is full which means that $D^\uparrow(E_F)$ is as low as in copper but the Fermi level in the $\downarrow$ spin band lies in the $d$ band and, therefore, $D^\downarrow(E_F)$ is much higher than $D^\uparrow(E_F)$. The situation for iron is somewhat different in that the density of states at $E_F$ is higher for $\uparrow$ spin electrons than for $\downarrow$ spin

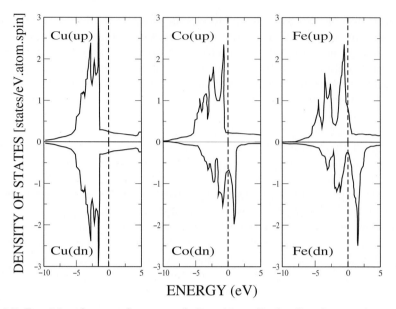

**Fig. 4.7.** Densities of states of copper, cobalt and iron. Broken line denotes the position of the Fermi level.

electrons. Also the spin asymmetry in the density of states is not so large for iron as for cobalt. However, in either case, the spin asymmetry of the density of states results in different scattering rates for ↑ and ↓ spin electrons, i.e. spin dependent scattering. It should be noted that this mechanism operates even if the scattering potential itself is independent of the spin, i.e. non-magnetic impurities, vacancies or stacking faults in a ferromagnetic metal all lead to spin dependent scattering. Since the Mott scattering mechanism is effective in bulk ferromagnetic metals, we shall refer to it as bulk spin dependent scattering.

The relative shift of ↑ and ↓ spin bands is simply a consequence of the fact that the potentials seen by ↑ and ↓ electrons in a ferromagnetic metal are different because of the exchange interaction. This provides another mechanism of spin dependent scattering which is specific to multilayers. In an infinite ferromagnet this effect does not, of course, lead to any spin asymmetry of the resistance since as long as the potentials seen by ↑ and ↓ electrons are periodic they do not result in any dissipation of the electron momentum. However, electrons in a multilayer entering the ferromagnet from the non-magnetic spacer see a spin dependent potential barrier which reflects differently electrons with ↑ and ↓ spin orientations. In the CPP geometry, even perfect interfaces thus result in spin dependent scattering [12]. In the CIP geometry, electrons propagate mainly along interfaces and, therefore, this mechanism is effective only if the interfaces are rough (intermixing of magnetic and non-magnetic atoms).

As opposed to the bulk Mott mechanism discussed earlier, spin dependent scattering due to spin dependence of the scattering potentials takes place only at the ferromagnet/non-magnet interface and is, therefore, called interfacial spin

dependent scattering. To gain better understanding of the interfacial spin dependent scattering, it is instructive to examine the band structures of the most
common combinations of magnetic and non-magnetic metals used in GMR multilayers. These are Co/Cu and Fe/Cr, and their band structures in the [001]
direction are shown in Figs. 4.8 and 4.9. It can be seen from Fig. 4.8 that there

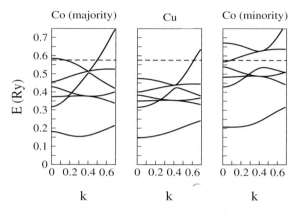

**Fig. 4.8.** Band structures of cobalt and copper along the [001] direction in the vicinity
of one of the Cu Fermi surface necks. Broken line denotes the position of the Fermi
level.

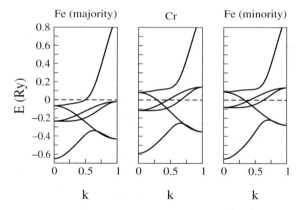

**Fig. 4.9.** Band structures of iron and chromium in the [001] direction. Broken line
denotes the position of the Fermi level.

is a very good match between the bands of Cu and the ↑ (majority) spin band
of Co. One can, therefore, conclude that ↑ spin electrons crossing the Cu/Co
interface experience only weak scattering, and this remains true even if Cu and
Co atoms are intermixed at the interface. On the other hand, there is a large
mismatch between the Cu and Co bands for the ↓ (minority) spin electrons re-

flecting a large difference between the atomic potentials of the two elements. It follows that ↓ spin electrons are strongly scattered at the Cu/Co interfaces. On the other hand, matching of the Fe and Cr bands is almost perfect for ↓ spin electrons but poor for the ↑ spin electrons. The spin asymmetry of scattering at the Fe/Cr interface has, therefore, a sign opposite to that for the Co/Cu interface.

The above discussion of spin dependent scattering based on the mismatch of bands of the magnetic and non-magnetic components of magnetic multilayers allows us also to understand which combinations of magnetic and non-magnetic metals should lead to optimum GMR. One clearly seeks as good a match as possible between the bands of the magnetic layers and those of the spacer layer in one spin channel and as large as possible mismatch in the other spin channel. It is clear from Figs. 4.8 and 4.9 that Co/Cu and Fe/Cr fulfil very well these requirements.

## 4.4   Resistor Network Theory of GMR

We now need to incorporate the effect of spin dependent scattering into the Boltzmann equation to determine the resistances of a magnetic multilayer in its ferromagnetic and antiferromagnetic configurations. Before we focus our attention on magnetic multilayers, it is useful to recapitulate a few basic facts about conduction in metals [16]. When an electric field $\mathbf{E}$ is applied to a metal, electrons experience a constant force $\mathbf{F} = -e\mathbf{E}$, where $e$ is the electron charge. Electrons in vacuum would accelerate in an electric field with a constant acceleration proportional to $\mathbf{E}$. However, electrons in a metal are scattered from imperfections and the scattering modifies the simple accelerated motion. Let us assume that on average each electron moves during a time interval $\tau$ without scattering. The time $\tau$ is called the mean free time and the distance covered during this time is the mean free path $\ell$.

During the time $\tau$ each electron with mass $m$ accelerates and acquires a velocity $\mathbf{v}$ in the direction of the electric field. When it is scattered from an imperfection, the direction of its velocity changes but the magnitude of the velocity remains unchanged (elastic collision). Every time an electron is scattered back it has to repeat the acceleration process to regain its velocity $\mathbf{v}$. As a result a steady state is reached where all the electrons move in the direction of the field with a velocity acquired during the accelerated motion between two scattering events

$$\mathbf{v} = -\frac{e\mathbf{E}\tau}{m}.$$ 
(4.2)

If in a constant electric field there are n electrons per unit volume, the electric current density is

$$\mathbf{j} = -ne\mathbf{v} = \frac{ne^2\tau\mathbf{E}}{m}.$$ 
(4.3)

Equation (4.3), which follows from the Boltzmann equation in the so called relaxation time approximation [17], is just Ohm's law and we can define the electrical conductivity $\sigma$ by $\mathbf{j} = \sigma\mathbf{E}$. Since later on we are going to discuss resistances, it is more convenient to introduce the resistiviy $\rho = 1/\sigma$. The resistivity is clearly given by

$$\rho = \frac{m}{ne^2\tau}. \tag{4.4}$$

Apart from the electron density $n$, the main factor that determines the resistivity is the mean free time $\tau$. The mean free time is clearly inversely proportional to the scattering probability. The scattering probability is, in turn, determined by two factors. The first is the strength of the scattering potential and the second the density of states at $E_F$ available for scattering. As already discussed, the first factor leads to interfacial spin dependent scattering in magnetic multilayers and the second is the Mott mechanism which results in spin dependent bulk scattering. We first consider the effect of the bulk spin dependent scattering. If follows from (4.3) and the above arguments that we can introduce a spin dependent resistivity for each ferromagnetic metal by

$$\rho_{FM}^{\uparrow} = 2\frac{\rho_{FM}}{1+\beta}; \ \rho_{FM}^{\downarrow} = 2\frac{\rho_{FM}}{1-\beta}, \tag{4.5}$$

where $\rho_{FM}$ is the total resistivity of the bulk ferromagnetic metal, $1/\rho_{FM} = 1/\rho_{FM}^{\uparrow} + 1/\rho_{FM}^{\downarrow}$ (assuming that the two spin channels remain independent). The parameter $\beta$ we have introduced will be referred to as the bulk scattering asymmetry. We shall treat $\beta$ as a phenomenological parameter but we expect from the discussion of the Mott scattering mechanism that $\rho^{\uparrow}/\rho^{\downarrow} \approx D^{\uparrow}(E_F)/D^{\downarrow}(E_F)$. In particular, it follows from Fig. 4.7 that $\beta < 0$ for cobalt and $\beta > 0$ for iron.

Similarly, we can introduce an interfacial scattering asymmetry assuming that there is a thin interfacial layer whose resistance $\rho_{F-N}^{\sigma}$ is spin dependent due to the presence of a spin-dependent potential barrier at the ferromagnet/nonmagnet interface. We, therefore, define an interfacial asymmetry parameter $\gamma$ by

$$\rho_{F-N}^{\uparrow} = 2\frac{\rho_{F-N}}{1+\gamma}; \ \rho_{F-N}^{\downarrow} = 2\frac{\rho_{F-N}}{1-\gamma}, \tag{4.6}$$

where $\rho_{F-N}$ is the total resistivity of the interfacial layer. We can again deduce from Figs. 4.8 and 4.9 that $\gamma < 0$ for cobalt and $\gamma > 0$ for iron. However, the actual magnitude of $\gamma$ is difficult to determine microscopically since it depends not only on the difference between the potentials seen by $\uparrow$ and $\downarrow$ spin electrons at the interface but also on the interfacial roughness and the thickness of the interfacial layer for which $\rho_{F-N}^{\sigma}$ is introduced. We shall, therefore, treat $\gamma$ as a free parameter.

We are now ready to calculate the GMR. The calculation will be described for bulk spin dependent scattering and CIP geometry. It is straightforward to include interfacial spin dependent scattering and this is the subject of the first exercise at the end of the chapter. The calculation of the CPP GMR is also left to the reader (second exercise).

Consider a periodic superlattice of alternating non-magnetic and magnetic layers with spin dependent scattering in the bulk of the ferromagnetic layers. Since the whole superlattice is made up of identical building blocks, superlattice unit cells, it is sufficient to calculate the resistances of a unit cell. In the antiferromagnetic configuration, the magnetic layers with antiparallel magnetizations are inequivalent and, therefore, the basic building block we have to consider (magnetic cell) consists of two magnetic layers containing $M$ atomic planes each and two non-magnetic layers of $N$ atomic planes each. The geometry for which the GMR is going to be calculated and the definition of a magnetic cell are illustrated in Fig. 4.10. It follows from (4.5) that an electron of a given

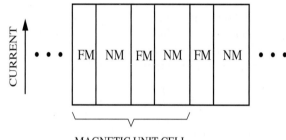

**Fig. 4.10.** Magnetic superlattice.

spin travelling in a superlattice sees regions of different local resistivities. The resistivity is high in those regions where there is a high density of states at $E_F$ available for scattering. There are, therefore, three different local resistivities in the superlattice unit cell: the resistivity of the non-magnetic spacer layer $\rho_{NM}$, which is the same for both spin orientations, and the high $\rho_{FM}^H$ and low $\rho_{FM}^L$ resistivities for the two different spin orientations in the ferromagnet. The low resistivity of the ferromagnet satisfies $\rho_{NM} \approx \rho_{FM}^L$ both for the Co/Cu and Fe/Cr systems. The distribution of such regions in a superlattice magnetic cell in its ferromagnetic and antiferromagnetic configurations is shown in Fig. 4.11. It is clear from Fig. 4.11 that the unit cell of a magnetic superlattice is equivalent to a system of eight resistors, with four resistors in each spin channel. To determine the magnetoresistance, we first need a rule for adding up the four resistors in the same spin channel. Once the total resistances in both spin channels are known, they can be simply added as resistors in parallel to give the total resistance of the magnetic unit cell. This needs to be done for the ferromagnetic ($\uparrow\uparrow$) and antiferromagnetic ($\uparrow\downarrow$) configurations. Following this prescription, we find that the resistances $R_{\uparrow\uparrow}$ and $R_{\uparrow\downarrow}$ are given by

$$\frac{1}{R_{\uparrow\uparrow}} = \left( \frac{1}{R_\uparrow} + \frac{1}{R_\downarrow} \right)_{\uparrow\uparrow} \; ; \; \frac{1}{R_{\uparrow\downarrow}} = \left( \frac{1}{R_\uparrow} + \frac{1}{R_\downarrow} \right)_{\uparrow\downarrow} , \qquad (4.7)$$

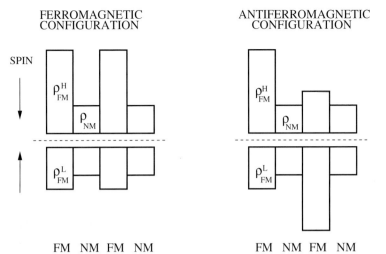

**Fig. 4.11.** Distribution of local resistivities in a magnetic unit cell.

where $R_\sigma$ is the resistance of the unit cell in a spin channel $\sigma$.

It is now necessary to determine the rules for adding up the four resistors in the same spin channel. It is clear from Fig. 4.11 that for the ferromagnetic configuration the problem reduces to the calculation of the resistance of a two-component superlattice with alternating regions of thicknesses $a$ and $b$ having resistances $\rho^a$ and $\rho^b$. For the antiferromagnetic configuration a four-component superlattice needs to be considered. To clarify the underlying physics, it is sufficient to investigate the two-component superlattice. Because the current flows in the direction of the layers forming the superlattice, one might be tempted to conclude that the resistances of the layers should always be added up as for resistors connected in parallel. However, it is easy to demonstrate that, in general, that would be incorrect.

Let us assume for simplicity that the resistivity $\rho^a$ is higher than $\rho^b$ because there is a higher density of scatterers in the layer $a$. This is illustrated in Fig. 4.12. Consider first the simplest case when there is a 'partition' between the layers $a$ and $b$ which prevents electrons crossing the $a/b$ interface. Such a system is clearly equivalent to two *independent* resistors because electrons remain confined to their respective layers. In this case the above argument applies, i.e. the two layers behave as ordinary resistors in parallel. However, there are no impenetrable partitions between the neighbouring regions in a superlattice. Electrons can cross easily the interface and undergo scattering in both layers. It follows that the two layers cannot be regarded as independent and, in general, the simple rules for a conventional network of resistors no longer apply. The reader might conclude that we can get no further without a detailed microscopic calculation. Fortunately this is not so and there are two physically important limits in which the total resistance of a superlattice can be easily evaluated.

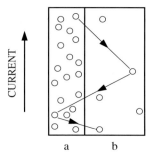

**Fig. 4.12.** Magnetic superlattice.

*Case A:* The mean free path in each layer of a superlattice is much shorter than the thickness of the layer. Because the mean free path is so short, very few electrons starting in one layer reach the neighbouring layer. It follows that electrons from different layers do not 'mix' and flow in their respective separate resistor channels as if the layers were separated by 'partitions'. All resistors then behave as resistors in a conventional resistor network and should be added in parallel. Inspection of Fig. 4.11 shows that there are exactly the same number of resistors of each type in the ferromagnetic and antiferromagnetic configurations, which means that $R_{\uparrow\uparrow} = R_{\uparrow\downarrow}$ and there is no magnetoresistance in this limit.

*Case B:* The mean free path in each layer is much longer than the thickness of the layer. For a metallic superlattice with a small number of atomic planes in each layer we are always close to this limit since typical mean free paths in metals are of the order of tens or even hundreds of interatomic distances. It is, therefore, the limit which is applicable to magnetic superlattices exhibiting the giant magnetoresistance.

Conduction electrons now sample equally layers with low and high resistivity and, therefore, experience an average resistivity. For a two-component superlattice this is given by

$$\bar{\rho} = \frac{a\rho^a + b\rho^b}{a + b} \tag{4.8}$$

The generalisation of (4.8) to a four-component superlattice is obvious.

The results for cases A and B obtained here from simple physical considerations can be derived as limits of a more general approach based on the Boltzmann equation [13]. In fact, such a microscopic calculation shows that the limit in which the simple averaging (4.8) applies is reached very rapidly and we are close to this limit already for a mean free path comparable with the thickness of the superlattice unit cell.

We can now apply (4.8) and the corresponding result for the four-component superlattice to evaluate the magnetoresistance $\Delta R/R$. In going from resistivities to the total resistance of a superlattice cell, care has to be taken of the dimensions of the cell. However, because the length of the sample in the direction of the current and the transverse area per atomic plane are the same in the ferromagnetic and antiferromagnetic configurations, they cancel out in the calculation of

$\Delta R/R$. It is, therefore, sufficient to determine for each spin channel the reduced resistance $R_\sigma$ per unit length and unit area. It is then clear from Fig. 4.11 that

$$\left(\frac{1}{R}\right)_{\uparrow\uparrow} = 2(M+N)^2 \left(\frac{1}{M\rho_{FM}^L + N\rho_{NM}} + \frac{1}{M\rho_{FM}^H + N\rho_{NM}}\right) \quad (4.9)$$

$$\left(\frac{1}{R}\right)_{\uparrow\downarrow} = \frac{8(M+N)^2}{M\rho_{FM}^L + M\rho_{FM}^H + 2N\rho_{NM}}. \quad (4.10)$$

Substituting from (4.9) and (4.10) into (4.1), we find that the magnetoresistance is given by

$$\frac{\Delta R}{R} = \frac{(1-\beta)^2}{4(1+N/M\mu)(\beta+N/M\mu)}, \quad (4.11)$$

where $\beta$ is the bulk scattering asymmetry and $\mu = \rho_{FM}^L/\rho_{NM}$.

It is now easy to pinpoint the main factors that determine the GMR. Clearly $\Delta R/R$ is a function of two variables, $\beta$ and $M\mu/N$. The most important requirement for large GMR is that the spin asymmetry ratio $\beta$ should be large. For a given $\beta$, the GMR increases with increasing $M\mu/N$ but saturates for a large value of this parameter. As a function of the spacer layer thickness $N$, the GMR decreases monotonically and falls off as $1/N^2$ for large $N$, which is as observed (see the broken line in Fig. 4.2). This can be viewed as 'shunting' of the cooperative effect of the magnetic layers by an 'inactive' spacer layer.

Experiments show [18] that CIP GMR also decreases with increasing thickness of the magnetic layers. This is not reproduced by (4.11). There are two possible reasons for the failure of (4.11) in the limit of thick magnetic layers. Since $\rho_{FM}^H$ is high, the mean free path in one of the spin channels may become shorter than the ferromagnetic layer thickness and the simple averaging (4.8) of the resistivities no longer applies. The other reason is our neglect of interfacial spin dependent scattering. When the simple averaging (4.8) fails, the only alternative is a numerical solution of the Boltzmann equation in the layer geometry [13]. On the other hand, the effect of interfacial scattering can be easily included in the resistor network formalism.

Until now we have assumed that bulk spin dependent scattering is dominant. We now adopt the other extreme point of view, i.e. assume that interfacial spin dependent scattering is so strong that bulk scattering can be neglected ($\beta = 0$). We shall further assume that there are $I$ interfacial atomic planes with $\rho_{F-N}^\sigma$ defined by (4.6). Finally we shall make a simplifying assumption $\rho_{NM} = \rho_{FM}^L$ (a good approximation for Co/Cu and Fe/Cr multilayers). It is then easy to show (exercise 1) that

$$\frac{\Delta R}{R} = \frac{(1-\gamma)^2}{4[1+(N+M)/2I\nu][\gamma+(N+M)/2I\nu]}, \quad (4.12)$$

where $\gamma$ is the interfacial scattering asymmetry defined in (4.6) and $\nu = \rho_{F-N}^L/\rho_{NM}$. The GMR now decreases both with increasing thicknesses $N, M$

of the spacer and the ferromagnet. The physical interpretation is that the magnetic layer first grows as a rough interface with strong spin dependent scattering and then turns into an 'inactive' shunting layer. It might seem that the model with dominant interfacial scattering explains better the dependence of the GMR on the ferromagnet thickness. However, numerical solution of the Boltzmann equation with dominant bulk spin dependent scattering leads also to a GMR which decreases with increasing thickness of the ferromagnetic layers, in good agreement with experiment [18]. One must, therefore, conclude that by analyzing experimental data in the CIP geometry, it is not possible to determine reliably the relative importance of bulk and interface spin dependent scattering. The situation is much clearer in the CPP geometry which will be now briefly discussed.

In applying the simple resistor model to the CPP GMR we need to make an assumption that the mean free path for spin-flip scattering is longer than the total length of the multilayer in the direction of the current. This is necessary for ↑ and ↓ spin channels to remain independent so that we can add up their total resistances $R_\uparrow$ and $R_\downarrow$ in parallel. We shall again introduce bulk scattering asymmetry using (4.5), but it is more convenient in the CPP geometry to characterise interfacial scattering by the total resistances of an interface for ↑ and ↓ spin channels, i.e. we define high $R_{F-N}^H$ and low $R_{F-N}^L$ interfacial resistances. Naturally, they are related to the total interfacial resistance $R_{F-N}$ and to the interfacial scattering asymmetry $\gamma$ via (4.6). Finally, we need to decide how to add up all the resistors in the same spin channel. This is simple in the CPP geometry since electrons move in the direction perpendicular to the layers and, therefore, sample individual layers one by one. All the layers thus behave as conventional resistors connected in series. We shall again consider a superlattice having $M$ atomic planes in each ferromagnetic layer and $N$ planes in each non-magnetic layer. To calculate the total resistances $R_{\uparrow\uparrow}$ and $R_{\uparrow\downarrow}$ of the superlattice in its ferromagnetic and antiferromagnetic configurations, we need to introduce also the total number $N_{MC}$ of magnetic unit cells. It is then straightforward to show that

$$(AR)_{\uparrow\downarrow} = \frac{N_{MC}}{2}[M(\rho_{FM}^L + \rho_{FM}^H) + 2\rho_{NM}N + 2A(R_{F-N}^H + R_{F-N}^L)], \quad (4.13)$$

where $A$ is the cross section area of the superlattice and the quantities $\rho_{FM}^L$, $\rho_{FM}^H$ have already been introduced in the CIP geometry.

Equation (4.13) can be used to test the validity of the series resistor model. It implies that the total resistance in the antiferromagnetic configuration increases linearly with the thickness of the superlattice ($N_{MC}$). It is found [19] that (4.13) is well obeyed for Co/Cu and Co/Ag multilayers.

One can easily obtain also the total resistance $R_{\uparrow\uparrow}$ of a magnetic superlattice in its ferromagnetic configuration and, hence, the GMR ratio $\Delta R/R$. However, it turns out that it is more useful to examine a closely related quantity

$$[(R_{\uparrow\downarrow} - R_{\uparrow\uparrow})R_{\uparrow\downarrow}]^{1/2} = \frac{N_{MC}}{2}(\beta\rho_{FM}^* M + 2\gamma AR_{F-N}^*), \quad (4.14)$$

where $\rho_{FM}^* = \rho_{FM}/(1-\beta^2)$, $R_{F-N}^* = R_{F-N}/(1-\gamma^2)$ and $\beta$, $\gamma$ are the bulk and interfacial scattering asymmetries. If we plot the left-hand side of (4.14) as a function of the thickness $M$ of the ferromagnetic layer keeping $N_{MC}$ fixed, we obtain a straight line with a slope $N_{MC}\beta\rho_{FM}^*/2$ and an intercept $N_{MC}\gamma AR_{F-N}^*$. The slope is thus determined entirely by bulk spin dependent scattering and the intercept by interfacial spin dependent scattering. It follows that the two types of scattering can be separated in the CPP geometry. An analysis of the CPP GMR experiments for Co/Ag and Co/Cu superlattices [20], [12] based on (4.14) shows that $\beta \approx 0.5$ and $\gamma \approx 0.6 - 0.8$ for both these systems. Bulk and interface scattering are, therefore, comparable.

The equivalent resistor theory of the GMR provides a correct semi-quantitative explanation of the effect and is particularly useful for analyzing experiments in the CPP geometry [12]. However, its main shortcomings are that the spin scattering asymmetries $\beta$ and $\gamma$ are introduced as phenomenological parameters and the differences between the band structures of the ferromagnetic and non-magnetic layers are ignored. An interesting recent development is a calculation [21] of the GMR assuming the Mott scattering mechanism (bulk spin dependent scattering) but using the Boltzmann equation in the layer geometry combined with a realistic tight-binding band structure. This approach provides a microscopic underpinning of the phenomenological resistor model described here.

## 4.5   Exercises

1. Derive (4.12) for the CIP GMR due to interfacial scattering.
2. Derive the formula (4.14) for the CPP GMR.

## References

1. P. Grünberg, R. Schreiber, Y. Pang, M.B. Brodsky, and H. Sower, Phys. Rev. Lett. **57**, 2442 (1986); G. Binasch, P. Grünberg, F. Saurenbach, and W. Zinn, Phys. Rev. B **39**, 4828 (1989).
2. M.N. Baibich, J.M. Broto, A. Fert, Van Dau Nguyen, F. Petroff, P. Etienne, G. Creuset, A. Friederich, and J. Chazelas, Phys. Rev. Lett. **61**, 2472 (1988).
3. S.S.P. Parkin, N. More, and K.P. Roche, Phys. Rev. Lett. **64**, 2304 (1990).
4. D.M. Edwards, J. Mathon, R.B. Muniz, and M.S. Phan, Phys. Rev. Lett. **67**, 493 (1991); P. Bruno and C. Chappert, Phys. Rev. Lett. **67**, 1602 (1991); J. Mathon, Murielle Villeret, A. Umerski, R.B. Muniz, J. d'Albuquerque e Castro, and D.M. Edwards, Phys. Rev. B **56**, 11797 (1997); P. Bruno, Phys. Rev. B **52**, 411 (1995).
5. R. Schad, C.D. Potter, P. Belien, G. Verbanck, V.V. Moshchalkov, and Y. Bruynseraede, Appl. Phys. Lett. **64**, 3500 (1994).
6. See the contribution by P. P. Freitas in this volume.
7. W.P. Pratt, Jr., S.F. Lee, J.M. Slaughter, R. Loloee, P.A. Schroeder, and J. Bass, Phys. Rev. Lett. **66**, 3060 (1991); M.A.M. Gijs, S.K.J. Lenczowski, and J.B. Giesbers, Phys. Rev. Lett. **70**, 3343 (1993).
8. A. Fert and I.A. Campbell, J. Phys. F: Metal Physics **6**, 849 (1976).

9. R.E. Camley and J. Barnas, Phys. Rev. Lett. **63** 664 (1989); J. Barnas, A. Fuss, R.E. Camley, P. Grünberg, and W. Zinn, Phys. Rev. B **42**, 8110 (1990).
10. S. Zhang and P.M. Levy, J. Appl. Phys. **69**, 4786 (1991); P.M. Levy, Solid State Phys. **47**, 367 (1994); P.M. Levy and S. Zhang, J. Magn. Magn. Mater. **151**, 315 (1995).
11. X.G. Zhang and W.H. Butler, Phys. Rev. B **51**, 10085 (1995); P. Zahn, I. Mertig, M. Richter, and H. Eschrig, Phys. Rev. Lett. **75**, 2996 (1995).
12. M.A.M. Gijs and G.E.W. Bauer, Adv. Phys. **46**, 285 (1997).
13. D.M. Edwards, J. Mathon, and R.B. Muniz, IEEE Trans. Magn. **27**, 3548 (1991).
14. T. Valet and A. Fert, Phys. Rev. B **48**, 7099 (1993); A. Fert, J.L. Duvail, and T. Valet, Phys. Rev. B **52**, 6513 (1995).
15. N.F. Mott, Adv. Phys. **13**, 325 (1964).
16. C. Kittel, *Introduction to Solid State Physics*, 4th edition (Wiley, New York, 1971).
17. W.A. Harrison, *Solid State Theory* (Dover Publications, New York, 1979).
18. D.M. Edwards, J. Mathon, R.B. Muniz, and S.S.P. Parkin, J. Magn. Magn. Mater. **114**, 252 (1992).
19. P.A. Schroeder, J. Bass, P. Holody, S.F. Lee, W.P. Pratt, Jr., and Q. Yang, Magnetic Ultrathin Films, Materials Research Society Proceedings **313**, Pittsburgh, 1993, p. 47.
20. J. Bass, P.A. Schroeder, W.P. Pratt, Jr., S.F. Lee, Q. Yang, P. Holody, L.L. Henry, and R. Loloee, Mater. Sci. Engng. **B31**, 77 (1995); W.P. Pratt, Jr., S.F. Lee, Q. Yang, P. Holody, R. Loloee, P.A. Schroeder, and J. Bass, J. Appl. Phys. **73**, 5326 (1993).
21. E.Y. Tsymbal and P.G. Pettifor, Phys. Rev. B **54**, 15314 (1996).

# 5 Electronic Structure, Exchange and Magnetism in Oxides

D. Khomskii

Laboratory of Solid State Physics, Groningen University,
Nijenborgh 4, 9722 AG Groningen,
The Netherlands

## 5.1 Introduction

Magnetic oxides are a curious class of materials with good prospects for applications in spin electronics. These materials exhibit a wide variety of magnetic properties (e.g. ferro-, ferri- and antiferromagnetic structures) as well as diverse transport properties including good insulators, systems with insulator–metal transitions, materials with "bad metallic" conductivity, good conductors and superconductors.

Typically in these materials the same electrons are responsible for both the magnetic and electric properties, so a strong interdependence of these characteristics should be seen. Indeed this is what is observed experimentally in many systems of this class. The most spectacular effect of this type is probably the "colossal magnetoresistance" (CMR) – the term coined recently, mostly for the effects observed in doped manganites [1]. This term may be also applied to some other compounds, e.g. EuO, see e.g. [2], in which the effect is sometimes much stronger than in the conventional manganites.

There are a number of textbooks, monographs [3,4] and review articles [5–7] in which the basic physics of the transition metal oxides are described. In this short chapter I will present a summary of the main concepts and notions used in describing the structure and properties of transition metal oxides. Special attention will be paid to the question when should *ferromagnetic* order be expected in oxides, and what is the relationship between the type of magnetic ordering and transport properties (notably resistivity) of magnetic oxides.

## 5.2 Transition Metal Ions in Crystals

Isolated ions with partially filled 3d-shells have 5-fold degenerate orbitals ($l = 2$, $(2l+1)$-degenerate levels), in which we can put up to 10 electrons ($2 \times (2l+1)$). The filling of these levels follows Hund's first rule: to minimize the Coulomb repulsion energy, electrons form a state with the maximum possible spin. Thus for example the ion $V^{3+}(d^2)$ should have spin $S = 1$; $Mn^{2+}(d^5)$ a spin $S = \frac{5}{2}$ etc.

When a transition metal (TM) ion is put into a crystal, the spherical symmetry of an isolated ion is reduced, and consequently some of the orbital degeneracy is lifted. This is called the splitting of levels due to a crystal field (CF).

The modification of the $3d$-level can be considered step by step by gradually reducing the symmetry of the surroundings. If for example a TM ion is put into a cubic crystal field (see Fig. 5.1), the 5-fold orbitally degenerate levels are split as shown in Fig. 5.2: three levels go down in energy, forming triply degenerate $t_{2g}$-levels, and two degenerate $e_g$-levels go up. The splitting of these levels $\Delta_{CF}$ is sometimes called $10\,Dq$, especially in chemical literature.

The splitting of these levels can be understood by examining the form of the $d$-wave functions for the $e_g$ orbitals:

$$\begin{aligned} d_{x^2-y^2} &= \tfrac{1}{\sqrt{2}}(x^2 - y^2) \\ d_{z^2} &= \tfrac{1}{\sqrt{6}}(2z^2 - x^2 - y^2)\,, \end{aligned} \qquad (5.1)$$

i.e. they have an electron density directed towards the negatively charged ions surrounding the TM ions, see Fig. 5.3. These ions are often called ligands, and the resultant crystal field splitting of $d$-levels is called a ligand field.

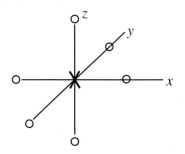

**Fig. 5.1.** × – transition metal ion;   ○ – negative ligand ions (e.g. oxygen)

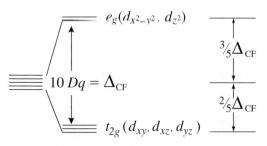

**Fig. 5.2.** Splitting of a 5-fold orbitally degenerate $d$-level of an isolated ion in an octahedral crystal field of Fig. 5.1.

In contrast the three $t_{2g}$ orbitals have lobes directed along diagonals in between the ligands, as in Fig. 5.4.

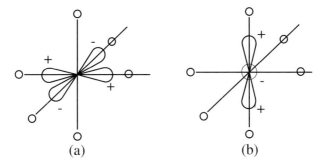

**Fig. 5.3.** The shape of the $e_g$-wave functions: (a) $d_{x^2-y^2}$-orbital; (b) $-d_{z^2}$-orbital

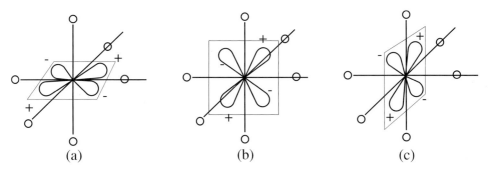

**Fig. 5.4.** The shape of the $t_{2g}$-wave functions: (a) $d_{xy}$; (b) $d_{xz}$ and (c) $d_{yz}$-orbitals

As seen from Figs. 5.3 and 5.4, the electron density in the $e_g$-orbitals is directed towards the negatively charged ligands, whereas those in $t_{2g}$-levels in between them. As a result the $e_g$-orbitals will experience a stronger Coulomb repulsion with ligands which raises their energies compared to those of the $t_{2g}$-levels. These simple considerations explain the crystal field splitting shown in Fig. 5.2. Note also that there are different signs of the corresponding lobes of $d$-functions, marked in Figs. 5.3 and 5.4. These are irrelevant for the Coulomb interaction with ligands giving a point-charge contribution to the CF, but play an important role further on.

There exists another contribution to the CF splitting besides the point charge contribution described above. This is the so called covalency contribution, due to a hybridization of the $d$-orbitals of the TM ion with the $p$-orbitals of the ligands (oxygen) as illustrated in Fig. 5.5. Due to this hybridization a mixing of these orbitals occurs, which causes the splitting of the $d$ and $p$ levels. It can easily be seen that the $e_g$-orbitals have a rather large overlap and hence a strong hybridization with the $p$-orbitals of oxygen occurs (directed towards the TM ion) leading to the so-called $\sigma$-orbitals, see Fig. 5.6. Consequently the mixing of $e_g$- and $p$-orbitals will be strong, and gives a corresponding upward shift of the $e_g$-levels,

$$\delta E_{e_g} \sim \frac{t^2_{pd\sigma}}{\Delta}. \tag{5.2}$$

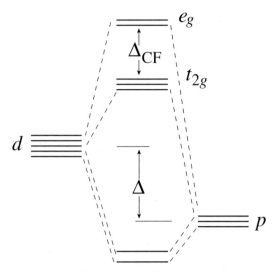

**Fig. 5.5.** The hybridization of $d$-levels of the TM ions and $p$-levels of the ligand leading to the repulsion of levels and splitting of $t_{2g}$ and $e_g$-levels.

This shift can be estimated by perturbation theory, assuming the $p$–$d$ hybridization (hopping matrix element) between the $e_g$- and $p$-orbitals $t_{pd\sigma}$ is small compared with the initial splitting of $d$- and $p$-levels $\Delta$, see Fig. 5.5.

Similar considerations show that the hybridization of the $t_{2g}$-orbitals with the corresponding $p$-orbitals of the ligands is smaller than that of $e_g$-orbitals. Indeed, as seen clearly from Fig. 5.4, the $t_{2g}$-orbitals are orthogonal to the $p$-orbitals and

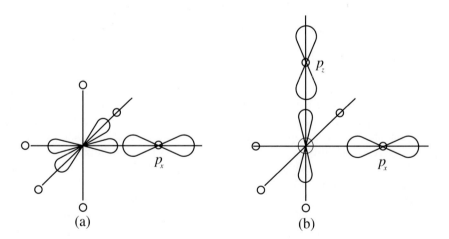

**Fig. 5.6.** Strong overlap and hybridization of $d_{x^2-y^2}$ (a) and $d_{z^2}$ (b) orbitals with the corresponding $p_x$ and $p_z$ orbitals ($\sigma$-orbitals of ligands)

directed towards the TM ions (here the signs of the $t_{2g}$-wave functions play a crucial role: by symmetry the overlap of $p_\sigma$-orbitals with the $t_{2g}$-orbitals is zero as shown in Fig. 5.6). The remaining overlap between $t_{2g}$ orbitals and $p$-orbitals shown in Fig. 5.7 is known as $\pi$-hybridization. This overlap is permitted by

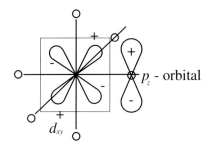

**Fig. 5.7.** An overlap and hybridization between one of the $t_{2g}$-orbitals ($d_{xz}$) with the corresponding $p$-orbitals ($p_z$) of a ligand ($\pi$-hybridization)

symmetry, but it is smaller than the $\sigma$-overlap of $e_g$-orbitals shown in Fig. 5.6. Since the $t_{2g}$–$p$ mixing is weaker and the upward shift of the $t_{2g}$-levels shown in Fig. 5.5 is smaller, then due to the point-charge contribution described above, the $e_g$ levels in a CF are higher in energy than the $t_{2g}$-levels. Thus both of these contributions to the CF, Coulomb repulsion with ligands and $p$–$d$ hybridization, in typical cases lead to the same consequence: the splitting of $d$-levels in a cubic (octahedral) crystal field with the form shown in Fig. 5.2. Typical values of the splitting between $t_{2g}$- and $e_g$-levels in TM oxides are $\Delta_{CF}(=10Dq) \simeq 1$–$2\,\mathrm{eV}$.

Now, using the rule (Hund's rule) formulated above, the ground state of a TM ion may be understood. From the formal valence of a TM in a given compound, the number of $d$-electrons left on the ion is found, and these electrons may be put in the CF-split levels of Fig. 5.2 one after another following Hund's rule, i.e. putting as many electrons with parallel spins as possible.

So, supposing the total number of $d$-electrons $n_d \lesssim 3$, then simply the total spin of the ion will be $S = n_d/2$ (we ignore for a while the question of remaining orbital degeneracy, see Sect. 5.3 below). However, if we have four $d$ electrons ($n_d = 4$), a problem may arise. If the fourth electron is placed with the spin parallel to those of the first three electrons, (i.e. according to Hund's rule), then we should place it on a higher-lying $e_g$-level, see Fig. 5.8a, which costs us an energy $\Delta_{CF}$. Alternatively, the fourth electron could be put on one of the lower $t_{2g}$-levels; but in this case because of the Pauli principle it should have the opposite spin, i.e. we have to violate Hund's first rule.

Both of these situations are met in practice. The first one leads to the so-called high-spin state of a TM ion, whereas the second one to the low-spin state. As seen from Fig. 5.8, the relative stability of one state with respect to another is determined by the ratio of the CF splitting $\Delta_{CF}$ and the Hund's rule stabilization energy (which may be described as an on-site ferromagnetic exchange interaction $-J_H \sum_{\alpha,\beta} \mathbf{S}_{i\alpha}\mathbf{S}_{i\beta}$, where $i$ is the site index and $\alpha$, $\beta$ are

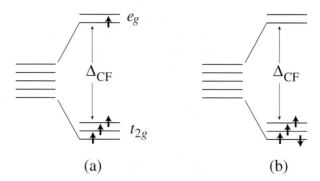

**Fig. 5.8.** High-spin state (a) and low-spin state (b) of an ion with 4 $d$-electron

indices of different $d$-orbitals[1]). If $\Delta_{CF} > J_H$ (or rather larger than the total Hund's rule stabilization energy, which for the case shown in Fig. 5.8 will be $3J_H$, the difference in the number of pairs of parallel spins between the configurations in Fig. 5.8a and 5.8b), then it would be favourable to form a low-spin state, occupying the lowest CF levels at the expense of Hund's rule exchange. In the opposite case the high-spin state will be stabilized.

In most cases the TM oxides have high-spin states (this is the case for $Mn^{3+}(d^4)$ whose configuration corresponds to that of Fig. 5.8a). However there are notable exceptions. The ionic states of $Co^{3+}(d^6)$, $Ni^{3+}$ $(d^7)$ and $Ru^{4+}$ $(d^4)$ are often low level spin ones. Also by changing the parameters such as tempera-

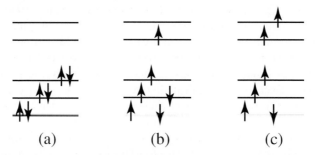

**Fig. 5.9.** Different possible electronic configurations of ions $Co^{3+}$ and $Fe^{2+}$ $(d^6)$: (a) low-level spin state; (b) intermediate spin state; (c) high-spin state

ture, pressure and composition of the material different spin states may appear, even real phase transitions as the material crosses over from one state to another may occur. This may be the situation in $LaCoO_3$ [3] (for which even a more complicated situation may exist, with the stabilization of an intermediate-

---

[1]  Strictly speaking, the Hund's rule stabilization energy is not an ordinary exchange interaction but is due to the difference of the direct Coulomb interaction of electrons on different orbitals; for our purposes this subtle difference is however irrelevant.

spin state [8]). Maybe the most interesting and the most important phenomena of this kind occur in some of the $Fe^{2+}$-containing compounds (including many biologically important ones): probably the low spin–high spin transition of $Fe^{2+}$ in such compounds plays an important role in functioning of such molecules and compounds, in particular in red blood cells.

## 5.3   Orbital Degeneracy and Jahn–Teller Effect

Now going one step further and consider what happens when the point symmetry of a TM ion is further reduced. It was already noted (although we put it for a while under the rug) that there may be situations when the detailed occupation of one or another crystal field level is not uniquely determined. For example, which particular $t_{2g}$-orbitals would be occupied in the $V^{4+}(d^1)$ ion in an octahedral coordination, or which $e_g$-orbital would the fourth $d$-electron go into in a high-spin state of $Mn^{3+}(d^4)$, see Fig. 5.8a.

There exists a very powerful and general theorem in quantum mechanics – the Jahn–Teller theorem – which states, crudely speaking, that the only degeneracy permitted in the ground state of any quantum system is the Kramers degeneracy. This is connected with the invariance with respect to time inversion. In simple terms it is the degeneracy of the spin up and down states (in systems without magnetic order). All the other types of degeneracy including orbital degeneracy (in Fig. 5.8a) are forbidden and should be lifted by the corresponding decrease of symmetry which lifts this degeneracy. The essence of this theorem (which, as Teller himself states in the preface to the book on the JT effect, [9], was actually formulated by Landau) is that: there is always a perturbation reducing the symmetry with a linear term representing the splitting of the degenerate levels (an energy gain) and a quadratic term representing the energy loss, see Fig. 5.10. Following from the standard perturbation theory of degenerate levels in quantum mechanics, the energy of the system as a function of perturbation $u$ has indeed the form

$$E(u) = -gu + \frac{Bu^2}{2} \qquad (5.3)$$

where the first term is the splitting of degenerate levels, and the second one is, e.g., the elastic energy of a deformation reducing the symmetry.[2]

For example a regular octahedron may be deformed into a tetragonal to lift the cubic symmetry, see Fig. 5.11. From similar considerations to the ones in Sect. 5.2 it can be seen that the local elongation of $O_6$-octahedra (Fig. 5.11a) decreases the Coulomb energy of the orbital $d_{z^2}$ in comparison with $d_{x^2-y^2}$, whereas a local compression of the octahedra (Fig. 5.11b) decreases the energy of the other orbital, $d_{x^2-y^2}$.

---

[2] There may appear important complications in case of isolated JT centers: due to quantum effects there may occur tunneling between the states in the right and the left minima in Fig. 5.10. This gives rise to so-called vibronic effects, considered in detail e.g. in [9]. In concentrated systems which we consider here, these effects are usually not very important, and we will not discuss them in what follows.

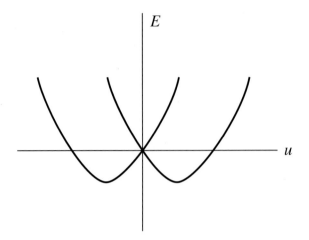

**Fig. 5.10.** The change of the total energy of a system with two degenerate levels as a function of perturbation (deformation) $u$ decreasing the symmetry and lifting the degeneracy

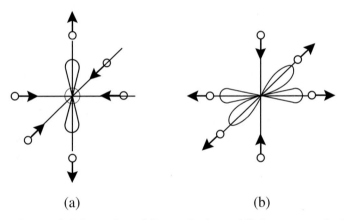

(a)                                    (b)

**Fig. 5.11.** Tetragonal deformation of $O_6$ octahedra stabilizing one particular orbital: (a) Elongation, stabilizing $d_{z^2}$-orbital; (b) Compression, stabilizing $d_{x^2-y^2}$-orbital

This is the essence of the Jahn–Teller theorem in application to the transition metal ions with an orbital degeneracy. As typically the neighbouring TM ions have common ligands (e.g. oxygen), a local JT deformation around one centre interacts with the corresponding deformation of its neighbours, see e.g. Fig. 5.12, giving rise to correlated displacements. Consequently the symmetry of the crystal as a whole is reduced. This usually occurs as a structural phase transition – one of very few types (maybe the only one) of structural phase transitions for which we know for sure their microscopic origin. This is known as the cooperative Jahn–Teller effect (CJTE) or as orbital ordering. Due to the distortion a particular orbital is occupied at each center, this may result in a ferrodistortion, or ferro-orbital ordering, e.g. when all the octahedra are elongated in one direction; this is the situation in Mn ferrites or in $Mn_3O_4$. Local deformations are often correlated in an antiferrodistortive fashion (Fig. 5.12); this is known as an antiferro-orbital ordering.

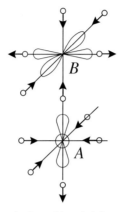

**Fig. 5.12.** An illustration of correlation of local deformations around neighbouring JT ions $A$ and $B$, with the corresponding orbital occupation.

Orbital ordering due to the JT effect takes place in many compounds. Typical ions displaying a strong JT effect are $Mn^{3+}(d^4)$, $Cr^{2+}(d^4)$, $Cu^{2+}(d^9)$. In these ions in octahedral surroundings there will be one electron ($Mn^{3+}$, $Cr^{2+}$) or one hole ($Cu^{2+}$) in a doubly degenerate $e_g$-level. Due to the strong overlap with the $p$-orbitals of ligands which in their turn strongly depend on the TM–O distance, the splitting of the $e_g$-levels with a shift of $nn$ oxygens is rather strong, which gives rise to a strong JT coupling. These ions are usually cited as typical JT ions. The JT effect for $Cu^{2+}$ is so strong that $Cu^{2+}$ is never found in a regular octahedron, but always in a strongly elongated one. Local distortion (elongation) around $Cu^{2+}$ can be so strong that one or two apex oxygens can "go to infinity" leaving $Cu^{2+}$ in a 5-fold (pyramid) or 4-fold (square) coordination. Such a coordination is indeed typical for many $Cu^{2+}$ compounds, the best known recent examples being high-$T_c$ cuprates like $YBa_2Cu_3O_7$ etc. Thus the JT nature of $Cu^{3+}$, even if it may not be directly responsible for high-$T_c$ superconductivity (although the person who discovered it, K. A. Müller, believes that it is), is at least very important for the stabilization of these rather unusual crystal structures.

The JT nature and corresponding orbital ordering is apparently very important in another nowadays popular class of compounds, namely the CMR manganites. The basic undoped compound $LaMnO_3$ contains a strong JT ion, $Mn^{3+}(d^4)$, and indeed there exists in $LaMnO_3$ a structural phase transition at $\sim 800\,K$ caused by the CJTE and orbital ordering. The crude orbital structure of $LaMnO_3$ at room temperature is shown in Fig. 5.13 (the actual occupied orbitals are slightly different). Thus the undoped $LaMnO_3$ has an antiferro-orbital ordering with locally elongated octahedra packed so that the long axes alternate in the basal plane. Such a distortion is rather typical: it helps to minimize the total strain of a crystal.

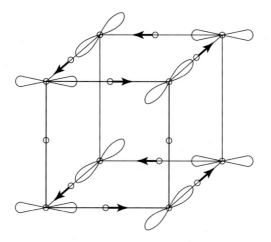

**Fig. 5.13.** Orbital structure and shifts of oxygens in undoped LaMnO$_3$

As in strong JT ions we are dealing with the double orbital degeneracy and it is convenient to describe the orbital structure by effective pseudospin operators $\tau$ so that e.g. the state $\tau^z = +\frac{1}{2}$ corresponds to orbital 1 (e.g. $d_{z^2}$), and $\tau^z = -\frac{1}{2}$ to orbital 2 ($d_{x^2-y^2}$). An arbitrary linear superposition of orbitals can always be formed, and written as

$$|\theta\rangle = \cos\frac{\theta}{2}|d_{z^2}\rangle + \sin\frac{\theta}{2}|d_{x^2-y^2}\rangle \qquad (5.4)$$

(coefficients are written in such a form to guarantee proper normalization of the wave function $|\theta\rangle$). The corresponding states can then be depicted in the $(\tau^z, \tau^x)$−, or $\theta$-plane, see Fig. 5.14. In Fig. 5.14 the state with $|\theta = 0\rangle$ would correspond to $|d_{z^2}\rangle$, $|\theta = \pi\rangle = |d_{x^2-y^2}\rangle$. But what is more interesting, the states obtained from $|d_{z^2}\rangle$, $|d_{x^2-y^2}\rangle$ by the rotation of axes can be marked on this diagram (as in a regular octahedron the directions $x$, $y$ and $z$ are equivalent). Of course, not only an orbital $d_{z^2}$ extended in the $z$-direction can be formed, but also equivalent orbitals $d_{x^2}$ and $d_{y^2}$ extended along the $x$ and $y$-axes (such orbitals are shown in Fig. 5.13). In Fig. 5.14 these orbitals would correspond to $|\theta = \pm\frac{2}{3}\pi\rangle$, and orthogonal orbitals $d_{z^2-x^2}$, $d_{z^2-y^2}$ to $|\theta = \pm\frac{1}{3}\pi\rangle$. Thus in this language the orbital structure of LaMnO$_3$ (shown in Fig. 5.13) would correspond to a "canted" $\tau$-antiferromagnetism, with one sublattice having pseudospins at an angle $\theta = \frac{2}{3}\pi$ and another at $\theta = -\frac{2}{3}\pi$ (actual angles of sublattices in LaMnO$_3$ are somewhat different, closer to $\simeq \pm 97°$).

Up to now only the strong JT ions with double $e_g$-degeneracy have been discussed. There are, however, many materials with triply-degenerate $t_{2g}$-orbitals:

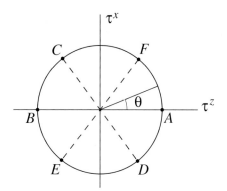

**Fig. 5.14.** $\tau^z$–$\tau^x$ plane in which all the possible $e_g$ states can be conveniently visualized. In this diagram the point $A$ $(\theta = 0)$ corresponds to the orbital $d_{z^2}$, $B$—to $d_{x^2-y^2}$, $C$ $(\theta = \frac{2}{3}\pi)$—to $d_{x^2}$, $D$—to $d_{z^2-y^2}$, $E$ $(\theta = -\frac{2}{3}\pi)$—to $d_{y^2}$ and $F$—to $d_{z^2-x^2}$

for example compounds of $V^{4+}(d^1)$ or $V^{3+}(d^2)$. Such materials also display features typical for the JT effect, however, there is one important difference between the $t_{2g}$-ions compared with the $e_g$-ions. For the latter the real orbital moment is quenched, for $t_{2g}$-ions there exists a nonzero orbital moment $l_{eff} = 1$ and correspondingly there is a spin–orbit interaction $\lambda l \cdot s$. This interaction can by itself lift the orbital degeneracy, and in typical situations it leads to a distortion (which in this case is usually called magnetostrictive distortion) opposite in sign to the one caused by the JT effect. Thus different compounds containing $t_{2g}$ ions may distort along different "routes": in some the ground state is determined by the JT distortion, but others develop "along the $ls$-route" and have a magnetostrictive distortion of the opposite sign. The second class of compounds typically contains materials with $Co^{2+}(d^7)$ and high-spin $Fe^{2+}(d^6)$ ions. The characteristic feature of compounds in which $ls$-coupling dominates and determines the ground state, is a decrease of symmetry and a lifting of the degeneracy occurring simultaneously with the magnetic ordering. In typical systems where the JT effect dominates, this happens independently and usually at higher temperatures than any magnetic ordering.

There are a number of interesting features in magnetic oxides containing JT ions, for further details see [6] and [10]. In the next section the mechanisms and the main features of the exchange interaction in oxides are discussed in which the orbital structure of corresponding ions plays a crucial role.

## 5.4  Exchange Interaction in Magnetic Insulators

Predominantly the electronic structure of isolated TM ions in crystals has been considered, with their attributes in concentrated systems arising from an interaction between ions. One such effect already mentioned above is the cooperative Jahn–Teller effect, or orbital ordering. More importantly in these systems are the magnetic interactions leading to some kind of long-range magnetic ordering, and the possibility of electron transfer from site to site. This effect is responsible for the transport properties seen in corresponding materials.

The simplest description of these properties should include the possibility of electron hopping as well as the effect of the Coulomb interaction between electrons. For a more complex model such details as the orbital structure of the corresponding ions should be taken into account. The $d$-electrons in crystals are described by the so-called Hubbard model:

$$H = -\sum t_{ij} c_{i\sigma}^{+} c_{j\sigma} + U \sum n_{i\uparrow} n_{i\downarrow} \equiv H' + H_0. \qquad (5.5)$$

Here the first term describes the hopping of $d$-electrons from site $j$ to site $i$, and the second term is the on-site Coulomb repulsion of $d$-electrons. This nondegenerate Hubbard model ignores such complications as a possible orbital degeneracy, but it is sufficient for the description of both insulating and metallic states of our system as well as for formulating the basics of the exchange interaction in magnetic insulators.

As is well known, for weak interaction $U \ll t$ the model (5.5) describes the metallic state, with the band dispersion

$$H' = \sum \varepsilon_k c_{k\sigma}^{+} c_{k\sigma}, \qquad \varepsilon_k = -2t(\cos k_x + \cos k_y + \cos k_z). \qquad (5.6)$$

(The simplest tight-binding approximation with only nearest neighbour hopping is used and the spectrum is written for a simple cubic lattice.) In this case the system would be metallic even for an exactly half-filled band with one electron per site, $n = 1$, independent of the distance between corresponding sites.

It is clear however that for large enough distance between sites, which means small hopping matrix element $t$, and for $n = 1$, the ground state should be insulating with electrons localized each at its site. This state is called a Mott or a Mott–Hubbard insulator, and it is due to the second term in the Hamiltonian (5.5), the on-site Coulomb repulsion $U$. If the overlap of the $nn$ electron wave functions is small enough so that $t \ll U$, care should be taken with the second term in (5.5) which should be minimized if there is exactly one electron per site and electrons are forbidden to hop onto the already occupied site. As a result the ground state electrons will be localized, and creation of charge excitations (transfer of an electron from its site to another one) would cost an energy $U$ (the repulsion of the transferred electron with the one already existing at this site). The energy gain in this process would be $\sim t$ (both the extra electron and the hole left at the first site can now move through the crystal and gain corresponding kinetic energy of the order of their bandwidth (5.6), i.e. $\sim t$), but if $U \gg t$ the energy loss of this process $\sim U$ exceeds the energy gain $\sim t$, so that the material would remain an insulator, with an energy gap $E_g \sim U - t$. Essentially, this is the physics of strongly correlated (strongly interacting) electron systems.

There are a number of interesting and to a large extent still unsolved problems in the physics of strongly correlated electron systems, see e.g. [11,12], but these cannot be discussed here in detail. Therefore, we will concentrate mostly on the properties of such systems for a simple case, i.e. for an integral number of electrons per site (e.g. $n = 1$) and for the case of a strong interaction $U \gg t$.

For $n = 1$ and $U \gg t$ the ground state of our system is an insulator. Thus there exists at each site a localized electron, i.e. a localized magnetic moment with $s = \frac{1}{2}$. These moments of course somehow interact with one another so that some kind of magnetic ordering should be established and spin degeneracy would be lifted at low temperatures; otherwise the Nernst theorem would be violated.

The main term of the Hamiltonian (5.5), the second term, leads to the formation of localized moments but it does not lift this spin degeneracy. However, the first term in (5.5), the electron hopping term, lifts this degeneracy (in second order of perturbation theory in $t/U \ll 1$) and leads to an antiferromagnetic exchange interaction between these localized magnetic moments. This result can be obtained rigorously, see e.g. [13,12], but here we will use a simple form which will be easily generalized later for more realistic cases of orbital degeneracy structure.

Consider two neighbouring sites with either parallel or antiparallel spins shown in Fig. 5.15. Electrons initially localized each at its site want to delocalize

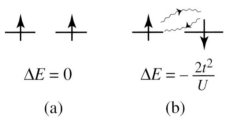

$$\Delta E = 0 \qquad\qquad \Delta E = -\frac{2t^2}{U}$$

(a)                  (b)

**Fig. 5.15.** Two neighbouring sites with parallel (a) and antiparallel (b) spins. The energy gain due to virtual hopping of an electron to a neighbouring site is shown

as much as possible by virtual hopping to neighbouring sites; by the Heisenberg uncertainty principle this would decrease their kinetic energy. Such delocalization may be caused by the first term of the Hamiltonian (5.5). In the situation of Fig. 5.15a this process is, however, forbidden by the Pauli principle; on the other hand for the antiparallel spins, Fig. 5.15b, this process is allowed. As a result the first electron hops to a neighbouring site (with the matrix element $t$), and then back. As usual in second order perturbation theory in quantum mechanics, this decreases the energy of the system, the energy gain being given by $\Delta E = -2t^2/U$ (the term $H'$ in (5.5) acts twice, therefore we have $t^2$ in the numerator; in the denominator, as usual, the energy of the intermediate state should stand and this is the energy of the repulsion of two electrons at the same site $U$). The factor of 2 comes from the fact that the left spin can make an "excursion" to the right and the right one to the left.

As a result the configuration with antiparallel spins is preferred relative to the one with the parallel spins. This can be described by the effective exchange interaction

$$H_{eff} = J \sum S_i S_j, \qquad J = \frac{2t^2}{U}. \qquad (5.7)$$

Virtual hopping of electrons leads to an antiferromagnetic Heisenberg exchange interaction and this type of exchange mechanism is usually called *superexchange* (sometimes also kinetic exchange). This is the main mechanism of exchange interaction in magnetic insulators like transition metal oxides.

## 5.5    Charge-Transfer versus Mott–Hubbard Insulators

Before discussing the complications introduced by the realistic crystal and orbital structure of materials, one more point should be discussed here. In contrast to the idealized model (5.5), in real materials, e.g. in oxides, usually there are ligands (oxygen ions) between TM ions. Consequently the hopping of $d$-electrons from site to site (first term in (5.5)) occurs not directly but via oxygen $p$-orbitals.

In many cases the oxygen $p$-states can be excluded and this reduces the description to the effective model (5.5); however, this is not always the case. Accordingly all magnetic insulators may be divided into two big groups: Mott–Hubbard insulators for which the description given above applies without any restriction, and charge-transfer insulators in which one should treat oxygen $p$-states in an apparent way.

The basic general Hamiltonian describing both the $d$-electrons of the TM and the $p$-electrons of oxygen has the form

$$H = \sum \varepsilon_d d_{i\sigma}^+ d_{i\sigma} + \varepsilon_p p_{j\sigma}^+ p_{j\sigma} + t_{pd}(d_{i\sigma}^+ p_{j\sigma} + \text{h.c.}) + U n_{di\uparrow} n_{di\downarrow}. \qquad (5.8)$$

Depending on the ratio of the charge-transfer excitation energy $\Delta = \varepsilon_d - \varepsilon_p$ and the Coulomb repulsion $U$, a division into two groups can be made.

If the oxygen $p$-levels lie deep enough, $\Delta \gg U$, the lowest charged excited states are those corresponding to the transfer of a $d$-electron from one TM site to another:

$$d^n + d^n \longrightarrow d^{n-1} + d^{n+1}.$$

This process as described above costs an energy $U$ and for $U \gg t$ gives the Mott–Hubbard insulating state. Still even in this case real hopping occurs via the oxygen $p$-states, but it can be excluded in perturbation theory, obtaining the effective $d$–$d$ hopping $t_{dd} = t = t_{pd}^2/\Delta$. This is the $d$–$d$ hopping $t$ which would enter the effective Hubbard model (5.5) and later the exchange integral (5.7). This is typically seen in oxides of the early transition metals, like Ti and V.

On the other hand there may be situations, where the charge-transfer energy $\Delta = \varepsilon_d - \varepsilon_p$ (the energy necessary to transfer an electron from the filled $2p$-level of $O^{2-}$ to a $d$-level of a neighbouring TM) is less than $U$. In this case the lowest charge-carrying excitations will be just these excitations: the transfer of an electron from oxygen to the TM, or the transfer of a hole to an oxygen:

$$d^n p^6 \longrightarrow d^{n+1} p^5 \equiv d^{n+1} \underline{L}$$

(the notation $\underline{L}$ is very often used nowadays and means "ligand hole" – the state with one electron on a ligand – here oxygen – missing).

According to this division we may draw a general phase diagram, the so called Zaanen–Sawatzky–Allen (ZSA) [14] diagram, shown in Fig. 5.16. Many oxides of late 3d-metals (Co, Ni, Cu) belong to this second category, the charge-transfer insulators.

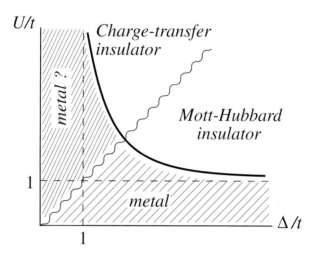

**Fig. 5.16.** Zaanen–Sawatzky–Allen diagram showing the regions of Mott–Hubbard ($\Delta > U$) and charge-transfer ($\Delta < U$) insulators

If $n = 1$ and the hopping $t$ is small, then even with $\Delta \ll U$ (but $\Delta \gg t$) we would have a similar situation to the one described above: the ground state will still be an insulator with electrons localized at the TM ions and with localized magnetic moments; this is similar to conventional Mott insulators. In the simplest cases such insulators will be antiferromagnetic, with the only difference being the exchange integral $J$ in (5.7) expressed as

$$J = \frac{2t_{pd}^4}{\Delta^2(2\Delta + U_{pp})}. \tag{5.9}$$

(Here we take into account the lowest excited states participating in the virtual electron hopping: the electrons that transfer from the oxygen $p$-shell to the TM $d$-levels, see Fig. 5.17; we also included the effective repulsion of two $p$-holes on the same oxygen $U_{pp}$ which contributes to the energy of one of the excited states of Fig. 5.17.) Thus from the point of view of magnetic properties, charge-transfer insulators do not significantly differ from the conceptually simpler Mott–Hubbard insulators, and for our purposes their difference could be ignored. However, it should be realized that there may be differences in these compounds in their excitation spectra, transport properties etc. Further effects

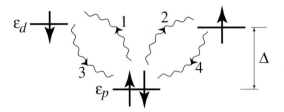

**Fig. 5.17.** Virtual processes with the hopping of $p$-electrons from oxygen to two neighbouring TM ions giving rise to an effective antiferromagnetic interaction in charge-transfer insulators. The numbers at the wavy lines denote the sequence of virtual transitions

are discussed in detail in Refs. [15,7] and used for the description of some real materials, e.g. $CrO_2$ in [16].

## 5.6    Goodenough–Kanamori–Anderson Rules

Until now when discussing the exchange interaction only the simplest case of a nondegenerate $d$-orbital containing one electron has been considered. In reality, as discussed in Sect. 2, $d$-electrons have a rather rich orbital structure: different orientation, different overlap between themselves and with the $p$-states of their ligands and possibly orbital degeneracy. All these details play an important role in determining the corresponding exchange interaction and determine finally the large variety of magnetic properties of TM insulators.

Rules called the Goodenough–Kanamori–Anderson (GKA) rules [3] were formulated in order to predict the observed phenomena.

There are many details and particular cases to consider, so only the main rules are formulated and the general approach explained; however, sometimes the outcome is not clear without detailed calculations. Nevertheless, the general trend is rather straightforward and results from the physics already described in previous sections.

So for the simplest case shown in Fig. 5.18a the localized electrons on two neighbouring TM ions occupy orbitals that are directed towards each other (or overlapping with the same $p$-orbital of the intermediate ligand, Fig. 5.18b). Thus the exchange interaction, according to (5.7) and (5.8) will be rather strong and antiferromagnetic. Note that in a real situation the $dd$-overlap occurs usually via

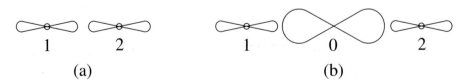

**Fig. 5.18.** Overlap of $d$-orbitals of two transition metal ions 1 and 2, direct one (a) or via $p$-orbital of the intermediate oxygen O (b)

$p$-orbitals of intermediate ligands, Fig. 5.18b. In this case the geometry of the corresponding bonds is crucial: the case of Fig. 5.18b represents what we call 180°-exchange (the angle TM1–O–TM2 is 180°). *This is the first GKA rule: the 180° exchange between filled orbitals (sometimes one speaks about half-filled orbitals, having in mind that there is* **one** *electron at each orbital, not two) is relatively strong and antiferromagnetic.*

Now consider the situation with filled orbitals but with a 90°-exchange path, Fig. 5.19. Likewise in this case we are also dealing with the orbitals (shaded one on TM1 ion and white on ion TM2) but interacting via the oxygen with a 90° exchange path 1–O–2. As discussed in the previous section, the actual electron transfer occurs between the $d$-orbitals of the TM and the $p$-orbitals of oxygen. It can be easily seen from Fig. 5.19 that two different orthogonal orbitals $p_x$ and $p_y$ overlap with the corresponding $d$-orbitals of sites 1 and 2. As a result the virtual electron hops as shown in Fig. 5.19b: one electron is transferred from the $p_x$-

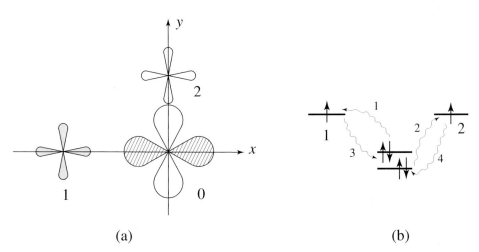

(a)                                           (b)

**Fig. 5.19.** The scheme of 90°-exchange illustrating the second GKA rule. (a) The relevant orbitals at the transition metals 1 and 2 (TM1 and TM2) and oxygen O; (b) Virtual processes with the hopping of $p$-electrons trom a "corner" oxygen into two neighbouring TM ions giving rise to the ferromagnetic interaction.

orbital to the TM1 ion, and another electron from a *different* orbital of the same oxygen, $p_y$, goes into the TM2 ion. Thus in the excited intermediate state there will be two electrons missing, or two $p$-holes present, on the oxygen. Depending on the relative spin orientation of TM1 and TM2 the remaining $p$-electrons will be either parallel (the case shown in Fig. 5.19b), or antiparallel. The energy of this intermediate state is in the denominator of the corresponding energy, and as usual the state with the lowest denominator is favoured. Accordingly, to Hund's rule (also valid for the oxygen ion), it is best to have the spins of the two oxygen electrons, or two $p$-holes, parallel: thus the spins of TM1 and TM2 should be parallel too. As a result of such an exchange process a *ferromagnetic* interaction between the moments of the TM ions 1 and 2 is favoured.

The energy difference between the parallel and antiparallel configurations is

$$J \sim -\frac{t_{pd}^4}{\Delta^2}\left(\frac{1}{2\Delta + U_p - J_H} - \frac{1}{2\Delta + U_p}\right) \simeq -\frac{t_{pd}^4}{\Delta^2(2\Delta + U_p)}\frac{J_H}{(2\Delta + U_p)} \quad (5.10)$$

(cf. (5.9)), i.e. the 90° ferromagnetic exchange would contain a small factor $\sim J_H/(2\Delta + U_p)$. *Thus the second GKA rule reads: 90°-exchange between (half)-filled orbitals is* **ferromagnetic** *and relatively weak.*

To illustrate the third and final GKA rule, consider an electron hopping between the $d$-states, as in a simple Hubbard model, not between occupied orbitals, as in Fig. 5.15b or Fig. 5.18, but between an *occupied* and an *empty* orbital. (In real life such hopping goes still through the intermediate ligands but we can skip this detail for a while.) Thus, imagine two TM ions with orbital ordering so that there is no overlap between occupied orbitals, and the only overlap is between an occupied and an empty one, see Fig. 5.20. In this figure the $d_{x^2-z^2}$-orbital (white

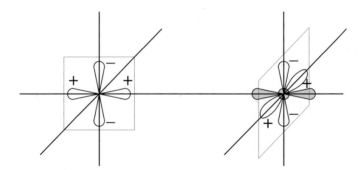

**Fig. 5.20.** The scheme illustrating the origin of the ferromagnetic exchange for the overlap between occupied and empty orbitals of neighbouring TM ions

one) is occupied at the site 1 and the $d_{y^2-z^2}$-orbital (white) at the site 2. (This orbital occupation is seen in $KCuF_3$, with the only difference being, the orbitals of one $d$-hole in $Cu^{2+}(d^9)$ and not of a $d$-electron are those concerned.) Due to the symmetry of the corresponding wave functions (note the signs in Fig. 5.20!) these orbitals are orthogonal, and there is no hopping between them. However there is a possibility of electron hopping from the occupied $d_{x^2-z^2}$-orbital of the site 1 into an *empty* $d_{x^2}$ (shaded orbital) at the site 2. Such virtual hopping is in principle allowed irrespective of the relative orientation of the spins of the sites 1 and 2. However, as in the previous case, the energies of the intermediate states entering the denominators in second order perturbation theory would differ, cf. Fig. 5.21. In this case, due to Hund's rule, the energy gain in the intermediate state is $J_H$ and the total exchange will be ferromagnetic. Again, as in (5.10), the total exchange constant will be given by the corresponding difference

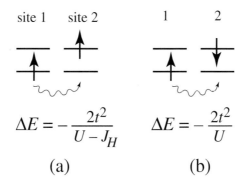

$$\Delta E = -\frac{2t^2}{U - J_H} \qquad \Delta E = -\frac{2t^2}{U}$$

(a)                        (b)

**Fig. 5.21.** Virtual hoppings from the occupied orbital on site 1 into an empty one on site 2 and corresponding energy gains for the parallel (a) and antiparallel (b) spin orientations

of the energies of the states 5.21a and 5.21b, i.e.

$$J \simeq -\frac{2t^2}{U}\frac{J_H}{U}. \tag{5.11}$$

(Taking into account $J_H < U$ and expanding the expression in Fig. 5.21 in $J_H/U$; typical values for the TM oxides are $J_H \sim 0.8\,\mathrm{eV}$ and $U \sim 3$–$5\,\mathrm{eV}$.) *Thus we justify our third GKA rule: when the exchange is due to an overlap between an occupied and an empty orbital, the resulting exchange is ferromagnetic and relatively weak.*

These main rules will be illustrated by a few examples, for instance, perovskite materials with the basic structure shown in Fig. 5.22. This is probably the simplest feasible structure: magnetic ions form a simple cubic lattice, and the ligands (e.g. oxygen ions) are sitting in between them. The TM ions in this struc-

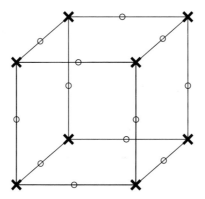

**Fig. 5.22.** Schematic crystal structure of perovskites; × — transition metal ions; ○ — oxygen (or other ligand) ions

ture are surrounded by oxygen octahedra, connected by the common corners, so that the superexchange paths TM–O–TM are 180° ones.[3]

Now according to the GKA rules formulated above, when both the $e_g$-orbitals of the TM ion are occupied, we have an antiferromagnetic $nn$ interaction, and as a result the magnetic ordering is of a simple two-sublattice type (the so-called $G$-type antiferromagnetism). This is the situation, e.g., in $KMnF_3$ ($Mn^{2+}$, $d^5$), $KNiF_3$ ($Ni^{2+}$, $d^8$), $LaFeO_3$ ($Fe^{3+}$, $d^5$) etc.

For another example, as already mentioned, $KCuF_3$ contains the typical JT ion $Cu^{2+}(d^9)$ with one hole in a doubly-degenerate $e_g$-level. In this material the orbital ordering shown in Fig. 5.23 is realized. As explained above, see Fig. 5.20, an exchange interaction in the basal $xy$-plane is relatively weak and *ferromag-*

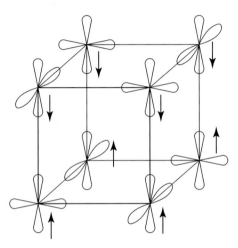

**Fig. 5.23.** One of two possible types of orbital ordering in $KCuF_3$ (the other one differs by the interchange of the occupied orbitals in the upper $xy$-plane)

*netic.* On the other hand, along the $z$-direction the lobes of occupied orbitals are directed towards each other, which, according to the first GKA rule, results in a strong antiferromagnetic interaction in this direction. As a result the magnetic structure shown in Fig. 5.23 is obtained: the ferromagnetic $xy$-planes are coupled antiferromagnetically in the $z$-direction (this magnetic structure is called $A$-type antiferromagnetism). Thus, due to a particular orbital ordering, the magnetic structure in a compound close to a cubic one is rather anisotropic. Moreover, as explained, the antiferromagnetic exchange in $z$-direction is much stronger than the ferromagnetic one in $xy$-plane, so in effect $KCuF_3$ has properties of a quasi-one-dimensional antiferromagnet (antiferromagnetic chains in the $z$-direction with weak ferromagnetic coupling between them), and, despite

---

[3] In reality often $MO_6$-octahedra are tilted so that the exact angle TM–O–TM is less than 180°. In many cases this has important consequences; for our general discussion however we ignore this complication

being structurally a nearly cubic compound, is one of the best examples of one-dimensional antiferromagnets.

Another very important example is given by the compound $LaMnO_3$, the basis of the CMR materials so popular nowadays. Its orbital structure has been already shown in Fig. 5.13. The application of the GKA rules to this compound is not as straightforward as above (see the detailed discussion in [7]), but the outcome is the same: the type-$A$ antiferromagnetic structure (ferromagnetic planes stacked antiferromagnetically) is seen, with the important difference from $KCuF_3$ being, the ferromagnetic exchange in the basal plane is stronger than the antiferromagnetic one between these planes.

The fact that due to the orbital ordering four of the nearest neighbours out of six have ferromagnetic coupling may be important for the formation of the ferromagnetic state in doped $LaMnO_3$, although usually this is ascribed to the double exchange mechanism, see below. This is definitely important, but it may well be that the factor mentioned above (ferromagnetic coupling due to orbital ordering) also plays some role in it.

In order to illustrate the second GKA rule, many materials with 90°-exchange which gives ferromagnetic interaction may be cited. Probably the simplest examples are provided by the one-dimensional structures containing $Cu^{2+}$, shown in Fig. 5.24. In these structures the angle Cu–O–Cu is often very close to 90°, and

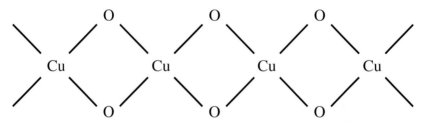

**Fig. 5.24.** The main structural motive of several compounds containing $Cu^{2+}$ coupled by the 90° Cu–O–Cu superexchange path

$x^2 - y^2$-like orbitals of Cu give rise to a (weak) ferromagnetic interaction. This is seen, in particular, in the "telephone number" compound $Ca_{14}Cu_{24}O_{41}$, in which spin ladders coexist with the spin chains having the structure of Fig. 5.24, and the exchange in these chains is known to be ferromagnetic.

Finally the compound $CuGeO_3$ with a similar structure may be mentioned: this is the first inorganic material showing a spin-Peierls transition. The angle Cu–O–Cu is slightly larger than 90° ($\sim 98°$) but is not enough to make the exchange antiferromagnetic (as is needed for the spin-Peierls transition). This presents a formidable problem, and special physical mechanisms are needed to overcome the second GKA rule and make the exchange antiferromagnetic [17].

## 5.7    Exchange Mechanism of Orbital Ordering

As discussed previously, even with a fixed lattice, superexchange leads not only to magnetic, but also to orbital ordering [18,6]. In Sect. 5.3 the Jahn–Teller theorem, the lifting of the orbital degeneracy and the resulting orbital ordering, the distortion of the lattice and the effective intersite interaction due to this mechanism was discussed. The general ideas developed in Sect. 5.6 will be applied to a particular question: what are the possible mechanisms of orbital ordering in systems with orbitally degenerate (Jahn–Teller) ions.

In order to shed light on this question, as shown in Fig. 5.25, imagine two neighbouring TM ions with one electron on each, occupying doubly-degenerate orbitals. Suppose for simplicity that only diagonal hoppings are allowed: those

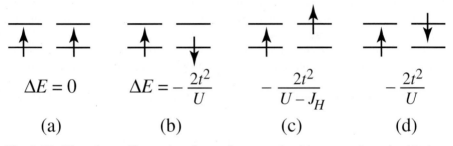

(a)          (b)          (c)          (d)

**Fig. 5.25.** The scheme illustrating the tendency to simultaneous spin and orbital ordering due to superexchange mechanism for a doubly-degenerate orbitals

between orbitals 1 in a $nn$ state and those between orbitals 2: $t^{11} = t^{22} = t$, $t^{12} = 0$. If the arguments are generalized as shown in Figs. 5.15 and 5.21, the energy gain due to virtual hopping of electrons onto neighbouring sites can easily be calculated (as shown in Fig. 5.25). Immediately it can be seen that due to the influence of the Hund's rule interaction (which decreases the energy of the intermediate state with two parallel spins), the state of Fig. 5.25c is favoured. Thus, whereas for the nondegenerate case of Fig. 5.15 the superexchange leads to an antiferromagnetic spin interaction, here the state with both the spin and orbital ordering is stabilized; in this particular case the one with the ferromagnetic spin ordering and "antiferro-orbital" one.

Again this treatment can be made rigourous and the effective exchange Hamiltonian in contrast to (5.7), would contain terms $S_i S_j$, terms describing orbital ordering $\tau_i \tau_j$ and terms describing the coupling of the spin and orbital degrees of freedom of the form $S_i S_j \cdot \tau_i \tau_j$.

Due to the complicated shape of real $e_g$-functions the actual form of the resulting effective Hamiltonian for real systems is also much more complex [18], but the main qualitative conclusions are similar: even with a fixed lattice (without real electron–lattice interaction usually invoked to explain cooperative Jahn–Teller ordering), in purely electronic terms the exchange interactions may cause both the magnetic (spin) and orbital (Jahn–Teller) ordering. One can even speak here about the "Jahn–Teller ordering without Jahn–Teller interaction". It should be noted that in this mechanism the orbital and magnetic orderings are strongly interrelated (although they can in general occur at different temperatures).

An interesting and important question arises as to the dominant mechanism leading to orbital ordering in real materials. Of course, generally speaking, both mechanisms (via electron–lattice interaction and via exchange interaction) are present simultaneously. Usually they both lead to the same orbital structure, so that experimentally it is very difficult to separate the contribution of different mechanisms. Novel possibilities to resolve this problem appeared recently with the development of a new calculation technique, the LDA + U method [19], which gives the possibility to obtain orbitally ordered structures by either fixing the undistorted lattice or taking into account the real lattice distortion. These calculations can be visioned as "numerical experiments" which permit to isolate the purely electronic contribution from the lattice one.

The experience gained so far [20] shows that indeed, correct orbital structures can be obtained even by fixing a symmetric lattice. A comparison of the energies from corresponding solutions shows that about 60–70% of the total energy gain due to the orbital ordering is provided by the electronic (exchange) contribution, the remaining 30–40% being gained when the lattice is released and permit to relax to a new equilibrium position corresponding to the correct orbital occupation. Thus from these calculations we can indeed conclude that the electronic (exchange) contribution gives at least a comparable, but maybe even dominant, contribution to the orbital ordering, as compared to the usually invoked electron–lattice interaction (although the latter is definitely also important).

## 5.8   Doping of Magnetic Insulators; Double Exchange

Up to now we have dealt exclusively with magnetic insulators with an integer number of electrons per TM ion. In this section the general case of doping (i.e. changing the average occupation of the $d$-levels) in magnetic insulators will be discussed.

A number of intriguing phenomena may be seen when doping: such as the coexistence of different valence states of a TM e.g. magnetite $Fe_3O_4$ (formally $Fe^{2+}$ and $Fe^{3+}$ ions) or $NaV_2O_5$ ($V^{4+}$ and $V^{5+}$). The different valence states can order ("crystallize") in the material, giving a charge-ordered (CO) state; usually these CO states are insulating. However, there may be situations when rapid exchange, or hopping, of electrons between different ions exists.

In contrast with Mott insulators with an integer number of electrons, here the hopping does not in general require any extra excitation energy and can occur quite freely, giving rise to a metallic conductivity. Thus a metallic state can be obtained with its magnetic properties differing strongly from the parent insulating compounds. Typically a ferromagnetic (or at least "more ferromagnetic") spin arrangement is obtained.

This phenomenon not only exists in doped or mixed-valence compounds, but also in oxides with integer valence. If there is a large enough electron hopping $t$ and corresponding bandwidth $W$ ($\sim t$) and if $W \gtrsim U$, a metallic state will be obtained. Sometimes these states behave magnetically as Pauli paramagnets (possibly with exchange enhancement) as in $LaNiO_3$. But in other systems of this type, especially those with small or negative charge-transfer gap (see Sect. 5.5), the corresponding metallic state is still characterized by strong electron correlations. In these cases magnetic ordering in such metallic states may exist; again usually a ferromagnetic one is seen. This is the case in $SrFeO_3$ or $SrCoO_3$.

For the rest of this section we will discuss the origin of ferromagnetism in doped Mott insulators. The most popular example of such a system nowadays is provided by doped $LaMnO_3$, e.g. $La_{1-x}Sr_xMnO_3$, where with increasing $x$ we go from the insulating state with $A$-type (layered) antiferromagnetism via a complicated intermediate state (its exact nature is not known, see below) to a ferromagnetic state for $x \gtrsim 0.18$. It is this last state which displays the property of Colossal Magnetoresistance (CMR) and which attracts now such attention.

The basic concept used to explain the appearance of ferromagnetism in these systems and its interplay with the metallic conductivity is through the double exchange model, first developed by Zener [21], later put on firm theoretical grounds by Anderson and Hasegawa [22] and by De Gennes [23]. See textbooks and review articles [5,7] for a detailed description; we give here only the general scheme without any details which can be found in these reviews.

The general idea for CMR is as follows. Suppose a lattice of localized electrons, and add a certain (small) number of extra electrons or holes which can in principle propagate through the crystal but these may also interact with the background of localized spins. For the system $La_{1-x}Sr_xMnO_3$ our initial ionic state for $x = 0$ is $Mn^{3+}(t_{2g}^3 e_g^1)$, and by substituting La by Sr we remove $x$ electrons, i.e. create $x$ holes in the $e_g$-band (or create $x$ $Mn^{4+}(t_{2g}^3)$ ions).

It may be easier to visualize (although more difficult to prepare experimentally) the opposite situation: start with the material $CaMnO_3$ containing $Mn^{4+}(t_{2g}^3)$ ions with three localized $t_{2g}$-electrons (the spin of such ions is $S = \frac{3}{2}$), and dope it with a rare earth, e.g. $Sm^{3+}$: by this we add a certain number of $d$-electrons in $e_g$-levels which in general can form a narrow band.

Such electrons in general can indeed move through the crystal (if we ignore the potential of the impurity itself, all the positions of the extra electron are equivalent). But if magnetic order of the background localized spins exists, it can influence and maybe hinder the motion of doped charge carriers.

$CaMnO_3$ is known to have a two-sublattice ($G$-type) antiferromagnetic ground state in which all the nearest neighbours of a given site have spins opposite to it. However, in $CaMnO_3$ a strong Hund's rule coupling exists between $d$-electrons, which (classically) forces the extra electron to have its spin parallel to the localized spin of the site. As a result the situation shown in Fig. 5.26 will take place: an extra electron at the site $i$ should have spin up, but it cannot hop to the neighbouring site $j$ having spin down.

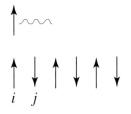

**Fig. 5.26.** An illustration of why the antiferromagnetic ordering hinders the motion of an extra electron in case of strong Hund's rule coupling

If localized spins are treated classically, it can be shown that for strong Hund's rule exchange $J_H > t$ (where $t$ is the hopping matrix element of the extra electron) the *effective* hopping is reduced,

$$t_{ij} \longrightarrow t_{eff} = t \cos \frac{\theta_{ij}}{2} \tag{5.12}$$

where $\theta_{ij}$ is the angle between spins of the sites $i$ and $j$. Thus for purely antiferromagnetic ordering $\theta_{ij} = \pi$, and $t_{eff} = 0$. On the other hand, if the system is made *ferromagnetic*, $\theta_{ij} = 0$, the electrons can move freely, $t_{eff} = t$.

The electrons hopping from site to site with the matrix element $t_{eff}$ (5.12) would form a band, with the spectrum (e.g. in cubic lattice)

$$\varepsilon(k) = -2t_{eff}(\cos k_x + \cos k_y + \cos k_z) \tag{5.13}$$

and a (small) number of doped electrons would occupy the states at the bottom of this band near $\varepsilon_{min} = -6t_{eff}$. Thus whereas in the undoped materials, as assumed, the system is antiferromagnetic due to the corresponding exchange interaction of localized (here $t_{2g}$) spins $J$, in the doped system a gain in energy (kinetic energy of the doped carriers) by increasing $t_{eff}$ can be obtained, i.e. by making the system "more ferromagnetic".

If now we assume that two sublattices form a canted structure with the angle between sublattices $\theta$, the total energy (per site) as a function of this angle may be written in the quasiclassical approximation as

$$E(\theta) = JS^2 \cos \theta - 6tx \cos \frac{\theta}{2} . \tag{5.14}$$

(Here we took into account the relations (5.7), (5.12), and assumed that $x$ is small so that all the electrons are at the bottom of corresponding band.) The minimisation of the energy (5.14) in $\theta$ gives

$$\cos \frac{\theta}{2} = \frac{3}{2} \frac{t}{JS^2} x , \tag{5.15}$$

i.e. under the influence of doping (increasing $x$) the original antiferromagnetic structure becomes *canted*, see Fig. 5.27, i.e. there will coexist both antiferromagnetic and ferromagnetic components of the magnetic order.

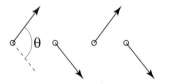

**Fig. 5.27.** The canted antiferromagnetic structure, with the antiferromagnetic component $\pm S^z = \pm S \sin\theta/2$ and the ferromagnetic one $S_x = S \cos\theta/2$

With increasing $x$ the canting angle $\theta$ would decrease and for

$$x > x_c = \frac{2}{3} \frac{JS^2}{t} \tag{5.16}$$

the magnetic order would become purely *ferromagnetic*. This is the standard explanation for the change of the magnetic structure with doping. It is used for example for the description of the appearance of ferromagnetism *together with the metallic conductivity* in doped manganites.

There are many subtle points in this model. First of all, even in this quasiclassical treatment, taking result (5.11) we cannot yet say that the resulting magnetic structure would be a two-sublattice canted antiferromagnetic one. Only the local pitch angle $\theta_{ij}$ is fixed by this treatment; one can however imagine an alternative magnetic structure, e.g. the helicoidal one, with the same pitch angle.

More important may be the neglect of quantum effects in the above treatment. The "ferromagnetic" on-site Hund's rule coupling does not necessarily require the spins of the extra electron and of localized spins to be parallel, as in Fig. 5.26. These should form a state with *maximum total spin*, e.g. a triplet state of the localized spin $s = \frac{1}{2}$, but this triplet can have a total $z$-projection 0, $\frac{1}{\sqrt{2}}(1\uparrow2\downarrow + 1\uparrow2\downarrow)$. It can be shown [24,25] that the motion of the extra electron is quantum-mechanically possible even on a purely antiferromagnetic background, albeit with a reduced bandwidth. The account of these factors modifies the resulting phase diagram, leading e.g. to the appearance of a lower critical concentration $x_{c1}$ [25] below which the original antiferromagnetic structure is undistorted.

But probably the most "dangerous" point in this treatment is the assumption of spatial homogeneity of the system. Using formulae (5.14) and (5.16) the total energy of the assumed homogeneous canted state can be calculated with the minimum energy being

$$E(x) = -JS^2 - \frac{9}{2} \frac{t^2}{JS^2} x^2 . \tag{5.17}$$

However, in this state the compressibility $\kappa^{-1} \sim d^2E/dx^2$ is *negative* [25] which indicates an absolute instability of such a state towards phase separation, i.e. creation of a state with an inhomogeneous distribution of both the extra charge carriers and magnetic order. For instance, the ferromagnetic metallic "droplets" may be formed in an antiferromagnetic insulating matrix [25].

A detailed analysis of the resulting state requires also an account of the long-range Coulomb interaction and lattice distortion; in addition in real manganites of the change of orbital ordering etc., and this is still not a tractable problem. However, generally speaking the creation of a spatially inhomogeneous state in lightly doped manganites, with the properties (e.g. in transport) of percolation systems should be expected and only at higher doping level the (more) homogeneous metallic and ferromagnetic state can be obtained, although some inhomogeneity can still be present.

Doped manganites can be connected to a specific feature of orbital degeneracy of $e_g$-levels or $e_g$-bands in which mobile electrons reside [26]. Probably this is relevant for an interesting and not yet completely understood asymmetry in the behaviour of doped manganites with $x > 0.5$ and $x < 0.5$. Whereas for overdoped manganites ($x > 0.5$) a very stable insulating state with regular stripe structure is formed, for $x < 0.5$ a ferromagnetic metallic state is obtained.

## 5.9    Concluding Remarks

In this short review the basic physical factors determining the electronic and magnetic structure of magnetic oxides were explained. Of course, this chapter cannot substitute a real textbook, but the main physical factors and mechanisms determining the type of magnetic ordering and its interplay with the insulating or metallic behaviour of corresponding materials were stressed.

There are always a lot of specific details important for particular materials. All these details, however, do not invalidate the general conclusion that, generally speaking, antiferromagnetism exists typically in insulating materials, and ferromagnetism is associated with metallic conductivity. There are of course some exceptions to this rule, i.e. there exist ferromagnetic insulators, but these are rare exceptions, and usually a special explanation is required, for further details see the discussion in [7]. The general trend is the one formulated above: antiferromagnetism prefers to coexist with the insulating state, and ferromagnetism with the metallic one. The underlying mechanism responsible for this trend is due to *virtual* hopping of electrons typically giving rise to a superexchange-generated antiferromagnetism, whereas *real* motion of electrons stabilizes ferromagnetism by the double-exchange mechanism. This strong interplay between magnetic structure and transport properties makes these systems so interesting and promising for possible applications in spin electronics.

## References

1. J. M. D. Coey, M. Viret, and S. von Molnar, Adv. Phys. **48**, 167 (1999).
2. Y. Shapira, S. Foner, R. L. Aggarwal, and T. B. Read, Phys. Rev. B **8**, 2316 (1973).
3. J. B. Goodenough, "Magnetism and chemical bond," Interscience Publ., N.Y.-Lnd., 1963.
4. A. Abraham and B. Bleney, "Electron paramagnetic resonance of transition ions," Clarendon Press, Oxford 1970.

5. P. W. Anderson, in "Solid State Physics", ed. H. Ehrenreich, F. Seitz, and D. Turnbull, Academic Press, N.Y. 1963, vol. 14, p. 99.
6. K. I. Kugel and D. I. Khomskii, Sov. Phys.—Uspekhi **25**, 231 (1982).
7. D. Khomskii and G. Sawatzky, Solid State Comm. **102**, 87 (1997).
8. M. A. Korotin, S. Yu. Ezhov, I. V. Solovyev, V. I. Anisimov, D. I. Khomskii, and G. A. Sawatzky, Phys. Rev. B **54**, 5309 (1996).
9. R. Englman, "The Jahn-Teller effect in molecules and crystals", Wiley–Interscience, Lnd.–N.Y. 1972.
10. M. Kaplan and B. Vekhter, "Cooperative phenomena in Jahn-Teller crystals", Plenum Press, 1995.
11. N. F. Mott, "Metal–Insulator transitions", Taylor and Francis, London 1974.
12. D. I. Khomskii, The Phys. of Metals and Metallography, **29**, 31 (1970).
13. P. W. Anderson, Phys. Rev. **115**, 2 (1959).
14. J. Zaanen, G. A. Sawatzky, and J. Allen, Phys. Rev. Lett. **55**, 418 (1985).
15. D. Khomskii, Lithuanian Journal of Physics, **37**, 65 (1997).
16. M. A. Korotin, V. I. Anisimov, D. I. Khomskii, and G. A. Sawatzky, Phys. Rev. Lett. **80**, 4305 (1997).
17. W. Geertsma and D. Khomskii, Phys. Rev. B **54**, 3011 (1996).
18. K. I. Kugel and D. I. Khomskii, Sov. Phys.—JETP **37**, 725 (1973).
19. V. Anisimov, A. Aryasetiawan, and A. Lichtenstein, J. Phys. C: Condens. Matter **9**, 767 (1997).
20. A. Lichtenstein, V. Anisimov, and J. Zaanen, Phys. Rev. B **52**, R5467 (1995).
21. C. Zener, Phys. Rev. **81**, 403 (1951).
22. P. W. Anderson and H. Hasegawa, Phys. Rev. **100**, 675 (1955).
23. P. G. De Gennes, Phys. Rev. **118**, 141 (1960).
24. E. L. Nagaev, Sov. Phys.—Uspekhi **166**, 833 (1996).
25. M. Kagan, D. Khomskii, and M. Mostovoy, Eur. Phys. J. B **12**, 217 (1999).
26. J. van den Brink and D. Khomskii, Phys. Rev. Lett. **82**, 1016 (1999).

# 6 Transport Properties
# of Mixed-Valence Manganites

M. Viret

Service de Physique de l'État Condensé, CEA-Saclay, 91191 Gif/Yvette, France

**Abstract.** This chapter reviews electronic transport properties of manganites of the type $(La_{1-x}Ca_x)MnO_3$. Resistivity variations with composition, temperature and field are presented in some detail and the different types of magnetoresistive effects measured in these systems are analysed. Particular attention is given to ferromagnetic phases which could potentially be used as magnetic sensors. A broader review paper "mixed valence manganites" has been published by J.M.D. Coey, M. Viret and S. von Molnár [1].

## 6.1 Electronic Structure

### 6.1.1 Ionic Model

The perovskite-structure oxides $ABO_3$ with La on A sites and a closed-shell ion on B sites (that of the $3d$ ions) are transparent insulators. Their Fermi level falls in a gap between the valence bands composed of $O_{2p}$ and the bottom of the empty conduction bands which are derived from the unoccupied $5d/6s$ orbitals of the $La^{3+}$. Usually filled $3d^{10}$ levels lie below the top of the $2p$ band whereas empty $d^0$ levels lie near the top of the gap. The $d$-levels fall progressively lower in energy on passing along the $3d$ series because of the increasing nuclear charge. For $3d^n$ cations, the Fermi level falls in the narrow $d$ band, and a lot of interesting physics flows from the strong electron correlations in this band. The traditional approach has been to regard transition-metal oxides as ionic compounds with well-defined $3d$ configurations incorporating an integral number of localized electrons per ion. This approach has been successful in accounting for many aspects of their behaviour, including magnetic moment formation. The model has considerable predictive power, and it is possible to relate the parameters in the model to spectroscopic measurements. Electronic properties of transition metal oxides are determined by an interplay of several interactions of comparable magnitude, all of order 1 eV. Schematically, these are (i) the Mott–Hubbard interaction $U_{dd}$, which is the cost of creating a $d^{n+1}d^{n-1}$ charge excitation in an array of $d^n$ ions; (ii) the charge transfer interaction, $U_{pd}$, which is the cost of transferring an oxygen $p$ electron to the neighbouring $d$ ion to create a $p^5 d^{n+1}$ charge excitation from $p^6 d^n$, (iii) the transfer integral $t$ which determines the $d$-electron bandwidth, $W$, (iv) the Hund's rule on-site exchange interaction $U_{ex}$ which is the energy required to flip a $d$-electron spin and (v) the crystal-field interaction $\Delta_{cf}$ and the Jahn–Teller effect $\delta_{JT}$. Besides the classical Bloch–Wilson insulators where the Fermi level falls in a gap in the one-electron density of states, transition-metal oxides may be Mott–Hubbard or charge-transfer insulators when the electron correlations are such that $U_{dd} > W$ or $U_{pd} > W$, respectively [2,3]. Most oxides of the early 3$d$ transition metals are Mott–Hubbard insulators and many oxides of the late 3$d$ transition metals are charge-transfer insulators. At the end of the 3$d$ series, the charge transfer gap may go to zero, and the oxides then become metals. In the middle of the series where $U_{dd} \approx U_{dp}$ the nature of the gap is less clear-cut. A summary of the magnitudes of these interactions in LaMnO$_3$ is given in table 6.1.

We consider further the electronic structure of manganese ions in B sites of the perovskite structure, where they are coordinated by an octahedron of oxygen neighbours, assuming for the moment that there is an integral number of $d$-electrons per site. It is useful to focus here on the one-electron energy levels rather than the multi-electron states, although they are closely related for the $d^4$ ion Mn$^{3+}$. The interelectronic correlations which give rise to Hund's rules for the free ion are perturbed by the crystalline electrostatic field due to the oxygen anions. The five $d$-orbitals, each of which can accommodate one electron of each spin, are split by the octahedral crystal field into a group of three $t_{2g} - d_{xy}, d_{yz}, d_{zx}-$ orbitals which have their lobes oriented between the oxygen neighbours and a group of two $e_g - d_{x^2-y^2}, d_{z^2}$ orbitals which are directed towards the oxygen neighbours. The former are obviously lower in energy because of the electrostatic repulsion of electrons on neighbouring sites, and the crystal field splitting $\Delta_{\text{cf}}$ ($\equiv 10\,Dq$) between the $t_{2g}$ and $e_g$ orbitals is of order 1.5 eV. The intra-atomic correlations which give rise to Hund's first rule (maximum $S$) is represented on a one-electron energy diagram by introducing an energy splitting of $\uparrow$ and $\downarrow$ orbitals $U_{\text{ex}}$, which is $> \Delta_{\text{cf}}$. Good evidence that $U_{\text{ex}}$ and $\Delta_{\text{cf}}$ are quite similar in magnitude in the perovskite-structure oxides is provided by the trivalent cobalt in LaCoO$_3$ which does not follow Hund's first rule; it is in a low-spin state, $3d^6$, $t_{2g}^6$ with $S = 0$. Trivalent nickel in Ni-substituted manganites is also low-spin [4]. Manganese ions are generally high-spin; the divalent ion, Mn$^{2+}$, has a very stable $3d^5$ configuration, a half-filled shell $t_{2g}^3 \uparrow e_g^2 \uparrow$ with $S = 5/2$ and a spherically-symmetric electron density. Trivalent manganese is $3d^4$, $t_{2g}^3 \uparrow e_g^1 \uparrow$ with $S = 2$, whereas quadrivalent manganese is $3d^3$, $t_{2g}^3 \uparrow$ with $S = 3/2$. The spin only moments of these ions are $5\mu_B$, $4\mu_B$ and $3\mu_B$, respectively.

A distortion of the oxygen octahedron lowers the symmetry of the cubic crystal field in such a way that the centre of gravity of the $t_{2g}$ levels and the centre of gravity of the $e_g$ levels is unchanged. There is therefore nothing to be gained by Mn$^{2+}$ or Mn$^{4+}$ from such a distortion, but Mn$^{3+}$ can lower its energy in proportion to the distortion, and the corresponding penalty in elastic energy will scale as the distortion squared, hence the marked tendency of $d^4$ ions to distort their octahedral environment in order to lower their energy. This is the Jahn–Teller effect. For example, the tetragonal elongation of the octahedron found in the O$'$-type structure will stabilize the $d_{z^2}$ orbital relative to the $d_{x^2-y^2}$ orbital. The $t_{2g}$ orbitals overlap relatively little with the orbitals of nearby oxygen or lanthanum ions, so these electrons tend to form a localized $t_{2g}^3 \uparrow$ ion core. However the $e_g$ orbitals overlap directly with the $p$ orbitals of the oxygen neighbours, so they tend to form a $\sigma^*$ antibonding band.

The end-member compounds such as $LaMnO_3$ have a distorted perovskite structure where the Fermi level falls in a gap between the two Jahn–Teller split $e_g$ bands. However, intermediate compositions such as $(La_{1-x}Ca_x)MnO_3$ with a cubic structure have a partly-filled $\sigma^*$ band, extending in three dimensions. These band electrons, which we refer to as the Zener electrons, hop from one Mn site to another with spin memory. They are both conduction electrons and mediators of the ferromagnetic exchange. Direct overlap of the $t_{2g}$ core electrons of adjacent manganese ions leads to antiparallel exchange coupling, since only the $\downarrow$ orbitals are empty [4].

The Hund's rule exchange splitting $U_{ex}$ between the states with total spin 2 and 1 is evaluated taking $s = 1/2$ and $S = 3/2$ as the spins of the $e_g$ electrons and the $t_{2g}^3$ ion core. The result $U_{ex} = J_H(S + 1/2)$ gives $2J_H$ as the splitting. The value of $U_{ex}$ is found from optical conductivity data to be 2.0 eV [5] which is slightly greater than $\Delta_{cf}$, as expected for a high spin ion. Hence $J_H = 1.0$ eV.

**Table 6.1.** Estimates of characteristic energies in $LaMnO_3$ (in eV)

| $U_{dd}$ | $J_H$ | $U_{pd}$ | $\Delta_{cf}$ | $U_{pp}$ | $\delta_{JT}$ | $W = 12t$ | $J_{ij}$ |
|------|------|------|------|------|------|------|------|
| 4.0 | 1.0 | 4.5 | 1.8 | 7.0 | 0.6 | 1.0 | 0.001 |

Despite the success of the ionic model, electron transfer and orbital admixtures are more properly treated in a band model. Band structure determinations involve extensive computations, and it is more difficult to build up an intuitive understanding of the relation between electron number and physical properties. Recent developments, particularly the local spin density approximation [6], offer access to the magnetic ground-state and provide insight into the extent of orbital mixing which is considerable in all transition-metal oxides. Most difficult to treat accurately are the interatomic correlations.

## 6.1.2   Band Model

Unlike the ionic $4f$ levels, which are unbroadened by overlap and hybridization, the $3d$ ionic levels acquire a substantial bandwidth, of order 1 eV, from overlap with the neighboring orbitals. The bandwidth $W = 2zt$, where $t$ is the transfer integral and $z$ is the number of manganese nearest neighbours is sensitive to Mn-O distances and Mn-O-Mn bond angles [7]. In the double exchange model it also depends on magnetic order, because the transfer integral depends on the angle between neighbouring spins $\theta$ via $t = t_0 cos(\theta/2)$.

Band structure calculations take account of intra-atomic correlations in an averaged way and they reproduce the one-electron energy level splittings of the crystal field theory by admixture of $p$ and $d$ electron orbitals. Before discussing the electronic structure of materials which exhibit large negative magnetoresistance, it is appropriate to consider first the end members. With this method, it is possible to take different crystallographic unit cells (orthorhombic or cubic) and different magnetic arrangements (ferromagnetic, antiferromagnetic or paramagnetic) in order to calculate the free energy associated with each of them and thereby evaluate which is the ground state. For each case, an electronic density of states is obtained and the character of the wave functions ($s$, $p$, $d$) can be evaluated. The different calculations agree very well on the main features. For both end member compounds, Satpathy $et$ $al.$ find that the ionic descriptions $La^{3+}Mn^{3+}O_3^{2-}$ and $Ca^{2+}Mn^{4+}O_3^{2-}$ are good approximations to reality. The calculated magnetic moments for $LaMnO_3$ are in quite good agreement with the experimental values of $3.7 - 3.9\mu_B$ [8,9].

For mixed La-Ca compositions, there is not an integral number of $d$ electrons per atom and the bands are not completely filled. One may therefore expect the compounds to become metallic in the absence of a distortion which creates distinct $Mn^{3+}$ and $Mn^{4+}$ sites. Pickett and Singh [10,11] modelled the $x = 0.33$ system by an ordered triple cell of general formula $La_2CaMn_3O_9$incorporating layers of (La-Ca-La) cations. The compound is found to be ferromagnetic with an average moment of $3.51\mu_B$ per formula unit. Two distinct Mn sites appear in such a cell where some Mn have an A-site cation environment with La neighbours only ($Mn_{La-La}$) and the others have one plane of La and one plane of Ca surrounding them ($Mn_{La-Ca}$). The moments for the two Mn sites are very close and the system should be viewed more in terms of hybridised spin-polarized Mn(3$d$)-O(2$p$) bands rather than strong mixed valence. Also the local densities of states are quite distinct for the two Mn sites. Near $E_F$, ($Mn_{La-Ca}$) bands constitute a nearly "half metallic" pair (i.e. the d states are fully spin-polarized) whereas ($Mn_{La-La}$) bands have an equal number of spin up and down states. The minority spin occupation is therefore determined by shifts in potential arising from nearby cation charges ($La^{3+}$ neighbours create a more favourable environment for an electron). The precise electronic structure is then expected to depend sensitively on local environment effects such as the cation disorder, deviations from stoichiometry and local strain. Pickett and Singh [11] also suggest that charge disorder could lead to localisation of minority states (in fact $Mn_{La-La}$ states) which would generate purely half metallic conduction without the majority carriers being localised. Above $T_c$, the absence of a net moment forces charge carriers to go through differently oriented regions which, due to minority state localisation, would induce very poor conductivity (thermally activated behaviour).

These calculations, and others [12], suggest that disorder plays a crucial role in the electronic state of the manganites. This can lead to Anderson localisation with a determinant role of the magnetisation [13,14]. In the partly-ferromagnetic state, the homogeneous compounds have mixed electronic spin character at the Fermi level. However, the interesting idea of preferential localisation of carriers in the minority spin subband provides support for the full spin polarisation of mobile charge carriers.

Experimentally, there is good evidence of electron localization by potential fluctuations. $(LaMn)_{1-\delta}O_3$ with $\delta = 0.01$ and $0.05$ are both insulators, although they exhibit a large linear term in the heat capacity which suggests a high density of states [15]. The Fermi level therefore lies below the mobility edge. It is also apparent from the conductivity of cation-deficient and doped $LaMnO_3$ that B-site vacancies are more effective than A-site vacancies or A-site cation disorder at localizing the electronic states.

### 6.1.3   Phase Separation

The conclusion that the presence of any small concentration of Zener electrons in a planar antiferromagnet inevitably leads to canting [16] has been criticized by Nagaev [17] and Mishra *et al.* [18]. The latter suggest that for realistic, non-infinite values of the Hund's rule coupling $J_H \approx 1eV$, the canting may be supressed for certain carrier concentrations. Furthermore, it has been shown by Arovas and Guinea that an inhomogeneous magnetic two-phase region is to be expected in the de Gennes model between the ferromagnetic and antiferromagnetic phase fields; the "canted" phase may actually be composed of finely-imbricated nanoscale ferromagnetic and antiferromagnetic domains with different electron concentrations [19]. Phase segregation has also been found in a Kondo lattice model [20,21] where compositions with average electron density around $0.9/Mn$ ($x \approx 0.1$) spontaneously segregate into antiferromagnetic regions with $x = 1$ and ferromagnetic regions with $x = 0.8$. Recent reviews of the experimental and theoretical situation can be found in Ref. [22,23]. A better awareness of the possibility that magnetic and crystallographic two-phase regions may separate single-phase fields could help to rationalize observations of magnetic and crystallographic inhomogeneity in manganites.

Experimentally, neutron studies in inelastic and small angle configurations [24,25] evidenced some degree of phase separation in lightly doped $La_{1-x}Ca_xMnO_3$. In single crystals of the 6% doped composition, entities 1-2 nm in size were observed but the small angle neutron scattering signal corresponds only to a weak contrast in magnetisation. Thus, it seems that instead of finely-imbricated ferromagnetic and antiferromagnetic regions, it is more appropriate to view the system as a canted antiferromagnet where the canting angle varies by about $20°$ on a nanometer scale. The susceptibility of the $x = 0.06$ compound shows a sharp peak at $T_N$ (like that of $LaMnO_3$ [26]) which is characteristic of canted antiferromagnetism. In ferromagnetic $La_{0.7}Ca_{0.3}MnO_3$ thin films, temperature dependent telegraphic noise was attributed to activation of ferromagnetic clusters of nanometre size [27].

## 6.2   Resistivity and Magnetoresistance

In the first study of the mixed-valence manganese perovskites, van Santen and Jonker reported resistivity measurements on ceramic samples of

$(La_{1-x}A_x)MnO_3$ (A = Ba, Ca, Sr) as a function of temperature and composition [28,29]. Their main result was the striking correlation between the magnitude of the resistivity and the magnetic state of the compounds. Outside the ferromagnetic concentration range, resistivities are high and thermally activated, but a resistance anomaly appears around $T_c$ for the ferromagnetic compositions, where there is a transition from thermally activated to metallic-like conductivity. Representative modern data on a thin film is shown in Fig. 6.1.

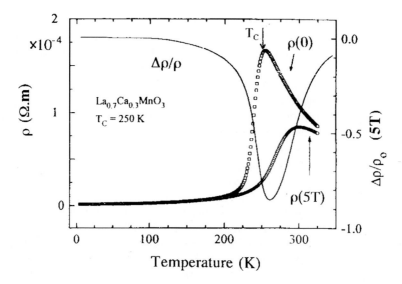

**Fig. 6.1.** Typical resistivity and magnetoresistance of a mixed-valence manganite showing a metal-insulator transition. Data are for a thin film of $(La_{0.7}Ca_{0.3})MnO_3$ for which $T_c = 250$ K (from [30])

When Volger [31] discovered the large negative magnetoresistance effect near $T_c$ in 1954, (Fig. 6.2) he showed it to be isotropic, i.e. independent of the relative orientation of the current and the field, and frequency-dependent (he also made the first measurements of Hall effect, thermopower and specific heat).

The single-crystal magnetoresistance measurements reported 15 years later by a Canadian group for the $(La_{1-x}Pb_x)MnO_3$ system [32] showed the effect to be quite substantial, with a 20% decrease in resistivity at 310 K in an applied field of 1 T, but these results and those of Volger made little impression on the broader scientific community, who were more aware of magnetic semiconductors (like EuO), where huge magnetoresistance effects were seen at low temperature. Twenty more years elapsed before the effect was rediscovered, at a time of intense interest in using magnetoresistance in can be above room-temperature for some compositions (e.g. La-Ba, La-Pb or La-Sr with $x \approx 0.3$). The resistivity peak temperature $T_m$ practically coincides with $T_c$ for crystals with $x \approx 0.3$. The

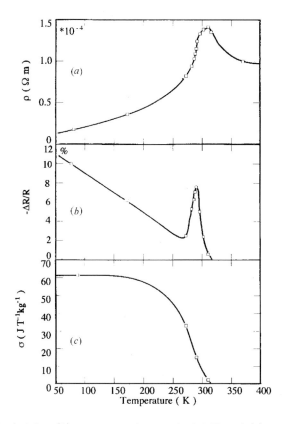

**Fig. 6.2.** (a) Resistivity, (b) magnetoresistance in 0.3 T and (c) magnetization of a $(La_{0.8}Sr_{0.2})MnO_3$ ceramic [31]

isotropic negative magnetoresistive effect requires an applied field in excess of 1T, to approach saturation, as illustrated for the thin film of $(La_{0.7}Ca_{0.3})MnO_3$ in Fig. 6.1. The saturation magnetoresistance in the vicinity of $T_c$ is bounded by the drop in resistivity below $T_c$, which depends in turn on $T_c$ itself [33,34], as shown in Fig. 6.3.

The field-induced change in resistivity is comparable to the high-field resistivity when the Curie point is close to ambient temperature, but it can be several orders of magnitude greater when $T_c$ lies at low temperature. The correlation is shown in Fig. 6.4, where $\rho(T_m)/\rho_0$ is plotted against $1/T_m$. The resistivity of any particular sample depends essentially on its magnetization.

A different type of magnetoresistance effect is the anisotropic magnetoresistance (AMR) which depends on the relative orientation of the current and magnetization. It is a small effect related to the orbital moment, which has been observed at low fields in films of tetragonal $(La_{0.7}Ca_{0.3})MnO_3$ [35].

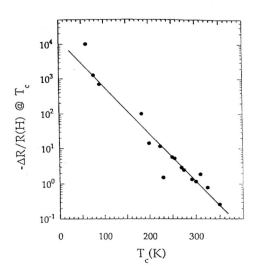

**Fig. 6.3.** Variation of the high-field giant magnetoresistance effect in mixed-valence manganites at their Curie temperatures as a function of $T_c$ , after [33]

## 6.2.1   Variations with Doping Level

The end-member $LaMnO_3$ is a semiconductor, with reported values of the activation energy ranging from 0.10 to 0.36 eV [36]. The sample-dependent bandgap which is small compared to the energy of the charge transfer excitation measured optically ($\approx$ 1eV) [37] or the bandgap measured by photoelectron spectroscopy ($\approx$ 1.3eV) [38] probably reflects doping of the material by defects and nonstoichiometry, and localisation of the resulting carriers. Fig. 6.5 sketches the variation of resistivity and magnetic ordering temperatures for some La manganites as a function of the proportion of $Mn^{4+}$, which is varied by cation doping (or cation deficiency). The correlation of the two phenomena is the basis for Zener's double exchange model. For antiferromagnetic phases, the Zener bandwidth is zero and the insulating character dominates the transport. As the material becomes ferromagnetic, the conduction electrons hop from one Mn to another with spin memory. The ferromagnetic interaction is therefore associated with high electronic mobilities and low resistivities. Electronic transport follows the magnetic transitions, with very high resistivity for the $x = 0$ end member, minimum resistivity obtained for $x = 0.3$ and higher resistivities for the other antiferromagnetic structures at $x = 0.5$. The number of conduction holes can be taken as the number of $Mn^{4+}$ introduced at low doping (small $x$) and the resistivity might be expected to follow this number. Instead, it is found experimentally that the conduction changes rather rapidly at a threshold corresponding to a hole concentration of $x \approx 0.15$, which suggests that localisation must play an important role in the transport properties at low doping. Charge ordering, even on a local scale, also promotes localisation; in the $(La_{1-x}Sr_x)MnO_3$ system, for example, charge ordering occurs around $x = 0.125$ [40]. The electrical resistivity

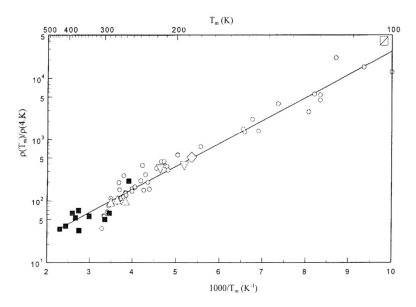

**Fig. 6.4.** Variation of the logarithm of the resistivity ratio $\rho_m/\rho_0$ in thin films of mixed-valence manganites at as a function of $1/T_m$, o($La_{0.7}Ca_{0.3}$)$MnO_3$; squares: ($La_{0.8}Sr_{0.2}$)$MnO_3$; other points (La,Pr,Ca,Sr)$MnO_3$ ; courtesy of K. Steenbeck

of a crystal with $x = 0.15$ is anisotropic, being a factor 2 - 5 lower for the current in-plane rather than along the c-axis [41].

### 6.2.2   Temperature– and Field–Induced Resistive Transitions

Magnetic transitions are commonly observed by varying the temperature. In mixed-valence manganites these can be of three types: Curie and Néel points and transitions to a canted state. A schematic magnetic phase diagram is shown in Fig. 6.6.

#### First Order Transitions

The largest magnetoresistive effects seem to accompany first order transitions. Here, we will give an overview of the different resistive effects associated with these magnetic transitions illustrated by a few examples.

Transitions at the W-W' line in Fig. 6.6 are mostly second order but compounds which have their resistivity peak at low temperature, $T_m \leq 100$ K, tend to exhibit thermal hysteresis [42–47]. In compounds with low Curie points where charge ordering takes place above $T_c$, the competition between Coulomb and ferromagnetic energies make the transition first order. Neutron studies [48] established that the ($Pr_{1-x}Ca_x$)$MnO_3$ system has charge-ordered phases over the entire composition range. The Czech group noted the absence of the cubic structure normally associated with ferromagnetism in these systems together with the

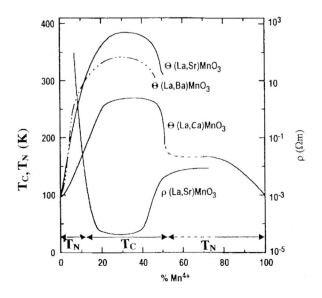

**Fig. 6.5.** Magnetic ordering temperatures and resistivities as a function of $Mn^{4+}$ content $x$ (schematic) after [39]

absence of a drop in resistivity below $T_c$. Low-temperature measurements have shown the existence of a first order transition between a state of high thermally-activated resistivity and metallic-like behaviour in $(Pr_{0.7}Ca_{0.26}Sr_{0.04})MnO_3$ [49], $(Nd_{0.7}Ca_{0.2}Sr_{0.1})MnO_3$ [50], $(Pr_{0.58}La_{0.12}Ca_{0.30})MnO_3$ [51], $(Pr_{1-x}Ca_x)MnO_3$ [52,53] and $(Nd_{0.7}Sr_{0.3})MnO_3$ [54], the latter in thin film form being most likely oxygen non-stoichiometric. Huge resistance ratios are related to cation size [50,55]. In fact, a good recipe for making compounds with a first-order transition triggered by a magnetic field consists of making an atomic mixture of two compounds, one of which presents charge ordering down to low temperature with activated resistivity over the entire range (for instance $Pr_{0.7}Ca_{0.3}MnO_3$), and the other behaving "normally" with a resistivity maximum near $T_c$ (for instance $Pr_{0.7}Sr_{0.3}MnO_3$). A judicious choice of the proportion of each leads to a compound with borderline behaviour in the thermally-activated regime where application of a strong magnetic field may induce a phase transition. Such effects are generally hysteretic and the removal of the applied field does not necessarily restore the initial state unless the temperature is increased again (Fig. 6.7). The transition to the metallic state can be sensitively changed by substituting $^{18}O$ by $^{16}O$ [56].

At a first-order irreversible transition from a semiconducting state, the resistivity of $(Pr_{0.70}Ca_{0.26}Sr_{0.04})MnO_3$, for example, was reduced by an impressive 11 orders of magnitude in 4 T at 30 K [49]. High-resolution electron microscopy reveals that some of these compounds are inhomogeneous on a microscopic scale [58]. Also, small angle neutron scattering indicates that magnetic inhomo-

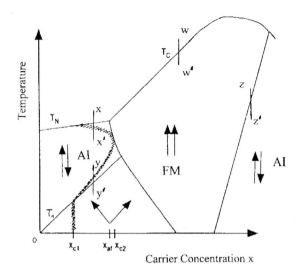

**Fig. 6.6.** Schematic magnetic phase diagram of $(La_{1-x}A_x)MnO_3$ modified from [39]. The insulator-metal transition is hatched

geneities exist on a nanometer scale. It seems here that the resistive transitions are due to the onset of tunneling between conducting grains through canted insulating material. Huge magnetoresistive aftereffects have also been observed in $Pr_{0.7}Ca_{0.3}MnO_3$ single crystals [57]. There, an insulating state was obtained with time when the sample, in the low-temperature zero field metallic state, is driven to a temperature close to a transition in the resistive phase diagram. With time, the magnetisation was observed to decrease slowly (like in magnetic viscosity experiments in spin glasses) while the resistivity jumps very suddenly, at some time, to unmeasurably high values as shown on Fig. 6.8. The explanation is that the sample is in a metastable ferromagnetic state where small clusters are temperature activated towards a antiferromagnetic or canted insulating state. The dramatic resistivity behaviour was taken as evidence for percolation effects in these compounds where time-induced resistive transitions reflect the breakage of the last conduction percolation path [57].

Large magnetic field effects on the resistance are also observed for compounds undergoing low-temperature magnetic transitions involving re-entrant antiferromagnetism or canting $(T_1)$ transitions Y-Y' or Z-Z' on Fig. 6.6. Resistance changes in the opposite sense accompany the re-entrant antiferromagnetic transitions near $x = 0.5$ (Z-Z' on Fig. 6.6). These are very pressure-sensitive and, in certain cases, can be triggered by applied pressures of the order of 0.5 GPa [59]. Under pressure, the resistivity becomes strongly hysteretic thereby underlining the first-order character of the transition. For example, $(La_{0.5}Ca_{0.5})MnO_3$ orders first as a canted antiferromagnet, and at lower temperature as an antiferromagnetic insulator, with huge magnetoresistance around the W - W' transition [60].

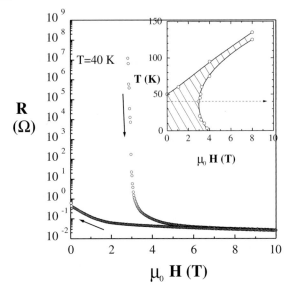

**Fig. 6.7.** Resistivity versus magnetic field at T = 40 K in (Pr$_{0.67}$Ca$_{0.33}$)MnO$_3$) cooled from 300 to 40 K in zero field. The phase diagram in the $(H, T)$ plane is shown in the inset. Notice the large hysteresis region in which the antiferromagnetic insulating phase and the ferromagnetic metallic one can coexist [57].

(R$_{0.5}$Sr$_{0.5}$)MnO$_3$ shows a transition from ferromagnetic or paramagnetic metal to paramagnetic insulator when R = Sm [61–63], but there are two magnetic transitions when R = Pr [64] or Nd [65,66] and the antiferromagnetic ground state may be charge-ordered. The Tsukuba group have studied the ferro - AF transition in (Nd$_{1-x}$Sr$_x$)MnO$_3$ and (Sm$_{1-x}$Sr$_x$)MnO$_3$ near $x = 0.5$ [67] and in ({Nd,Sm}$_{0.5}$Sr$_{0.5}$)MnO$_3$ single crystals [59]. The transition is accompanied by charge ordering, cell distortion and re-entrant thermally activated conduction. It is so strongly hysteretic that it can be completely suppressed by applied fields (Fig. 6.9). In the case of {Nd$_{0.25}$Sm$_{0.75}$}$_{0.5}$Sr$_{0.5}$MnO$_3$, a relatively low field (1 T) was enough to wipe out the re-entrant behaviour and generate a 6 order of magnitude resistivity decrease. At 4 K there is a similar irreversible field-induced transition from a canted semiconducting state to a metastable ferromagnetic metallic state, where the resistivity of (Pr$_{0.6}$Ca$_{0.4}$)MnO$_3$ changes by more than 12 orders of magnitude [69]. Caignaert *et al.* showed that the passage to a canted state through an antiferromagnetic state greatly enhances the magnitude of the magnetoresistive effect at the transition [70]. The effect of many B-site substitutions on the magnetoresistance at transitions occurring in manganites with $x \approx 0.5$ has been reported by the Caen group [71,72,62,73].

Another interesting observation concerns the variation of the magnetic transition temperature $T_1$ between the antiferromagnetic and canted states (Y-Y' in Fig. 6.6) with applied fields. Application of a large magnetic field can greatly shift $T_1$ to extend the conducting temperature range while affecting the para-

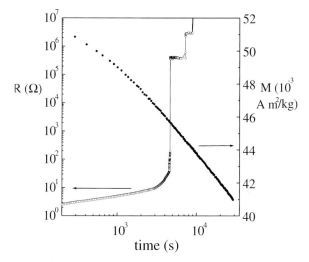

**Fig. 6.8.** Comparison between the time dependences of magnetic moment and resistance in $Pr_{0.67}Ca_{0.33}MnO_3$. The system was initially driven in the ferromagnetic metallic at $T = 45$ K and $H = 0.5$ Tesla, as indicated in the text. Notice the absence of any singularity in magnetic moment at the time $\tau_0 = 4600s$, where the resistance jumps by several orders of magnitude

magnetic Curie temperature very little. The Caen group reported a shift from 75 K to 130 K in a 5 T field, while the paramagnetic Curie temperature remained constant at 170 K in their $(Pr_{0.7}Sr_{0.05}Ca_{0.25})MnO_3$ sample [74]. In other magnetic compounds of the same series, the cell parameters were found to vary at $T_c$ so as to minimise the distortion of the octahedron. This effect is more important in compounds which do not present an intermediate magnetic structure (only one magnetic ordering temperature in Fig. 6.6) [70], and is almost invisible when the activated behaviour survives at low temperature. This is consistent with the Jahn-Teller effect becoming dynamic for the conventional manganites below $T_c$ as electrons become more mobile.

**Reversible Transitions**

There are two types of reversible magnetic transitions which give rise to very different resistive behaviours:

Néel points (X-X' in Fig. 6.6) are generally not accompanied by major resistivity changes, and in any case, activated behaviour persists through the transition to the antiferromagnetic state.

At the Curie point, where ferromagnetic order sets in for compounds with $x$ around 0.3, the resistivities present large anomalies with associated giant magnetoresistance (Z-Z' on Fig. 6.6). A typical curve was shown in Fig.6.1 where a high maximum is obtained near $T_c$. Electric and magnetic transitions are close to one another but may not be completely superimposed. The peak in the resistivity, especially of poorly-crystallized ceramics lies at the temperature $T_m$

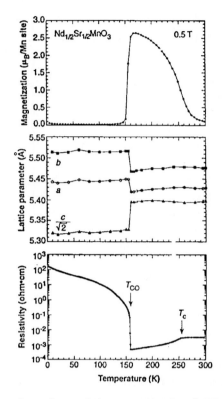

**Fig. 6.9.** Temperature-dependence of the magnetisation, lattice parameters and resistivity of $(Nd_{0.5}Sr_{0.5})MnO_3$ [68]. $T_c$ and $T_{CO}$ are the ferromagnetic and charge ordering transitions respectively

somewhat below $T_c$ [75,76], but in crystals these temperatures almost coincide [77]. An interesting property is that they are both pressure and magnetic field dependent, with increases of order 30 K/GPa and 10 K/T, respectively.

The metal-insulator transition in these manganites is atypical in that it is the low-temperature phase which is metallic, and the high-temperature phase which is insulating, rather than the reverse. This is due to the magnetic nature of the transition; the double exchange setting in just below $T_c$ drives the material metallic. As an external field is applied, the magnetic order is enhanced and the resistivity decreases. The field is most effective near $T_c$ where the magnetic susceptibility is maximum.

Numerous recent papers present magnetoresistance measurements at the Z-Z' transition in different manganese oxide compounds. A large effect is essentially linked to the presence of an adequate amount of $Mn^{4+}$, around 30%, which can be introduced by cation substitution or cation deficiency [78]. The effect is inhibited by charge order. It increases as the resistivity $\rho_{max}$ in the localised state just above $T_c$ increases (Fig. 6.3), but the residual resistivity $\rho_0$ is usually a lower bound on the minimum resistivity that can be achieved. Large applied fields are needed to achieve colossal magnetoresistance effects in bulk and thin-film crystals. Kusters *et al.* [79] decreased the resistance of their $(Nd_{0.5}Pb_{0.5})MnO_3$ crystal by two orders of magnitude near $T_c = 180$ K by applying an 11 T field. Von Helmolt *et al.* reported large effects at room-temperature in oriented $(La_{0.67}Ba_{0.33})MnO_3$ films [80] grown by pulsed laser deposition on $SrTiO_3$ substrates, followed by post-annealing treatments at $900°C$ in order to raise $T_c$ above 300 K. The resistive transition becomes sharper and the magnetoresistance ratio defined as $R(0)/R(H)$ at room-temperature reached 2.5 at $\mu_0 H = 6$ T. The effect can be much larger in compounds having a lower Curie temperature as illustrated in Fig. 6.3. Jin *et al.* [81] at Bell grew $(La_{0.67}Ca_{0.33})MnO_3$ thin films by pulsed laser deposition on $LaAlO_3$ substrates which showed $R(0)/R(H)$ of 5.6 at 100 K in an applied field $\mu_0 H = 6$ T. Here a suitable post-annealing treatment ($900°C$ under 3 atm of $O_2$ for 1/2 hour) reduces the transition temperature and greatly enhances the magnetoresistance ratio (defined as the ratio of the resistance difference divided by the high-field resistance) to 'colossal' values of order 10000 at 110 K and 6 T [82]. It is reported that there is also an optimum film thickness of about 100 nm [83] for the magnetoresistive effect, possibly related to strain. The magnitude of the magnetoresistance effect is increased by substituting some Y for La, but the Curie temperature is decreased [84,85,42,86,87]. Again, the effect is optimized by appropriate annealing [88]. At constant doping, the decrease in Curie temperature when a heavier rare earth is substituted for La depends on the rare earth cation radius [89]. The high-field magnetoresistance changes sign and becomes positive above $2T_c$ [87].

A comprehensive study of the influence of deposition conditions and post-annealing treatments on the magnetoresistance was reported by the Maryland group [90,54]. Thin films of $(La_{0.67}Ba_{0.33})MnO_3$ and $(Nd_{0.7}Sr_{0.3})MnO_3$ were synthesised by pulsed laser deposition in a 40 Pa $N_2O$ atmosphere at temperatures ranging from $615°C$ to $815°C$. The films grown at lower temperatures exhibited better crystallographic quality. The temperature at which the resistivity of the films reaches its maximum is found to decrease with increasing deposition temperature, and the resistance of that maximum is pushed to higher values. For the post-annealing, both temperature and time of the treatment tend to increase the temperature of the maximum and lower the peak resistivity. For the lower Curie temperatures, $R(0)/R(H)$ of the order of 100 can be reproducibly achieved in 8 T by the post-deposition treatments. The largest negative magnetoresistance effects are correlated with substrate-induced lattice distortion [91,92]. Reducing the oxygen content in $(La_{0.67}Ba_{0.33})MnO_{3-\delta'}$ extends the temperature range in which the large magnetoresistive effect is observed, and eventually makes the compound insulating when $\delta' = 0.2$ [93]. Defects induced by irradiation with $Ar^+$ ions have a similar effect [94]. Prolonged annealing of $(La_{1-x}Sr_x)MnO_3$ first decreases, then increases the resistivity [95].

At first sight, the decrease in resistivity with applied magnetic fields seems to be different above and below $T_c$. Fig. 6.10 shows a representative set of $R(H)$ curves at different $T$. Above $T_c$, the magnetoresistive effect is not very large in small fields, and the curves are bell-shaped initially with $\Delta\rho \propto \mu_0 H^2$ [89,96,97]

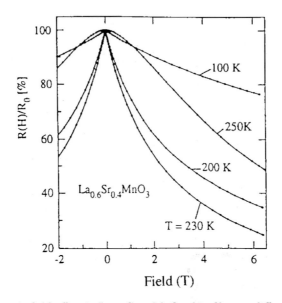

**Fig. 6.10.** Magnetic field effect in $La_{0.7}Ca_{0.3}MnO_3$ thin films at different temperatures above and below $T_c = 240$ K [80]

or $\Delta\rho \propto \mu_0 M^2$ [98]. In the vicinity of $T_c$, the resistivity drops rather quickly at low fields and saturates at higher fields, but at low temperature, the effect is again small. Just below $T_c$, $\Delta\rho$ varies roughly as $\sqrt{H}$ [89], but others report that the resistivity is nearly linear in $H$ [99] or that conductivity is linear in $H$ [100]. In fact, what seems to be qualitatively different behaviour may just reflect the effect of the magnetization. In the paramagnetic state, $M \propto H$, hence $\Delta\rho \propto \mu_0 H^2$. On the critical isotherm, $M \propto H^{1/\delta}$ where $\delta \approx 3 - 5$, hence $\Delta\rho \propto \mu_0 H^{2/\delta}$. Below $T_c$, $M$ depends nonlinearly on $H$; an exponential variation $\Delta\rho \propto \exp[-H/H_0]$ has been reported [101]. A single scaling function $\Delta\rho \propto \exp[(-M/M_0)^2]$ can reproduce the field and temperature dependence both above and below $T_c$ [102]. The field-dependence may be best rationalized in terms of a straight resistor in series with a magnetoconductor, which varies as $H^2$ above $T_c$ and as $H$ below $T_c$ [103].

Most magnetoresistance measurements have been made on thin films with the field $H$ applied in the film plane, so there is no demagnetizing effect. Nevertheless, it is important to understand which of the magnetic vectors, $B$, $H$ or $M$ produces the magnetoresistance phenomena. Insofar as conduction is by spin-polarized electrons, the important quantity appears to be the magnitude of the magnetization $M_{av}$ averaged over the spin-diffusion length $\lambda_s$, which is expected to be a multiple of the mean free path $\lambda$, probably a few tens of interatomic spacings. Well above $T_c$, where the magnetisation follows a Brillouin function which is linear in field at low fields, $M_{av} \approx M$ ($= 0$ in zero field). Magnetization varies very quickly around $T_c$, where small fields have a large effect (high magnetic susceptibility). Below $T_c$ in the multidomain ferromagnetic state

$M_{av} \approx M_s$, since the domain size $\gg \lambda_s$ and the domain wall width is also likely to be greater than $\lambda_s$. At temperatures well below $T_c$, the spontaneous magnetisation in compounds with $x$ around 0.3 is close to the saturation value and the field effect on the magnetisation at the scale of the spin-diffusion length is minimal, the saturation of $M$ being only due to domain wall motion. However, the magnetization is sharply inhomogeneous near grain boundaries in a ferromagnetic polycrystal, and there $M_{av}$ can be much less than $M_s$ for those electrons crossing from one grain to the next. Also, effects of spin injection through grain boundaries can greatly affect the low-temperature MR. This effect is discussed below.

**Ferromagnetic State**

The ferromagnetic state in compounds with $x \approx 0.3$ is metallic in the sense that the resistivity is practically temperature-independent, with a slight positive temperature coefficient. Experimental evidence for metallic behaviour comes from power-law dependences of $R$ at low-temperature where $T^2$ [104,105] or $T^{2.5}$ [106] increases have been reported. These could be ascribed to correlation effects in a degenerate electron gas [107], but there may also be a contribution from spin-wave scattering. In fact, the coefficient of the $T^2$ term derived from the measurements of Urushibara on single crystals, which is a measure of the correlation strength, is comparable to values for other highly correlated electron metals such as $CeAl_3$ and $CeCu_2Si_2$ [108]. However, only in a few compounds such as $(La_{0.7}Sr_{0.3})MnO_3$ does the magnitude of the residual resistivity have the value expected of a metal, $\rho_0 < 1.5 10^{-6} \Omega m$, applying the criterion that the mean free path should exceed the interatomic spacing [109]. In other compounds, the residual resistivity can be up to ten orders of magnitude greater (as shown in Fig. 6.11), making them extremely peculiar metals. In some cases, a slight upturn in $\rho$ is reported below 10-20 K [31,110,33,111] which can be suppressed by applied field or pressure. The resistivity in the 'bad' metallic state is unusually sensitive to pressure, with variations of $\rho_0$ of one order of magnitude per GPa [113,110]. It is also dependent on the substrate and preparation conditions for thin films which may distort the lattice [114,91], and depends critically on the grain size [115–117] and grain-boundary angle [118]. A variation in $\rho_0$ from $3 \times 10^{-6}$ to $2 \times 10^{-3} \Omega m$ for $(La_{0.67}Ca_{0.33})MnO_3$ has been ascribed by Gupta *et al.* [116] to spin-dependent scattering at grain boundaries as the crystallite size decreases from bulk to 3 μm. Resistivities as high as $8 \times 10^3 \Omega m$ are ascribed by Coey *et al.* to intrinsic magnetic barriers between misaligned ferromagnetic regions which are in the nanometer size range [112]. Large reductions in $\rho_0$ are associated with increases in reduced magnetization [119], and it is likely that canted $Mn^{4+}$ spins in the low temperature ferromagnetic state may cause the lack of magnetic saturation [120]. Very direct evidence of spin-dependent scattering in the ferromagnetic metallic state has been provided by the resistive behaviour of bilayers and heterostructures described below.

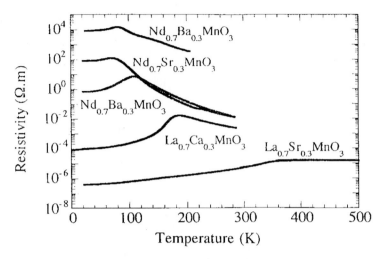

**Fig. 6.11.** Resistivity as a function of temperature for films of mixed-valence manganites with $x = 0.3$. after [112]

## Grain Boundary and Tunneling Effects

Size effects are important in fine-grained polycrystals. The high-field, colossal magnetoresistance is independent of grain size [121], but there is a low-field magnetoresistance at all temperatures below $T_c$ in polycrystalline material which is absent in single crystals (Fig. 6.12).

The low-field effect is distinctly different from colossal magnetoresistance. It increases with decreasing grain size, and grows rapidly at low temperature. The effect, which has been observed in polycrystalline ceramics [117,116,123,124] and thin films [125,126,103,39] is attributed to spin-dependent grain boundary scattering. The boundaries are about 1 nm wide. The grain boundary scattering has been isolated in experiments on manganite strips on a bicrystal substrate [127,128]. Domain wall scattering varies as $\delta_w^{-2}$, hence the narrow "domain walls" that occur at grain boundaries may be much more efficient at scattering electrons than the wide Bloch walls that occur in the bulk [129]. Unlike the high-field colossal magnetoresistance which is an intrinsic property, usually greatest near $T_c$, the low-field response reflects the micromagnetic state of the sample and peaks at the coercive field. The effect can be most directly seen in pressed powder compacts of manganites (Fig. 6.13) [130] and other half-metallic ferromagnets [131,132], where the powder magnetoresistance (PMR) is entirely due to interparticle contacts. It is associated with the degree of alignment of the magnetization of the grains or particles, but it may be sensitive to paramagnetic or quenched spin disorder at grain boundaries [116,123,133]. Surface spin disorder can arise because absent bonds weaken the exchange interactions of atoms in the interface, or alter the balance of ferromagnetic and antiferromagnetic interactions. A different explanation of the low-field magnetoresistance of tetragonal $(La_{0.6}Ca_{0.3})MnO_3$ films in terms of domains has been given by O'Donnell [134].

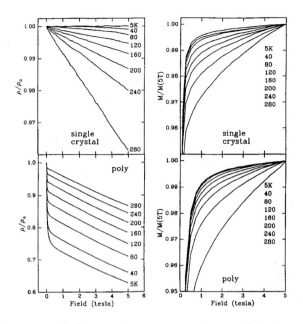

**Fig. 6.12.** Comparison of the negative magnetoresistance of single-crystal and poly-crystalline $(La_{1-x}Sr_x)MnO_3$ [122]

Whatever the explanation, the low-field effect could be more significant for applications than the intrinsic colossal magnetoresistance, but the effects reported at room temperature so far are very small [126].

Sandwich structures and superlattices have been prepared from the ferromagnetic manganites with nonmagnetic spacer layers, that may be insulating, metallic or superconducting. In superlattices of $(La_{0.67}Ca_{0.33})MnO_3$ and the ferromagnetic metallic oxide $SrRuO_3$ ($T_c = 162$ K), enhancement of the magnetoresistance effect in the ferromagnetic state at low temperatures in the superlattice as compared to the pure film is ascribed to spin-dependent scattering at the interfaces [135].

Tunnel spin valves, where the current is forced through an insulating barrier that separates two ferromagnetic layers, are appropriate systems for an enhanced magnetoresistive response since small applied fields can modify the magnetic configuration of the device. The barrier decouples the ferromagnetic electrodes so that their relative orientation can easily be changed. Trilayers can be fabricated for current perpendicular to the plane (cpp) geometry using conventional UV lithography. In tunnel junctions with fully-polarized electrodes, electron tunneling - supposedly without spin flip - should be forbidden when the electrodes have antiparallel magnetization. The IBM group first reported magnetoresistive effects as large as 80 % in $(La_{0.7}Sr_{0.3})MnO_3/SrTiO_3/(La_{0.7}Sr_{0.3})MnO_3$ or $(La_{0.7}Ca_{0.3})MnO_3/SrTiO_3/(La_{0.7}Ca_{0.3})MnO_3$ trilayers at 4.2 K [136,137]. A similar structure where a 3 nm thick $SrTiO_3$ layer separates 25 and 33 nm thick manganite layers, has been measured by Viret *et al.* [138]. Charge carriers were

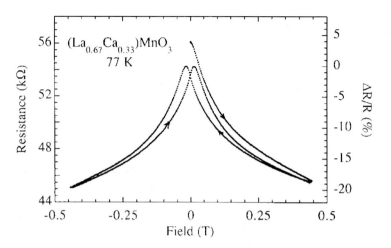

**Fig. 6.13.** Magnetoresistance of a $(La_{0.67}Ca_{0.33})MnO_3$ powder compact at 77 K. The initial magnetization curve is shown, together with the hysteresis which arises from the hysteresis of the ferromagnetic powder [130]

forced to tunnel between a $(La_{0.7}Sr_{0.3})MnO_3$ bottom stripe into a 6 μm side $(La_{0.7}Sr_{0.3})MnO_3$ square through the 3 nm thick insulator. Figure 6.14 shows a magnetoresistance curve at 4.2 K where the nominal resistance is multiplied by 5.5 when the top square flips into the antiparallel configuration. This magnetoresistive effect at such low fields (of the order of 10 mT) is the highest yet reported in any magnetoresistive system. Effects of this magnitude demonstrate the high degree of spin polarization of electrons at the Fermi level in mixed-valence manganites. A quantitative estimate of the spin polarization can be made using the relation [139] $(\sigma^\uparrow - \sigma^\downarrow)/(\sigma^\uparrow + \sigma^\downarrow) = (1 - 2a)^2$, where $\sigma^\uparrow$ and $\sigma^\downarrow$ are the conductances in the parallel and antiparallel states, and $a$ is the proportion of spin-up carriers. A 450% magnetoresistive effect gives $a = 0.91$ and a polarization of 83%. This is much higher than in any of the $3d$ metals or alloys, where the largest polarization of tunnelling electrons is 40%, for Fe [139].

The magnetoresistance of tunnel spin valves falls dramatically as the temperature is raised (Fig. 6.15). This was explained by the presence of an oxygen-deficient layer at the $SrTiO_3/(La_{0.7}Sr_{0.3})MnO_3$ interface which has a lower Curie temperature than the bulk of the $(La_{0.7}Sr_{0.3})MnO_3$ and induces spin flipping which decreases the magnetoresistive effects. Measurements of neutron reflectivity with polarisation analysis confirm that interfaces between manganite electrodes and both the substrate and the barrier are magnetically affected [140,141]. A reduced magnetisation has been measured down to several nm inside the manganite. A reduced magnetisation leads to decreased spin polarisation at the interface with the insulating barrier and hence a reduced magnetic effect on the tunneling. Presumably, the oxygen is off-stoichiometry in this interface which affects magnetic order and Curie temperatures. Tunnel magnetoresistance effects are greatly suppressed above the interface ordering temperature leading to a

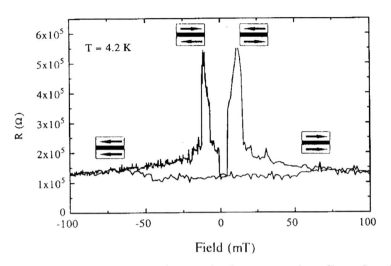

**Fig. 6.14.** Magnetoresistance ratio for a sandwich structure of two $(La_{0.67}Sr_{0.33})MnO_3$ films separated by a 3 nm layer of $SrTiO_3$. The magnetization of the two magnetic layers switch at different fields [138]

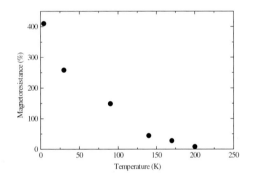

**Fig. 6.15.** Evolution of the magnetoresistive effect with temperature in tunnel spin valves. The dramatic decrease of the TMR amplitude is attributed to the effect of a magnetically reduced interface [138].

maximum in the $R(T)$ characteristic of the junctions. Therefore, it appears crucial to be able to better control the magnetisation of the interface layer in order to achieve a large effect at room temperature. Another problem in these systems is the reproducibility of the magnetic switching [140]. Because the magnetoresistive effect is so large, any spin misalignment in the electrodes in the antiparallel configuration produces a parallel conduction channel which shortcuts the large induced resistance. This effect is particularly important in micron-size electrodes where micromagnetic configurations are quite complex. This could also greatly enhance the problem of noise of magnetic origin in potential devices.

## Paramagnetic State

As in the case of the high-temperature superconductors, the physical properties of the manganites in the region above the transition temperature are anomalous. Transport above $T_c$ is still a matter of controversy as numerous groups have reported different behaviour. Data on compounds with $x \approx 0.3$ were first fitted with a purely activated law, $\rho = \rho_\infty \exp(E_0/kT)$, where the gap $E_0$ is typically 0.1 eV [79,30,100]. There is also evidence for a $\rho \propto T \exp(E_0/kT)$ behaviour over an extended temperature range [105,114,92]. Others find that Mott's variable range hopping (VRH) expression, $\rho = \rho_\infty \exp((T_0/T)^{1/4})$ is appropriate [142,112,96,143,144,14,106]. The simple activation law could indicate the opening of a gap at the Fermi level above $T_c$. Photoemission data support the view that a small band gap appears at $T_c$ in $(La_{0.7}Ca_{0.3})MnO_3$ [145,146] (although the manganite surface probed in photoemission is unrepresentative of the bulk [147]), and there is evidence from tunnelling spectroscopy [148] and Hall effect [149] for a change in the density of states at $T_c$. However, it is difficult to justify a gap over a range of $Mn^{4+}$ concentrations from $x = 0.2$ to $x = 0.4$ in the absence of any change in structure. Moreover, Hundley et al. [30] could not interpret their activated behaviour as excitation to extended states, since the measured mobility was so small that a mean free path could not be defined. Others who found $\rho = \rho_\infty \exp(E_0/kT)$ activated behaviour did not have Hall or other transport results to identify the form of transport [150,100].

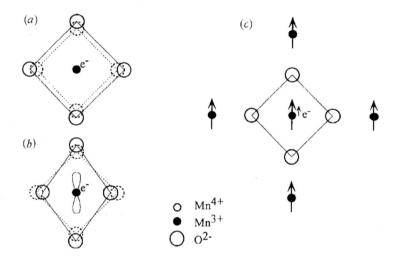

**Fig. 6.16.** Schematic picture of (a) a small dielectric polaron, (b) a Jahn-Teller polaron and (c) a magnetic (spin) polaron

A common view is that the carriers form small dielectric polarons (Fig. 6.16) [151,152,79]. There is some direct evidence of small polaron formation in the distribution of Mn - O bond lengths contained in the pair distribution function of $(La_{1-x}Ca_x)MnO_3$ with $x = 0.12$ [153]. The hopping motion of polarons leads to a resistivity of the form $\rho = (kT/ne^2D)\exp(E_0/kT)$ where $n$ is the carrier density and $D$ is the polaron diffusion constant. Here, there may be contributions to $E_0$ of magnetic, elastic (Jahn-Teller) or coulombic origin. Park $et$ $al.$ [145] suggest that strong polaron effects lead to a charge fluctuation energy of about 1.5 eV above $T_c$, but that the Jahn-Teller effect is less significant than the normal small-polaron contribution. Good agreement with data was reported for the simple form $\rho \propto T\exp(E_0/kT)$, with $E_0 = c + E_{trap}$, where $E_{trap}$ is the trapping energy [154,105]. Another related form of polaron conduction, in which the prefactor is also dependent on the state of magnetization, was used to extract the lattice polaron trapping energy $\approx 0.35$ eV [114]. Remarkably, this value remained constant for a variety of films having different magnetic and transport properties. Perhaps the most extended high temperature measurements on single crystals and thin films of the Ca and Sr doped La-manganites are those of Snyder $et$ $al.$ [105]. Their resistivity data, shown in Fig. 6.17, follow the small polaron hopping law very well. These authors show that at least part of the hopping energy must come from lattice distortions, as can be seen in the abrupt change from temperature independent to activated transport near the 750 K structural transition in the inset of Fig. 6.17.

Another plausible view is that the presence of magnetic disorder above $T_c$ together with the intrinsic variations in the Coulomb potential due to the presence of $A^{3+}$ and $A^{2+}$ ions in the lattice leads to the formation of a mobility edge

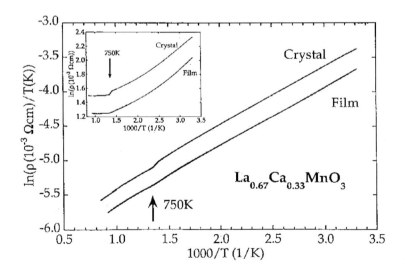

**Fig. 6.17.** High temperature resistivity (warming and cooling) of $(La_{0.67}Ca_{0.33})MnO_3$ film and crystal in zero field after [105]

[155]. At high temperatures, carriers will be excited from the Fermi energy $E_F$ to the mobility edge $E_m$, giving an activated conductivity. At lower temperatures it may be possible to discern a nearest-neighbour hopping process with a lower activation energy, which transforms into uncorrelated variable range hopping $\ln\rho \propto T^{-1/4}$ when the available phonon energy is so small as to make longer-range hops necessary to find a site sufficiently close in energy for hopping to occur. This view is not incompatible with small dielectric polaron formation since variable-range hopping of small polarons also leads to $\ln\rho \propto T^{-1/4}$ [156]. For highly-correlated electron systems a Coulomb gap appears at $E_F$ and the hopping law is then $\ln\rho \propto T^{-1/2}$ at temperatures below the correlation gap [157]. Experimental evidence for VRH behaviour above $T_c$ is presented in Fig. 6.18 in a range of ferromagnetic manganites with $x \approx 0.3$. Resistivity data there can usually be fitted with a power law between 1/2 and 1/4. There is also evidence of VRH behaviour when $x \approx 0.5$ [64].

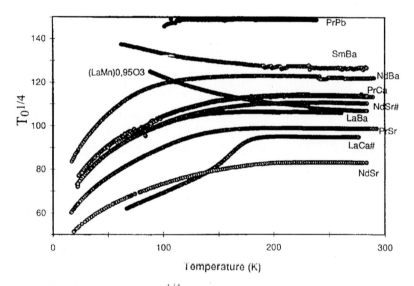

**Fig. 6.18.** Plot of the parameter $T_0^{1/4}$ versus temperature showing variable range hopping behaviour above $T_c$ in a range of ceramic samples of $x = 0.3$ compounds (# indicates thin films)

An important question in the manganites is the existance of a magnetic contribution to the localising random potential. A Coulomb random potential may be caused by substitution or vacancies on A or B sites. For instance, the electron occupancy corresponding to 30% $Mn^{4+}$ can be achieved in three ways, by divalent cation substitution with $x = 0.3$, by lanthanum deficiency with $z = 0.077$ or by nonstoichiometry with $\delta = 0.048$. All three compounds are ferromagnets, but only the first two are metals. Activated conduction is measured down to the lowest temperatures in ferromagnetic $(LaMn)_{0.95}O_3$ [158,159]. This demonstrates that the Coulomb random potential due to the B-site (manganese) vacancies is well able to produce localisation even in the absence of a fluctuating magnetic potential. In cation-deficient $La_{1-z}MnO_3$, where there are only A-site vacancies, the carriers are localised when $z < 0.05$, whether in the antiferromagnetic phase ($z < 0.03$) or in the ferromagnetic phase ($0.03 < z < 0.05$) [160]. However, the ferromagnet with $z \approx 0.05$ is a metal [78]. Therefore, for similar $Mn^{3+}/Mn^{4+}$ ratios [161,162], B-site vacancies localise the $e_g$ electrons more effectively than A-site vacancies. In the substituted lanthanum compounds with A-site disorder ($La_{1-x}^{3+}A_x^{2+})MnO_3$, the random potential variations $\Delta V$ due to divalent and trivalent A-site ions are insufficient to produce localisation in the ferromagnetic state when $x = 0.3$. This indicates that the ratio of $\Delta V$ to the occupied $\sigma^*$ bandwidth $W$ is less than the critical value for 'diagonal' or Anderson localisation. However when a smaller rare earth, such as Pr or Sm replaces La, a ferromagnetic insulating state is sometimes found. This may be associated with a decreased Mn-O-Mn bond angle which reduces the transfer integral and bandwidth, hence tipping $\Delta V/W$ above the value necessary for localisation [109].

## Polarons

We have seen that the manganites with $x \approx 0.3$ which are ferromagnetic metals below $T_c$ generally exhibit activated conduction above $T_c$. Setting aside some minor changes in lattice parameters at $T_c$, the metal-nonmetal transition must evidently be magnetically driven. It is very likely that the carriers polarise their immediate environment creating short-range magnetic correlations. The large effective mass and small activation energy of the carriers indicate polaron formation. We have discussed the possibility that these polarons are normal small dielectric polarons where the electron bears with it a dilatation of the $MnO_6$ octahedron. The possibility that these polarons form bound pairs (bipolarons) above $T_c$ has been discussed in ref. [163].

Other possibilities are Jahn-Teller polarons where the electron carries with it an axial distortion of the $MnO_6$ octahedron, and magnetic polarons where there is a ferromagnetic polarisation of the surrounding Mn core spins (Fig. 6.16). The influence of an applied field on the resistivity and thermal expansion above $T_c$ indicates that the polarons have magnetic character [79,80,164,106,143]. Although small-angle neutron scattering [106,143,165,166] confirms the presence of nanometer-scale magnetic coherence and fluctuating short-range order in the vicinity of $T_c$ which persists up to $\approx 1.3T_c$, proper analysis of the spin pair-correlation function does not support the existence of magnetic polarons with a well-defined size, at least in $(La_{0.75}Sr_{0.25})MnO_3$ single crystals [166]. The magnetic coherence is similar to that encountered in critical scattering from Fe or Ni. Viret *et al.* introduced a spin-pair correlation function of a form consistent with magnetic exchange interactions that extend beyond nearest neighbours. As regards the dynamics of the magnetic fluctuations, muon spin relaxation indicates unusual spin dynamics below $T_c$ [167]; NMR shows that two neighbouring moments remain parallel for more than $10^{-5}$ s [168]. More direct evidence of spin correlations on the scale of 1 nm at $T_c$ has been deduced from analysis of the quasi-elastic neutron peak in the Ca compounds [165]. In the Sr crystal, a picture of slowly-fluctuating moments where ferromagnetic interactions are mediated by Zener electron hopping seems preferable to one of fast-moving spin polarons with a definite size.

Jahn-Teller polarons are local distortions of the lattice around $Mn^{3+}$ ions. When these polarons are frozen, the Jahn-Teller distortion becomes cooperative as in the O' structure of compounds with small values of $x$. The more conducting rhombohedral compounds with $x \approx 0.3$ have equal Mn - O bond lengths imposed by symmetry, but dynamically-fluctuating local distortions may still stabilise the Jahn-Teller polarons [169]. Both static and dynamic distortions have been evidenced by neutron diffraction. The Jahn-Teller (J-T) polaron can form in a solid when the J-T stabilisation energy $\delta_{JT}$ is comparable to the conduction electron bandwidth $W$. Unlike the dielectric polaron where a charge polarisation decorates the carriers and increases their effective mass, the J-T polaron carries with it a local distortion which removes the degeneracy of the electronic ground state. Some evidence for the J-T polaron is seen in the temperature-dependence of the lattice parameters of orthorhombic samples with $x \approx 0.3$. As these manganites with $x \approx 0.3$ are cooled towards $T_c$, the ratio $c/\sqrt{2}a$ in the orthorhombic cell decreases sharply [170,70,7,171] reflecting an increasing deformation of the $MnO_6$ octahedron in the basal plane which splits the $e_g$ band. The greater the distortion, the more localised are the charge carriers. This deformation disappears in the metallic state, below $T_c$. The dynamic Jahn-Teller effect has been detected through the variation of the Debye-Waller factor for samples with $x = 0.3$ in the vicinity of $T_c$ [170,172,173] and in the temperature-dependence of the optical conductivity [120]. Ion channeling experiments also seem to provide direct evidence for a dynamic Jahn-Teller distortion around $T_c$ [174].

More dramatic evidence for polarons, which are probably of the Jahn-Teller variety, is the isotope effect on the Curie temperature discovered in $(La_{0.8}Ca_{0.2})MnO_3$ by Zhao *et al.* [175](Fig. 6.19). $T_c$ is 21 K higher in samples made with $^{16}O$ than in those made with $^{18}O$. An isotope effect on the EPR signal is also associated with a larger exchange integral in $^{16}O$ than in $^{18}O$ samples [176].

The composition $x = 0.2$ is one where the dynamic Jahn-Teller effect for $Mn^{4+}$ is most pronounced [177]. No such isotope effect is found in the ferromagnetic perovskite $SrRuO_3$, where there is no strong Jahn-Teller ion. The effective polaron bandwidth $W_{eff}$ depends on the characteristic frequency $\omega$ of the optical phonons involved: $W_{eff} = W \exp(-\gamma\delta_{JT}/\hbar\omega)$ where $W$ is the bare bandwidth, $\delta_{JT}$ is the Jahn-Teller stabilization energy which is about 0.5 eV [178,179,42] and $\gamma$ is a parameter depending on $\delta_{JT}/W$, which is in the range 0 - 1 [175]. Another estimate of $\delta_{JT}$ is 0.35 eV [114]. The phonon frequency varies as $M^{-1/2}$, where $M$ is the isotope mass. Hence a significant reduction in $W_{eff}$ is expected on passing from $^{16}O$ to $^{18}O$. This translates into a reduction in Curie temperature since $T_c \propto W_{eff}$ in the limit where the Hund's rule coupling $J_H \gg W_{eff}$ [180,169]. There is also an effect on the resistivity in both paramagnetic and ferromagnetic

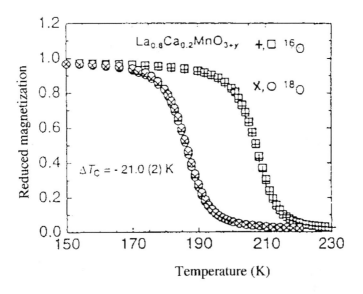

**Fig. 6.19.** Temperature-dependence of the magnetization of $(La_{0.8}Ca_{0.2})MnO_3$ samples made with different oxygen isotopes [175]

states on replacing $^{16}O$ by $^{18}O$. In the case of $(\{La,Pr\}_{0.7}Ca_{0.3})MnO_3$ the isotope substitution is sufficient to convert a low-temperature metallic phase into an insulator [56].

The absence of polaron conduction and colossal magnetoresistance effects in mixed-valence cobalt perovskites is indirect evidence in favour of J-T polarons in manganites, since low-spin $Co^{4+}$ is not a strong Jahn-Teller ion.

### 6.2.3   Models for Electronic Transport

Spin-disorder scattering has been repeatedly invoked to explain the transport properties of GMR systems. The first models were developed in the late fifties and sixties by de Gennes and Friedel [181], Kasuya [182], van Peski-Tinbergen and Dekker [183] and Fisher and Langer [184]. These early calculations deal with spin scattering of conduction electrons in ferromagnetic metals and degenerate semiconductors. Resistivity anomalies are expected at the Curie point because of the onset of magnetic order. Friedel and de Gennes have considered the effects of long-range order on the scattering near the critical point. Above $T_c$, the magnetic scattering is constant and temperature independent. They deduced, however, that the resistivity should present a singularity at the Curie temperature with a smooth decrease in scattering below $T_c$. Fisher and Langer argued that short-range fluctuations make the dominant contribution to the temperature-dependent part of the resistivity, so that the singularity should be in the derivative of the resistivity at $T_c$ (which is expected to follow the heat capacity). These theories apply to ferromagnetic metals where the densities of conduction electrons with up and down spins are comparable ($N^\uparrow \approx N^\downarrow$). However, for double exchange (DE) materials, Searle and Wang [32] pointed out that this approximation is invalid. In fact, $(N^\uparrow - N^\downarrow)/(N^\uparrow + N^\downarrow)$ is of order $DE/E_F$ where $DE$ is the energy of the spins interacting with the applied field and $E_F$ is the Fermi level. In the double exchange case, we have $DE/E_F > 1$. In most conventional models, the scattering by spin disorder enters the expression of the conductivity via the scattering cross section. To calculate it, theories use the Born approximation or take $DE/E_F$ as an expansion parameter, both of which require $DE/E_F \ll 1$. Kasuya's [182] expression, however, was used successfully in lightly doped EuO [185], although $DE/E_F \gg 1$, and the results were confirmed independently by later work [186].

Searle and Wang [32] constructed a simple model from the molecular field approach considering a strong coupling between the Mn cores and the conduction electrons spins. They consider two kinds of possible excitations: the first consists in flipping one Mn spin along with the spin of any Zener electron located at the site (because of Hund's coupling). This effectively removes that electron from those contributing to the global DE interaction since it can no longer hop from its site to the others (its spin is no longer parallel to that of the other Mn cores). The second is the spin flip of a manganese core alone which does not remove an electron from the DE interaction. Hence, electrons can be separated in two groups: those contributing to the global DE interaction and those which do not. Using the expression for the decrease in energy due to the DE interaction derived by Zener [122] it is possible to express the total energy density of the electron system as a function of the average spin polarisations associated with the Mn cores and the DE electrons. This can be expressed as a function of the applied field by using a molecular field expression. The scattering probabilities of the electrons are then calculated following Zener's assumptions that an electron can only travel between two Mn sites if they have parallel spins. The scattering probabilities ($P^\uparrow$ and $P^\downarrow$ respectively for up electrons and down electrons) can then be expressed as a function of the average spin polarisation i.e. the reduced magnetization, $M(T)/M(0)$. Conductivities of up and down bands are inversely proportional to the scattering probabilities and the expression for the total conductivity obtained by Searle and Wang is $\sigma_t = (N/C)\{[1 + (M(T)/M(0))^2]/[1 - (M(T)/M(0))^2]\}$ with $C$ an adjustable parameter and $N$ the total carrier density. Fig. 6.20 compares the result of fits from different models to the resistivity curve obtained by the Canadian group in their single crystals of (La,Pb)MnO$_3$. The model gives a fairly good fit to their data.

Kubo and Ohata [187] obtained essentially the same result from a more rigorous approach starting from the double exchange Hamiltonian (also known as the $s - d$ model or Kondo lattice Hamiltonian, although the $s - S$ coupling is positive, whereas in the Kondo effect it is negative) $H = -t_{\text{eff}} \sum_{<ij>\sigma} c_{i\sigma}^\dagger c_{j\sigma} - J_H \sum_i \boldsymbol{S}_i \cdot \boldsymbol{s}_i$. They detailed the resulting conductivity further by calculating that the resistivity should vary as $T^{9/2}$ at low temperature in the spin-wave approximation. Although, as already indicated, temperature exponents of the

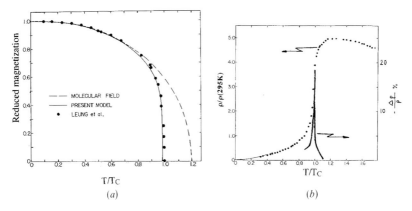

**Fig. 6.20.** (a) Reduced magnetization of a $(La_{0.69}Pb_{0.31})MnO_3$ crystal as a function of temperature. The dashed line is the result of the standard molecular field approximation. (b) Normalised resistivity and magnetoresistance in 1 Tesla as a function of $T/T_c$. In both figures, the dots are experimental data and the solid line is based on the model of Searle and Wang [32]

order of 2 were measured at low temperature, $T^{9/2}$ corrections have been seen by Snyder *et al.* [105]. Furukawa includes a term $-\mu \sum_{j\sigma} n_\sigma$ where $\mu$ is the chemical potential, and points out that large shifts of $\mu(\sim 0.1W)$ are to be expected as a function of temperature and magnetization [188].

It is noticeable that the extremely steep slope of the $R(T)$ curve just below $T_c$ (Fig. 6.1) cannot be accounted for by the above models. Consequently, several groups have developed other models to account for the temperature and magnetic field dependence of the resistivity, and have proposed scaling relationships between $\rho$ and $M$ [189,30,14,101,190]. Pierre *et al.* [191] proposed an exchange-induced band crossing model, like EuO. Zhang considers a spin-polaron model with clustering where the transfer integral is treated as a perturbation [192]. Another approach is a qualitative percolation model based on transport of Ising-like spins on a resistor network [193]. The work of Hundley *et al.* [30], based on the observation that transport occurs by hopping above $T_c$, resulted in the following empirical relationship $\rho(H,T) = \rho_m \exp\{-M(H,T)/M_0\}$ where $\rho_m$ and $M_0$ are fitting parameters. Although, in general, $\rho_m$ is linear in $T$ for hopping processes near 250 K, Hundley *et al.* were able to fit their data over a wide temperature range using $\rho_m = 21~\mu\Omega cm$ and $\mu_0 M_0 = 0.20$ T (Fig. 6.21). This suggested a direct link between resistivity and the state of magnetization of the sample. Furthermore, the relationship is exponential leading to a picture in which the binding energy of the carrier is magnetization dependent. Nonetheless, quantitative agreement depends on the fitting parameters ($M_0$ is one third of the saturation magnetization) and their interpretation is unclear. Dionne proposed a similar model where the thermally-activated hopping energy depends on magnetization [190]. Another form of phenomenology incorporates a two current model [194,195], where the conductivity of one channel is purely dependent on

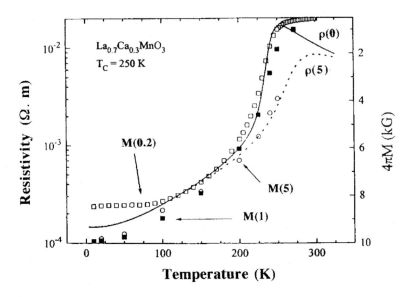

**Fig. 6.21.** Resistivity data on $(La_{0.7}Ca_{0.3})MnO_3$ at different temperatures and applied fields scaled in terms of the magnetization [30]

magnetic order, and the other is due to excitation of carriers across a gap, $E_0$. The form chosen [195], $\sigma = \alpha[M/M_s]^2 + \beta\exp\{-E_0/kT\}$ where $\alpha$, $\beta$ and $E_0$ are fitting parameters, mimics the observed transport reasonably well. At low temperatures, the first term is large and short-circuits the sample. At $T_c$, the magnetization becomes zero, and only the second term remains. Snyder $et$ $al.$ [103] suggest that the magnetoresistance above and below $T_c$ are best fitted by a resistance in series with a magnetoconductor giving the empirical expression $\rho = \rho_\infty + 1/\{\sigma_0 + F(H)\}$ where $F(H) \approx H$ for $T < T_c$ and $F(H) \approx H^2$ for $T > T_c$. Several attempts have been made to incorporate both magnetic and lattice effect within one theoretical framework. Apart from any shortcomings in fitting the data to a magnetic term alone, there is experimental evidence from near edge X-ray absorption [172] and ion channeling [174] for local lattice distortions, especially above $T_c$, in Ca or Ba-doped $LaMnO_3$. Furthermore, a more rigorous quantitative calculation by Millis $et$ $al.$ [179] based on the double exchange Hamiltonian was unable to give the right order of magnitude for the Curie temperature, and could not account for the experimental resistivity behaviour. Their main results for electrical transport were:

-$\rho$ only has a derivative discontinuity at $T_c$ while the resistivity peak is found around $T_c/2$,

-below $T_c$, there are two terms in the expression for the resistivity, the amplitude of the spin fluctuations and an additional term proportional to $M^2$. Their effect is opposite and, at least near $T_c$ the latter dominates and makes the resistivity increase with an applied field.

-The calculated resistivity above $T_c(\sim 1.5\ \mu\Omega cm)$ is far too small.

This is due to the Hund's rule coupling, $J_H$, being much greater than $t$, the average hopping energy (proportional to the bandwidth), which makes local fluctuations of $S_i \cdot S_j$ scatter electrons. When $J_H \ll t$, fluctuations in $S_i \cdot S_i$ are most important and lead to the Fisher-Langer [184] predictions which yield decreasing resistivity with decreasing temperature. Since $J_H > t$ in manganites (Table 6.1) Millis *et al.* are critical of any mean field approach, including Searle and Wang's [32] and Furukawa's [196,197]. They also point out that their result differs from the work of Kubo and Ohata [187] because of their inclusion of the $S_i \cdot S_j$ fluctuations. Millis *et al.* argue that the magnetic fluctuations do not significantly reduce the electron bandwidth in the double exchange model, so that a Fermi liquid picture of weakly scattered bandlike electrons follows [179]. Their calculation demonstrates that double exchange alone cannot account quantitatively for the properties of the manganites and they suggest that a complementary mechanism, namely the Jahn-Teller distortion, should be included in a description of the motion of the $e_g$ charge carriers. Millis and colleagues [169,198] have also calculated the effects of combining Jahn-Teller distortions with magnetic interactions to obtain the resistivity and magnetic transition temperature. Qualitative agreement with experimental data is obtained. In addition, they have examined in detail the Fermi liquid to polaron crossover [199,200], which was alluded to in the original work [179] to explain the low value for the conductivity.

Röder *et al.* [201,202] have also studied a model which includes Jahn-Teller coupling and double exchange. This model does not account for the resistivity in detail but finds that the charge is dressed both by lattice distortion and local magnetic order. The authors show that for $T \ll T_c$, the polarons are very large, extending over many lattice sites and, consequently, overlap one another to form bands as envisioned by Zhou *et al.* [203]. For $T \approx T_c$, the carrier becomes self-trapped by the lattice distortion with spin polarization around the position of the hole. A similar conclusion has been drawn by Lee and Min [204]. This picture is very similar to the bound magnetic polaron of Kasuya [205], except that the trapping potential is the substitutional impurity; for $T > T_c$, in the dilute limit, the polaron is a localized charge surrounded by a nearest neighbour spin cloud.

An argument that Jahn-Teller effect is inessential has been advanced by Varma [13] who points out that similar insulator-metal transitions occur in mixed-valent $Tm(Se_x Te_{1-x})$ compounds where no Jahn-Teller distortion exists. He explains the insulating state by carrier localization due to magnetic disorder, which creates nondiagonal disorder in the hopping matrix elements $t_{ij}$ connecting near neighbour sites where the spins are randomly distributed and slowly fluctuating. The result is a band where at least one half of the states are localized. If the carrier is localized, it will also tend to form a spin polaron whose motion in the presence of an electric field will be governed by the slow spin fluctuations. With increasing magnetic order, near and below $T_c$, these localization effects become smaller. Similarly, an applied magnetic field decreases the magnetic disorder and increases the localization length, thereby decreasing the resistivity.

A theory based on the concept of bipolarons has been developed by Alexandrov *et al.* [163]. The idea is that two dielectric polarons with opposite spins form pairs in the paramagnetic state. This is possible in the case of strong electron-phonon coupling materials. Above the Curie point, these bipolarons have a very small mobility and transport is by thermal activation. Upon cooling through $T_c$ the exchange interaction of the $p$ polaronic holes (the charge carriers) with $d$ electrons competes with the binding energy. The pairs are broken and single polarons in a ferromagnetic background can conduct current like in a metal. Hence, the transition to the paramagnetic state is accompanied by a "current carrier density collapse", where both mobility and number of charge carriers must drop. In the paramagnetic state, thermally activated pair breaking releases polarons which can participate in the transport. The effect of the onset of magnetisation and a possible applied field are described in detail in ref. [163]. The resistivity is predicted to peak at $T_c$ and the field effect leads to CMR. This is an elegant and interesting theory but no unambiguous experiment has been reported so far demonstrating the existence of bound polarons above $T_c$. Also, the role of disorder on transport has not been presented but seems to be of minor importance which is not in perfect agreement with the experimental situation.

The magnetic character of the metal-insulator transition and the variable range hopping behaviour observed above $T_c$ (Fig. 6.18) call for a theory that combines these elements. The problem here is that localisation lengths $(1/\alpha)$, inferred from $T_0$ using the expression $kT_0 = 18\alpha^3/N(E)$ appropriate for Anderson localization in doped semiconductors where $N(E)$ is the density of states, are of order 0.05 nm which is clearly unphysical [112]. The Dublin group [14,206] developed the idea of carrier localization by magnetic rather than charge disorder to explain the transport properties over the entire range of temperature and field. The metal-insulator transition is ascribed to a modification of the spin-dependent exchange potential $-J_H s \cdot S$ associated with the onset of magnetic order at $T_c$. Here $J_H$ is the on-site Hund's rule exchange coupling of an $e_g$ electron with $s = 1/2$ to the $t_{2g}$ ion core with $S = 3/2$. Since $J_H \approx 1$ eV, there is a band of states of width $U_m \approx 2$ eV into which the electron may hop (Fig. 6.22) and the $e_g$ electrons may be localized by the random spin-dependent potential above $T_c$, where conduction is by variable-range hopping. When a magnetic field is applied to the manganite or when there is an internal molecular field, the random distribution of spin directions is narrowed, and the average magnetic potential decreases. Over the whole temperature range, the resistivity is expected to vary as [14]: $\ln(\rho/\rho_\infty) = \{T_0[1 - (M/M_s)^2]/T\}^{1/4}$ where $M/M_s$ is the reduced magnetization. Also, considering that the number of available sites for the hopping electron may be limited by the Jahn-Teller distortions and other factors, the parameter $T_0$ is different from Mott's original value, given above; the expression becomes $kT_0 = 171\alpha^3 U_m V$, where $V$ is the cell volume. Taking $U_m = 2$ eV, the corresponding localization lengths are typically in the range 0.4 nm, the average hopping range at room temperature is 1.5 nm and the hopping energy at room temperature is $\Delta E = 0.1$ eV. These numbers are physically plausible since the localization length exceeds the ionic radius of $Mn^{3+}$ and the hopping distances

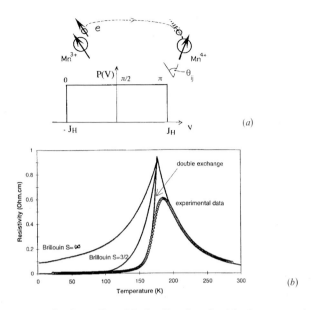

**Fig. 6.22.** Resistivity of a $La_{0.7}Ca_{0.3}MnO_3$ film fitted with the magnetic localization model with different expressions for the magnetization. The upper panel shows the spin-dependent potential distribution $P(V)$ experienced by a hopping Zener electron in the paramagnetic state [80]

are 3 - 4 times the Mn-Mn separation. Moreover, the resistivity and magnetoresistance curves are well reproduced by the above modified VRH formula [14], as shown in Fig. 6.22.

Nagaev [17,207] has developed a theory where the resistivity peak and magnetoresistance are explained by considering localization and scattering from an exchange potential that is caused by a difference in the local magnetization close to a divalent impurity and far from it.

An alternative to considering the effects of magnetization on the localization length is to construct an electronic density of states for the $e_g$ band (Fig. 6.23) where the density of states itself is greatly modified at $T_c$. For simplicity we assume that the $e_g$ band is unsplit by the Jahn-Teller effect and can accomodate two electrons. The exchange splitting remains in the paramagnetic state, since the local moment persists. A decrease in band width with increasing temperature reflects the effect of magnetic order on $t$ or $W$. For $T \ll T_c$, localized states are due only to Coulomb disorder, including possible Jahn-Teller distortions. Since the total number of states remain the same, the area under the curve has to be conserved. The Fermi level $E_F$ which, in the metallic regime, is within the extended states above the mobility edge $E_\mu$ that separates the localized from the extended states, can find itself in the localized region of the density of states (hatched) as magnetic order decreases. At very high temperatures, the bandwidth reduction is greatest because the bandwidth is $\propto t_0 cos(\theta/2)$ in

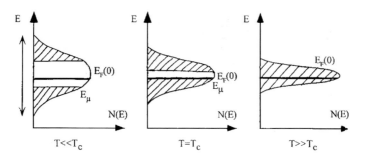

**Fig. 6.23.** Band picture for manganites with $x = 0.3$ as a function of temperature.

the double-exchange model and the magnetic disorder includes even nearest neighbours. Most (possibly all) states are localized because of a combination of the influence of Coulomb disorder on the mobility edge and magnetic disorder on the bandwidth.

It remains a challenge to clarify and modify these pictures. Very likely both magnetic and charge disorder, and lattice distortion effects play some role in the transport above $T_c$. It is a question of relative magnitude. Evidence for Jahn-Teller and lattice polaronic effects in the electronic transport has been presented, but magnetic interactions clearly dominate the potential near $T_c$. The respective roles of the two lattice effects, the magnetic and Coulomb disorder in the localisation process in the different regimes of carrier concentration and temperature needs to be better understood.

## 6.3   Applications

Physical properties that may be exploited include the temperature dependence of resistivity and the magnetoresistance. The rapid variation of resistivity in the vicinity of $T_c$, with relative changes as high as 10-20%/K suggest uses as a bolometer. The ability to modify the composition of the perovskite oxides so as to place the Curie temperature anywhere in the range from 50-380 K, and thereby tune the temperature variation of the resistivity gives flexibility. Drawbacks are the temperature-dependent sensitivity with the highest values available only in a limited range near $T_c$. Furthermore, the maximum sensitivity falls rapidly as $T_c$ increases; also when $T_c$ is low, ($< 100$ K) the resistivity shows thermal hysteresis.

Prospects are brighter for exploiting the magnetoresistance. Potential applications include magnetic sensors, magnetoresistive read heads and magnetoresistive random access memory. In ferromagnetic compositions with $x \approx 0.3$, the maximum high-field magnetoresistance is associated with the resistivity peak near $T_c$. One method to broaden the magnetoresistive response is to use materials with a canted ferromagnetic structure having a composition $0.4 < x < 0.5$ [78,208]. A direct application of magnetoresistance is to use the materials as magnetic field sensors. The isotropy of the magnetoresistance means that a spherical crystal with a demagnetizing factor $N = 1/3$ may be used to sense the magnitude of a magnetic field regardless of its direction [209]. Other shapes, such as films or cylinders may be used to measure the field in a particular direction or plane but the appropriate demagnetizing factor should be taken into account. A prototype position sensor based on a thin film with permanent magnet bias has been built [210]. It is based on $(La_{0.67}Ca_{0.33})MnO_3$ which is sprayed or screen-printed onto a ceramic substrate and fired at $1400°C$. A Wheatstone's bridge configuration is used to improve sensitivity and compensate for thermal drift. Sensitivity is around $\sim 10 \%/T$.

Another sensor application is a magnetoresistive microphone [82]. When the manganite sensor is presented to the field $B$ directly it is possible to detect relatively large fields, of order 1 T, which are required to modify significantly the ferromagnetic order and thereby induce magnetoresistance. However the sensitivity $1/\rho(d\rho/dB_0)$ can be enhanced by flux concentration using a soft, anhysteretic ferrite to guide the flux to a small cross section occupied by the manganite sensor [211]. The magnetoresistive response can be amplified a thousand-fold in a limited field range (Fig. 6.24). Segments of $YBa_2Cu_3O_7$ have also been used to

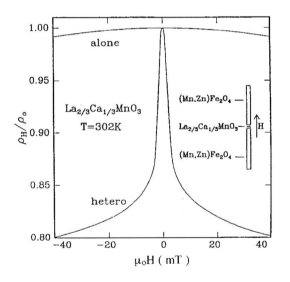

**Fig. 6.24.** Resistivity as a function of applied field at room temperature for a thin $(La_{0.67}Ca_{0.33})MnO_3$ film with and without flux concentration using long rods of Mn-Zn ferrite [211]

enhance the flux below the superconducting transition temperature $T_{sc} = 90$ K [212].

A more promising approach to achieving a good low-field magnetoresistive response is to focus on the micromagnetic structure rather than to depend on the applied field to modify the intrinsic magnetic structure within each domain, i.e. exploit the low-field magnetoresistance rather than the colossal magnetoresistance. Here, the low-field response is favoured by a high density of grain boundaries [116,211,123,213,128,127], and in tunnel spin valve structures [137,136,138]. A first step towards miniaturizing the tunnel spin valve is the current-perpendicular-to-plane sandwich device consisting of a micron-size diameter pillar composed of two layers of ferromagnetic manganite separated by an insulating spacer layer. A 450 % change of resistance was observed according to whether the magnetization of the two manganite layers is parallel or antiparallel [138]. The switching field for the device was 100 Oe at 4.2 K. Besides showing the potential for generating large resistivity changes in low fields with manganite heterostructures, this prototype device demonstrates how it is possible to use films of ferromagnetic manganites as electronic spin polarizers and analysers. A problem here is that the low-field magnetoresistance, whether in spin polarised tunnel junctions, polycrystalline ceramics or powder compacts is strongly temperature dependent in all circumstances where it has been observed to date. Huge effects are seen at cryogenic temperatures, but the effect at room temperature does not exceed 1%. Furthermore, the Curie temperature of manganites is too low for most applications. Read heads, for example, have to be able to operate up to 100°C, which is uncomfortably close to the maximum Curie temperature in Sr-doped manganites.

The ferromagnetic metallic manganites have an almost completely spin-polarized $3d$ band. They are therefore a potential source of spin-polarized electrons for a variety of spin electronic devices. The problem here is how best to inject these electrons across the manganite interface at room temperature while still retaining their spin polarization. Compatibility with existing Si-based structures may need to be addressed. Three-terminal devices based on manganite/superconductor heterostructures operating at liquid-nitrogen temperature [214] may open new perspectives in high-speed electronics. It has been suggested that manganite/superconductor layer structures could be useful for ultrasonic wave amplification, thermal switching and thermocouple infrared detection [215].

An issue which is considered elsewhere in this book is noise in manganite devices. There are reports of exceptionally large $1/f$ noise [216 218] or thermal noise in the vicinity of $T_c$. The huge $1/f$ noise has a non-Gaussian character associated with magnetic inhomogeneities [218]. An implication is that there may be problems exploiting the low-field magnetoresistance effect in low-frequency applications, but the $1/f$ noise should not be a problem in high-frequency applications such as read heads. In the area of thin films and heterostructures, an important task is to uncover the role of the grain boundary in polycrystalline films (and ceramics) and to understand the contributions of intrinsic and/or interface effects to the temperature-dependence of the low-field magnetoresistance. Generally, the temperature range for useful magnetoresistive effects needs to be extended. Then, a knowledge of the exact degree of spin polarization of the conduction electrons in different materials and characterization and control of the interfaces in heterostructures will be essential to achieve effective spin injection in layered or planar devices. Changes in the electronic structure induced by photons, electric field or other means which are long-lived (hysteretic phase transitions) need to be explored and controlled, especially in thin films, for potential storage applications.

There is much work to be done, but it is sufficiently challenging and potentially rewarding in terms of interesting physics and novel magnetoelectronic devices that the subject of mixed-valence manganites and related compounds will likely flourish for another number of years. The manganite research has directed the attention of the community more generally towards electron localisation of magnetic origin and half-metallic ferromagnetic oxides, where near-perfect spin polarization opens prospects for novel devices based on magnetic control of the electron stream.

# References

1. J. M. D. Coey, M. Viret, and S. von Molnár, Advances in Physics **48**, 167 (1999).
2. J. Zaanen, G. Sawatzky, and J. Allen, Physical Review Letters **55**, 418 (1985).
3. J. Zaanen and G. Sawatzky, Journal of Solid State Chemistry **88**, 8 (1990).
4. J. B. Goodenough, Wiley-Interscience, New York (1966).
5. Y. Okimoto et al., Physical Review Letters **75**, 109 (1995).
6. D. Sarma et al., Physical Review Letters **75**, 1126 (1995).
7. J. L. Garcia-Munoz, J. Fontcuberta, M. Suaadi, and X. Obradors, Journal of Physics: Condensed Matter **8**, (1996).
8. J. Elemans, B. vanLaar, K. v. d. Veen, and B. Loopstra, Journal of Solid State Chemistry **3**, 238 (1971).
9. F. Moussa et al., Physical Review B **54**, 15149 (1996).
10. W. Pickett and D. Singh, Europhysics Letters **32(9)**, 759 (1995).
11. W. E. Pickett and D. J. Singh, Physical Review B **53**, 1146 (1996).
12. F. Shi, M. Ding, and T. Lin, Solid State Communications **96**, 931 (1995).
13. C. M. Varma, Physical Review B **54**, 7328 (1996).
14. M. Viret, L. Ranno, and J. Coey, Physical Review B **55**, 8067 (1997).
15. L. Ghivelder et al., Physical Review B **60**, 1 (1999).
16. P. G. de Gennes, Physical Review **118**, 141 (1960).
17. E. L. Nagaev, Physics Uspekhi **39**, 781 (1996).
18. S. K. Mishra, S. Satpathy, F. Aryasetiawan, and O. Gunnarasson, Physical Review B **55**, 2725 (1997).
19. D. Arovas and F. Guinea, Physical Review B **58**, (1998).
20. S. Yunoki et al., Physical Review Letters **80**, 845 (1998).
21. E. Dagotto, S. Yunoki, M. A.L., and A. Moreo, Physical Review B **58**, (1998).
22. S. vonMolnár and J. Coey, Current Opinion in Solid State and Material Science **3**, 171 (1998).
23. A. Moreo, S. Yunoki, and E. Dagotto, Science **283**, 2034 (1999).
24. M. Hennion et al., Physical Review B **56**, R 497 (1997).
25. M. Hennion et al., Physical Review Letters **81**, 1957 (1998).
26. V. Skumryev et al., European Physical Journal B **11**, 401 (1999).
27. B. Raquet, A. Anane, S. Wirth, P. Xiong and S. von Molnár, Physical Review Letters **84**, 4485 (2000).
28. J. H. vanSanten and G. H. Joncker, Physica **16**, 599 (1950).

29. G. H. Jonker, Physica **20**, 1118 (1954).
30. M. F. Hundley *et al.*, Applied Physics Letters **67**, 860 (1995).
31. J. Volger, Physica **20**, 49 (1954).
32. C. W. Searle and S. T. Wang, Canadian Journal of Physics **48**, 2023 (1970).
33. K. Khazeni *et al.*, Physical Review Letters **76**, 295 (1996).
34. M. F. Hundley *et al.*, Journal of Applied Physics **79**, 4535 (1996).
35. J. Eckstein *et al.*, Applied Physics Letters **69**, 1312 (1996).
36. G. H. Jonker, Journal of Applied Physics **37**, 1424 (1966).
37. T. Arima and Y. Tokura, Journal of The Physical Society of Japan **64**, 2488 (1995).
38. A. Chainani, M. Mathew, and D. D. Sarma, Physical Review B **47**, 15397 (1993).
39. R. von Helmolt, J. Wecker, K. Samwer, and K. Baerner, Journal of Magnetism and Magnetic Materials **151**, 411 (1995).
40. Y. Yamada *et al.*, Physical Review Letters **77**, 904 (1996).
41. A. Anane *et al.*, Journal of Magnetism and Magnetic Materials **377**, 377 (1997).
42. H. Y. Hwang *et al.*, Physical Review Letters **75**, 914 (1995).
43. S. K. Singh, S. Palmer, D. M. Paul, and M. R.Lees, Applied Physics Letters **69**, 263 (1996).
44. W. Archibald, J.-S. Zhou, and J. B. Goodenough, Physical Review B **53**, 14445 (1996).
45. R. Thomas, L. Ranno, and J. Coey, Journal of Applied Physics **81**, (1997).
46. R. vonHelmolt, J. Wecker, T. Lorenz, and K. Samwer, Applied Physics Letters **67**, 2093 (1995).
47. M. Kasai *et al.*, Japanese Journal of Applied Physics **35**, L489 (1996).
48. E. Pollert, S. Krupicka, and E. Kuzwicova, Journal of Physics and Chemistry of Solids **43**, 1137 (1982).
49. A. Maignan, C. Simon, V. Caignaert, and B. Raveau, Solid State Communications **96**, 623 (1995).
50. F. Millange *et al.*, Zeitschrift fur Physik B **101**, 169 (1996).
51. A. Maignan, C. Simon, V. Caignaert, and B. Raveau, Comptes Rendus de l'Academie des Sciences Serie II **321**, 297 (1995).
52. J. Barratt, M. Lees, G. Balakrishnan, and D. McPaul, Applied Physics Letters **68**, 424 (1996).
53. M. R. Lees, J. Barratt, G. Balakrishnan, and D. M. Paul, Physical Review B **52**, 14303 (1995).
54. G. Xiong *et al.*, Applied Physics Letters **67**, 3031 (1995).
55. H. Kuwahara *et al.*, Science **272**, 80 (1996).
56. N. Babushkina *et al.*, Journal of Applied Physics **83**, 7369 (1998).
57. A. Anane *et al.*, Physical Review B **59**, 256 (1999).
58. M. Hervieu *et al.*, Physical Review B **53**, 14274 (1996).
59. Y. Tokura *et al.*, Physical Review Letters **76**, 3184 (1996).
60. G. Gong *et al.*, Journal of Applied Physics **79**, 4538 (1996).
61. F. Damay *et al.*, Solid State Communications **98**, 997 (1996).
62. F. Damay, A. Maignan, N. Nguyen, and B. Raveau, Journal of Solid State Chemistry **124**, 385 (1996).
63. M. Kasai, H. Kuwahara, Y. Tomioka, and Y. Tokura, Journal of Applied Physics **80**, 6894 (1996).
64. P. Wagner *et al.*, Physical Review B **55**, 3699 (1997).
65. P. Laffez *et al.*, Materials Research Bulletin **31**, 905 (1996).
66. H. Kawano, R. Kajimoto, M. Kubota, and H. Yoshizawa, Physical Review B **53**, 14709 (1996).

67. Y. Tomioka *et al.*, Applied Physics Letters **70**, 3609 (1997).
68. H. Kuwahara *et al.*, Science **270**, 961 (1995).
69. M. R. Lees, J. Barratt, G. Balakrishnan, and D. M. Paul, Journal of Physics: Condensed Matter **8**, 2967 (1996).
70. V. Caignaert *et al.*, Journal of Magnetism and Magnetic Materials **153**, L260 (1996).
71. A. Maignan, C. Martin, and B. Raveau, Zeitschrift fur Physik B **102**, 19 (1997).
72. A. Maignan and B. Raveau, Zeitschrift fur Physik B **102**, 299 (1997).
73. F. Damay, A. Maignan, C. Martin, and B. Raveau, Journal of Applied Physics **81**, 1372 (1997).
74. A. Maignan, C. Simon, V. Caignaert, and B. Raveau, Journal of Magnetism and Magnetic Materials **152**, L5 (1996).
75. R. Mahesh, R. Mahendiran, A. Raychaudhuri, and C. Rao, Journal of Solid State Chemistry **120**, 2045 (1996).
76. R. Mahendiran, R. Mahesh, A. K. Raychaudhuri, and C. N. R. Rao, Solid State Communications **99**, 149 (1996).
77. J. Z. Liu *et al.*, Applied Physics Letters **66**, 3218 (1995).
78. R. Mahendiran *et al.*, Physical Review B **53**, 3348 (1996).
79. R. M. Kusters *et al.*, Physica B **155**, 362 (1989).
80. R. von Helmolt *et al.*, Physical Review Letters **71**, 2331 (1993).
81. P. Ganguly, P. A. Kumar, P. Santhosh, and I. Mulla, Journal of Physics: Condensed Matter **6**, 533 (1994).
82. S. Jin, Materials Transactions JIM **37**, 888 (1996).
83. S. Jin *et al.*, Applied Physics Letters **67**, 557 (1995).
84. S. Jin *et al.*, Applied Physics Letters **66**, 382 (1994).
85. J. Fontcuberta *et al.*, Solid State Communications **97**, 1033 (1996).
86. A. Nossov, J. Pierre, V. Vassiliev, and V. Ustinov, Solid State Communications **101**, 361 (1996).
87. R. Mahendiran, R. Mahesh, A. K. Raychaudhuri, and C. N. R. Rao, Physical Review B **53**, 12160 (1996).
88. A. Nossov, J. Pierre, V. Vassiliev, and V. Ustinov, Journal of Physics: Condensed Matter **8**, 8513 (1996).
89. N. Sharma *et al.*, Journal of Magnetism and Magnetic Materials **166**, 65 (1997).
90. H. L. Ju *et al.*, Applied Physics Letters **65**, 2108 (1994).
91. N.-C. Yeh *et al.*, Proceedings of Materials Research Society Meeting, San Francisco 145 (1997).
92. N.-C. Yeh *et al.*, Journal of Applied Physics **81**, 5499 (1997).
93. H. L. Ju *et al.*, Physical Review B **51**, 6143 (1995).
94. C. H. Chen, V. Talyansky, C. Kwon, and M. Rajeswari, Applied Physics Letters **69**, (1996).
95. N. Sengoku and K. Ogawa, Japanese Journal of Applied Physics **35**, 5432 (1996).
96. Y. Jia *et al.*, Solid State Communications **94**, 917 (1995).
97. C. L. Canedy *et al.*, Journal of Applied Physics **79**, 4546 (1996).
98. J. Fontcuberta *et al.*, Physical Review Letters **76**, 1122 (1996).
99. K. Li *et al.*, Journal of Physics D: Applied Physics **29**, 14 (1996).
100. G. C. Xiong *et al.*, Solid State Communications **97**, 599 (1996).
101. B. Martinez *et al.*, Physical Review B **54**, 10001 (1996).
102. J. O'Donnell *et al.*, Physical Review B **54**, R6841 (1996).
103. G. J. Snyder *et al.*, Applied Physics Letters **69**, 4254 (1996).
104. A. Urushibara *et al.*, Physical Review B **51**, 14103 (1995).

105. G. Snyder *et al.*, Physical Review B **53**, 1 (1996).
106. J. M. DeTeresa *et al.*, Physical Review B **54**, 1187 (1996).
107. N. Ashcroft and N. Mermin, "Solid State Physics" (Holt-Saunders Japan, Tokyo, 1976).
108. G. Stewart, Z. Fisk, and J. Williams, Physical Review B **28**, (1984).
109. N. Mott, "Metal insulator transitions (2nd edition)" (Taylor and Francis, London, 1985).
110. J. Zhou, W. Archibald, and J. Goodenough, Physical Review B **57**, R2017 (1998).
111. J. Heremans, M. Carris, S. Watts, and K. Dahmen, Journal of Applied Physics **81**, 4967 (1997).
112. J. M. D. Coey, M. Viret, L. Ranno, and K. Ounadjela, Physical Review Letters **75**, 3910 (1995).
113. K. Khazeni *et al.*, Journal of Physics: Condensed Matter **8**, 7723 (1996).
114. N.-C. Yeh *et al.*, Journal of Physics: Condensed Matter **9**, 3713 (1997).
115. R. Sanchez, J. Rivas, C. Vasquez, and A. Lopez-Quintela, Applied Physics Letters **68**, 134 (1996).
116. A. Gupta *et al.*, Physical Review B **54**, R15629 (1996).
117. H. Y. Hwang, S. W. Cheong, N. P. Ong, and B. Batlogg, Physical Review Letters **77**, 2041 (1996).
118. K. H. Kim *et al.*, Physical Review B **55**, 4023 (1997).
119. J. L. Garcia-Munoz *et al.*, Physical Review B **55**, R668 (1997).
120. S. Kaplan *et al.*, Physical Review Letters **77**, 2081 (1996).
121. R. Mahesh, R. Mahendiran, A. Raychaudhuri, and C. Rao, Applied Physics Letters (1996).
122. C. Zener, Physical Review **81**, 440 (1951).
123. L. Balcells, J. Fontcuberta, B. Martinez, and X. Obradors, Journal of Physics: Condensed Matter **10**, 1889 (1998).
124. H. L. Ju and H. Sohn, Solid State Communications **102**, 463 (1997).
125. R. Shreekala *et al.*, Applied Physics Letters **71**, 282 (1997).
126. J. Y. Gu *et al.*, Applied Physics Letters **70**, 1763 (1997).
127. N. Mathur *et al.*, Nature **387**, 266 (1997).
128. K. Steenbeck *et al.*, Applied Physics Letters **71**, 968 (1997).
129. G. Gehring and D. Coombes, Journal of Magnetism and Magnetic Materials **177**, 873 (1998).
130. J. M. D. Coey, Philosophical Transactions of the Royal Society A **356**, 1519 (1998)
131. J. M. D. Coey *et al.*, Physical Review Letters **80**, 3815 (1998).
132. J. M. D. Coey, A. Berkowitz, L. Balcells, and F. Putris, Applied Physics Letters **72**, 734 (1998).
133. B. Martínez, Ll. Balcells, J. Fontcuberta, X. Obradors, C. H. Cohenza and R. F. Jardim, Journal of Applied Physics **83**, 7058 (1998).
134. J. O'Donnell *et al.*, Physical Review B **55**, 5873 (1997).
135. G. Q. Gong *et al.*, Physical Review B **54**, R3742 (1996).
136. Y. Lu *et al.*, Physical Review B **54**, R8357 (1996).
137. J. Z. Sun *et al.*, Applied Physics Letters **69**, 3266 (1996).
138. M. Viret *et al.*, Europhysics Letters **39**, 545 (1997).
139. R. Meservey and P. Tedrow, Physics Reports **238**, 173 (1994).
140. M. Viret *et al.*, Journal of Magnetism and Magnetic Materials **198-199**, 1 (1999).
141. F. Ott *et al.*, Journal of Magnetism and Magnetic Materials **211**, 200 (2000).
142. R. von Helmolt, L. Haupt, K. Baerner, and U. Sondermann, Solid State Communications **82**, 693 (1992).

143. J. M. deTeresa *et al.*, Nature **386**, 256 (1997).
144. M. Jaime *et al.*, Applied Physics Letters **68**, 1576 (1996).
145. J.-H. Park *et al.*, Physical Review Letters **76**, 4215 (1996).
146. J.-H. Park *et al.*, Journal of Applied Physics **79**, 4558 (1996).
147. J. Choi *et al.*, Journal of Applied Physics **81**, (1998).
148. A. Biswas and A. K. Raychaudhuri, Journal of Physics, Condensed Matter **8**, L739 (1996).
149. W. Westerburg *et al.*, European Physical Journal B **14**, 509 (2000).
150. I. Troyanchuk, Sov. Phys. JETP **75**, 132 (1992).
151. P. Gerthsen and K. Hardtl, Z. Naturforschung **17**, (1962).
152. R. Heikes, R. Miller, and M. R., Physica **30**, 1600 (1964).
153. S. Billinge *et al.*, Physical Review Letters **77**, 715 (1996).
154. G. F. Dionne, MIT Technical Report (1996).
155. N. Mott and E. Davies, "Electronic processes in noncrystalline materials" (Oxford University Press, Oxford, 1971).
156. G. Triberis and L. Friedman, Journal of Physics C **18**, 2281 (1985).
157. B. Shlovskii and Efros, "Electronic properties of doped semiconductors" (Springer, Berlin, 1984).
158. L. Ranno *et al.*, Journal of Physics: Condensed Matter **8**, L33 (1996).
159. J. Toepfer and J. Goodenough, Journal of Solid State Chemistry **130**, 117 (1997).
160. F. Lotgering, Philips Research Reports **25**, 8 (1970).
161. S. S. Manoharan *et al.*, Journal of Solid State Chemistry **117**, 420 (1995).
162. A. Gupta *et al.*, Applied Physics Letters **67**, 3494 (1995).
163. A. Alexandrov and A. Bratkovsky, Physical Review Letters **82**, 141 (1999).
164. J. Fontcuberta *et al.*, Europhysics. Letters. **34**, 379 (1996).
165. J. Lynn *et al.*, Physical Review Letters **76**, 4046 (1996).
166. M. Viret *et al.*, Europhysics Letters **42**, 301 (1997).
167. R. H. Heffner *et al.*, Physical Review Letters **77**, 1869 (1996).
168. C. Kapusta *et al.*, Journal of Physics **11**, 4079 (1999).
169. A. Millis, B. Shraiman, and R. Mueller, Physical Review Letters **77**, 175 (1996).
170. P. Radaelli *et al.*, Physical Review B **54**, 8992 (1996).
171. D. Argyriou *et al.*, Physical Review Letters **76**, 3826 (1996).
172. T. Tyson, J. de Leon, S. Conradson, and B. A.R., Physical Review B **53**, 13985 (1996).
173. P. Dai, J. Zhang, H. Mook, and S. Liou, Physical Review B **54**, R3694 (1996).
174. R. P. Sharma *et al.*, Physical Review B **54**, 10014 (1996).
175. G. Zhao, K. Conder, H. Keller, and K. Muller, Nature **381**, 676 (1996).
176. A. Shengelaya, G. Zhao, H. Keller, and K. Mueller, Physical Review Letters **77**, 5296 (1996).
177. G. Matsumoto, Journal of The Physical Society of Japan **29**, 615 (1970).
178. R. Burns, "Mineralogical Applications of Crystal Field Theory", 2nd Edition (Cambridge University Press, London, 1992).
179. A. J. Millis, P. B. Littlewood, and B. I. Shraiman, Physical Review Letters **74**, 5144 (1995).
180. P. W. Anderson and H. Hasegawa, Physical Review **100**, 675 (1955).
181. P. G. de Gennes and J. Friedel, Journal of Physics and Chemistry of Solids **4**, 71 (1958).
182. T. Kasuya, Progress of Theoretical Physics **22**, 227 (1959).
183. T. V. Peski-Tinbergen and A. Dekker, Physica **29**, 917 (1963).
184. M. Fisher and J. Langer, Physical Review Letters **20**, 665 (1968).

185. S. vonMolnár and M. Shafer, Journal of Applied Physics **41**, 1093 (1970).
186. T. Penney, M. Shafer, and J. Torrance, Physical Review B **5**, 3669 (1972).
187. K. Kubo and N. Ohata, Journal of The Physical Society of Japan **33**, 21 (1972).
188. N. Furukawa, Journal of the Physical Society of Japan **63**, 3214 (1997).
189. Y. Tokura *et al.*, Journal of The Physical Society of Japan **63**, 3931 (1994).
190. G. F. Dionne, Journal of Applied Physics **79**, 5172 (1996).
191. J. Pierre *et al.*, Physica B **225**, 214 (1996).
192. S. Zhang, Journal of Applied Physics **79**, 4542 (1996).
193. P. J. M. Bastiaansen and H. J. F. Knops, Computational Material Science **10**, 225 (1998).
194. J. Z. Sun, L. Krusin-Elbaum, S. Parkin, and G. Xiao, Applied Physics Letters **67**, 2726 (1995).
195. J. Nunez-Regueiro and A. Kadin, Applied Physics Letters **68**, 2747 (1996).
196. N. Furukawa, Journal of the Physical Society of Japan **63**, 3214 (1994).
197. N. Furukawa, Journal of the Physical Society of Japan **64**, 2734 (1995).
198. A. J. Millis, Physical Review B **53**, 8434 (1996).
199. A. J. Millis, R. Mueller, and B. Shraiman, Physical Review B **54**, 5405 (1996).
200. A. J. Millis, R. Mueller, and B. Shraiman, Physical Review B **54**, 5389 (1996).
201. H. Roeder, J. Zang, and A. Bishop, Physical Review Letters **76**, 1356 (1996).
202. J. Zang, A. R. Bishop, and H. Röder, Physical Review B **53**, R8840 (1996).
203. J. Zhou, W. Archibald, and J. Goodenough, Nature **381**, 770 (1996).
204. J. D. Lee and B. I. Min, Physical Review B **55**, 12454 (1996).
205. T. Kasuya and A. Yanase, Reviews of Modern Physics **40**, 684 (1968).
206. M. Viret, L. Ranno, and J. Coey, Journal of Applied Physics **81**, (1997).
207. E. L. Nagaev, Physical Review B **54**, 16608 (1996).
208. J. Lawler, J. Coey, J. Lunney, and V. Skumryev, Journal of Physics: Condensed Matter **8**, 10737 (1996).
209. L. Ranno *et al.*, Journal of Magnetism and Magnetic Materials **157/158**, 291 (1996).
210. L. Balcells *et al.*, Applied Physics Letters **69**, 1486 (1996).
211. H. Y. Hwang, S.-W. Cheong, and B. Batlogg, Applied Physics Letters **68**, 3494 (1996).
212. Z. Dong *et al.*, Applied Physics Letters **68**, 3432 (1996).
213. J. O'Donnell *et al.*, Applied Physics Letters (pp97).
214. V. A. Vas'ko *et al.*, Physical Review Letters **78**, 1134 (1997).
215. Y. Tang *et al.*, Physica Status Solidi B **182**, 509 (1994).
216. M. Rajeswari *et al.*, Applied Physics Letters **69**, 851 (1996).
217. G. Alers, A. Ramirez, and S. Jin, Applied Physics Letters **68**, 3644 (1996).
218. H. Hardner, M. Weissman, M. Jaime, and R. Treece, Applied Physics Letters **68**, 272 (1997).

# 7 Spin Dependent Tunneling

F. Guinea, M. J. Calderón and L. Brey

Instituto de Ciencia de Materiales, Consejo Superior de Investigaciones Científicas, Cantoblanco, E-28049, Madrid, Spain

**Abstract.** A pedagogical introduction to tunneling processes in magnetic junctions is presented. Different effects which influence the magnetoresistance of the junction are reviewed.

## 7.1 Introduction

The magnetic field dependence of the current through magnetic junctions is a subject of great interest. Spin dependent tunneling poses many interesting scientific questions [1], and the number of applications for magnetic junctions continues to grow. The present work intends to give a pedagogical introduction to the physical processes which are responsible for the wide variety of phenomena observed in magnetic junctions. There are many types of junctions and materials, and a fraction of them are not yet completely understood. We will mostly discuss those mechanisms which can be considered reasonably understood, and of wide applicability. No attempt is made to cover in full the large and growing bibliography published on the subject. We will try to use whenever possible examples of current interest.

In a field as vast as this one, a selection of topics is unavoidable. We first give a rough clasification of magnetic junctions. Then, we discuss how the basic models of tunneling need to be modified in the presence of a spontaneous magnetization of the electrodes. This is done in Sect. 7.2. We next discuss the changes that the interface can induce in the basic elastic processes mentioned earlier, in Sect. 7.3. Sect. 7.4 deals with inelastic tunneling, mediated by magnetic excitations at the electrodes or at the surface. In Sect. 7.5, we analyze the changes in the magnetic structure of the surface which can be expected. In Sect. 7.6 charging effects are presented. Some conclusions are discussed in Sect. 7.7.

## 7.2 Magnetic Junctions

### 7.2.1 Types of Junctions

Transport between two bulk metallic electrodes can be roughly classified into two types, schematically shown in Fig. 7.1:

– When the separation between the electrodes exceeds a few angstroms, electrons move between the electrodes by tunneling. The probability that any one

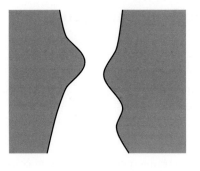

(a)

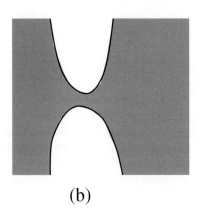

(b)

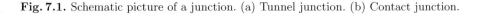

**Fig. 7.1.** Schematic picture of a junction. (a) Tunnel junction. (b) Contact junction.

electron tunnels through a barrier of height $V$ and length $l$ is given by:

$$\mathcal{T} \sim \exp\left[-c\sqrt{\frac{2mV}{\hbar^2}}\,l\right] \tag{7.1}$$

where $c$ is a constant of order unity, which depends on the detailed shape of the barrier and on the electronic wavefunctions. The barrier can be the vacuum created between the two electrodes, in which case the height of the barrier is given by the work function of the electrodes. More commonly, a barrier is created by inserting an insulating layer between the two electrodes. In this case, the barrier height depends on the position of the edges of the gap of the insulating material with respect to the Fermi level of the electrodes.

As tunneling depends exponentially on the distance between the electrodes, we expect that, in a junction of macroscopic size, the current will be due to tunneling events at protrusions of the interface. A change of a few angstroms can greatly modify the tunneling probability. The conductance at any of these points is given by :

$$g \sim \frac{e^2}{h} \mathcal{T} \tag{7.2}$$

– Alternatively, the two electrodes can be in contact in some points. Then, the conductance of each contact is given by $e^2/h$ times the number of electron channels through the contact. This number is given, roughly, by the cross section of the contact expressed in units of $k_{\mathrm{F}}^{-2}$, where $k_{\mathrm{F}}$ is the Fermi wavevector, which gives the size of the electronic wavefunctions. Then:

$$g \sim \frac{e^2}{h} k_{\mathrm{F}}^2 A \tag{7.3}$$

where $A$ is the area of the junction. A number of interesting experiments have been recently performed in magnetic junctions where with a single contact of atomic dimensions [2–4].

### 7.2.2  Magnetic Properties

The previous discussion ignores the possible magnetization of the electrodes. If the number of electrons of the two spin polarizations is not equal, we must define a spin dependent conductance, at each of the points where electrons move from one electrode to the other. The expressions defined earlier are modulated by the density of states of each type of electrons. For a given bias voltage $V$, the electrons which participate in the conduction come from the levels located, at most, at a distance $eV$ from the Fermi energy, $E_{\mathrm{F}}$. Thus, in order to understand the transport at small bias voltages we need to know density of states at the Fermi level, $D_\uparrow(E_{\mathrm{F}})$ and $D_\downarrow(E_{\mathrm{F}})$. As mentioned earlier, the tunneling amplitude can also depend on the electronic wavefunctions, which, in a magnetic system, will be spin dependent. A sketch is given in Fig. 7.2.

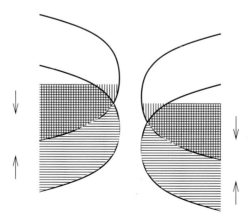

**Fig. 7.2.** Sketch of the density of states at the two sides of a magnetic junction

An applied field modifies the polarization of the electrodes, leading to changes in the densities of states described above. Typically, magnetic fields are too weak to alter significantly the energy barrier at the interface. In many cases, we are only interested in the changes in the transport properties induced by the field, i. e., the magnetoresistance of the junction. Then, we do not need to consider the details of the barrier, which can be described in terms of an energy dependent transmission coefficient.

Another reasonable approximation which allows us to greatly simplify the analysis of the junction is to assume that the direct magnetic coupling between the electrodes is negligible. As mentioned earlier, in a typical junction only a few regions are in close contact. Even in these areas, exchange couplings between atoms in different electrodes will be reduced, as the distances are larger than the interatomic spacings. Thus, the orientation of the magnetization in each electrode will be determined by bulk effects and by its previous history. On the average, we can assume that the relative orientation of the magnetization of the two electrodes can take any value.

On the other hand, under an applied magnetic field, fluctuations in the magnetization of each of the electrodes away from the applied field will be greatly suppressed. Thus, in order to compute the magnetoresistance of the junction, we just need to compare the conductance with a random orientation of the magnetization of the electrodes and that when both magnetizations are aligned.

Let us assume further that:

$$D_\uparrow(E_F) \propto N_\uparrow$$
$$D_\downarrow(E_F) \propto N_\downarrow \tag{7.4}$$

where $N_\uparrow$ and $N_\downarrow$ are the number of electrons with up and down spins. Then, in the unpolarized situation, we expect that:

$$G_0 \propto \frac{1}{2}(N_{\uparrow L}N_{\uparrow R} + N_{\uparrow L}N_{\downarrow R} + N_{\downarrow L}N_{\uparrow R} + N_{\downarrow L}N_{\downarrow R})$$
$$= N_L N_R \tag{7.5}$$

where the indices $L$ and $R$ stand for the right and left electrodes, and $N_L$ and $N_R$ are the total number of electrons.

In the polarized case, on the other hand, we have:

$$G_H \propto N_{\uparrow L}N_{\uparrow R} + N_{\downarrow L}N_{\downarrow R} \tag{7.6}$$

Hence, we have:

$$\frac{G_H - G_0}{G_0} = \frac{(N_{\uparrow L} - N_{\downarrow L})(N_{\uparrow R} - N_{\downarrow R})}{N_L N_R} \tag{7.7}$$

Hence, the magnetoresistance is directly proportional to the polarization of the electrodes. This simple analysis roughly explains the pioneering experiments in spin tunneling [5,6].

A more realistic theory should, at least, include two additional effects:

– The density of states at the Fermi level needs not be proportional to the total polarization.

– The wavefunctions of the majority and minority electrons near the barrier need not be the same. Then, the transmission coefficient acquires a spin dependence, which influences the magnetoresistance.

These two effects were first analyzed using a plane wave description for the electronic wavefunctions [7]. However, many magnetic materials include transition elements, so that a more complicated description is required. In some cases, like Fe alloys, the polarization of the density of states at the Fermi surface is the opposite to the total polarization [8]. Near the surface, the s electrons are more delocalized such that they play a major part in the transport process. When this is the case, the junction magnetoresistance is mostly determined by the polarization of the s electrons, which can be different from the bulk polarization, dominated by the d orbitals [9]. A detailed analysis of the role of the atomic orbitals near the junction for many situations of practical interest can be found in [10]. In addition, some orbitals of the magnetic electrodes can make bonds with the atoms in the insulating barrier. If that happens, the magnetoresistance becomes quite independent of the bulk properties of the electrodes, and interesting possibilities for tailoring the properties of the junction arise [11]. The density of states at a surface can also be modified by the existence of surface states. These states will be polarized by the bulk magnetization, leading to resonances in the magnetoresistance as function of bias voltage [12].

### 7.2.3   Problems.

A three dimensional electron gas has the two spin subbands split by an exchange potential, $\Delta$. The dispersion relation is:

$$\epsilon_{k,\uparrow} = \frac{\hbar^2 k^2}{2m} - \Delta$$

$$\epsilon_{k,\downarrow} = \frac{\hbar^2 k^2}{2m} + \Delta \tag{7.8}$$

Calculate the values of $[D_\uparrow(E_F) - D_\downarrow(E_F)]/[D_\uparrow(E_F) + D_\downarrow(E_F)]$ and $(N_\uparrow - N_\downarrow)/(N_\uparrow + N_\downarrow)$, assuming that $\Delta \ll E_F$.

Answer:

$$\frac{D_\uparrow(E_F) - D_\downarrow(E_F)}{D_\uparrow(E_F) + D_\downarrow(E_F)} = \frac{\Delta}{2E_F}$$

$$\frac{N_\uparrow - N_\downarrow}{N_\uparrow + N_\downarrow} = \frac{3\Delta}{2E_F} \tag{7.9}$$

Calculate the same quantities for a two dimensional magnetic layer.

Answer:

$$\frac{D_\uparrow(E_F) - D_\downarrow(E_F)}{D_\uparrow(E_F) + D_\downarrow(E_F)} = 0$$

$$\frac{N_\uparrow - N_\downarrow}{N_\uparrow + N_\downarrow} = \frac{\Delta}{E_F} \tag{7.10}$$

## 7.3   Magnetic Impurities

In the previous section, we have assumed that an applied magnetic field aligns the magnetization of the electrodes. Isolated paramagnetic impurities, however, may need a much higher applied field to become polarized. The existence of these misaligned impurity levels can lead to transport processes not allowed within the framework discussed in the previous section.

To understand why it is the case, let us consider resonant tunneling through a single paramagnetic impurity located between the two electrodes, as sketched in Fig. 7.3.

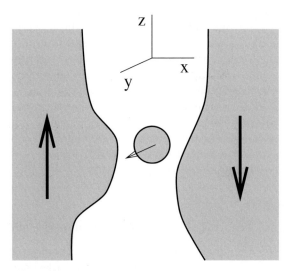

**Fig. 7.3.** Paramagnetic impurity in a magnetic junction (see text).

We assume that Hund's rule freezes the electronic spin of the impurity. We simplify things further by assuming that the tunneling processes mediated by the impurity do not involve exchange of electrons. Hence, the same electron which moves from one electrode to the impurity continues, elastically and without loss of coherence, to the second electrode. As the problem does not involve electron-electron interactions, it can be reduced to a problem of independent electrons moving in a spin dependent potential:

$$\mathcal{H} = \mathcal{H}_\mathrm{L} + \mathcal{H}_\mathrm{R} + \mathcal{H}_\mathrm{tunn} + \sum_i V_{\mathrm{imp}\ ss'}(\boldsymbol{r}_i - \boldsymbol{R}_\mathrm{imp}) \qquad (7.11)$$

where $\mathcal{H}_\mathrm{L}$ and $\mathcal{H}_\mathrm{R}$ stand for the Hamiltonians which describe the left and right electrodes, and $\mathcal{H}_\mathrm{tunn}$ takes into account the remaining tunneling processes.

We first assume that the magnetization in the two electrodes is aligned (high field case). The direct tunneling processes included in $\mathcal{H}_\mathrm{tunn}$ include only tunneling from a state in the majority (minority) band to another state in the majority (minority) band of the other electrode. Let us take the direction of the magnetization as the z-axis. A possible electronic state in a paramagnetic impurity whose moment is oriented along the x-axis has the form:

$$|\Psi\rangle = \frac{1}{\sqrt{2}}\left(|\uparrow\rangle + |\downarrow\rangle\right) \tag{7.12}$$

The impurity potential in (7.11) mixes the two spin states of the electrodes. In terms of states with polarization along the z-axis, we can write:

$$V_{\text{imp}\uparrow\uparrow}(\mathbf{r} - \mathbf{R}) = V_{\text{imp}\uparrow\downarrow}(\mathbf{r} - \mathbf{R})$$
$$= V_{\text{imp}\downarrow\downarrow}(\mathbf{r} - \mathbf{R}) = \frac{1}{2}V_{\text{imp}}(\mathbf{r} - \mathbf{R}) \tag{7.13}$$

Hence, an electron with a given polarization in one electrode can tunnel through the impurity, with equal probability, into the two polarization states of the other electrode.

We can estimate these effects by averaging over all impurity polarizations, taking into account the influence of the applied field [13]. A low field can align the polarization of the electrodes, but will have a negligible influence on the impurities. Then, resonant tunneling through the impurities gives a contribution to the conductance of the junction which is independent of the applied field, lowering the magnetoresistance. When the field is strong enough to align the impurities, a magnetoresistance close to that of a clean junction will be recovered. The crossover between these two regimes takes place when $\mu_0 SH/k_B T \approx 1$. For an impurity with total spin $S = 3/2$, for instance, the typical fields are 60 T at 300 K and 0.8 T at 4 K.

We have, so far, ignored the possibility that the impurity contains a partially filled energy shell, so that the electron which tunnels in needs not be the same as the one which tunnels out. In these processes, the total spin of the impurity remains unchanged, but its orientation is modified. They are responsible for the Kondo effect in bulk systems. If the electrodes are not magnetic, tunneling through a Kondo impurity leads to a sharp resonance at the Fermi level [14], which is strongly dependent on temperature and bias voltage. The generalization of this study to magnetic electrodes is yet to be done. In principle, one would expect that the formation of the Kondo resonance will be suppressed. If it is the case, the inclusion of these exchange processes will not modify significantly the independent electron tunneling model discussed above.

Finally, we can extend the previous analysis to incoherent tunneling through the impurities. Transport of electrons can be separated into two independent processes. The tunneling electrons lose memory of their spin during the passage through the impurity, leading, again, to a reduction of the magnetoresistance. Incoherent tunneling can be important if the impurity is actually a magnetic cluster with many internal degrees of freedom.

## 7.4   Magnetic Excitations

So far, we have considered elastic processes which do not involve the creation, or absorption, of excitations in the junction. This has greatly simplified the analysis, as the different interactions between quasiparticles need not to be taken into account.

There are many inelastic processes possible in a junction. We are interested in those which can influence the magnetoresistance. This will happen if the excitations which are created during the tunneling process are themselves magnetic. A clear signature of inelastic processes is their temperature dependence. When the temperature is increased, there is more thermal energy available, $k_\mathrm{B}T$, and the number of excitations in the system also increases. The rate of inelastic processes is enhanced. In the case of a junction, an alternative way of providing energy to the system is through the applied voltage. The associated energy scale is $eV$. Thus, the contribution to the conductance of inelastic processes scales in the same way with temperature or voltage.

As mentioned in Sect. 7.2, electron transport involves processes which take place in regions of mesoscopic dimensions, at the surface of the junction. Let us assume that this scale is $a$. An electron localized in a region of length $d$ can only couple to excitations with wavelengths greater than $d^{-1}$. The inelastic contribution to the conductance will be proportional to the number of excitations with energies $\omega_{\boldsymbol{k}} \leq \max(k_\mathrm{B}T, eV)$, and wavevectors such that $|\boldsymbol{k}| \leq d^{-1}$. At zero temperature and finite bias voltage, we find that:

$$
\begin{aligned}
I(V) \propto & \int_0^{eV} d\epsilon\, N_{\mathrm{L},s}(\epsilon) f(\epsilon) \\
\times & \int_0^{\min(\epsilon,\omega_d)} d\epsilon'\, N_{\mathrm{R},s'}(\epsilon')[1 - f(\epsilon - \epsilon' - eV)] D(\epsilon')
\end{aligned}
\tag{7.14}
$$

We assume that one electron tunnels from the left electrode, where the density of states is $N_\mathrm{L}(\epsilon)$, to the right electrode, creating an excitation of energy $\epsilon'$. $D(\epsilon')$ is the density of states of excitations with energy $\epsilon'$. $\omega_d$ stands for the energy of a magnetic excitation of wavevector comparable to $d^{-1}$. If the excitation has spin one, like a magnon in a ferromagnet, then $s' = -s$. A similar formula gives the differential conductance at zero bias and finite temperature.

The dispersion relation of bulk magnons in a ferromagnet is:

$$
\omega_{\boldsymbol{k}} = J|\boldsymbol{k}a|^2
\tag{7.15}
$$

where $J$ is the exchange coupling, and $a$ is a length of the order of the size of the unit cell. Then, $D(\epsilon') \propto \epsilon'^{1/2}$, and, from (7.14), we obtain:

$$
I(V) \propto
\begin{cases}
V \left(\frac{eV}{J}\right)^{3/2} & eV \ll J\left(\frac{a}{d}\right)^2 \\
V \left(\frac{a}{d}\right)^3 & eV \gg J\left(\frac{a}{d}\right)^2
\end{cases}
\tag{7.16}
$$

Using an analogous argument, the contribution of inelastic processes to the differential conductance at zero temperature is given by:

$$G(T) \propto \begin{cases} \left(\frac{k_B T}{J}\right)^{3/2} & k_B T \ll J \left(\frac{a}{d}\right)^2 \\ \left(\frac{a}{d}\right)^3 & k_B T \gg J \left(\frac{a}{d}\right)^2 \end{cases} \tag{7.17}$$

The temperature dependence follows the decrease in the magnetization of the electrodes, $M(T=0) - M(T) \propto T^{3/2}$.

The analysis discussed here can be extended to magnetic excitations localized at the interface between the electrodes (see next section). For instance, ferromagnetic spin waves in two dimensions have a density of states independent of energy. That changes the exponent $3/2$ in (7.16) and (7.17) to 1.

Magnon assisted tunneling leads to spin flip processes, and reduces the magnetoresistance of the junction. Recent experiments in different types of junctions confirm the picture presented here [15–17] (see also [18]).

## 7.5  Magnetic Properties of the Interface

Atoms at the surface of the junction are surrounded by an environment quite different from that at the bulk of each electrode. Changes in the magnetic structure of the surface are to be expected. The decrease in the number of nearest neighbours typically leads to a shift of the density of magnetic excitations towards lower energies [19]. The characterization of the magnetic properties of a surface is, however, a very difficult task, and there is not a unique recipe to solve the problem. First principles calculations of surface magnetic properties are scarce [20]. In fact, it is very difficult to estimate, from first principles, the magnetic properties of bulk materials.

We can get a simple estimate of the changes of the magnetic structure at the surface by analyzing a semiinfinite chain of spins coupled ferromagnetically. In the bulk, the spin waves are given by:

$$\omega_k = 2J[1 - \cos(ka)] \tag{7.18}$$

so that the density of excitations is:

$$D_{\text{bulk}}(\omega) = \frac{1}{\pi\sqrt{\omega(4J - \omega)}} \tag{7.19}$$

which has the characteristic $\omega^{-1/2}$ divergence of one dimensional models. Assuming that the couplings near the surface are the same as in the bulk, it can be shown [19] that the density of states at the last atom of the surface is given by:

$$D_{\text{surf}}(\omega) = \frac{2}{\pi} \frac{1}{\omega + \sqrt{\omega(4J - \omega)}} \tag{7.20}$$

The divergence at low energies is now twice the value of the divergence found in the bulk density of states. As both densities are normalized, this increase in

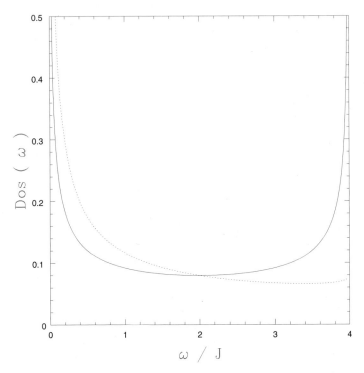

**Fig. 7.4.** Density of spin wave excitations inside a ferromagnetic one dimensional system (full line), and at the surface (broken line).

low energy modes implies a reduction in the weight of the high energy modes at the surface, as shown in Fig. 7.4. At low temperatures, the spins at the surface fluctuate more than in the bulk. In addition, the surface excitations can mediate spin flip processes, reducing the magnetoresistance of the junction, as discussed in the previous section.

The effects can be more dramatic in double exchange materials [21]. In these systems, the effective coupling between spins is proportional to the kinetic energy of the conduction electrons, which is reduced near the surface [22]. In addition, the lattice can be distorted at the surface, leading to different charge states of the magnetic ions. As in these compounds there are residual antiferromagnetic couplings, besides the ferromagnetic double exchange mechanism, the surface can become antiferromagnetic.

### 7.5.1   Problems

The surface of a magnetic junction is covered by a thin antiferromagnetic layer. The spin wave dispersion relation is:

$$\omega_k = J_{\mathrm{AF}} \mathbf{k} a \qquad (7.21)$$

where $J_{AF}$ is the antiferromagnetic coupling, $k$ is a two dimensional vector, and $a$ is a length of the order of the unit constant. The scale of the roughness of the interface junction is $d$. Calculate, using the method described in Sect. 7.4, the contribution of inelastic processes involving spin-wave excitations, to the conductance as function of voltage and temperature.

Answer:

$$G(V) \propto \begin{cases} \left(\frac{eV}{J_{AF}}\right)^2 & eV \ll J_{AF}\left(\frac{a}{d}\right) \\ \left(\frac{a}{d}\right)^2 & eV \gg J_{AF}\left(\frac{a}{d}\right) \end{cases} \tag{7.22}$$

$$G(T)|_{V=0} \propto \begin{cases} \left(\frac{k_B T}{J_{AF}}\right)^2 & k_B T \ll J_{AF}\left(\frac{a}{d}\right) \\ \left(\frac{a}{d}\right)^2 & k_B T \gg J_{AF}\left(\frac{a}{d}\right) \end{cases} \tag{7.23}$$

## 7.6 Charging Effects in Granular Systems

Metallic junctions play a major role in determining the transport properties of granular systems. If the grains are magnetic, the magnetoresistance of these samples is usually controlled by the motion of the electrons across the interfaces between the grains, and not by intrinsic effects related to the bulk properties of the materials [23,24].

Tunneling between small metallic grains shows additional effects besides those considered so far. The increase of the charge of a grain by one electron charge leads to a change in the electrostatic energy of the grain approximately equal to $e^2/\epsilon_0 R$, where $R$ is the average radius of the grain and $\epsilon_0$ is the dielectric constant of the surrounding material. When this energy is larger than the thermal energy, $k_B T$, tunneling of electrons is suppressed. This phenomenon is called Coulomb Blockade [25]. For particles of radius $\sim 100$ nm, the corresponding charging energies correspond to $T = 30$ K.

The interplay of charging effects and magnetism leads to a wide variety of phenomena. The resistance of granular materials shows an upturn at low temperatures [23,24], which can be interpreted in terms of an average charging energy, assuming a given distribution of grain sizes and barrier heights [26]. In the presence of an applied field, the effective charging energy changes, typically decreasing [27].

The dependence of the effective charging energy on applied field can be understood by noting that the field increases the intergrain conductances. Transport between grains is more efficient when the magnetization is aligned. Thus, the field enhances the delocalization of the electrons, which will spend more time on neighboring grains. The effective charging energy is reduced, in agreement with observations [27].

Charging effects have been invoked to explain upturns in the resistance of conventional junctions, which may have almost disconnected protrusions or even metallic grains within the barrier [28].

### 7.6.1   Problems

A granular material has grains of radius $R \approx 50$ nm. The dielectric constant of the medium is $\epsilon_0 = 5$. Calculate at which temperature Coulomb blockade effects become appreciable.

Answer: $T \approx 90$ K.

## 7.7   Conclusions

The present review analyzes some processes which determine spin dependent transport at magnetic junctions. We have not tried to give an exhaustive description of this rapidly growing field. We have not discussed some topics indirectly related to the subject, like transport across grain boundaries [29], or across boundaries between magnetic domains [30,31]. Non equilibrium effects, like the spin accumulation due to the slowness of the spin equilibration have also not been discussed [32]. On the other hand, we have tried to use examples of current interest, and we have discussed in some detail those processes which can influence most the magnetoresistance of the junction.

## References

1.  R. Meservey and P. M. Tedrow, Phys. Rep. **238**, 173 (1994).
2.  J. L. Costa-Krämer, Phys. Rev. B **55**, R4875 (1997).
3.  F. Ott, S. Barberan, J. G. Lunney, J. M. D. Coey, P. Berthet, A. M. de Leon-Guevara and A. Revcolevschi, Phys. Rev. B **58**, 4656 (1998).
4.  N. García, M. Muñoz and Y.-W. Zhao, Phys. Rev. Lett. **82**, 2923 (1999).
5.  M. Jullière, Phys. Lett. **54A**, 225 (1975).
6.  M. B. Stearns, J. Magn. Magn. Mater. **5**, 167 (1977).
7.  J. C. Slonczewski, Phys. Rev. B **39**, 6995 (1989).
8.  B. Drittler, N. Stefanou, S. Blügel, R. Zeller, and P. H. Dederich, Phys. Rev. B **40**, 8203 (1989).
9.  E. Yu. Tsymbal and D. G. Pettifor, J. Phys.: Condens. Matter **9**, L411 (1997).
10. M. Villeret, J. Mathon, R. B. Muniz, and J. d'Albuquerque e Castro, Phys. Rev. B **57**, 3474 (1998). See also J. Mathon, Phys. Rev. B **56**, 11810 (1997).
11. J. M. De Teresa, A. Barthélémy, A. Fert, J. P. Contour, R. Lyonnet, F. Montaigne, P. Seneor, and A. Vaurès, Phys. Rev. Lett. **82**, 4288 (1999). J. M. De Teresa, A. Barthélémy, A. Fert, J. P. Contour, F. Montaigne and P. Seneor, Science **286**, 507 (1999).
12. R. Wiesendanger, M. Bode and M. Getzlaff, Appl. Phys. Lett. **75**, 124 (1999).
13. F. Guinea, Phys. Rev. B **58**, 9212 (1998).
14. S. Hershfield, J. H. Davies, and J. W. Wilkins, Phys. Rev. Lett. **67**, 3720 (1991).
15. S. Zhang, P. M. Levy, A. C. Marley, and S. S. P. Parkin, Phys. Rev. Lett. **79**, 3744 (1997).
16. J. Moodera, J. Nowak and R. J. M. van de Veerdonk, Phys. Rev. Lett. **80**, 2941 (1998).
17. C. H. Shang, J. Nowak, R. Jansen, and J. S. Moodera, Phys. Rev. B **58**, R2917 (1998).

18. A. M. Bratkovsky, Appl. Phys. Lett. **72**, 2334 (1998).
19. D. C. Mattis "The Theory of Magnetism I: Statics and Dynamics", Springer, New York (1988).
20. T. Asada, G. Bihlmayer, S. Handschuh, S. Heinze, Ph. Kurz, and S. Blügel, J. Phys.: Condens. Mater. **11**, 9347 (1999).
21. J. M. D. Coey, M. Viret, and S. von Molnár, Adv. Phys. **48**, 167 (1999).
22. M. J. Calderón, L. Brey, and F. Guinea, Phys. Rev. B **60**, 6698 (1999).
23. J. M. D. Coey, A. E. Berkowitz, Ll. Balcells, F. F. Putris, and A. Barry, Phys. Rev. Lett. **80**, 3815 (1998).
24. Ll. Balcells, J. Fontcuberta, B. Martinez, and X. Obradors, Phys. Rev. B **58**, R14697 (1998).
25. "Single Electron Tunneling", M. H. Devoret and H. Grabert eds., Plenum Press, New York (1992).
26. J. S. Helman and B. Abeles, Phys. Rev. Lett. **37**, 1429 (1976).
27. M. García-Hernández, F. Guinea, A. de Andrés, J. L. Martínez, C. Prieto, and L. Vázquez, Phys. Rev. B **61**, 9549 (2000).
28. J. Z. Sun, D. W. Abraham, K. Roche, and S. S. P. Parkin, Appl. Phys. Lett. **73**, 1008 (1998).
29. N. D. Mathur, G. Burnell, S. P. Isaac, T. J. Jackson, B. S. Teo, L. L. MacManus-Driscoll, L. F. Cohen, J. E. Evetts, and M. G. Blamire, Nature **266**, 387 (1997).
30. P. M. Levy and S. Zhang, Phys. Rev. Lett. **79**, 5110 (1997).
31. L. Brey, preprint cond-mat/9905209.
32. A. Brataas, Y. U. Nazarov, J. Inoue, G. E. W. Bauer, Phys. Rev. B **59**, 93 (1999).

# 8 Basic Semiconductor Physics

H. J. Jenniches

Department of Physics and Astronomy, University of Leeds, Leeds LS2 9JT, United Kingdom

## 8.1 Introduction

### 8.1.1 What is a Semiconductor?

Semiconductors were originally defined as materials with a conductivity between that of metals and that of insulators in the range between $10^2$ and $10^{-9}$ $(\Omega \mathrm{cm})^{-1}$. This definition emphasises electronic transport and shows that the conductivity in semiconductors can be varied over an impressive eleven orders of magnitude. It does not, however, indicate the microscopic mechanism for this behaviour. A more advanced definition aiming at the explanation of the conductivity variation reads as follows:

A semiconductor is a solid state material, which is insulating at low temperatures and has a measurable electronic conductance at higher temperature. The electronic conductivity is due to the well-defined chemical composition, which does not change in high electric fields or due to some influence from outside the solid-state material.

Although this definition seems to leave the problem of defining conductivity, specifically electronic conductivity, this term is readily explained in solid state physics textbooks where classical and quantum mechanical free electron models are introduced. As the focus here is on semiconductors, it is sufficient to know that the conductivity of a material is determined by the charge carrier concentration and the mobility of these charge carriers. In the case of electrons as charge carriers it is, thus, necessary to find out, what determines the number of electrons in a solid and which of those are available to conduct electrical current (free or nearly free electrons) as opposed to electrons which are tied up in the bonds within the solid. Therefore, conductivity must be explained by the kind of structural bonds on an atomic level and the presence of free (delocalised) charge carriers due to these bonds.

The periodic system of elements orders the elements in terms of their electronic and nuclear structure and is one way of finding out, how many electrons an element can make available, either for conducting electrical current or for bonding to other elements. Electrons in incomplete (outer) electron shells are called valence electrons and the number of valence electrons defines the type of bond the element can undergo and it determines to some degree the atomic radii of the atoms as this is related to the fact how loosely or tightly the outer electrons are bound to the atom. In the middle of the periodic system – between the metals on the left and the insulators on the right – resides the group of elemental semiconductors (group IV). The elements in this group have four valence electrons per atom, i.e. four partly filled electron shells that need to be filled. Therefore, elements of this group are reactive and can undergo bonds with many other elements. On the other hand, elements with four valence electrons tend to form *covalent bonds* when bonding to other elements of the same group. Covalent bonds involve two electrons per bond and are fairly strong. The electrons involved in these bonds are not free but localised. However, if energy, e.g. light or heat, is absorbed by the material, the bond might be broken and an electron is set free and can contribute to conduction. And, that is precisely, how semiconductors work.

The most important group IV semiconductors are silicon (Si) and germanium (Ge). But semiconductors are not only from group IV, but can be binary combinations of group III and V elements, like gallium arsenide (GaAs) and indium phosphide (InP), or binary combinations of group II and VI elements, like zinc selenide (ZnS) and cadmium telluride (CdTe) or even ternary alloys, having the same number of electrons in total, "tied up" in covalent bonds. As the same number of electrons is involved in all these materials they tend to have the same or very similar crystal structures in their solid state. As one would expect with carbon being in the same group of the periodic system, the crystal structure of Si and Ge is in fact the diamond structure. The crystal structure of the binary semiconductors, such as GaAs or InP, is the zincblende structure, which is the same structure for binary compounds with the two different elements occupying alternating positions [1].

At absolute zero temperature ($T = 0$ K), all electrons are bound to their parent atoms. There are no free electrons left that would enable electric current to flow. Therefore, semiconductors are insulators at low temperatures.

### 8.1.2   Simple Band Structure

It is now easy to redraw the structural view to introduce a simple band structure. Consider Fig. 8.1, where a simple bond model is shown on the left hand side and a simple schematic band structure on the right hand side. All valence electrons, which are tied up in covalent bonds on the left are represented by the valence band on the lower right.

At absolute zero temperature, in this energy diagram, all states in the valence band are occupied and the conduction band is completely empty. The semiconductor is insulating. At higher temperature or rather if energy in some form is absorbed by the electrons, i.e. lattice vibrations, photons, etc., an electron-electron bond can break and the electron becomes a free charge carrier capable of conducting electrical current, the hole left in its place, however, is immediately filled with other valence electrons from the surrounding bonds and thus is also considered to be a free charge carrier, but of positive charge. Holes, therefore, also contribute to the electric conductivity of the semiconductor. This process of free electron formation is called *electron-hole pair generation*.

Fig. 8.1 illustrates the electron-hole pair generation at label 1. This mechanism is producing a larger amount of thermally generated electron-hole pairs,

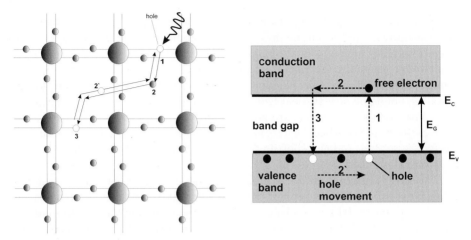

**Fig. 8.1.** Bond model (left) and simple band structure (right). After [2].

thus increasing the electrical conductivity of the semiconductor. Moving through the crystal (label 2), i.e. after some time, the free electron will jump into another broken bond somewhere in the crystal, cancelling the hole existing there at that precise moment. This process is called *electron-hole recombination* (label 3). The introduction of the hole as a positive charge carrier simplifies the analysis; instead of observing the movement of a very large amount of electrons in the valence band, our view is focused on a relatively small number of holes that move in the opposite direction.

The generation of a free electron is represented in the energy diagram on the right hand side of Fig. 8.1 by an electron jumping from the valence into the conduction band (which represents the energy states of the electrons which contribute to conduction). The electron at the bottom of the conduction band has only potential energy. Under the influence of the electric field it will gain some kinetic energy, which will enable electric current to flow. The increase of the kinetic energy is represented by the electron moving upward from the bottom of the conduction band. On the other hand, the increase of the kinetic energy of the hole is represented by the hole moving downward in the valence band. The energy difference between the valence and the conduction band, the forbidden energy zone, is called *energy gap* and is the single most important parameter for semiconductors, as it determines which energy needs to be absorbed to generate charge carriers.

In pure semiconductors, free carriers are generated exclusively by the process of electron-hole pair generation described above. Therefore, the concentration of electrons in equilibrium equals the concentration of holes. Such semiconductors are called *intrinsic semiconductors.*

But, whereas our simple band structure model only assigns an energy to the electrons, real band structures look somewhat different due to the fact that electrons also have momentum, which means that the direction of travel matters for the "fine" structure of the band. The real structure is based on the wave picture for the properties of the electrons, and this is derived from wave mechanics, i.e. quantum mechanics. The wave provides a way of calculating the effects an electron can produce. In quantum mechanics the electron energy is given by $E = \hbar\omega$, its momentum is $p = \hbar k$ and its group velocity is $v = d\omega/dk = dE/dp$, where $\hbar$ is $h/2\pi$, $\omega$ is the radian frequency and $k$ is the wave number (the number of radians of phase change in unit distance). These equations are derived from a simple one-dimensional model of a solid, consisting of a row of $N$ atoms with a distance $a$ apart, each with two free electrons. When evaluating what values of $k$ can be used to make allowed wave functions or states, the highest wavelength is equal to $2Na$ and the lowest $k$, therefore, becomes $\pi/(Na)$. Further $k$-values are equally spaced $2\pi/Na$ ... until $2N\pi/Na$ which is $2\pi/a$. The regular spacing of allowed values of $k$ occurs in three directions when a three-dimensional lattice of atoms in ordinary space is analysed – that is why $k$-space or momentum space is useful to obtain a view of electron states. The $2N$ waves with $k$ running from $\pi/Na$ to $2\pi/a$ are all the waves one needs to make a band. Note, that they are just enough for the number of electrons in the crystal, and that the wavelength goes from the size of the crystal to the distance between the atoms. Band structures are periodic along the k axis because the real crystal is periodic in space. Usually only one interval of $k$ is shown, conventionally the first Brillouin zone, which is called a reduced zone diagram. Energy diagrams against $k$ ($E$-$k$ diagrams) for real semiconductors are specific for certain crystallographic directions and are also of a more complex structure than our simple band structure in Fig. 8.1.

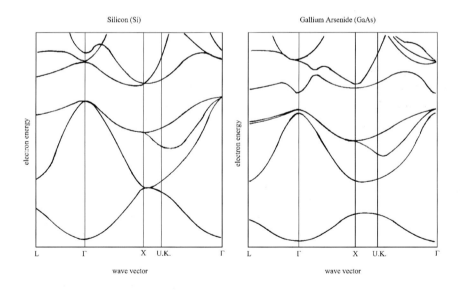

**Fig. 8.2.** Real band structures of Si (left) and GaAs (right). After [3].

Figure 8.2 shows the real band structures of the most important semiconductors, Si and GaAs with their four valence bands on the bottom of the respective diagrams and the four conduction bands on the top. The $\Gamma$ point is the zone centre [000] with the L point denoting the [111] zone boundary and X denoting the [100] zone boundary, i.e. the shapes are not symmetric about the zone centre. It is now much more difficult to find the energy gap or band gap because of the irregularity of the structures. The band gap is given by the smallest distance between the conduction band and the valence band, i.e. the gap between the minimum in the conduction band and the maximum in the valence band. It is apparent that for GaAs the minimum in the conduction band and the maximum in the valence band occur at the same $k$ value (here in the zone centre), whereas for Si there is a k-shift between the minimum and the maximum. Semiconductors like GaAs are called *direct band-gap semiconductors* (the transition is a purely optical transition, the material can be used for light-emitting diodes), those semiconductors with the minimum of the conduction band at a different value of k from the maximum of the valence band are called *indirect band-gap semiconductors* (the indirect transition involves an additional "step", i.e. one or more phonons.

The curvature of the bands is also of significance. As the velocity is $v = dE/dp$ the acceleration $a$ of an electron due to some external force can be expressed as $a = dv/dt = (d/dt)(dE/dp) = (dp/dt)(d^2E/dp^2)$. But $dp/dt$ is the rate of change of momentum, and hence equals the applied force $F$. Thus $d^2E/dp^2$ replaces the mass in the equation of motion $F = ma$ and we can describe the response of a carrier to a force by using $(d^2E/dp^2)^{-1}$ instead of the mass. This new term is known as the *effective mass* $m^*$ of a carrier and summarises the way the interaction with the lattice affects the carrier motion. The effective mass of a free electron is $m_e{}^*$ and the effective mass of an electron in the bottom of the conduction band is usually less than the free-electron mass, and may be much less. Thus the real band structures can be described in terms of their value of the curvature, and, therefore, in terms of their effective masses. A sharply curved band has a large value of $d^2E/dp^2$ and hence a small effective mass and is, therefore, called *light-hole band*. As can be seen from the diagrams the effective mass may vary with direction (anisotropy).

## 8.2    Charge–Carrier Concentration, Band Gap and Fermi Energy

### 8.2.1    Intrinsic Semiconductors

The first section introduced semiconductors and the important terms and characteristic parameters. For an evaluation of the conductivity of semiconductors and their behaviour, for instance with temperature, the number of charge carriers has to be calculated. The electron and hole densities, $n$ and $p$, are represented on the right hand side of Fig. 8.3 (top) as the area under the curve of the electron and hole distributions, respectively. The charge-carrier distribution is the product of the number of electron states per unit energy per unit volume, which is known as the density of states, and the fraction of occupied states at each energy, which is given by the Fermi function, as known from standard solid state textbooks.

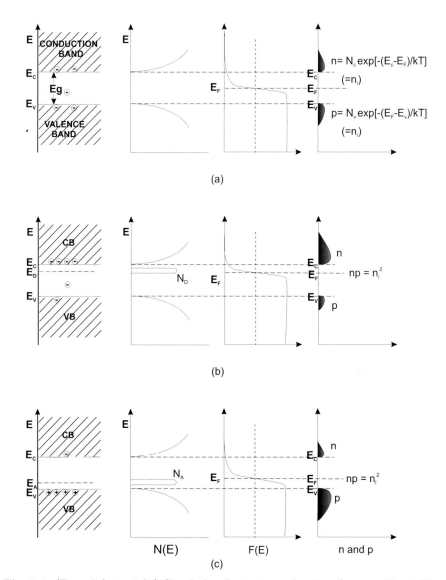

**Fig. 8.3.** (From left to right) Simple band structure, density of states, Fermi–Dirac distribution and charge carrier concentration for (a) intrinsic, (b) n-type and (c) p-type semiconductors at thermal equilibrium. After ref. [1].

The density of states for a semiconductor as a function of electron energy is shown in Fig. 8.3 (top). The density of states is zero in the band gap and has a calculable shape elsewhere. Electrons and holes occupy states near the band edges with parabolic dispersion relations having effective masses $m_c^*$ and $m_v^*$, respectively. For electrons near the lower edge of the conduction band the density of states $D(E)$ is found to be

$$D_c(E) = \frac{8\sqrt{2}\pi(m_c^*)^{3/2}}{h^3}\sqrt{E - E_c} \qquad (8.1)$$

and, accordingly, for holes near the valence-band edge

$$D_v(E) = \frac{8\sqrt{2}\pi(m_v^*)^{3/2}}{h^3}\sqrt{E_v - E} \qquad (8.2)$$

Since electrons are fermions, they obey the Pauli–exclusion principle and Fermi–Dirac statistics, provided the system is in thermal equilibrium. The Fermi–Dirac distribution functions for electrons and holes are given by

$$f_n(E) = \frac{1}{1 + \exp\left[\frac{E - E_{F_n}}{kT}\right]} \qquad (8.3)$$

$$f_p(E) = \frac{1}{1 + \exp\left[\frac{E_{F_p} - E}{kT}\right]}. \qquad (8.4)$$

The Fermi energy or Fermi level $E_F$ is the energy where the probability of a state being occupied by an electron or hole is one half. It is apparent from Fig. 8.3 (top) that the Fermi function is symmetrical about the Fermi level and tends to zero for large positive energies and to one for large negative energies. $E_{F_n}$ and $E_{F_p}$, the Fermi energies for electrons and holes, are equal under equilibrium conditions, i.e. $E_{F_n} = E_{F_p} = E_F$. In that case

$$f_p(E) = 1 - f_n(E). \qquad (8.5)$$

If the energy is much larger than the Fermi level, then the Fermi–Dirac distribution function can be approximated by the Maxwell–Boltzmann distribution, i.e. the electrons behave like classical billiard balls:

$$f_n(E) = \exp\left[-\frac{E - E_F}{kT}\right]. \qquad (8.6)$$

Accordingly, for an evaluation of the number of holes the fraction of states that are unoccupied is given by

$$f_p(E) = \exp\left[\frac{E - E_F}{kT}\right], \qquad (8.7)$$

provided only holes are taken into account and the relevant energy range is well below $E_F$. Note that the equations are equivalent with the sign reversed – another indication that the idea of holes carrying positive charge is correct.

With the distributions of electrons in the conduction band and holes in the valence-band being described by

$$dn(E) = D_c(E)f_n(E)dE \tag{8.8}$$

$$dp(E) = D_v(E)f_p(E)dE, \tag{8.9}$$

the electron and hole densities are obtained by integration.

Therefore, the total number of electrons per unit volume is found as

$$n = \frac{1}{4}\left(\frac{2m_c^* kT}{\pi\hbar^2}\right)^{3/2} \exp\left[-\frac{E_c - E_F}{kT}\right] \tag{8.10}$$

and accordingly the number of holes per unit volume becomes

$$p = \frac{1}{4}\left(\frac{2m_v^* kT}{\pi\hbar^2}\right)^{3/2} \exp\left[-\frac{E_F - E_v}{kT}\right] \tag{8.11}$$

Collecting the terms outside the exponentials into single symbols $N_c$ and $N_v$ yields a more compact expression for $n$ and $p$ which is

$$n = N_c \exp\left[-\frac{E_c - E_F}{kT}\right] \tag{8.12}$$

$$p = N_v \exp\left[-\frac{E_F - E_v}{kT}\right]. \tag{8.13}$$

$N_c$ and $N_v$ are called the effective density of states for the conduction band and for the valence band, respectively. They can be thought of as the number of states that would be required to give the same value for $n$ (or $p$) if all the states were at a single energy, that of the respective band edge. $N_c$ and $N_v$ vary with temperature while the actual density of states varies very little.

The product of the concentrations of electrons and holes in equilibrium

$$np = N_c N_v \exp\left[-\frac{E_c - E_v}{kT}\right] \tag{8.14}$$

gives an expression for the band gap $E_G = E_c - E_v$, the single most important parameter for semiconductors. Therefore,

$$E_G = kT \ln\left[\frac{N_c N_v}{np}\right] \tag{8.15}$$

The product of $n$ and $p$ depends only on the temperature and the kind of semiconductor, and not on hole or electron densities. In intrinsic semiconductors without added impurities, the electron density equals the hole density and the name *intrinsic carrier density* $n_i$ is given to this carrier density, such that

$$np = n_i{}^2 \tag{8.16}$$

This equation is very important in the study of semiconductors but seems to be difficult to emphasise adequately, such that this equation is sometimes called the *semiconductor equation*. It is an example of the chemical law of mass action.

Dividing the two equations for $n$ and $p$ yields

$$p/n = \frac{N_v}{N_c} \exp\left[\frac{E_c + E_v - 2E_F}{kT}\right] \qquad (8.17)$$

which can be rewritten to determine the Fermi energy

$$E_F = \frac{E_c + E_v}{2} - \frac{kT}{2} \ln\left[\frac{p}{n}\right] - \frac{3}{4}kT \ln\left[\frac{m_c^*}{m_v^*}\right]. \qquad (8.18)$$

This equation shows that the Fermi level for intrinsic material ($\ln(p/n) = 0$) is at the average of $E_c$ and $E_v$ with a small correction when $m_c^* \neq m_v^*$. When $p \neq n$ then the Fermi level shifts towards the band with the majority carriers.

It is worth remembering that the semiconductor equation holds for any values of $p$ or $n$ as long as a thermal equilibrium situation is being described. So it often is valid. However, when $p$ and $n$ are controlled in some device by the external conditions, then one may be far from thermal equilibrium. From (8.14) it becomes clear that the variation of the intrinsic carrier concentration with temperature can be caused by

1. a variation of the effective masses of the carriers,
2. the pre-exponential term $T^{3/2}$,
3. a variation of the band gap $E_G$, and
4. the $kT$ term in the denominator of the exponential function argument.

The variation of the effective masses of the carriers for small temperature changes can be neglected. Therefore, these are assumed to be constant. The variation of the band gap for small temperature changes can be described by a linear function. This linearisation is fairly common whenever a simple expression for the relation between the band gap and temperature is needed for a certain temperature range. By substitution a linear coefficient in a logarithmic plot can be obtained. This coefficient has the same order of magnitude for all common semiconductors. A different kind of band gap variation is found with the composition when intrinsic semiconductors are considered. There is some relation between the band gap and the lattice constant, such that a kind of phase diagram can be plotted. Within certain limits the band gap widens with decreasing lattice constant. The band gap can be custom engineered by picking two binary semiconductors with band gaps on either side of the required band gap and gradually substituting the constituent elements to meet the required band gap. For instance, the band gap for $AlAs_{50}Sb_{50}$ will be half way between the band gaps for AlAs and AlSb.

## 8.2.2    P and N Type Doping

Another way to engineer the band gap and all electrical properties of a semiconductor is called doping. Pure semiconductor materials (say 99.999%) are called intrinsic semiconductors as their properties are intrinsic to the material itself. By adding some impurities to a semiconductor its electrical properties are changed. Doped semiconductors are called *extrinsic semiconductors*. Impurities that cause the increase of electron concentration are called *donors*. A semiconductor in which the concentration of electrons is higher than the concentration of holes is said to be an (extrinsic) *n-type* semiconductor. The concentration of electrons in Si and Ge can be increased by doping with penta-valent elements, such as phosphorus (P) and arsenic (As). These elements have five valence-electrons, but only four are necessary to form a covalent bond with the host semiconductor. The extra electron will be loosely bound to its parent atom and a very small amount of energy (referred to as ionisation energy) will be sufficient to tear it off. Naturally, when the fifth electron leaves, the donor atom becomes a positive (donor) ion. At room temperature, this ion is "frozen" in the semiconductor crystal, and does not contribute to the current flow. The presence of a donor impurity is represented by discrete states within the energy gap, very close to the conduction band.

A *p-type* semiconductor, in which the hole concentration is higher than the electron concentration, is obtained by adding *acceptor* impurities. In Si and Ge, acceptors are usually tri-valent elements, such as boron (B). Since these elements have only three valence electrons, one is missing to complete covalent bonds with host semiconductor atoms. Therefore, such an atom will bind an electron that would otherwise jump from the valence band into the conduction band, thus preventing the formation of an electron-hole pair. By catching an electron, the acceptor impurity will become a negatively charged ion. Acceptors, therefore, remove an electron from the valence band, leaving a mobile hole. Acceptor impurities introduce energy states within the energy gap, very close to the valence band. Since ionisation energies of typical donor and acceptor impurities are rather small (10 to 50 meV), at room temperature ($T$=300 K, $kT$=25 meV) almost all impurities are ionised.

Compound semiconductors consist of a tri-valent and a penta-valent element (III–V compounds), such as gallium arsenide (GaAs) or gallium phosphide (GaP) or a di-valent and a hexa-valent element (II–VI compounds), such as zinc sulfide (ZnS). The chemical bond is formed by the component with higher valence lending some electron(s) to the component with lower valence. Donor impurities in compound semiconductors are elements with valence higher than that of the component they substitute, and acceptor impurities are elements with valence lower than that of the component they substitute. It is interesting to notice that a tetra-valent element, such as Si and Ge, in III–V compound can be both, a donor impurity (if it substitutes a tri-valent component) or an acceptor impurity (if it substitutes a penta-valent component).

Shallow donors or acceptors are impurities which really increase the concentration of electrons or holes in the semiconductor, while deep level impurities which are reducing the concentration of carriers due to the nature of their bond to the host crystal which is then a significant modification/perturbation to the crystal structure and the binding.

### 8.2.3   Impurity Bands

When a semiconductor is doped with donor or acceptor impurities, impurity energy levels (or impurity bands) are introduced in the band gap. A donor level is defined as being neutral if filled by an electron and positive if empty. An acceptor level is neutral if empty and negative if filled by an electron.

The calculation of impurity energy levels is based on calculating the ionisation energy for a donor replacing the mass by the conductivity effective mass in relation to the mass and ionisation energy of the hydrogen atom. The ionisation energy is the energy necessary to free the fifth valence electron from the donor atom, or to capture the fourth electron onto the acceptor atom. Accordingly, for the case of a donor impurity the ionisation energy equals the distance of the donor impurity level from the top of the band gap. For the case of an acceptor impurity it equals the distance of the acceptor impurity level from the bottom of the band gap.

Usually, every impurity introduces several energy levels into the band gap, for instance, Au in Ge has three acceptor levels and one donor level in the band gap. Any impurity modifies the electrical properties of the semiconductor material by creating additional energy levels. The highest purity is, therefore, required in the manufacturing of semiconductor devices.

### 8.2.4   Charge–Carrier Concentration and Fermi Energy of Extrinsic Semiconductors

Again, the knowledge of the position of the Fermi level allows a determination of the concentration of electrons and holes in the system. When impurity atoms are introduced the Fermi level must adjust itself to preserve charge neutrality. Under the conditions of electrical neutrality the total of positive charges (free holes and fixed ionised donor or acceptor atoms) must equal the total of negative charges (free electrons and fixed ionised acceptor atoms) in a semiconductor, therefore

$$p + N_d^+ = n + N_a^- , \tag{8.19}$$

where $p$ and $n$ are the intrinsic hole and electron concentrations, $N_d^+$ and $N_a^-$ are the concentrations of ionised donors and acceptors, respectively. If the concentrations of the impurities are not too high, almost all of them will be ionised at room temperature. Consequently, it can be assumed that the concentrations of ionised impurities is equal to the total concentration.

For n-type semiconductors, as shown in Fig. 8.3, middle section, for a known concentration of donors $N_d$ that introduce an energy level $E_d$ within the band gap, the concentration of ionised donors is given as

$$N_d^+ = N_d \left[ 1 - \frac{1}{1 + \frac{1}{2} \exp\left[ (E_d - E_F)/kT \right]} \right] \tag{8.20}$$

Similarly for a p-type semiconductor (Fig. 8.3, bottom), for a known concentration of acceptors $N_a$ that introduce an energy level $E_a$ within the band gap, the concentration of ionised acceptors is given as

$$N_a^+ = \frac{N_a}{1 + 4 \exp\left[(E_a - E_F)/kT\right]} \tag{8.21}$$

The factor in front of the exponential function in the denominator of the expression for $N_d^+$ equals 2 since a donor level can accept one electron with either spin or can have no electron. The equivalent factor in $N_a^-$ equals 4 because each acceptor impurity level can accept one hole of either spin and the impurity is doubly degenerate as a consequence of the two degenerate valence bands at wave vector $k$=0 in Si, Ge and GaAs [1].

At very low temperatures $E_F - E_d \gg kT$ and $E_a - E_F \gg kT$, respectively, which results in exponential functions in the denominators of the above formulas being much larger than one, and, consequently, $N_d^+$ and $N_a^-$ approaching zero. As the temperature increases, the exponential function diminishes and the ionised impurity concentrations approach the total concentrations.

Thus, in an n-type semiconductor the charge neutrality equation becomes

$$n = p + N_d^+ \tag{8.22}$$

and for a set of given $N_c$, $N_d$, $N_v$, $E_c$, $E_d$, $E_v$ and $T$ the Fermi level $E_F$ can be determined.

At elevated temperatures (above room temperature) the charge neutrality condition can be approximated by

$$n + N_a = p + N_d \tag{8.23}$$

and together with (8.14) one can evaluate the electrons and holes in an n-type semiconductor as

$$n_{n_0} = \frac{1}{2}\left[(N_d - N_a) + \sqrt{(N_d - N_a)^2 + 4n_i^2}\right], \tag{8.24}$$

which is approximately $N_d$, if $N_d - N_a \gg n_i$ and $N_d \gg N_a$ and

$$p_{n_0} = \frac{n_i^2}{n_{n_0}} \approx \frac{n_i^2}{N_d} \tag{8.25}$$

and therefore

$$E_c - E_F = kT \ln(N_c/N_d) \quad \text{or} \quad E_F - E_{Fi} = kT \ln(n_{n_0}/n_i). \tag{8.26}$$

$E_{Fi}$ denotes the Fermi energy in the intrinsic case.

Equivalently for p-type semiconductors

$$p_{p_0} = \frac{1}{2}\left[(N_a - N_d) + \sqrt{(N_d - N_a)^2 + 4n_i^2}\right], \tag{8.27}$$

which is approximately $N_a$, if $N_a - N_d \gg n_i$ and $N_a \gg N_d$ and

$$n_{p_0} = \frac{n_i^2}{p_{p_0}} \approx \frac{n_i^2}{N_a} \tag{8.28}$$

and therefore

$$E_F - E_v = kT \ln(N_v/N_a) \quad \text{or} \quad E_{Fi} - E_F = kT \ln(p_{po}/n_i) \quad (8.29)$$

The subscripts $n$ and $p$ refer to the type of semiconductor, the subscript 0 refers to the thermal equilibrium condition.

For n-type semiconductors the electron is referred to as majority carrier and the hole as minority carrier, since the electron concentration is the larger of the two. The roles are reversed for p-type semiconductors. These simple solutions can be obtained if $|N_a - N_d| \gg n_i$. It is seldom necessary to deal with situations when this is not the case, but if this does occur, approximate values for $p$ and $n$ can be obtained by first ignoring the minority carrier in (8.23) and then using the semiconductor equation and $n + N_a$ alternately to obtain more accurate solutions. NB: In doped material, the minority carrier density is several orders of magnitude smaller than $n_i$.

In highly doped semiconductors, only a fraction of impurities replaces atoms in the crystal lattice of the host semiconductor, which results in part of them remaining electrically inactive. At lower temperatures (typically below 200 K), the energy of crystal lattice vibrations is not sufficient to ionise all impurities. The temperature range where this takes place is referred to as partial ionisation range or freeze out region.

## 8.3    Carrier Transport

### 8.3.1    Introduction

The next step and a first, simple application of doped semiconductors would be to join p-type and n-type semiconductor in order to make a p-n junction, i.e. a basic semiconductor diode. But when considering an interface between an n- and a p-doped semiconductor it is obvious that there would be an enormous particle-density gradient around this interface. This gradient will lead to diffusion of one type of charge-carriers into the other type of semiconductor, thus creating a diffusion current. Therefore, firstly basic transport phenomena have to be discussed.

The reasons for a net flow of holes or electrons [4], i.e. current, are

1. an electric potential gradient $dV/dx$,
2. a particle-number density gradient $dn/dx$,
3. a temperature gradient $dT/dx$

Current due to an electrical potential gradient is called drift current while current due to a particle number density gradient is a diffusion current. The third point will not be discussed here, but is significant for such useful devices as thermoelectric cooling systems and power generators.

## 8.3.2   Drift Current and Mobility

If an electric field $E$ accelerates carriers of charge $q$ and effective mass $m^*$ their acceleration $a$ is given by

$$a = \frac{qE}{m^*} \tag{8.30}$$

The distance each carrier is travelling in the direction of $E$ by this acceleration is $\frac{1}{2}a\tau^2$, where $\tau$ is the time between collisions. The average drift velocity $v_d$ is found by taking an average value of the distance, and dividing by the average value of $\tau$, such that it can be written

$$v_d = \frac{q\tau_d E}{m^*}, \tag{8.31}$$

where $\tau_d$ is $\overline{\tau^2}/\overline{\tau}/2$, making $\tau_d$ the mean time between collisions as appropriate for carrier drift. With $q\tau_d/m^* = \mu$, known as the mobility for the particular carriers in the material one obtains

$$v_d = \mu E \tag{8.32}$$

From this equation it is apparent that the drift produced by the same field on holes and on electrons is in opposite directions. The drift currents carried by both holes and electrons, however, add and the current density $j$ carried by drifting carriers is

$$j = nqv_d = nq\mu E, \tag{8.33}$$

where $n$ is the total number of drifting charge carriers. If there are different kinds of carriers present they contribute separately to the current density [4].

## 8.3.3   Diffusion Current

The flux $F$ of particles down a density gradient $dn/dx$ is described by

$$F = -D\frac{dn}{dx} \tag{8.34}$$

where $D$ is the diffusion coefficient. This equation is known as Fick's law and applies to any example of diffusion. The electric current density is

$$J = -qD\frac{dn}{dx} \tag{8.35}$$

where $q$ is the charge of the carrier.

When considering the diffusion of holes and electrons in the same density gradient, holes and electron fluxes are in the same direction, and, thus, the conventional electrical currents tend to cancel – in fact the opposite of the situation when drift was examined [4]. In fact, $D$ and $\mu$ are related by

$$\frac{D}{\mu} = \frac{kT}{e} \tag{8.36}$$

a relation found by Einstein. Both, $D$ and $\mu$ are each an effect of collisions of carriers with the same lattice defects. Consequently a formula for $D$ is

$$D = \frac{kT\tau_{\mathrm{d}}}{m^*} \tag{8.37}$$

The average distance that a carrier can diffuse before recombining is known as the *diffusion length* $\lambda_{\mathrm{d}}$. If the mean distance between collisions is $\lambda$, then in a time $\tau_{\mathrm{r}}$, the recombination lifetime, a carrier can make $\tau_{\mathrm{r}}/\tau_{\mathrm{d}}$ collisions and can diffuse a distance $\lambda(\tau_{\mathrm{r}}/\tau_{\mathrm{d}})^{1/2}$. With the average thermal velocity being $(kT/m^*)^{1/2}$, the diffusion length is obtained to

$$\lambda_{\mathrm{d}} = \lambda(\tau_{\mathrm{r}}/\tau_{\mathrm{d}})^{\frac{1}{2}} = (D\tau_{\mathrm{r}})^{\frac{1}{2}}. \tag{8.38}$$

Unfortunately $\tau_{\mathrm{r}}$ has to be determined by experiment, as recombination can occur in several ways, for instance, via traps in the middle of the band gap.

In the evaluation of devices it is sometimes more appropriate to consider both types of charge carriers in a combined parameter and therefore assume *ambipolar diffusion*. The ambipolar diffusion coefficient is then given by

$$D_{\mathrm{a}} = \frac{n_n + p_n}{n_n/D_p + p_n/D_n} \tag{8.39}$$

and the ambipolar lifetime is

$$\tau_{\mathrm{a}} = \frac{p_n - p_{n_0}}{U} = \frac{n_n - n_{n_0}}{U}, \tag{8.40}$$

where $U$ is the net recombination rate [1].

The continuity equation balancing the numbers of particles entering and leaving a region by considering diffusion, drift, generation and recombination of charge carriers is a useful starting point for further analysis.

### 8.3.4  Mobility and Conductivity

The *mobility* $\mu$ has already been given in (8.32). Perhaps the easiest way of thinking of mobility is merely as the constant relating $v_{\mathrm{d}}$ and $E$. Another way of putting the same idea is that the mobility is the velocity for unit field. Notice that for high mobility a large value of $\tau_{\mathrm{d}}$ (the mean time between collisions of charge carriers) and a small value for the effective mass are necessary.

Mobility is a useful generalising concept when discussing the effect of an electric field on carriers. A very useful first approximation is to say that $\mu$ is a constant for a given carrier and material. But, in fact $\mu$ depends on temperature, and on doping density when this is high, because of the effect on the collision time. The mobility also depends on the electric field: it falls at high fields, so that the drift velocity in many materials tends to a maximum limit. For Si the maximum drift velocity is $10^5$ m/s for both holes and electrons. The field needed for electrons to reach this speed is about $2 \cdot 10^6$ V/m, but holes require fields above $10^7$ V/m, where avalanche ionisation is beginning to set in. Nevertheless a constant mobility is usually assumed for nearly all devices.

As with other parameters the mobility can be taken from plots of mobility vs. impurity concentration in conjunction with the rough temperature dependence $\mu \propto m^{*-\frac{3}{2}}T^{-\frac{1}{2}}$ [1].

The *conductivity* $\sigma$ of a sample may be defined as the current that flows across a unit cube when the field and hence voltage between opposite faces is unity. Thus for a semiconductor where both holes and electrons are present, with e the electronic charge

$$\sigma = ne\mu_e + pe\mu_h, \tag{8.41}$$

where the subscripts $e$ and $h$ are referring to electrons and holes, respectively. Again, for simplicity, one can take values of the conductivity or resistivity from appropriate plots for n- and p-type semiconductors which are readily available in the relevant literature [1].

### 8.3.5 Band Bending

So far only equilibrium conditions were considered – but, what is happening in terms of the energy diagrams, when a semiconductor is subject to an electric field? Also, energy diagrams do not really have an x-axis, but from now on $x$ means the position and here $x = 0$ is the surface of the semiconductor. When dealing with p-n junctions it will be the location of the metallurgical junction.

Both holes and electrons are essentially mobile charges $q_h$ and $q_e$ (not to be misunderstood as the donor or acceptor atoms, which are "fixed" within the normal atomic diffusion limits) and these positive and negative charges will move when placed into an electric field. To visualise this it is useful to consider a simple plate capacitor [4]: On the positive plate of the charged capacitor there is a positive surface-charge density and on the negative plate an equal negative charge. By considering what states are available for theses charges to occupy it is possible to establish how thick the surface layer is which they cause. In a metal there are plenty of states into and from which the extra electrons can be transferred, and the electric field falls from a high value outside the metal towards zero in a few atomic layers. In a semiconductor there are only a small number of states which, at acceptable energies, can be filled or emptied. If a large charge has to be accommodated then the charge on the "semiconductor" capacitor extends much further into the crystal involving a surface layer of a thickness that is no longer negligible.

In Fig. 8.4 (middle) a p-type semiconductor forms the negative part of a capacitor. The usual electrostatic convention that surface charges exist in negligibly thin layers is on this occasion being refined by a more detailed analysis. Poisson's equation is used to analyse the layer in consideration. The Poisson equation relates the charge density and the electric field intensity and takes the general form

$$\nabla E = \frac{\rho(\boldsymbol{r})}{\varepsilon}, \tag{8.42}$$

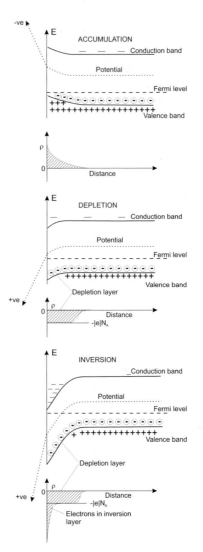

**Fig. 8.4.** Band bending – the three types of surface layer for a p-type semiconductor: accumulation (top), depletion (middle) and inversion (bottom). After [4].

where $\rho(\boldsymbol{r})$ is the charge density as a function of position and $\varepsilon = \varepsilon_0 \varepsilon_r$ is the permittivity of the medium with $\varepsilon_0$ being the permittivity of free space and $\varepsilon_r$ the relative permittivity of the material under consideration. In a one-dimensional model one has

$$\frac{d^2 V}{dx^2} = -\frac{\rho}{\varepsilon},\tag{8.43}$$

where $V$ is the potential in Volts. $\rho$ can be expressed as the product of the charge carrier density and the charge, i.e. $eN_a$ for a p-type semiconductor. Provided one can define the boundary conditions for the region in question, it is possible to obtain the electric field intensity $E$ after integration of the above expression. The charge density at the geometrical surface is zero, so that the electric displacement, $D = \varepsilon E$, is continuous in the mathematical sense on entering the semiconductor surface. As $\varepsilon$ is different inside the semiconductor, the electric field inside is not the same as the electric field outside. The displacement falls to zero deep in the semiconductor from its value $D$ outside. Hence the total charge density per unit area, in the whole surface layer is, by Poisson's equation, equal to $D$.

The bands bend in this case to lower energies near the surface. This removes the valence band from the Fermi level so that there are no holes in the valence band near the surface, though the acceptors stay ionised. The volume-charge density is that of the acceptors, and a useful estimate of the thickness $t$ of the surface layer, if the acceptor concentration is $N_{\mathrm{a}}$, is

$$ teN_{\mathrm{a}} = \varepsilon E = D = \varepsilon \frac{dV}{dx}. \tag{8.44} $$

This sort of surface layer is called *depletion layer*, and, for instance, occurs on one side of a p-n junction, which is discussed in the next section. The relation between the voltage $V$ across the depletion layer and the thickness of the layer then becomes

$$ V = \frac{t^2 eN_{\mathrm{a}}}{2\varepsilon_{\mathrm{r}}\varepsilon_0}. \tag{8.45} $$

If a p-type semiconductor is made the positive plate of a capacitor (see Fig. 8.4, top) and hence the electric field is directed out of the surface, the bands are bent to more negative energies near the surface and there is an *accumulation* of holes near the surface. The charge density increases near the surface as the valence band approaches the Fermi-level, so that the formula describing the thickness of the layer has to include the integral of a charge density which varies with position. The formula is thus more complex than that for the depletion layer.

A strong field into the surface of a p-type semiconductor (Fig. 8.4, bottom) can cause the conduction band to become the nearer of the two bands to the Fermi level. The surface is then n-type even though the bulk is p-type, and electrons occupy states in the conduction band near the surface in this so-called *inversion layer*. Such a layer forms the conducting channel in many practical devices.

If, instead of a p-type semiconductor, n-type is used, corresponding effects occur for the opposite electric field direction. Thus, a field into the surface produces enhancement, a weak field out of the surface produces depletion, and a strong field out of the surface produces inversion. In the inversion layer the majority charge carriers would be holes [4].

## 8.4   P–N Junction

### 8.4.1   Barrier Potential

A p-n junction is formed when p-type and n-type semiconductor regions are adjacent to each other. The p-n junction is a semiconductor diode (a diode is a two-terminal device that has a high resistance to electric current in one direction but a low resistance in the other direction). The conductance properties of the p-n junction depend on the direction of the voltage, which can in turn be used to control the electrical nature of the device. The behaviour of the p-n junction is the basic characteristic of many semiconductor devices, which makes the p-n junction a fundamental device.

As discussed previously, joining p-type and n-type semiconductors to make a p-n junction will lead to diffusion of one type of charge-carriers into the other type of semiconductor. This leads to electron-hole recombination throughout a certain width from the initial metallurgical junction with the formation of a depletion layer within that width. As the depletion layer is depleted of charge carriers it represents a high electrical resistance compared to the n- and p-type parts of the semiconductor device. On both ends of the p-n junction charge neutrality still holds, but within the depletion layer it does not. Therefore, this is a non-equilibrium situation and the current balance has to be considered. It is useful to analyse a simple situation: a p-n junction in equilibrium with no applied voltage or net current as shown in Fig. 8.5. Without examining the detailed variation of carrier density or potential across the junction three statements can be made [4]:

1. The net current of electrons across the junction is zero.
2. The net current of holes across the junction is zero.
3. At all points $pn = n_i^2$ is valid.

The zero total current density of electrons can be thought of as two cancelling components caused by drift and diffusion. These two components can be written as

$$-n(x)e\mu_e\frac{dV(x)}{dx} + eD_e\frac{dn(x)}{dx} = 0, \qquad (8.46)$$

where $V(x)$ is the potential and $n(x)$ the electron density at a distance $x$ from one side of the junction. Rearranged this becomes

$$\mu_e\frac{dV(x)}{dx} = D_e\frac{1}{n}\frac{dn(x)}{dx}. \qquad (8.47)$$

Integrating (8.47) from $x = x_n$, well on the n-side of the junction, to $x = x_p$, safely on the p-side one obtains

$$\int_{x_p}^{x_n}\mu_e\frac{dV(x)}{dx}dx = \int_{x_p}^{x_n}D_e\frac{1}{n}\frac{dn(x)}{dx}dx. \qquad (8.48)$$

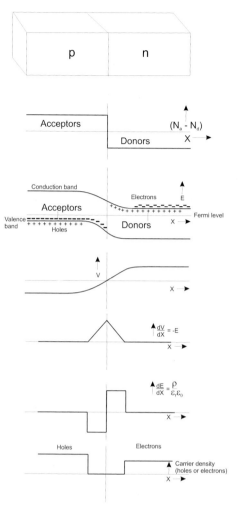

**Fig. 8.5.** A p-n junction diode without bias, from top to bottom: sketch of a p-n junction, dopant density throughout the abrupt junction, electron energy diagram, potential variation, electric field variation with peak at the boundary between p and n, differential of electric field variation and carrier density variation. After [4].

The left hand side becomes an integral in $V$ with limits $V_n$ and $V_p$ and the right hand side an integral in $n$ with limits $n_n$ and $n_p$, in each case well away from the junction. Then

$$V_n - V_p = \frac{D_e}{\mu_e} \ln \left[ \frac{n_n}{n_p} \right]. \tag{8.49}$$

If the junction was made by doping the n-side with $N_d$ donors and the p-side with $N_a$ acceptors, then a good approximation is $n_n = N_d$, and $n_p = n_i^2/N_a$.

Hence

$$V_n - V_p = \frac{D_e}{\mu_e} \ln \left[ \frac{N_d N_a}{n_i^2} \right].$$  (8.50)

Thus, there is a potential difference between the two sides, the so-called barrier potential $V_b$, which curbs the tendency of the electrons to diffuse away from the place where they are densest. Unless the doping is very heavy, the barrier potential is less than the band gap of the semiconductor. Typical values of barrier potential for junctions in Ge and Si are 0.4 V and 0.8 V [4], respectively. The barrier potential results from the fact that large amounts of electrons and holes (the respective majority charge carriers) diffuse from their sides into the depletion zone and recombine and thus annihilate the charge connected with them. What is left are the fixed donor and acceptor ions in the n- and p-type semiconductor material, which represent positive and negative immobile charges which therefore change the charge distribution in the p-n junction.

In an equilibrium situation the Fermi level is constant right through the system. The difference of the potential energy on the two sides is equal to $(E_c - E_F)$ on the p-side minus $(E_c - E_F)$ on the n-side.

With (8.12) and (8.13) this gives

$$V_b = V_n - V_p = \frac{kT}{e} \ln \left[ \frac{N_c}{n_n} \right] - \frac{kT}{e} \ln \left[ \frac{N_c}{n_p} \right],$$  (8.51)

but as before $n_n = N_d$ and $n_p = \frac{n_i^2}{N_a}$, so that

$$-V_b = \frac{kT}{e} \ln \left[ \frac{N_d N_a}{n_i^2} \right].$$  (8.52)

If this is compared to the former equation for $V_b$, it is apparent that

$$\frac{kT}{e} = \frac{D_e}{\mu_e}.$$  (8.53)

This equation is identical with (8.36), the Einstein relation. The derivation can be repeated for holes; the barrier potential comes out accordingly, and there is a corresponding version of (8.53) which is

$$\frac{kT}{e} = \frac{D_h}{\mu_h}.$$  (8.54)

Thus, if either mobility or the diffusion constant is known, the other quantity can be calculated.

### 8.4.2   Depletion Zones

As mentioned above, a region close to the metallurgical p-n junction tends to be depleted of holes and electrons, and is known as a depletion layer. Fig. 8.6 shows p-n junctions with forward and reverse bias. The externally applied bias

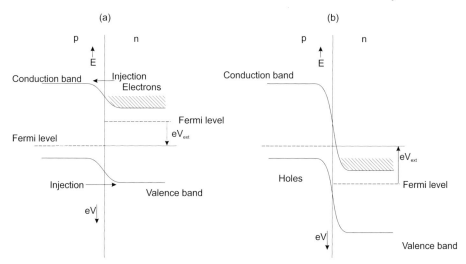

**Fig. 8.6.** Energy bands for a p-n junction with (a) forward bias and (b) reverse bias. After [4].

shows up as a difference between the Fermi levels on the two sides – this is a fundamental point in reading or constructing diagrams. The applied voltage occurs mostly across the junction/depletion layer.

The reverse bias adds to the barrier potential and results in a wider depletion layer. The forward bias subtracts from the barrier potential – in general the forward bias never reverses the usual situation to make the p-side more positive. In terms of device properties this means that the reverse bias decreases or possibly "blocks" the current flowing from one side of the device to the other side, whereas the forward bias leads to enhanced current flow. This is exactly, how the p-n junction works as a basic rectifying diode.

In order to understand real devices an investigation of the detailed structure of the depletion layer is required. Two assumptions are made: the first is that the density of doping atoms changes rapidly from one value $N_a$ on the p-side to another steady value $N_d$ on the n-side. Such a junction is known as a step junction or an abrupt junction. In practice, as long as the cross-over occurs in a distance that is much less than the full width of the depletion layer, this assumption will be a good approximation. The second assumption is that the distance over which the electron or hole density falls to a small value is short enough to be ignored. This is equivalent to saying that the depletion layer must be much thicker than the Debye length (the maximum distance at which the Coulomb fields of charged particles are expected to interact), or that the total potential difference $V_j$ across the junction is much greater than $kT/e$. In the depleted parts of the p-region, the only charges are fixed acceptor ions, which are negative. Thus using Poisson's equation again

$$\frac{dE}{dx} = \frac{\rho}{\varepsilon_r \varepsilon_0} = \frac{-eN_a}{\varepsilon_r \varepsilon_0}. \tag{8.55}$$

Hence

$$E = -\frac{eN_a x}{\varepsilon_r \varepsilon_0} + C. \tag{8.56}$$

When $E = 0$, $x = x_p$ at the boundary of the depletion layer in the p-region, so that

$$E = -\frac{eN_a(x - x_p)}{\varepsilon_r \varepsilon_0}. \tag{8.57}$$

Because of the choice of the direction of the x axis in Fig. 8.6, $x_p$ is a negative number, so $E$ reaches its most negative value $E_{\max}$ when $x = 0$,

$$E_{\max} = \frac{eN_a x_p}{\varepsilon_r \varepsilon_0}. \tag{8.58}$$

In the depleted part of the n-region, the only charges are the donor ions, which are positive,

$$\frac{dE}{dx} = \frac{eN_d}{\varepsilon_r \varepsilon_0}, \tag{8.59}$$

$$E = \frac{eN_a(x - x_n)}{\varepsilon_r \varepsilon_0}, \tag{8.60}$$

where $x_n$ is the boundary of the depletion layer in the n-region. From the analysis of the n-layer,

$$E_{\max} = -\frac{eN_a x_n}{\varepsilon_r \varepsilon_0}. \tag{8.61}$$

The total potential difference $V_j$ across the junction is the sum of the voltages across the two parts,

$$V_j = \frac{e}{2\varepsilon_r \varepsilon_0}(N_d x_n^2 + N_a x_p^2). \tag{8.62}$$

The two equations for $E_{\max}$ must give the same answer, so $N_a|x_p| - N_d|x_n|$ Hence there will be a thicker depletion layer where the doping is lighter and a thinner one where there is heavy doping. Equation (8.62) also means that most of the voltage appears across the lightly doped side. In an extreme case one can ignore the thickness and voltage on the heavily doped side. Such a junction may be indicated by a plus sign ($p^+ - n$), and the total thickness of the depletion layer is proportional to $\sqrt{V_j}$. Accordingly the total depletion layer thickness or width $W$ is

$$W = \sqrt{\frac{2\varepsilon_r \varepsilon_0 (N_d + N_a)}{eN_d N_a} V_j}. \tag{8.63}$$

### 8.4.3 Varicap$^{TM}$ Diode or Varactor Diode

This is an example of a device that is basically a simple p-n junction but refers to a diode in which the capacitance varies with the applied voltage. It is, in fact, the capacitance of the depletion layer, which depends on the bias voltage. In the analysis of transistors the depletion-layer capacitance is a factor which limits performance (it affects the frequency response). In most diodes the capacitance is unwanted, but in varactor diodes or Varicaps$^{TM}$, this capacitance is exploited. The capacitance $C(V_j)$ of the depletion layer is

$$C(V_j) = \frac{dQ}{dV}(V_j) = \frac{dQ}{dW}(V_j)\frac{dW}{dV}(V_j), \tag{8.64}$$

where $Q$ is the charge on each side of the capacitor, $W$ is the total thickness of the depletion layer and $V_j$ is the total difference in potential between the p- and n-sides. If we take a $p - n^+$ diode, then the major part of $V_j$ and of $W$ is in the p-region. As a result $W \approx x_p$ and $Q = eN_a x_p A$ with $A$ being the area of the junction. Hence

$$\frac{dQ}{dW} = \frac{dQ}{dx_p} = eN_a A. \tag{8.65}$$

From (8.62), when $x_n \ll x_p$

$$\frac{dW}{dV_j} = \frac{2\varepsilon_r \varepsilon_0}{2eN_a A x_p}. \tag{8.66}$$

Eliminating $x_p$ from (8.66) by using (8.62) again and then writing out the expressions for the two terms in (8.64), gives

$$C(V_j) = A\sqrt{\frac{eN_a \varepsilon_r \varepsilon_0}{2}}\frac{1}{\sqrt{V_j}}. \tag{8.67}$$

Thus $C(V_j)$ decreases as the bias becomes more negative. If $C(V_j)$ is expressed in terms of the depletion-layer thickness, then

$$C = \frac{A\varepsilon_r \varepsilon_0}{W}. \tag{8.68}$$

So the capacitance is identical with that of an ordinary capacitor of the same size, shape, and permittivity as the depletion layer. This is true for all p-n junctions, no matter what the variation of doping with position is.

### 8.4.4 Light Emitting Diodes

A light emitting diode (LED) is another example for a device that is basically a simple p-n junction made up from particular materials. LEDs belong to the group of optoelectronic components like semiconductor laser diodes and photodiodes, that have revolutionised communication technology. LEDs can convert electrical energy into optical radiation by means of electroluminescence, i.e. the generation of light by passing an electrical current through the material under an applied electric field. Electroluminescent light differs from thermal radiation or incandescence (radiation as a result of the temperature of the material) in the relatively narrow range of wavelengths contained within its spectrum. For LEDs, the spectral line width is typically 10 to 50 nm. The electroluminescent light is obtained by injecting minority carriers into the region of a p-n junction where radiative transitions can take place [1]. This is achieved by forward biasing the p-n junction, such that many holes are pushed from the p-region into the junction region and many electrons are pushed from their n-region into the junction region [5]. In the junction region the electrons fall into holes, i.e. recombine. The recombination of electrons and holes leads to the release of energy – in the band theory this corresponds to an electron falling from the conduction band or one of the donor levels just beneath it in the n-region to the valence band or one of the acceptor levels in the p-region [6]. There are two main competing processes for the recombination: (a) a simple radiative transition with a lifetime $\tau_{rr}$ resulting in the emission of a photon, and (b) a non-radiative transition via recombination centres with a lifetime $\tau_{nr}$ resulting in the release of a phonon. The quantum efficiency is the fraction of the excited carriers that combine radiatively to the total recombination [1]. For an LED to be efficient, the ratio $\tau_{rr}/\tau_{nr}$ must be small, i.e. radiative transitions have to be highly probable. For indirect band-gap semiconductors (see Sect. 8.1.2 and Fig. 8.2) like Si and Ge this is not the case and the energy released is all dissipated as heat, but for direct semiconductors like GaAs the transition is purely by emission of a photon.

In general, the recombination rate is proportional to the surplus carrier density in the junction region, i.e. the intensity of the light emission is depending on the current through the junction. Typical current/voltage values for an LED are 20 mA at 2-3 Volts. As LEDs are responding very quickly to changes in the current, the light emission can be modulated at frequencies of several MHz.

The energy of the emitted photons is approximately equal to the band gap, i.e. the photon wavelength can be varied by using materials with different band gaps. The variation of the band gap with composition for ternary semiconductor compounds was briefly discussed at the end of Sect. 8.2.1, whereas band gap engineering via the addition of impurities is explained in Sects. 8.2.2 and 8.2.3. For instance, GaAs emits infra-red radiation and GaP doped with Zn and O emits red light at 1.7 eV. Particular emphasis has to be given to the design and construction of the p-n junction in the LED: a photon is only useful when it has emerged from the diode [4]. Absorption in the semiconductor and total internal reflection at the surface reduce the effective output by a factor of up to one hundred in some devices.

## 8.5    Haynes–Shockley Experiment

The Haynes–Shockley experiment is an example of a typical experiment in semiconductor physics, i.e. a measurement of characteristic semiconductor parameters. The classic experiment by Haynes and Shockley (1951) apparently lead to the development of the transistor. The experiment is described in a paper entitled "The Mobility and Life of Injected Holes and Electrons in Germanium" [7]. Alternative contact types or injection methods are now preferred, but this is the way it was first done. Minority carrier mobility and lifetime, as measured in this experiment, are vital parameters for materials from which, for instance, bipolar (minority carrier junction) transistors are made.

In this experiment electron-hole pairs are produced by a (hot) electron injection, i.e. a pulsed signal applied through probes on the semiconductor. These carriers are then allowed to drift in opposite directions in an applied electric field. The minority carriers are detected by a reverse-biased metal point contact. The arrival of the minority carriers is observed on an oscilloscope to be delayed by a time which is proportional to the distance the carriers have travelled, so the minority carriers have a definite drift velocity. The drift velocity is found to be proportional to the drift field, so the minority-carrier mobility can be calculated. The value of the mobility for a specific carrier (e.g. electrons in Si) turns out to be the same whether they are majority or minority carriers, and the experiment shows evidently that minority carriers are needed for a complete explanation of semiconductor phenomena [4]. An alternative injection technique for this experiment could be the photo-production of electron-hole pairs, i.e. using a light source focussed onto a small area, demonstrating the importance of minority carriers and their mobility. Complementing Hall measurements can then give the mobility of majority carriers.

As this classic experiment initialised the development of conventional semiconductor devices, one might expect similar experiments studying spin coherent transport in semiconductors to promote the new technology of spin electronics. The first very promising studies in this field have been performed in recent years and are reviewed in the chapter on "Spin transport in Semiconductors" by M. Ziese.

## 8.6   Exercises

1. What are the equilibrium concentrations of holes and electrons at 300 K in silicon doped with
   (a) $N_d = 3 \cdot 10^{20} \mathrm{m}^{-3}$
   (b) $N_a = 2 \cdot 10^{22} \mathrm{m}^{-3}$ and
   (c) $N_d = 7.5 \cdot 10^{16} \mathrm{m}^{-3}$, $N_a = 5.0 \cdot 10^{16} \mathrm{m}^{-3}$.
   Assume $n_i = 1.45 \cdot 10^{16} \mathrm{m}^{-3}$.
   What would you expect to happen to the carrier concentrations as the temperature was raised to 500 K?

2. Calculate the resistance at 300 K between opposite faces of a bar of silicon of length 1 cm with a cross-sectional area of 6 mm². Assume
   (a) the silicon to be intrinsic
   (b) the silicon to be doped with $10^{22}$ donors/m³ and
   (c) the silicon to be doped with $10^{22}$ acceptors/m³.
   Assume $n_i = 1.45 \cdot 10^{16} \mathrm{m}^{-3}$, $\mu_e = 0.135 \mathrm{m}^2 \mathrm{V}^{-1} \mathrm{s}^{-1}$ and $\mu_h = 0.043 \mathrm{m}^2 \mathrm{V}^{-1} \mathrm{s}^{-l}$.

3. What donor doping density is required to form a resistor of 1.5 k$\Omega$ in an integrated circuit? The length of the resistor is 75 $\mu$m and its width is 5 $\mu$m. You may assume that the depth of the diffusion is 4 $\mu$m. Assume $T = 300$ K.

4. An abrupt Si p-n junction is doped with $10^{21}$m$^{-3}$ B atoms in the p-region and $10^{20}$m$^{-3}$ P atoms in the n-region. Calculate
   (a) the barrier potential $V_b$,
   (b) the depletion layer thickness without applied voltage,
   (c) the depletion layer thickness when -10V is applied and
   (d) the depletion layer capacitance with -10V bias,
   if the area is $10^{-8}$ m$^2$.
   Assume $T = 300$ K, $n_i = 1.45 \cdot 10^{16}$m$^{-3}$ and $\varepsilon_r = 11.9$.

5. In the Haynes–Shockley experiment a narrow, pulsed beam of light generates electron-hole pairs in an n-type semiconductor at a distance of 50 mm from a rectifying probe which acts as a detector of minority carriers. If the electric field along the line joining the point of incidence of the light beam and the probe is 1000 V/cm and the detected signal is observed 10 $\mu$s after the light pulse calculate the mobility of the minority carrier.

Answers:

1. (a) $n = 3 \cdot 10^{20}$ m$^{-3}$, $p = 7 \cdot 10^{11}$ m$^{-3}$
   (b) $p = 2 \cdot 10^{22}$ m$^{-3}$, $n = 1.05 \cdot 10^{10}$ m$^{-3}$
   (c) $n = 3.16 \cdot 10^{16}$ m$^{-3}$, $p = 6.64 \cdot 10^{15}$ m$^{-3}$
2. (a) $R_i = 3.92 \cdot 10^6$ $\Omega$
   (b) $R_n = 7.72$ $\Omega$
   (c) $R_p = 21.7$ $\Omega$
3. $N_d = 1.16 \cdot 10^{23}$ m$^{-3}$ or $N_a = 3.36 \cdot 10^{23}$ m$^{-3}$
4. (a) $V_b = 0.51$ V
   (b) W=2.7 $\mu$m
   (c) W=12.3 $\mu$m
   (d) C=0.085 pF
5. $\mu = 0.05$ m$^2$/Vs

# References

1. S. M. Sze, "Physics of Semiconductor Devices", Wiley and Sons (1981).
2. J. Sribar and J. Divkovic Puksec, "Physics of Semiconductor Devices" (in Croatian), Element (1994).
3. J. R. Chelikowsky and M. L. Cohen, Phys. Rev. B **14**, 556 (1976).
4. D. A. Fraser, "The Physics of Semiconductor Devices", Oxford University Press (1986).
5. H. D. Young and R. A. Freedman, "University Physics", Addison Wesley (2000).
6. R. Muncaster, "A Level Physics", Stanley Thomas (1993).
7. J. R. Haynes and W. Shockley, Phys. Rev. **81**, 835 (1951).

# 9   Metal–Semiconductor Contacts

D. I. Pugh

Department of Physics, University of York, York YO1 5DD, United Kingdom

Metal–semiconductor contacts display a range of electrical characteristics from strongly rectifying to ohmic, each having its own applications. The rectifying properties of metal points on metallic sulphides were used extensively as detectors in early radio experiments, while during the second world war the rectifying point contact diode became important as a frequency detector and low level microwave radar detector [1]. Since 1945 the development of metal semiconductor contacts has been stimulated by the intense activity in the field of semiconductor physics and has remained vital in the ohmic connection of semiconductor devices with the outside world. The developments in surface science and the increased use of Schottky barriers in microelectronics has lead to much research with the aim of obtaining a full understanding of the physics of barrier formation and of current transport across the metal-semiconductor interface. Large gain spin electronic devices are possible with appropriate designs by incorporating ferromagnetic layers with semiconductors such as silicon [2]. This inevitably leads to metal-semiconductor contacts, and the impact of such junctions on the device must be considered. In this section we aim to look simply at the physical models that can be used to understand the electrical properties that can arise from these contacts, and then briefly discuss how deviations of these models can occur in practical junctions.

The simplified energy band diagrams for a metal and $n$-type semiconductor are shown in Fig. 9.1. Assuming the metal work function $\phi_m$ is much greater than the semiconductor work function $\phi_s$, then when brought into intimate contact under conditions of thermal equilibrium, electrons pass from the conduction band of the semiconductor into the metal until the Fermi levels equalise. This leaves behind a depletion region in the semiconductor causing band bending and a barrier $\phi_{bn}$.

The diffusion potential $V_i$, or amount by which the bands are bent upwards, is given by

$$V_i = \phi_m - \phi_s \tag{9.1}$$

The bending upwards of the bands in an $n$-type semiconductor produces a barrier to electrons from semiconductor to metal. The barrier height $\phi_b$ as viewed from the metal is usually quoted:

$$\phi_{bn} = V_i + (E_c + E_F) = \phi_m - \chi_s \tag{9.2}$$

Where $\chi_s\{= \phi_s - (E_c - E_F)\}$ is called the electron affinity of the semiconductor. Equation 9.2 is known as the Schottky limit, and was developed independently by Schottky and Mott in 1938 [3]. The drift and diffusion of majority

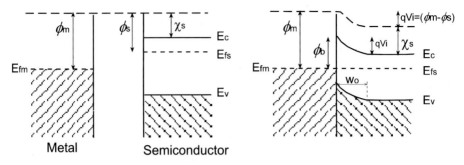

**Fig. 9.1.** Simplified band diagram of a metal and a $n$-type semiconductor, $\phi_m \gg \phi_s$ before (left) and after (right) contact

carriers govern the actual current voltage characteristics of the contact across the depletion region and emission over the barrier. These two processes are essentially in series and the resulting characteristics are governed by which ever causes the most impediment to majority carrier flow [4]. At thermal equilibrium, the rate at which the electrons diffuse across the barrier from the semiconductor to the metal is balanced by the rate at which electrons drift across the barrier in the opposite direction due to the junction electric field; there is no net current (Fig. 9.2a). Applying a forward bias voltage $V_F$ across the contact reduces the depletion region width as the depletion voltage reduces from $V_i$ to $(V_i - V_F)$. The electrons in the semiconductor see a reduced barrier and flow to the metal increases. Negligible voltage appears across the low resistance metal, and so $\phi_b$ and electron flow into the semiconductor remains unchanged. As a result there is a net flow of electrons from the semiconductor to the metal (Fig. 9.2b).

Applying a negative bias $-V_R$ across the contact increases the potential drop across the region by $(V_i + V_R)$ and increases the barrier seen from the semiconductor, while the flow from the metal to the semiconductor remains unchanged. The net effect is a small current flow from the semiconductor to the metal and rectification has occurred (Fig. 9.2c). The resulting contact is known as a Schottky barrier and is firmly established as a diode device in microelectronic technology It allows devices with higher conductances than is possible with $p$-$n$ structures, has a lower turn on voltage (0.2 V for Al/$n$-Si) and is also a majority carrier device enabling faster recovery times and higher frequency applications [5].

The case for an $n$-type semiconductor with $\phi_m \ll \phi_s$ is shown in Fig. 9.3. After contact electrons flow from the metal into the conduction band of the semiconductor, causing a small surface accumulation of electrons on the semiconductor side of the boundary. There is no potential barrier to electrons flowing in either direction. The region of the contact is of low resistance, the highest resistivity region being the bulk semiconductor and any applied voltage appears across this region and does not affect the contact band diagram. The bulk semiconductor resistance determines current flow, and the contact is known as ohmic. These are an important group of metal-semiconductor contacts and are central

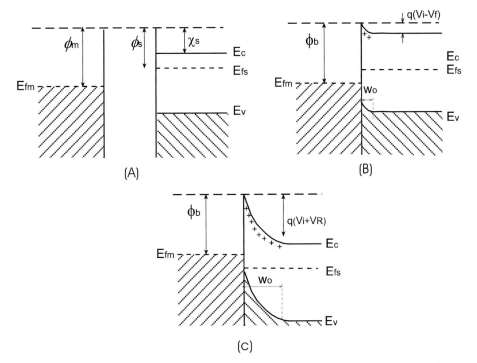

**Fig. 9.2.** (a) metal–semiconductor contact at thermal equilibrium and (b) under a forward applied bias $V_F$ and (c) under applied reverse bias $V_R$.

in silicon and gallium arsenide technologies as interconnects between semiconductor devices and the outside world [5].

In practice, metal semiconductor contacts do not follow the Schottky–Mott model and the barrier height is not proportional to the work function of the metal. The barrier height is instead more complicated and related to one or a combination of the work function of the metal, the electron affinity and resistivity of the semiconductor, barrier reduction due to image force lowering and the nature and density of semiconductor surface states. A high density of surface states, as found in covalent semiconductors such as Si, Ge and GaAs can effectively pin the barrier height and make it completely independent of the metal work function [6]. Barrier heights are also relatively insensitive to the doping level of the semiconductor provided it is below $10^{17}$ cm$^{-3}$ [5]. Knowledge of the microscopic structure of the metal-semiconductor interface and of the interfacial reactions of the metal semiconductor atoms is also necessary to fully characterise metal-semiconductor junctions. When brought into intimate contact the semiconductor may react with the metal to form one or more chemical compounds which affect barrier height. Examples of this type of interface are contacts consisting of metal-silicide-silicon, which are becoming increasingly important in the manufacture of repeatable and reproducible Schottky devices, and are find-

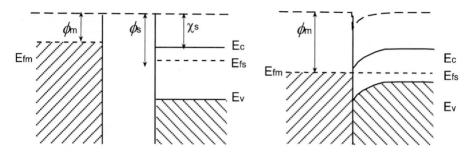

**Fig. 9.3.** Band diagram for a metal and $n$-type semiconductor with $\phi_m \ll \phi_s$ before (left) and after (right) contact. The resulting contact is ohmic.

ing widespread use in modern microelectronics [7]. The reactive interface causes the contact to move into the interior of the silicon lattice, away from surface imperfections and contaminants. As a result the barrier height becomes very stable and reproducible. Silicides can also form low resistance ohmic contacts [5]. In another common type of interface, the natural stable oxide of the semiconductor of up to 20 Å in thickness may be present before and after metal contact is made. This oxide barrier is assumed to be an ideal insulator devoid of any charge, and the potential drop across it is negligible compared with that across the semiconductor depletion region. The oxide is usually thin enough for electrons to easily tunnel through, so this oxide layer does not act as a barrier for electron flow. This interfacial layer effectively decouples the metal from the semiconductor and so each can be treated as a separate system. The surface states are regarded as a property of the particular semiconductor-insulator combination and we can ignore any modification in surface dipole contributions to the metal work function or semiconductor affinity.

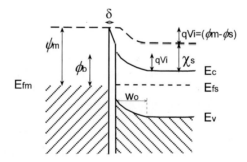

**Fig. 9.4.** The effect of an interfacial layer between the metal and semiconductor.

In summary, the resulting metal-semiconductor interface can have a variety of electrical characteristics from strongly ohmic to highly rectifying. The resulting barrier depends upon the materials used and the nature of the interface between them. This has great implications within any magnetic spin device that aims to combine magnetic metals with semiconductors. The quality of the interface and the effect of parameters such as heat and time on this junction and their effect on the electrical characteristics may define the potential applications of such junctions. The integration of magnetic layers with semiconductors to produce high gain spin electronic devices will inevitably produce metal-semiconductor contacts. Integration with silicon may form silicides, and the magnetic and electrical nature of these layers may be vital when considering spin injection and transport through the structures.

# References

1. B. L. Sharma, "Metal–Semiconductor Schottky Barrier Junctions and their Applications", Plenum Press (1984).
2. J. F. Gregg, W. Allen, R. Kirchmann, C. Sirisathitkul, J. P. Schille, N. Viart, M. Gester, S. M. Thompson, P. Sparks, V. Da Costa, K. Ounadjela, M. Skvarla, J. Magn. Magn. Mater. **175**, 1 (1997).
3. E. H. Rhoderick and R. H. Williams, "Metal–Semiconductor Contacts", Second Edition, Oxford Science Publications (1988).
4. E. H. Rhoderick, "Metal–Semiconductor Contacts", IEE Proc. Vol. **129**, p.1-24 (1982).
5. S. K. Ghandhi, "VLSI fabrication Principles, Silicon and Gallium Arsenide", Second Edition, Wiley and Sons (1994).
6. C. A. Mead, "Solid State Eletronics", Vol. 9 (1966).
7. G. Ottaviani, K. N. Tu, and J. W. Mayer, Phys. Rev. Lett. **44**, 284 (1980).

# 10 Micromagnetic Spin Structure

R. Skomski

Department of Physics and Astronomy and Center for Materials Research and
Analysis, University of Nebraska, Lincoln NE 68588, USA

**Abstract.** Magnetization inhomogenities, associated for example with grain boundaries, give rise to local spin-dependent potentials and affect the magnetoresistance.
The local magnetization $M(r)$ depends on both intrinsic and extrinsic factors. Intrinsic properties, such as spontaneous magnetization and anisotropy, are determined
on an atomic scale and are basically independent of the material's real structure and
history. Extrinsic properties, such as remanence and coercivity, are linked to magnetic hysteresis, realized on mesoscopic or macroscopic length scales, and are strongly
real-structure dependent. The local magnetization $M(r)$, which determines the magnetoresistance, is determined from a nonlinear and nonlocal micromagnetic energy
functional containing the intrinsic properties as parameters. This chapter focuses on
basic micromagnetic effects and on the spin structure at grain boundaries. Continuum
and layer-resolved analytic calculations yield a quasi-discontinuity of the magnetization
between misaligned and in-completely exchange-coupled grains and a disproportionally
large grain-boundary magnetoresistance.

## 10.1 Introduction

Electron scattering in advanced magnetoresistive materials depends on the spin-
dependent potential associated with the local magnetization $M(r)$. In order
to abstract from the atomic origin of the magnetoresistance, which is different
for GMR [1–5], CMR [6–9], and PMR materials [10], we introduce the term
spin-projecting magnetoresistance (SMR). The basic assumption of SMR is that
the magnetoresistance is a unique though generally difficult-to-calculate function of $M(r)$. SMR must be distinguished from ordinary magnetoresistance
and anisotropic magnetoresistance, which reflect Lorentz forces in typically non-
magnetic metals and spin-orbit coupling in transition metals, respectively. Physically, SMR means that the magnetic field alters the mean free paths for ↑ and
↓ channels by modifying the local potential felt by the conduction electrons.

The Bloch character of one-electron wave functions in perfect crystals implies
zero resistivity, but thermal or structural disorder yield finite mean free paths
$\lambda$, finite relaxation times $\tau$, and nonzero resistivities $\rho \propto 1/\tau$. Due to the Pauli
principle, the interaction between electrons depends on the relative spin orien-
tation, so that the local potential and the electron scattering is spin-dependent
(see Chap. 4). Subject to the availability of electronic states – as epitomized by
the density of states at the Fermi level – this mechanism leads to an explicit
magnetization dependence of the resistivity.

In the case of weakly inhomogeneous materials the resistivity is proportional to the square of the gradient of the spin-dependent local potential [11] and therefore proportional to the square of the magnetization gradient. Since magnetization inhomogenities are most pronounced in low and moderate magnetic fields $\boldsymbol{H}$, the resistance may be very large at low fields, whereas high magnetic fields reduce the resistance by aligning the spins.

The determination of the local magnetization (mesoscopic spin structure) is a micromagnetic problem. The traditional term micromagnetic [12] is somewhat unfortunate, because most micromagnetic phenomena – such as magnetic hysteresis – are nanostructural, realized on deep-submicron length scales. Micromagnetic or extrinsic properties reflect the *real structure* (defect structure, morphology, metallurgical microstructure) of a material. By contrast, intrinsic properties, such as spontaneous magnetization and magnetocrystalline anisotropy, refer to perfect crystals.

Micromagnetic problems are usually solved on a continuum level [12–14]. Narrow-wall phenomena, which have been studied for example in rare-earth cobalt permanent magnets [15], involve individual atoms and atomic planes and lead to comparatively small corrections to the extrinsic behavior. However, in the context of spin electronics, grain-boundary related scattering is generally non-negligible [4,8,9,16,17] and involves quite small length scales of about 1 nm [8]. This may lead to a disproportionally strong spin scattering and calls for a comparison of continuum and layer-resolved calculations.

This chapter elaborates basic ideas of magnetism and, in a sense, considers thin films, paramagnetic gases, bulk magnets, small particles, and wires on an equal footing. Sect. 10.2 is a brief summary of the atomic origin of magnetism, Sect. 10.3 deals with fundamental aspects of micromagnetism, and Sect. 10.4 is devoted to grain-boundary and narrow-wall phenomena.

## 10.2   Intrinsic Properties

Intrinsic properties refer to the atomic origin of magnetism and involve quantum phenomena such as exchange, crystal-field interaction, interatomic hopping, and spin-orbit coupling [18–22]. Quantities describing the mesoscopic spin structure, such as the coercivity $H_c$ and the remanence $M_r$, are extrinsic (real-structure related) [14,23–25], but intrinsic properties enter micromagnetic equations as local *micromagnetic parameters*. Table 10.1 shows the magnetic moment $m$, the spontaneous magnetization $M_S$, the Curie temperature $T_C$, and first uniaxial anisotropy constant $K_1$ for some magnetic materials. Not included are antiferromagnets, such as NiO, $GdFeO_3$, and $Ti_2O_3$, whose long-range magnetic order vanishes above the Néel temperature $T_N$, and oxides such as $CrO_2$ (FM), and $Y_3Fe_5O_{12}$ (FIM) (see Chaps. 12 and 6).

### 10.2.1   Magnetic Moment, Exchange, and Magnetization

Magnetic solids contain atoms characterized by a quantum-mechanical *magnetic dipole moment* $\hat{\mathbf{m}} = -\mu_B(\hat{\mathbf{l}} + 2\hat{\mathbf{s}})/\hbar$. Often one considers the net magnetic mo-

**Table 10.1.** Intrinsic and structural properties of some magnetic materials (FM = ferromagnet, FIM = ferrimagnet).

| | $m$ | $\mu_0 M_S$ | $T_C$ | $K_1$ | Comment |
|---|---|---|---|---|---|
| | $\mu_B$/f.u. | T | K | MJ/m$^3$ | |
| Fe | 2.23 | 2.15 | 1044 | 0.05 | Cubic FM |
| Co | 1.73 | 1.81 | 1390 | 0.53 | Hexagonal FM |
| Ni | 0.62 | 0.62 | 628 | -0.005 | Cubic FM |
| SmCo$_5$ | 8.0 | 1.07 | 1020 | 17.2 | Hexagonal FM |
| Nd$_2$Fe$_{14}$B | 37.6 | 1.61 | 585 | 4.9 | Tetragonal FM |
| BaFe$_{12}$O$_{19}$ | 19.9 | 0.47 | 742 | 0.33 | Hexagonal FIM |
| Fe$_3$O$_4$ | 4.0 | 0.63 | 860 | -0.012 | Cubic FIM |

ment $m$ per formula unit, which is measured in $\mu_B$. An alternative way of characterizing a material's net moment is to consider the spontaneous *magnetization* $M_S = \mathrm{d}m/\mathrm{d}V$, measured in A/m, or its flux-density equivalent $\mu_0 M_S$, measured in T. Here $\mathrm{d}V$ is a small volume element containing at least one unit cell. Since thermal excitations tend to disalign the atomic moments, the spontaneous magnetization is temperature-dependent. The zero-temperature spontaneous magnetization $M_S(T = 0)$ is determined by the atomic moments and often denoted by $M_0$.

There are two sources of magnetic moment: currents associated with the orbital motion of the electrons (orbital moment) and the electron spin (spin moment). Solid-state magnetism originates from the partly filled inner electron shells of transition-metal atoms. Of particular importance are the $3d$ iron-series elements, in particular Fe, Co, and Ni, and the $4f$ rare-earth elements, such as Nd, Sm, Gd, and Dy. On the other hand, $4d$ palladium-series elements, $5d$ platinum series elements, and actinide elements, such as U, have a magnetic moment in suitable crystalline environments.

The magnetic moment of iron-series transition-metal atoms in metals (Fe, Co, Ni, YCo$_5$) and non-metals (Fe$_3$O$_4$, NiO) is given by the *spin*, so that the moment, measured in $\mu_B$, is equal to the number of unpaired spins. The reason is that the orbital moment is largely quenched (destroyed) by the crystal field, although the small residual orbital moment (of the order of $0.1\mu_B$) is important in the context of magnetic anisotropy. Rare-earth atoms keep their orbital moments in metals and non-metals, because their partly filled shells lie deep inside the atoms and are not very much affected by the crystal field.

Figure 10.1 illustrates that the net magnetic moment depends on the type of zero-temperature magnetic order. In ferromagnets the atomic moments add, whereas ferrimagnets and antiferromagnets are characterized by two (or more) sublattices with opposite moments. This amounts to a reduction of the net moment (ferrimagnetism) or to the absence of a net moment (antiferromagnetism).

In order to understand moment formation and magnetic order one has to start from the many-electron Schrödinger equation (see Chaps. 2 and 5). The solution of that equation is complicated by Coulomb interactions of the type $1/|\boldsymbol{r} - \boldsymbol{r}'|$, where $\boldsymbol{r}$ and $\boldsymbol{r}'$ are the positions of the interacting electrons. For non-interacting electrons, the many-electron wave function factorizes, but the Coulomb repulsion makes such a separation impossible and gives rise to a variety of intra- and interatomic exchange contributions.

In the case of two electrons and two atomic sites, the problem reduces to the discussion of three parameters: the *hopping parameter t*, the *Coulomb* energy $U$ necessary to add a second electron into an atomic orbital, and the *direct exchange* $J_D$ [14]. The direct exchange is always positive, but for interatomic distances of interest it is not larger than about 0.1 eV, that is smaller than $U$ and $t$ by at least one order of magnitude. Comparing the energies of the lowest-lying $\uparrow\uparrow$ and $\uparrow\downarrow$ states yields the *effective exchange*

$$J_{\text{eff}} = J_D + \frac{U}{4} - \sqrt{t^2 + \frac{U^2}{16}} \qquad (10.1)$$

From this equation we see that the Coulomb repulsion $U$ and the direct exchange $J_D$ favor ferromagnetism ($J_{\text{eff}} > 0$), whereas interatomic hopping ($t$) tends to destroy ferromagnetism. The reason is that $\uparrow\downarrow$ electron pairs in an atomic orbital are unfavorable from the point of view of Coulomb repulsion, whereas parallel spin alignment $\uparrow\uparrow$ is favorable, because the Pauli principle implies that the two electrons are in different orbitals. However, this energy gain has to compete against a hopping-related increase in one-electron energies.

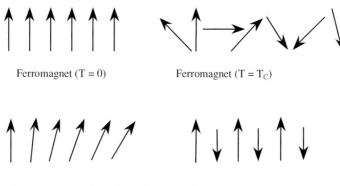

Ferromagnet (T = 0)      Ferromagnet (T = T$_\text{C}$)

Micromagnetic configuration (T = 0)    Ferrimagnet (T = 0)

**Fig. 10.1.** Magnetic order (schematic). Magnetoresistance involves both zero-temperature and finite temperature magnetic ordering. Much of the fascination of advanced magnetoresistive phenomena is based on the intriguing interplay between microscopic and mesoscopic physics (micromagnetism).

In oxides, $t \ll U$ and (10.1) yields $J_{eff} = J_D - 2t^2/U$. Due to the small-ness of the direct exchange, oxides are often antiferromagnets, but when $t = 0$ by symmetry, as in $CrO_2$, then $J_D$ gives rise to ferromagnetism (Goodenough-Kanamori rules, see Chaps. 5 and 12). In $3d$ metals, $t \gg U$ and (10.1) yields $J_{eff} = J_D + U/4 - t$. Since $U$ is a largely atomic property, metallic ferromagnetism is realized for not-too-large hopping, that is for narrow bands (Stoner criterion, see Chap. 2).

In order to discuss magnetic order, it is necessary to distinguish between intra-atomic and interatomic exchange interactions. Intra-atomic exchange is responsible for the formation of atomic moments, whereas interatomic exchange favors a ferromagnetic (or antiferromagnetic) alignment of neighboring spins. Typically, the intra-atomic exchange $J_{intra}$ is much larger than the interatomic exchange $J_{inter} = J$, and atomic moments tend to be quite stable. By compari-son, it is comparatively easy to disalign neighboring spins by thermal excitation and – to a lesser extent – by inhomogeneous magnetic fields and polycrystalline random-anisotropy contributions.

A widely-used approach to discuss interatomic exchange is the *Heisenberg* interaction $-J\hat{s}_1 \cdot \hat{s}_2$ between neighboring spins $\hat{s}_1$ and $\hat{s}_2$, where $J$ is an inter-atomic exchange constant. The derivation of the spontaneous magnetization $M_S$ of a solid from the corresponding Heisenberg Hamiltonian is a very complicated problem, but a number of approximations (normalized classical spins, restriction to nearest-neighbor interactions, mean-field approximation) lead to the simple result that $M_S$ vanishes above the *Curie temperature* $T_C = zJ/3k_B$, where z is the number of nearest neighbors.

Strictly speaking, the applicability of the Heisenberg model is limited to local-moment magnets, such as insulating transition-metal oxides and rare-earth met-als. In $3d$ metals, the magnetic moment is a band-structure property, involving at least a few neighboring atoms [14,26]. This leads to non-integer moments per atom, may yield moment and exchange-constant corrections at grain boundaries and interfaces, and means that quantities such as $J$ and $m$ should be considered as atomic parameters.

As indicated in Fig. 10.1, the vanishing of the spontaneous magnetization at $T_C$ reflects the thermally activated rotational misalignment of the atomic mo-ments. By contrast, the magnitude of the atomic moments remains largely un-changed [27]. The reason is that atomic moments are supported by intra-atomic exchange energies of the order of 1 eV ($10^4$ K), whereas the interatomic exchange does not exceed about 0.1 eV. This scenario is realized in both metals and non-metals, although interatomic hopping in itinerant metals, such as iron, may yield short-range correlations at and above $T_C$. The magnetization $M_S$ considered in micromagnetism is usually *averaged* over a few interatomic distances and can be regarded as a temperature-dependent but field-independent material constant (micromagnetic parameter). This means, in particular, that micromagnetic phe-nomena, such as domain formation and hysteresis, are realized by magnetization *rotations*.

Aside from long-range critical fluctuations in the vicinity of $T_C$, the spontaneous magnetization is caused by atomic-scale exchange interactions. However, there is also an exchange energy associated with micromagnetic magnetization rotations, such as domains. The local magnetization can be written as $M(r) = M_S(T)s(r)$, where $s(r)$ is the unit vector giving the local magnetization direction. Heisenberg exchange means that spin misalignment in ferromagnets $(J > 0)$ costs exchange energy. On a continuum level, the normalized magnetization $s_{1/2} = s(r) \pm b\, \partial s(r)/\partial x$ of two neighbouring atoms located at $r_{1/2} = \pm be_x$ correspond to the exchange energy

$$-Js_1 \cdot s_2 = -J + Jb^2 \left(\frac{\partial s}{\partial x}\right)^2 \tag{10.2}$$

More generally, any magnetization inhomogenity is punished by an exchange energy density

$$\frac{dE_{ex}}{dV} = A\,(\nabla s)^2 \tag{10.3}$$

where the *exchange stiffness* $A$ is of the order of 10 pJ/m ($10^{-11}$ J/m) for typical ferromagnets, see Table 10.2.

## 10.2.2   Anisotropy

The energy of a magnetic solid depends on the orientation of the magnetization with respect to the crystal axes, which is known as magnetic *anisotropy*. The anisotropy of permanent magnets is high in order to keep the magnetization in a desired direction, whereas soft magnets are characterized by a very low anisotropy. Materials with moderate anisotropy are often used as magnetic-recording media. In the field of magnetoresistance, anisotropy is a double-edged issue: high anisotropies enhance the magnetization gradient and the magnetoresistance, but they also make the material more difficult to magnetize.

It is convenient to write the magnetization as

$$M = M_S\left[\sin(\theta)\sin(\varphi)e_x + \sin(\theta)\cos(\varphi)e_y + \cos(\theta)e_z\right]. \tag{10.4}$$

The simplest anisotropy-energy expression is then

$$E_a = K_1 V \sin^2(\theta), \tag{10.5}$$

where $K_1$ is the *first uniaxial anisotropy constant* and $V$ is the magnet volume [28]. Equation (10.5) is widely used to describe uniaxial magnets (hexagonal, tetragonal, and rhombohedral crystals) and small ellipsoids of revolution (fine particles). For $K_1 > 0$ the easy magnetic direction is along the c- (or z-) axis, which is called *easy-axis anisotropy*, whereas $K_1 < 0$ leads to *easy-plane anisotropy*, where the easy magnetic direction is anywhere in the a-b- (or x-y-) plane. In cubic magnets there is no unique z-axis, but (10.5) can be used for small angles $\theta$ (see below). For very low symmetry (orthorhombic, monoclinic, and triclinic), the first-order anisotropy energy can be written as

$$E_{\rm a} = K_1 V \sin^2(\theta) + K_1' V \sin^2(\theta) \cos(2\varphi) \,, \qquad (10.6)$$

where $K_1$ and $K_1'$ are, in general, of comparable magnitude. This expression must also be used for magnets having a low-symmetry shape, such as ellipsoids having three unequal principal axes, and for a variety of surface anisotropies, such as that of bcc (011) surfaces.

An expression including second order anisotropy constants is [28]

$$\frac{E_{\rm a}}{V} = K_1 \sin^2(\theta) + K_2 \sin^4(\theta) + K_2' \sin^4(\theta) \cos(4\varphi) \,. \qquad (10.7)$$

This equation describes tetragonal, hexagonal, rhombohedral and cubic crystals. Hexagonal and rhombohedral crystals are characterized by $K_2' = 0$ (fourth-order uniaxial anisotropy), whereas in the tetragonal case $K_2$ and $K_2'$ are of the same order of magnitude.

The anisotropy of cubic crystals is often written as

$$\frac{E_{\rm a}}{V} = K_1^{\rm c} \left( \alpha_1^2 \alpha_2^2 + \alpha_2^2 \alpha_3^2 + \alpha_3^2 \alpha_1^2 \right) + K_2^{\rm c} \, \alpha_1^2 \alpha_2^2 \alpha_3^2 \,, \qquad (10.8)$$

where $\alpha_1 = \cos(\theta)$, $\alpha_2 = \sin(\theta) \cos(\varphi)$, and $\alpha_3 = \sin(\theta) \sin(\varphi)$ are the direction cosines of the magnetization direction. Analysis of (10.8) shows that $K_1^{\rm c} > 0$ favors the alignment of the magnetization along the (001) cube edges, which is called iron-type anisotropy, whereas $K_1^{\rm c} < 0$ corresponds to an alignment along the (111) cube diagonals referred to as nickel-type anisotropy. Comparison of (10.7) and (10.8) yields $K_2 = -7K_1^{\rm c}/8 + K_2^{\rm c}/8$ and $K_2' = -K_1^{\rm c}/8 - K_2^{\rm c}/8$. These relations mean (i) that the constant $K_1$ in cubic materials reflects fourth-order crystal-field interactions [14] and (ii) that there are only two independent constants when (10.7) is applied to cubic magnets [14]. Typical $K_2^{\rm c}$ values are 0.015, 0.05, and 0.28 MJ/m$^3$ for Fe, Ni, and Fe$_3$O$_4$, respectively.

By definition, there are no odd-order terms in (10.6)–(10.8). Odd-order anisotropies may be caused by relativistic Moriya-Dzialoshinskii interactions, exchange biasing, or particular micromagnetic regimes [14,29]. This refers in particular to uni-*directional* anisotropies of the type $K_{\rm ud} \cos(\theta)$, which correspond to a hysteresis-loop shift.

With respect to the *physical origin of anisotropy* it is necessary to distinguish between magnetostatic and magnetoelectric anisotropies. Magnetostatic interactions give rise to shape anisotropy, which is illustrated in Fig. 10.2: the magnetostatic energy of the spin configuration (a) is lower than that of the configuration (b), so that the easy magnetization corresponds to the lowest magnetostatic energy. For fine particles (see below), the shape-anisotropy contribution to $K_1$ is

$$K_{1,\rm sh} = \frac{\mu_0}{4} \left( 1 - 3N \right) M_{\rm s}^2 \,, \qquad (10.9)$$

where $N$ is the demagnetizing factor of the particle ($N = 0$ for long cylinders, $N = 1/3$ for spheres, and $N = 1$ for plates). Note that (10.9) and the simplified picture Fig. 10.2 do not apply to large particles, where the exchange stiffness $A$ is not able to ensure a uniform (coherent) spin orientation throughout the magnet (Sect. 10.3.3).

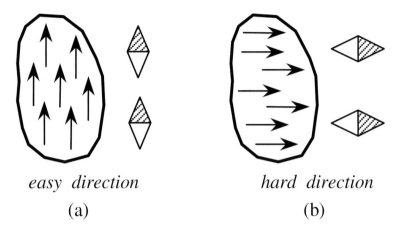

*easy direction*

(a)

*hard direction*

(b)

**Fig. 10.2.** Shape anisotropy of cubic magnets (schematic). The configuration (a) is energetically more favorable than the configuration (b), as one can deduce from the compass-needle analogy.

In non-cubic magnets there is also a magnetostatic contribution to the bulk anisotropy. However, in most materials the bulk or *magnetocrystalline* anisotropy reflects the competition between the spin-orbit coupling and the electrostatic crystal-field interaction (magnetoelectric anisotropy) [30]. The crystal field reflects the local symmetry of the crystal and acts on the orbits of the electrons in the partly filled inner shells. The anisotropy is then realized by the coupling of the orbital moments to the spins by the relativistic spin-orbit coupling $H_{SO} = \lambda_{SO}\hat{\mathbf{l}} \cdot \hat{\mathbf{s}}$. The spin-orbit coupling has two consequences: (i) it couples the magnetization (the spin) to the orbital motion of the electrons and (ii) it creates a small orbital moment in largely quenched magnets. Quenched wave functions correspond to standing waves of the type $\cos(2\varphi)$ and are favorable from the point of view of electrostatic crystal-field interaction, because they are able to adapt to the crystal field, but due to the standing-wave character of the quenched wave function the orbital moment and the anisotropy are zero. By contrast, unquenched wave functions, such as $\exp(2i\varphi)$, do not split in the crystal field benefit from the spin-orbit coupling, because their running-wave character amounts to a circular current.

Depending on the relative strengths of the crystal-field and spin-orbit interactions there are two limits of interest. Rare-earth $4f$ electrons are close to the atomic core and exhibit a strong spin-orbit coupling, whereas the crystal field felt by the $4f$ electrons is rather small. This means a rigid coupling between spin and orbital moment, and the magnetocrystalline anisotropy is given by the electrostatic interaction of the generally aspherical $4f$ charge cloud with the crystal field [14]. Although the $4f$ crystal field is much smaller than the crystal-field acting on iron-series $3d$ electrons, it creates a high rare-earth anisotropy contribution (Table 10.1). The much smaller anisotropy of $3d$ magnets is explained by the quenching of the orbital moment due to the crystal field. In the limit of complete quenching, $\langle \hat{l} \rangle = 0$ and $K_1 = 0$, but in reality the weak $3d$ spin-orbit coupling acts as a perturbation and yields some admixture of running-wave character, a small residual orbital moment, and some anisotropy.

In order to illustrate the origin of the $3d$ anisotropy we consider two $d$ orbitals, such as $|\Psi_1\rangle = |xy\rangle$ and $|\Psi_2\rangle = |x^2 - y^2\rangle$. The Hamiltonian is

$$H = \begin{pmatrix} A_0 & 0 \\ 0 & -A_0 \end{pmatrix} + 2\lambda_{SO}\cos(\theta)\begin{pmatrix} 0 & i \\ -i & 0 \end{pmatrix}, \tag{10.10}$$

where the crystal-field parameter $A_0$ describes the electrostatic energy of the two orbitals in the crystal field, $\cos(\theta)$ is the angle between spin direction and z-axis and the factor 2 is the magnetic quantum number of the $d$ orbitals. Diagonalization of (10.10) yields the energy eigenvalues

$$E_\pm = \pm\sqrt{A_0^2 + 4\lambda_{SO}^2 \cos^2(\theta)}. \tag{10.11}$$

By expanding $E_-$ into powers of the small quantity $\lambda_{SO}^2/A_0^2$ we obtain the second-order anisotropy energy

$$E_a = \frac{2\lambda_{SO}^2}{A_0}\sin^2(\theta). \tag{10.12}$$

An equation of this type was first derived by Bloch and Gentile [30]. The corresponding orbital moment scales as $\lambda_{SO}\mu_B/A_0$ [14]. Note that the qualitative result (10.12) applies to both metallic and non-metallic $3d$ magnets, but in metals the crystal-field splitting must be replaced by the band width [21]. To make quantitative predictions one has to extend (10.10) by including all occupied $3d$ orbitals and all unperturbed crystal-field or band-structure states.

The magnetocrystalline anisotropy is closely related to the *magnetoelastic* anisotropy, because strained crystals can be regarded as unstrained crystals having slightly different atomic positions. Magnetoelastic anisotropy is particularly important in cubic magnets, where uniaxial stress gives rise to uniaxial anisotropy contributions. The magnetoelastic contribution to the first anisotropy constant is (see e.g. [14])

$$K_{1,me} = \frac{3\lambda_S\sigma}{2}, \tag{10.13}$$

where $\sigma$ is the uniaxial stress and $\lambda_S$ is the saturation magnetostriction. Experimental room-temperature values of $\lambda_S$ are $-7 \times 10^{-6}$ for iron, $-33 \times 10^{-6}$ for nickel, $40 \times 10^{-6}$ for $Fe_3O_4$, $-1560 \times 10^{-6}$ for $SmFe_2$, $75 \times 10^{-6}$ for FeCo, and practically zero for $Fe_{20}Ni_{80}$ (permalloy).

As the spontaneous magnetization, anisotropy constants are temperature dependent: atomic excitations lead to the occupation of excited levels, and in the limit of very high temperatures all levels are occupied with equal probability (zero anisotropy). Note that the temperature equivalent of anisotropy energies per atom does not exceed about 1 K, but the switching of individual spins into states with reduced anisotropy is largely suppressed by the strong inter-atomic exchange.

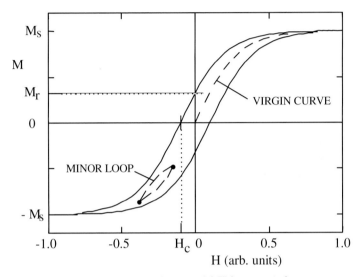

**Fig. 10.3.** Typical major M-H hysteresis loop.

## 10.3   Basic Micromagnetism

As mentioned in the introduction, magnetic properties derived from the hysteresis loop are *extrinsic* properties, because they describe the real structure of the magnet rather than the atomic (intrinsic) behavior. Figure 10.3 shows a typical *M-H* hysteresis loop. Note that hysteresis loops are usually corrected for the demagnetizing field $-NM$ by plotting the magnetization as a function of the internal field $H - NM$. In general, this skewing (shearing) correction makes the hysteresis loops more rectangular. *Major* or limiting hysteresis loops are obtained by starting from a fully aligned magnet where $M(r) = M_S e$. This is achieved by applying a large positive field. The loop is then obtained by monitoring the volume-averaged magnetization as a function of the external magnetic field $H$. *Minor loops* are obtained if the maximum applied field $\pm H$ is insufficient for complete saturation. They lie inside the major loop and therefore include a smaller area than the major loop. *Virgin curves* (initial curves) are obtained on increasing $H$ from zero after thermal demagnetization, that is after heating beyond $T_C$. *B-H* hysteresis loops are used, for example, to determine the energy product of permanent magnets [14].

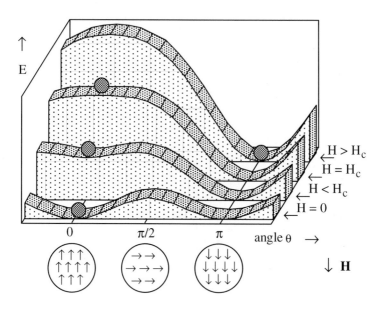

**Fig. 10.4.** Origin of coercivity: hysteresis involves metastable energy minima.

The most important extrinsic properties are the remanent magnetization or *remanence* $M_r$ which remains in a magnet after switching off a large magnetic field and the coercive force or *coercivity* $H_c$ that is the reverse field at which the average magnetization vanishes. Coercivity describes the stability of the remanent state and gives rise to the classification of magnets into hard magnetic materials (permanent magnets), semi-hard materials (storage media), and soft magnetic materials. Modern permanent magnets exhibit broad hysteresis loops with coercivities of order 1 T (0.8 MA/m), whereas semi-hard materials used in storage media exhibit narrow but rectangular hysteresis loops having coercivities of the order 0.05 T (40 kA/m). The coercivity of storage media is sufficient to assure the remanence of the stored information without requiring powerful and bulky writing facilities. Other extrinsic properties, such as the permanent-magnet energy product and loop squareness, go beyond the exclusive consideration of $M_r$ and $H_c$. The strong real-structure dependence of extrinsic properties is seen, for example, from the fact that the coercivity of technical iron doubles by adding 0.01 wt.% nitrogen [25]. The reason is that the interstitial nitrogen yields a local modification of $K_1$ which has a disproportionally strong impact on the motion of domain walls.

The explanation and determination of extrinsic properties is generally very complicated, and only in a few cases it is possible to use simple hysteresis models. One example is the coherent-rotation or Stoner–Wohlfarth model, which describes the hysteretic behavior of a uniformly magnetized particle (Fig. 10.4). However, truly one-dimensional energy landscapes, such as $E(\theta)$ in Fig. 10.4, are rarely encountered in practice. Most magnetization processes of interest are *incoherent*, and the associated energy landscape is multidimensional. For example, the applicability of the Stoner–Wohlfarth theory is limited to very small particles, and even in the case of single-domain particles (Sect. 10.3.3) the magnetization reversal may be incoherent. A further complication is that the involved micromagnetic equations are nonlocal and nonlinear, and only in a few cases it has been possible to obtain physically transparent solutions.

## 10.3.1   Coherent Rotation

The *Stoner–Wohlfarth* model [31] assumes that the magnetization remains coherent (uniform) throughout the magnet, as in Figs. 10.2 and 10.4. This is justified for very small particles or very thin films or wires, where the interatomic exchange is able to keep the spins parallel throughout the magnet (see Sect. 10.3.3). Incorporating the shape anisotropy into $K_1$, the magnetic energy of an aligned uniaxial Stoner–Wohlfarth particle is

$$\frac{E}{V} = K_1 \sin^2(\theta) + K_2 \sin^4(\theta) - \mu_0 M_S H \cos(\theta), \qquad (10.14)$$

where $H$ is the external magnetic field, applied in the z-direction, and $M_z = M_S \cos(\theta)$. The last term in this equation is the *Zeeman energy* $-\mu_0 \, \boldsymbol{m} \cdot \boldsymbol{H}$, which describes the interaction of a magnetized body with the external field.

Putting $H = 0$ in (10.14) yields a variety of zero-field spin configurations. When both $K_1$ and $K_2$ are positive, then minimization of (10.14) yields easy-axis anisotropy ($\theta = 0$). On the other hand, when both $K_1$ and $K_2$ are negative, then the magnetization lies in the basal plane: easy-plane anisotropy, $\Theta = \pi/2$. A particularly interesting regime is the *easy-cone* magnetism occurring if the conditions $K_1 < 0$ and $K_2 > -K_1/2$ are satisfied simultaneously [14,29]. The tilt angle between the z-axis and the easy magnetization direction is given by

$$\theta_c = \arcsin\left(\sqrt{\frac{|K_1|}{2K_2}}\right). \qquad (10.15)$$

Since the temperature dependences of $K_1$ and $K_2$ are generally different ($K_2$ is often negligible at high temperatures), the preferential magnetization direction may change upon heating (spin-reorientation transition). A similar film-thickness dependent transition is observed in films where surface and bulk anisotropy contributions compete.

For $K_2 = 0$, stability analysis of (10.14) yields the coherent-rotation *nucleation field*

$$H_N = \frac{2K_1}{\mu_0 M_S} \qquad (10.16)$$

at which the $\theta = 0$ state ($M_z = M_S$) becomes unstable. In terms of Fig. 10.4, this instability refers to the vanishing of the local energy minimum at $H_c = H_N$ and leads to a rectangular hysteresis loop.

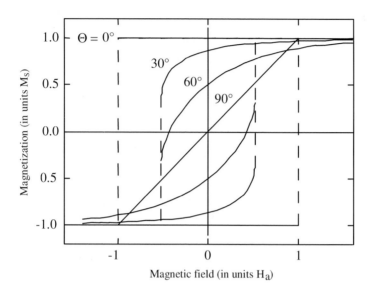

**Fig. 10.5.** Dependence of the magnetization on the angle $\theta$ between field and easy axis for a uniaxial magnet. Dashed lines indicate magnetization jumps.

Equation (10.16) translates the anisotropy constant $K_1$ into a field quantity, namely the *anisotropy field* $H_a = H_N$. It may be used as a coercivity estimate, although it almost invariably overestimates the coercivity by one order of magnitude. This discrepancy, known as Brown's paradox, is explained by the prevalence of incoherent magnetization processes in real magnets. For anisotropy fields in more complicated magnets see Ref. [14].

Many materials of interest in spin electronics are polycrystallites (nanocrystallites) or powders. Ignoring interparticle interactions, which are discussed in the following sections, we can describe those materials as ensembles of Stoner–Wohlfarth particles, characterized by a coherent rotation of the magnetization. Figure 10.5 shows hysteresis loops of uniaxial Stoner–Wohlfarth particles for different angles $\theta$ between the applied magnetic field and the crystallite's c-axis. The magnetic behavior of the material is then obtained as a superposition of Stoner–Wohlfarth loops. For uniaxial magnets the resulting remanence $M_r = M_S/2$, whereas for iron-type ($K_1 > 0$) and nickel-type ($K_1 < 0$) cubic magnets, $M_r/M_S$ equals 0.832 and 0.866, respectively. In the case of uniaxial magnets, the coercivity is equal to $0.479 H_a$.

## 10.3.2   Domains and Domain Walls

Until now we have neglected the mutual magnetostatic dipole interaction between atomic moments. The magnetostatic dipole field created by a magnet's own magnetization is given by

$$\boldsymbol{H}_{\mathrm{d}}(\boldsymbol{r}) = \frac{1}{4\pi} \int \mathrm{d}V' \frac{3(\boldsymbol{r} - \boldsymbol{r}')(\boldsymbol{r} - \boldsymbol{r}') \cdot \boldsymbol{M}(\boldsymbol{r}') - |\boldsymbol{r} - \boldsymbol{r}'|^2 \boldsymbol{M}(\boldsymbol{r}')}{|\boldsymbol{r} - \boldsymbol{r}'|^5} . \qquad (10.17)$$

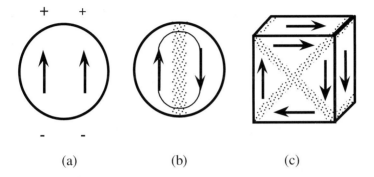

(a)                    (b)                    (c)

**Fig. 10.6.** Magnetostatic self-energy and flux closure (schematic). As implied by (10.17), the spin configurations (b) and (c) are more favorable than the configuration (a). The transition between domains is realized by a domain wall (grey area).

Due to a self-interaction contribution, this field differs by $\boldsymbol{M}/3$ from the internal magnetostatic field obtained from Maxwell's equations. However, magnetic fields couple as $\boldsymbol{M} \cdot \boldsymbol{H}$ to the magnetization, so that any term proportional to $\boldsymbol{M} \cdot \boldsymbol{M} = M_S^2$ amounts to a physically irrelevant shift of the zero-point of the self-interaction energy $-(1/2)\mu_0 \int \boldsymbol{H}_{\mathrm{d}} \cdot \boldsymbol{M} \mathrm{d}V$ [32], and the physics of magnetostatic self-interaction is fully contained in (10.17).

By expressing (10.17) in terms of the magnetic charge density $-\nabla \cdot \boldsymbol{M}$ it can be shown that the magnetostatic self-interaction energy is particularly low when there are no magnetic charges at the magnet's surface. From Fig. 10.6 we see that the absence of surface charges is linked to flux closure in the magnet. More generally, magnetostatic interactions tend to yield magnetic *domains* of opposite magnetization directions [23,24,33,34]. This explains why the net magnetization $\langle \boldsymbol{M}(\boldsymbol{r}) \rangle$ of many magnets is equal to zero, despite $\boldsymbol{M}(\boldsymbol{r})^2 = M_S^2$ throughout the magnet. For example, two pieces of soft iron do not attract each other, and to exert a force on a soft magnet one needs to destroy the domains by an external field. There are many different domain patterns of interest in the context of magnetoresistance (see particularly Chaps. 14 and 15).

A common feature of all domain structures is that the domains are separated by comparatively sharp *domain walls* [23,24]. The reason for the formation of

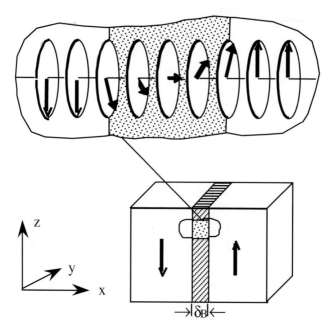

**Fig. 10.7.** Bloch wall lying in the y-z-plane (schematic). The arrows show the local magnetization direction.

domain walls is that the magnetization inside the domains lies along easy directions, whereas the transition between two easy magnetization directions involves energetically unfavorable spin orientations. Magnetocrystalline anisotropy favors narrow domain walls, but (10.5) shows that narrow walls, that is large magnetization gradients, are unfavorable from the point of view of exchange.

The domain-wall width is estimated very easily from dimensional arguments [23]. The domain-wall width is determined by the anisotropy constant $K_1$ and the exchange stiffness $A$, which are measured in $J/m^3$ and $J/m$, respectively, so that the only length and the only wall energy derivable from these parameters are the wall-width parameter $\delta_0 = (A/K_1)^{1/2}$ and the wall-energy parameter $\gamma_0 = (K_1 A)^{1/2}$, respectively. This means that the domain-wall thickness tends to be much larger than the interatomic spacing but is much smaller than typical domain sizes.

In order to make quantitative predictions one has to consider specific wall geometries. Figure 10.7 shows a $K_1$-only Bloch wall, where the *wall thickness* $\delta = \pi\delta_0$ and the wall energy $\gamma = 4\gamma_0$ [14,35]. The 180° Bloch wall shown in Fig. 10.7 is frequently encountered in uniaxial magnets. Other important wall configurations are thin-film 180° Néel walls, where the magnetization vector remains in a plane (in the z-x-plane in Fig. 10.7), and 90° walls observed in cubic crystals. Typical domain-wall widths are 5 nm and 100 nm for hard and soft magnetic materials, respectively.

The magnetostatic dipole interaction favors domain formation, but since the creation of domain walls costs energy, there are no walls if the gain in magnetostatic energy is smaller than the wall energy. For example, the wall in Fig. 10.7b – indicated by the dotted line – corresponds to a wall energy $\gamma \pi R^2$. The competing gain in magnetostatic energy is roughly equal to half the single-domain energy, that is $\mu_0 M_S^2 V / 12$, so that domain formation is favorable for particles whose radius exceeds a *critical single-domain radius*

$$R_{\mathrm{sd}} \simeq \frac{36\sqrt{AK_1}}{\mu_0 M_S^2} . \tag{10.18}$$

This value varies between a few nm in soft magnets and about 1 μm in hard magnets.

It is important to note that the critical single-domain radius is an equilibrium property and therefore largely unrelated to hysteresis. As illustrated in Fig. 10.4, hysteresis involves energy barriers and metastable states, and in hard magnetic materials, where $K_1$ is large, the structural length scales associated with hysteresis and coercivity are much smaller than $R_{\mathrm{sd}}$.

The critical single-domain radius can also be written as $R_{\mathrm{sd}} = 36\kappa l_{\mathrm{ex}}$, where

$$\kappa = \sqrt{\frac{K_1}{\mu_0 M_S^2}} \tag{10.19}$$

is the magnetic *hardness* parameter and

$$l_{\mathrm{ex}} = \sqrt{\frac{A}{\mu_0 M_S^2}} \tag{10.20}$$

is the *exchange length*. The exchange length $l_{\mathrm{ex}}$ is the length below which atomic exchange interactions dominate typical magnetostatic fields. For example, we will see that $l_{\mathrm{ex}}$ determines the coherence radius $R_{\mathrm{coh}}$ below which interatomic exchange is able to ensure coherent rotation. It also determines the thickness of soft-magnetic films below which Néel walls are energetically more favorable than Bloch walls and the grain size of two-phase magnets below which the hysteresis loops look single-phase like.

Table 10.2 shows typical micromagnetic parameters. Note that magnetically very hard and very soft materials are characterized by $\kappa \gg 1$ and $\kappa \ll 1$, respectively, whereas $l_{\mathrm{ex}} = 3$ nm for a broad range of magnetic materials.

## 10.3.3   Hysteresis and Coercivity

In order to explain the hysteresis loop of magnetic materials one needs to trace the local magnetization $\boldsymbol{M}(\boldsymbol{r}) = M_S \boldsymbol{s}(\boldsymbol{r})$ as a function of the applied field $H$. The starting point is the magnetic energy functional $E_{\mathrm{m}}$ obtained by adding the exchange energy (10.3), the anisotropy energy, the magnetostatic self-energy, as

**Table 10.2.** Micromagnetic parameters at room temperature. (The values for Fe and Ni are uniaxial estimates).

| Material | $\mu_0 M_S$ | $A$ | $K_1$ | $\delta$ | $\gamma$ | $l_{ex}$ | $\kappa$ | $R_{sd}$ | $H_0$ |
|---|---|---|---|---|---|---|---|---|---|
|  | T | pJ/m | MJ/m$^3$ | nm | mJ/m$^2$ | nm |  | nm | T |
| Fe | 2.15 | 8.3 | 0.05 | 40 | 2.6 | 1.5 | 0.12 | 6 | 0.06 |
| Co | 1.76 | 10.3 | 0.53 | 14 | 9.3 | 2.0 | 0.46 | 34 | 0.76 |
| Ni | 0.61 | 3.4 | -0.005 | 82 | 0.5 | 3.4 | 0.13 | 16 | 0.03 |
| BaFe$_{12}$O$_{19}$ | 0.47 | 6.1 | 0.33 | 14 | 5.7 | 5.9 | 1.37 | 290 | 1.8 |
| SmCo$_5$ | 1.07 | 22.0 | 17 | 3.6 | 77 | 4.9 | 4.35 | 764 | 40 |
| Nd$_2$Fe$_{14}$B | 1.61 | 7.7 | 4.9 | 3.9 | 25 | 1.9 | 1.54 | 107 | 7.6 |

implied by (10.17), and the Zeeman energy $-\mu_0 \boldsymbol{M} \cdot \boldsymbol{H} V$ describing the interaction with the applied field. For $K_1$-only uniaxial magnets we obtain

$$E_\mathrm{m} = \int \left[ A \left( \frac{\nabla \boldsymbol{M}}{M_S} \right)^2 - K_1 \left( \frac{\boldsymbol{n} \cdot \boldsymbol{M}}{M_S} \right)^2 - \frac{1}{2} \mu_0 \boldsymbol{M} \cdot \boldsymbol{H}_\mathrm{d}(\boldsymbol{M}) - \mu_0 \boldsymbol{M} \cdot \boldsymbol{H} \right] \mathrm{d}V \tag{10.21}$$

where $\boldsymbol{n}(\boldsymbol{r})$ is a unit vector denoting the crystallite's easy axis.

As illustrated in Fig. 10.4, hysteresis indicates difficulties in reaching the global (free) energy minimum. As a crude rule – and aside from the Stoner–Wohlfarth-like reversal in weakly interacting particle ensembles – there are two main coercivity mechanisms: nucleation and pinning. *Nucleation* determines the coercivity of nearly homogeneous magnets and means that the magnetization reversal occurs immediately after the original magnetization state becomes unstable. Examples are the Stoner–Wohlfarth nucleation field (10.16), which describes the reversal of the magnetization of an isolated small particle, and the localized nucleation [38] in submicron particles. *Pinning* governs the magnetization reversal in strongly inhomogeneous magnets and means that the coercivity is determined by the interaction of domain walls with structural inhomogenities. To realize magnetization reversal in pinning-controlled magnets, the reverse external field must be larger than some (de)pinning (or propagation) field. One typical pinning mechanism involves inhomogenities whose anisotropy constant is higher than that of the main phase: since high anisotropies yield high domain-wall energies, the penetration of the wall into the highly anisotropic regions is energetically unfavorable. This mechanism is also known as repulsive pinning, whereas the capturing of a wall in a low-anisotropy region is referred to as attractive pinning.

The trapping of walls by a small number of powerful pinning centers is called *strong pinning*. A simple strong-pinning expression is $H_\mathrm{p} =$

$(\mathrm{d}\gamma(x)/\mathrm{d}x)/(2\mu_0 M_\mathrm{S})$, where $\gamma(x)$ is the average wall energy as a function of the wall position [14,36]. By contrast, pinning caused by a large number of very small pinning centers, such as atomic defects, is called *weak pinning*. In the case of weak pinning, the wall energy is averaged over a distance of order $\delta_\mathrm{B}$, so that the density of pinning centers determines the pinning strength. Another mechanism involves many nucleation centers, so that the magnetization reversal is realized by the pinning-controlled growth and coalescence of a large number of domains.

In the hysteresis loop, the difference between nucleation and pinning is seen most easily from the virgin curves, which are obtained by thermal demagnetization. Figure 10.8 illustrates this distinction. After thermal demagnetization, domain walls in nucleation-controlled particles are very mobile, so that saturation is achieved in very low fields. By contrast, pinning centers impede the domain wall motion in both the virgin-curve and major-loop regimes.

For structurally (morphologically) homogeneous ellipsoids of revolution having the easy axis parallel to the axis of revolution, the nucleation problem can be solved exactly [12–14]. This is of some practical importance, because acicular (wire- or needle-like) magnets, fine particles, and thin films can be approximated by prolate, spherical, or oblate ellipsoids of revolution. The calculation consists of two steps: (i) the linearization of (10.21), as discussed in Sect. 10.4, and (ii) solving the resulting stability problem by eigenmode analysis. The corresponding eigenmodes $\boldsymbol{m}(\boldsymbol{r}) = \boldsymbol{M}(\boldsymbol{r}) - M_\mathrm{S}\boldsymbol{e}_z$ are known as *nucleation modes*, and the

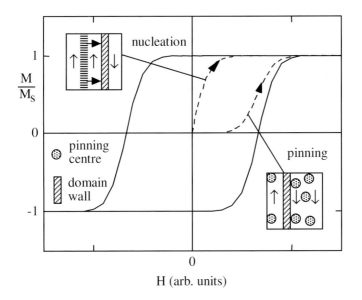

**Fig. 10.8.** Virgin curves for pinning-controlled and nucleation-controlled permanent magnets.

field at which the instability of the $m(r) = 0$ state occurs is the nucleation field.

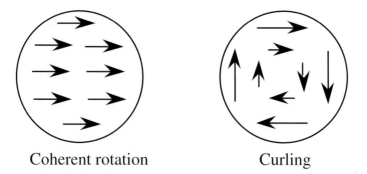

**Fig. 10.9.** Nucleation modes in homogeneous ellipsoids of revolution (top view on the equator plane). The arrows show $m(r)$.

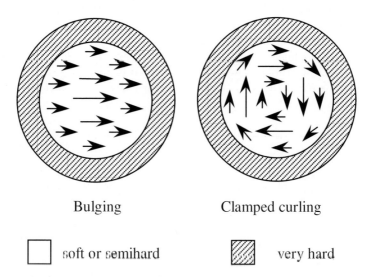

**Fig. 10.10.** Nucleation modes in spheres surrounded by a hard-magnetic shell (top view on the equator plane). The arrows show $m(r)$ for the core phase; in the surrounding shell, $m(r) = 0$. In both cases, the radial dependence of $m$ is given by spherical Bessel functions and localized in the soft region.

The nucleation field is [13,14]

$$H_N = \frac{2K_1}{\mu_0 M_S} + \frac{1}{2}(1 - 3N)M_S \tag{10.22}$$

for coherent rotation and

$$H_N = \frac{2K_1}{\mu_0 M_S} - NM_S + \frac{cA}{\mu_0 M_S R^2} \tag{10.23}$$

for the so-called curling mode. Here the radius $R = R_x = R_y$ refers to the two degenerate axes of the ellipsoid, and $c = 8.666$ for spheres ($N = 1/3$) and $c = 6.678$ for needles ($N = 0$). Coherent rotation and curling are realized in small magnets ($R < R_{\text{coh}}$) and large magnets ($R > R_{\text{coh}}$), respectively, where the coherence radius $R_{\text{coh}}$ is of the order of $5l_{\text{ex}}$ [37]. Note that these radii are *independent* of $K_1$. Figure 10.9 compares the coherent-rotation and curling modes. Coherent rotation is favorable from the point of view of exchange, but the exchange energy necessary to realize curling competes with the gain in magnetostatic energy associated with the flux-closure clearly visible in Fig. 10.10. We also deduce that the flux-closure contribution dominates the exchange in large magnets.

The coherent-rotation and curling modes are delocalized, that is the nucleation mode extends throughout the magnet. In general, inhomogenities lead to a localization of the nucleation mode [38]. An exactly solvable case is a soft or semi-hard magnetic sphere surrounded by a hard-magnetic shell. Eigenmode analysis then yields a *bulging* mode characterized by the symmetry of the coherent-rotation mode but incoherent due to the radial dependence of $m$ [39].

Figure 10.10 compares the bulging mode, realized for small particles, with the corresponding modified curling mode realized in large particles. The ultimate reason for the incoherent character of the bulging mode are the boundary conditions at the interface between the two magnetic phases. This yields not only an increase of the nucleation-field coercivity, as compared to Fig. 10.9, but also a singularity at the interface. In Sect. 10.4 we will see under which circumstances grain boundaries are sources of magnetoresistance.

### 10.3.4   Time Dependence of Magnetic Properties

The non-equilibrium character of magnetization processes means that magnetic properties are time-dependent. There are two basic types of time-dependent magnetic phenomena. Fast atomic processes lead to equilibrium on a local scale and realize intrinsic properties on sub-nanosecond time scales. For this reason, intrinsic properties can be regarded as equilibrium properties, and the energy functional (10.21) is also known as the micromagnetic *free* energy. Micro-magnetic processes are much slower, because atomic thermal excitations have to compete against many-atom energy barriers. For example, permanent magnetism relies on the fact that typical energy barriers are much larger than $k_B T$ [14]. Intermediate time scales are used, for example, to explain phenomena such as spin-wave resonance (see Chap. 15).

In a strict sense, ferromagnetism is limited to infinite magnets, because thermal excitations in finite magnets cause the net moment to fluctuate between opposite directions and yield – ultimately – a zero spontaneous magnetization. However, the corresponding equilibration time may be very large, and in practice it is often difficult to distinguish the magnetism of small particles from true ferromagnetism. In structurally inhomogeneous magnets (two-phase magnetism), each non-equivalent site $i$ exhibits a local spontaneous magnetization $M_i(T)$, and there is only one common Curie temperature. However, when the size of the inhomogenities is larger than about 1 nm, it is quite difficult to distinguish the inhomogeneous ferromagnet from a mixture of two phases [40].

A manifestation of *extrinsic* dynamics is that freshly magnetized permanent magnets loose a small fraction of their magnetization within the first few hours, which is known as *magnetic viscosity*. Typically, the magnetization loss is logarithmic, $\Delta M = -S \ln(t)$, where $S$ is the magnetic-viscosity constant [14,36,41]. A related effect is that coercivity depends on the sweep rate $dH/dt$ used to measure the loops: $H_c$ is largest for high sweep rates, that is for fast hysteresis-loop measurements.

Small energy barriers, realized for example in fine-particle ensembles, give rise to *superparamagnetism*. First, in particles whose radius is smaller than about 1 nm the external magnetic field is unable to produce saturation, because it cannot compete against thermal excitations. Secondly, there is a blocking radius below which thermal excitations are able to overcome anisotropy-energy barriers. Blocking radii scale as $(T/K_1)^{1/3}$ and are of the order of 5 nm for semi-hard materials.

## 10.4   Grain–boundary Magnetism

Real polycrystalline (nanocrystalline) magnets exhibit intergranular exchange coupling and magnetostatic interactions between grains. Strong intergranular interactions lead to the breakdown of the picture of individual grains and the magnetic reversal becomes a cooperative effect involving many grains.

A simple model is the *Preisach model*, where the interactions appear as random magnetic fields acting on the individual crystallites [36], but internal interaction fields are unable to give an appropriate description of cooperative magnetization processes. In fact, the validity of the internal-field approach is restricted to the non-cooperative ensembles, where the width of the switching-field distribution $P(H_c)$ of the (non-interacting) crystallites is larger than the magnitude of the interaction fields [42].

A better approach is the *random-anisotropy* theory [43–46], which focuses on the competition between interatomic exchange and random anisotropy. There are two main random-anisotropy effects: (i) the exchange favors parallel spin alignment throughout the magnet, and the remanence is exchanged-enhanced and (ii) in the limit of strong exchange interactions the coercivity of isotropic magnets vanishes. The relative strength of the intergranular exchange can be expressed in terms of the dimensionless parameter $A/K_1 R^2$, where $R$ is an average grain radius. This parameter shows that intergranular exchange is most effective in the limit of small grain sizes.

However, the original random-anisotropy theory cannot be used when two or more structural length scales are involved. This refers in particular to the effect of sharp grain boundaries [42,47]. Here we present a linear grain-boundary theory, which applies to weakly misaligned grains and is compatible with the scattering mechanism mentioned in the introduction.

## 10.4.1    Model

Consider an ensemble of exchange-coupled misaligned grains characterized by the local exchange stiffness $A(r)$ and the local easy direction $n(r)$. The starting point of the calculation is (10.21). Since SMR scales as $(\nabla M)^2$, we can restrict ourselves to short length scales, where magnetostatic interactions are of secondary importance (Sect. 10.3.2), and incorporate the self-interaction field into $H$. To linearize the problem we consider weakly misaligned grains and small deviations from perfect spin alignment. Since $|M(r)| = M_S$ and $|n(r)| = 1$, we can then write

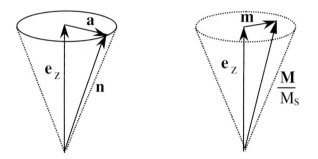

**Fig. 10.11.** Unit vectors $n(r)$ and $M(r)/M_S$ describing the polycrystalline easy axes and the local magnetization, respectively.

$$n(r) = \left(1 - \frac{a(r)^2}{2}\right) e_z + a(r) \tag{10.24}$$

and

$$M(r) = M_S \left(1 - \frac{m(r)^2}{2}\right) e_z + M_S m(r), \tag{10.25}$$

where $a \ll 1$, $m \ll 1$, abd $a \cdot e_z = m \cdot e_z = 0$. Figure 10.11 illustrates the meaning of $a$ and $m$.

Putting (10.24), (10.25) into (10.21) and taking $H = -H e_z$ yields

$$E = \int \left[A(r)(\nabla m)^2 + K_1 (m - a(r))^2 - \frac{\mu_0 M_S H}{2} m^2\right] dV. \tag{10.26}$$

Minimizing this equation with respect to $m$ we obtain

$$-\nabla (A(r)\nabla \cdot m) + (K_1 - \mu_0 M_S H/2) m = K_1 a. \tag{10.27}$$

Next we consider a grain boundary in the y-z-plane, as shown in Fig. 10.12. This is reasonable, because the perturbation $m(x)$ caused by the grain boundaries decays quite fast in the interior of the grains [42]. Since $a = a e_y$ and

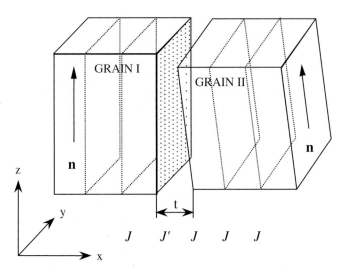

**Fig. 10.12.** Two neighboring grains and grain boundary. In this section, both continuum and layer-resolved configurations are discussed.

$m = me_y$, (10.27) now simplifies to

$$-\frac{\partial}{\partial x}\left[A(x)\frac{\partial m}{\partial x}\right] + \left[K_1 - \frac{\mu_0 M_S H}{2}\right]m = K_1 a(x) \qquad (10.28)$$

where $a(x)$ is equal to $a_I$ and $a_{II}$ in the respective grains I and II.

### 10.4.2   Boundary Conditions

As one can see from Fig. 10.10, grain or phase boundaries lead to singularities in the magnetization ($M$ or $m$) and may be potential sources of a pronounced magnetoresistance. The boundary conditions implied by (10.27) and (10.28) have been discussed in [48]. For an interface located at $x_0$, integration from $x_0 - \varepsilon$ to $x_0 + \varepsilon$ yields

$$\left[A(x)\frac{\partial m}{\partial x}\right]_{x_0 - \varepsilon} = \left[A(x)\frac{\partial m}{\partial x}\right]_{x_0 + \varepsilon}. \qquad (10.29)$$

This means that a jump in $A(\boldsymbol{r})$ changes the *slope* of the perpendicular magnetization component $m$.

Figure 10.13 shows that the boundary conditions (10.29) give rise to a variety of scenarios. In Fig. 10.13a, a phase I characterized by a large exchange stiffness is coupled to a phase II having a low $A$. When the anisotropy of phase I is very high, then the mode $\boldsymbol{m}(\boldsymbol{r})$ is localized in the phase II characterized by low or moderate anisotropy, regardless of the value of $A$ in the two phases. This regime is illustrated in Fig. 10.13b and realized, for example, in the composite Fig. 10.10. Of particular interest in spin electronics is the case where two grains

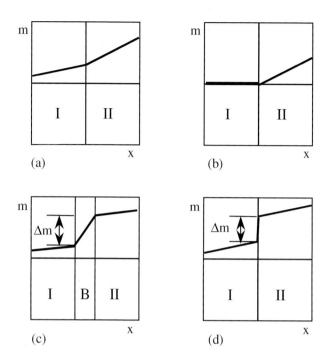

**Fig. 10.13.** Boundary conditions at interfaces: (a) interface between two phases of different exchange stiffness, (b) interface between hard and soft regions, (c) effect of grain-boundary region characterized by reduced exchange, and (d) reduced exchange coupling at interface. Note that (d) may be interpreted as a special case of (c) where $B = 0$.

are separated by a grain-boundary region (B) of reduced exchange stiffness. As shown in Figs. 10.13c and d, this leads to a *quasi-discontinuity* $\Delta m$ of the magnetization with a strong potential for SMR.

In order to calculate the magnitude of the quasi-discontinuity, we assume a thin grain-boundary region of thickness $t$ whose exchange stiffness $A'$ is smaller than the bulk exchange stiffness $A$ (Fig. 10.13c). Putting, for simplicity, $H = 0$ in (10.28) [49], we see that $m = a$ well inside the grains. It is therefore useful to consider the quantity $\Delta = \Delta m / |a_{II} - a_I|$, that is the fraction of the magnetization variation concentrated in the grain-boundary region. In the bulk, the difference $m - a$ decays exponentially [42], so that the calculation of $\Delta$ amounts to incorporating the boundary condition (10.29). After short calculation we obtain

$$\Delta = \frac{1}{1 + \dfrac{2A'\delta}{\pi A t}} . \qquad (10.30)$$

For $t = 0$, the quasi-discontinuity vanishes ($\Delta = 0$), whereas zero intergranular exchange ($A' = 0$) leads to $\Delta = 1$. On the other hand, since $t$ tends to be much smaller than the wall width $\delta$, an exchange enhancement at the grain boundary ($A' > A$) has no major impact on $\Delta$.

### 10.4.3   Layer–Resolved Spin Structure

Equation (10.28) describes ferromagnets on a continuum level, so that magnetization processes cannot be resolved on an atomic scale. For example, as indicated in Fig. 10.13d, grain boundaries tend to be atomically thin, and it is difficult to judge whether (10.30) remains valid in this limit.

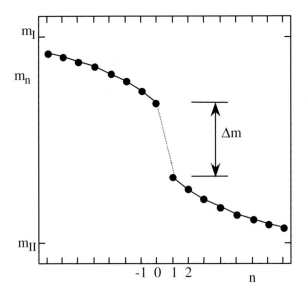

**Fig. 10.14.** Spin structure in the vicinity of the grain boundary. The jump $\Delta m$ means a quasi-discontinuity of the magnetization at the grain boundary.

In a layer-resolved analysis, as envisaged in Fig. 10.12, (10.28) must be replaced by

$$E = \sum_{n=-\infty}^{\infty} \left[ J_{n,n+1} \frac{(\boldsymbol{M}_n - \boldsymbol{M}_{n+1})^2}{M_S^2} - K_1 S_0 t_0 \frac{(\boldsymbol{n}_n \cdot \boldsymbol{M}_n)^2}{M_S^2} - \mu_0 \boldsymbol{M}_n \cdot \boldsymbol{H} S_0 t_0 \right]$$

(10.31)

where $S_0$ is the interface area, $J_{n,n+1} \simeq A(\boldsymbol{r}) t_0$ is the interlayer exchange coupling between adjacent atoms in the n-th and (n+1)-th layers, and each layer (index n) has a thickness $t_0 = t$. Restricting ourselves to the remanent state ($H = 0$) and using the approximation (10.25) we obtain, aside from a physically irrelevant zero-point energy,

$$E = \sum_{n=-\infty}^{\infty} \left[ J_{n,n+1}(m_n - m_{n+1})^2 + K_1 S_0 t_0 (m_n - a_n)^2 \right].$$

(10.32)

Here the interatomic *interface exchange* $J_{0,1} = J'$ is smaller than the bulk exchange $J_{n,n+1} = J$ for $n \neq 0$ (Fig. 10.12). Minimizing (10.32) yields the set of equations

$$J_{n,n+1}(m_n - m_{n+1}) + K_1 S_0 t_0 m_n = K_1 S_0 t_0 a_n \qquad (10.33)$$

subject to the boundary conditions $m_n = a_{I/II}$ for $n = \pm\infty$.

A typical solution is shown in Fig. 10.14. For $n < 0$ and $n > 1$, the dependence of the layer-resolved perpendicular magnetization contribution $m_n$ on $n$ is exponential, $(m_n - a_{I/II}) \sim \exp(\pm n/\lambda)$, and the decay length is

$$\lambda = \frac{1}{\mathrm{arcosh}(1 + K_1 S_0 t_0/2J)} \qquad (10.34)$$

In the interesting limit $K_1 S_0 t_0 \ll J$ this reduces to the bulk-type wall-width expression $(J/K_1 S_0 t_0)^{1/2}$. As in the continuum limit, the reduced interface exchange yields a quasi-discontinuity $\Delta m = m_1 - m_0$. For $K_1 S_0 t_0 \ll J$, we obtain

$$\Delta = \frac{\sqrt{K_1 S_0 t_0 J}}{\sqrt{K_1 S_0 t_0 J + 2J'}} . \qquad (10.35)$$

Aside from the use of atomic parameters, this result is very similar to (10.30).

Due to the quadratic dependence of SMR on the magnetization gradient, most of the magnetoresistive scattering by weakly coupled grains is associated with the quasi-discontinuity $\Delta$. From (10.30) and (10.35) we deduce that $\Delta = 1$ for $A' = 0$ and $J' = 0$, respectively, that is for zero intergranular exchange. Compared to ordinary domain-wall scattering, this corresponds to an increase of the magnetoresistance by a factor of order $\delta/t_0$, that is of the order of 100 for many materials of interest in spin electronics. However, a comparatively weak intergranular exchange is sufficient to yield a strong reduction of $\Delta$. Taking $t = t_0$ (one layer of reduced exchange) we find that $\Delta = 1/2$ when $A'/A$ and $J'/J$ are about $t_0/\delta$, that is of the order of 0.01. This means that SMR materials are not very forgiving with respect to residual intergranular exchange.

## 10.5   Concluding Remarks

A key finding of this chapter is that reduced exchange at grain boundaries yields a quasi-discontinuity of the magnetization, corresponding to a disproportionally strong domain-wall type magneto-resistance. By comparison, enhanced exchange in a thin grain-boundary region has no major effect on the spin structure. The same is true for anisotropy changes in the grain-boundary region, because the effect of anisotropy inhomogenities averages over at least a few nm.

The atomic origin of the grain-boundary exchange is of secondary importance in micromagnetism, because it is treated as a parameter. In any case, there is no conduction without some rudimentary exchange, so that the intergranular exchange may be small but is always nonzero.

The results obtained in Sect. 10.4 do not depend very much on whether one uses continuum or layer-resolved models. However, this does not mean that the relation between intrinsic and extrinsic properties at grain-boundaries is trivial, and a thorough and comprehensive description of the magnetic and magnetoresistive phenomena at grain boundaries remains a challenge.

Since the wall-width parameter $\delta$ scales as $(A/K_1)^{1/2}$, the domain-wall scattering is particularly strong in hard magnets, but in that case one needs undesirably large magnetic fields to change the spin configuration. Reduced intergranular exchange has a much more favorable effect on the magnetoresistance. From the point of view of spin-projecting magnetoresistance (SMR), the ideal magnetoresistive material is a hard-soft nanocomposite characterized by very weak intergranular exchange [50]. Of course, the realization of such a material remains a real challenge to atomic theory and magnet processing.

Note, finally, that SMR requirements are similar to the situation encountered in magnetic recording, where pronounced intergranular exchange between semi-hard grains leads to 'interaction domains' and reduces the storage density [51]. By contrast, two-phase permanent magnetism relies on a strong exchange coupling between hard and soft regions [42].

## Acknowledgement

Thanks are due to Prof. D. J. Sellmyer for discussing the manuscript. This work was supported by DOD, DOE, OXSEN, and CMRA.

## References

1. M. N. Baibich, J. M. Broto, A. Fert, F. Nguyen Van Dau, F. Petroff, P. Etienne, G. Creuzet, A. Friederich, and J. Chazelas, Phys. Rev. Lett. **61**, 2472 (1988).
2. A. E. Berkowitz, J. R. Mitchell, M. J. Carey, A. P. Young, S. Zhang, F. E. Spada, F. T. Parker, A. Hutten, and G. Thomas, Phys. Rev. Lett. **68**, 3745 (1992).
3. J. Q. Xiao, J. S. Jiang, and C. L. Chien, Phys. Rev. Lett. **68**, 3749 (1992).
4. J. F. Gregg, W. Allen, K. Ounadjela, M. Viret, M. Hehn, S. M. Thompson, and J. M. D. Coey, Phys. Rev. Lett. **77**, 1580 (1996).
5. P. M. Levy and S. Zhang, Phys. Rev. Lett. **79**, 5110 (1997).
6. R. von Helmolt, J. Wecker, B. Holzapfel, L. Schultz, and K. Samwer, Phys. Rev. Lett. **71**, 2331 (1993).
7. J. M. D. Coey, M. Viret, L. Ranno, and K. Ounadjela, Phys. Rev. Lett. **75**, 3910 (1995).
8. P. Schiffer, A. P. Ramirez, W. Bao, and S.-W, Cheong, Phys. Rev. Lett. **75**, 3339 (1995).
9. N. D. Mathur, G. Burnell, S. P. Isaac, T. J. Jackson, B.-S. Theo, J. L. MacManus Driscoll, L. F. Cohen, J. E. Evetts, and M. G. Blamire, Nature **387**, 266 (1997).
10. J. M. D. Coey, A. E. Berkowitz, Ll. Balcells, F. F. Putris, and A. Barry, Phys. Rev. Lett. **80**, 3815 (1998).
11. N. F. Mott and H. Jones, "The Theory of the Properties of Metals and Alloys", (University Press, Oxford, 1936).
12. W. F. Brown, "Micromagnetics", (Wiley, New York, 1963).
13. A. Aharoni, "Introduction to the Theory of Ferromagnetism", (University Press, Oxford, 1996).

14. R. Skomski and J. M. D. Coey, "Permanent Magnetism", (Institute of Physics, Bristol, 1999).
15. H. R. Hilzinger and H. Kronmüller, Phys. Lett. **51A** 59 (1975).
16. M. Viret, D. Vignoles, D. Cole, J. M. D. Coey, W. Allen, D. S. Daniel, and J. F. Gregg, Phys. Rev. B **53**, 8464 (1996).
17. A. Gupta, G. Q. Gong, G. Xiao, P. R. Duncombe, P. Lecoeur, P. Trouilloud, Y. Y. Wang and V. P. Dravid, and J. Z. Sun, Phys. Rev. B **54**, 15629 (1996).
18. E. Ising, Z. Phys. **31**, 253 (1925).
19. W. Heisenberg, Z. Phys. **49**, 619 (1928).
20. F. Bloch, Z. Phys. **57**, 545 (1929).
21. H. Brooks, Phys. Rev. **58**, 909 (1940).
22. J. C. Slater, Rev. Mod. Phys. **25**, 199 (1953).
23. F. Bloch, Z. Phys. **74**, 295 (1932).
24. L. Landau and E. Lifshitz, Phys. Z. Sowjetunion **8**, 153 (1935).
25. M. Kersten, Z. Phys. **44**, 63 (1943).
26. F. Cyrot-Lackmann, J. Phys. Chem. Solids **29**, 1235 (1968).
27. A well-known exception is, for example, the 'very weak intinerant ferromagnet' $ZrZn_2$.
28. Note that first, second, and third anisotropy constants, respectively, correspond to – at least – second-, fourth-, and sixth-order anisotropy contributions [14].
29. R. Skomski, H.-P. Oepen, and J. Kirschner, Phys. Rev. B **58**, 11138 (1998).
30. F. Bloch and G. Gentile, Z. Phys. **70**, 395 (1931).
31. E. C. Stoner and E. P. Wohlfarth, Phil. Trans. Roy. Soc. **A240**, 599 (1948).
32. In this equation, the factor $1/2$ is necessary to avoid double-counting interacting pairs of atomic moments.
33. C. Kittel, Rev. Mod. Phys. **21**, 541 (1949).
34. D. J. Craik and R. S. Tebble, Rep. Prog. Phys. **24** 116 (1961).
35. S. Chikazumi, "Physics of Magnetism", (Wiley, New York, 1964).
36. R. Becker and W. Döring, "Ferromagnetismus", (Springer, Berlin, 1939).
37. The coherence radius $R_{coh}$ is obtained by equating (10.22) and (10.23). Note that $R_{coh}$ is independent of $K_1$.
38. R. Skomski, J. Appl. Phys. **83**, 6503 (1998).
39. R. Skomski, J. P. Liu, and D. J. Sellmyer, Phys. Rev. B **60**, 7359 (1999).
40. R. Skomski and D. J. Sellmyer, J. Appl. Phys. **87**, 4756 (2000).
41. E. Kneller, "Ferromagnetismus" (Springer, Berlin, 1962).
42. R. Skomski, J. P. Liu, J. M. Meldrim, and D. J. Sellmyer, in "Magnetic Anisotropy and Coercivity in Rare-Earth Transition Metal Alloys", eds., L. Schultz and K.-H. Müller, (Werkstoffinformationsgesellschaft, Frankfurt/M., 1998), p. 277.
43. R. Harris, M. Plischke, and M. J. Zuckermann, Phys. Rev. Lett. **31**, 160 (1973).
44. K. Moorjani and J. M. D. Coey, "Magnetic Glasses", (Elsevier, Amsterdam, 1984).
45. E. M. Chudnovsky, W. M. Saslow, and R. A. Serota, Phys. Rev. B **33**, 251 (1986).
46. R. Skomski, J. Magn. Magn. Mater. **157-158**, 713 (1996).
47. T. Schrefl and J. Fidler, J. Magn. Magn. Mater. **175-181**, 970 (1998).
48. R. Skomski and J. M. D. Coey, Phys. Rev. B **48**, 15812 (1993).
49. Positive and negative fields reduce and enhance the spin inhomogenity, respectively, but a detailed analysis of the field dependence goes beyond this work.
50. The hard phase, which does not switch in small fields, could be ferrimagnetic to reduce the magnetostatic interactions between the grains.
51. D. J. Sellmyer, M. Yu, R. A. Thomas, Y. Liu, and R. D. Kirby, Phys. Low-Dim. Struct. **1-2**, 153 (1997).

# 11   Electronic Noise
# in Magnetic Materials and Devices

B. Raquet

Laboratoire de Physique de la Matière Condensée de Toulouse,
LPMCT-LNCMP-INSA, France

**Abstract.** With the development of magnetic devices and new materials for spin electronics on the sub-micron scale, we consider the relevant properties of electronic noise in magnetic solid-state microstructures. We review the most common types of electronic fluctuations in materials, namely, thermal noise, shot noise, $1/f$ noise and random telegraph noise. In each case, the discussion is illustrated by recent reports on electronic noise in magnetic materials and devices. We show that the resistance fluctuation measurement is an unique tool to probe the dynamics of magnetic instabilities and their coupling to the charge carriers via spin dependent scattering processes on a nanometric scale. We finally consider electronic noise in promising materials and devices for spin electronic applications like half metallic oxides, CMR perovskites and GMR-based magnetic sensors. Comments on recent results point out fundamental properties of the electronic and magnetic ground states which can be extracted from noise measurements. Special attention is paid to the noise behaviour and the signal-to-noise ratio in magneto-electronic applications.

## 11.1   Introduction

*A nuisance, an unwanted signal mixed in with the desired signal...* are common descriptions of electronic noise in measurement and instrumentation. It is true that voltage or current fluctuations impose practical limits on the performance of an electronic circuit or a measuring device. As a function of the measured entity, the noise level is usually defined in $nV/\sqrt{Hz}$, $nA/\sqrt{Hz}$, $nT/\sqrt{Hz}$..., which corresponds to the magnitude of the fluctuating quantity normalised to a 1 Hz-frequency bandwidth. The obvious challenge for any form of application or laboratory experiment is to increase the minimum detectable value of the desired variable; this naturally requires a reduction of the different sources of electrical noise interfering with the measurement.

However, for more than fifty years, physicists have understood that noise is not always a nuisance: the electronic noise inherent in materials or devices conveys fundamental information on the system dynamics near equilibrium. On many occasions, it has been demonstrated that intrinsic fluctuations are fingerprints of the internal dynamics. A "dirty" signal with random electrical fluctuations may reveal intrinsic properties of the conductivity, such as the microscopic behaviour of the charge carriers and their coupling to lattice defects, electronic traps, magnetic momentum.(for reviews see [1–6]). For example, very recently, noise studies have provided experimental evidence for macroscopic quantum tunnelling in the magnetisation reversal processes of mesoscopic systems [7]. In doped calcium manganites, the observation of a giant random telegraph noise has been interpreted in terms of intrinsic dynamic phase separation in the ferromagnetic ground state of the perovskite [8].

With the development of spin electronics [9–11] and nanoscale devices, various materials like magnetic oxides [12], half metallic ferromagnets [13], ferromagnetic semiconductors [14], giant magnetoresistance structures and tunnel junctions [15] must be reconsidered with respect to spin polarisation of the charge carriers and spin dependent conduction processes. Electrical noise studies are therefore of major interest both for estimating the intrinsic noise level and for a basic understanding of electronic mechanisms. Besides, the size reduction toward nanoscale devices tends to strongly enhance the apparent noise magnitude of the system. Noise studies as a function of the lateral size and the design of the devices provide an insight into the limiting factors for any application. It has been shown that, with the realisation of tiny samples, some active sources of noise may induce colossal electrical fluctuations with a lack of Gaussianity [16]. The question of a limiting volume below which a device may become too noisy to use unfortunately has to be considered.

This chapter is focused on the description of electronic noise in magnetic materials and devices, mainly considering electronic fluctuations in the range between very low frequencies up to few tens of MHz. After recalling the statistical tools usually employed to deal with stochastic processes (Sect. 11.2), we present the most common types of electronic noise in condensed matter, namely, thermal noise, shot noise, $1/f$ noise and random telegraph noise (Sect. 11.3). The text refers to recent reports on noise studies in magnetic materials and identification of the intrinsic sources of noise. The last section is devoted to electrical noise studies in leading materials for spintronic applications (Sect. 11.4). Final comments concern the limiting noise factors in magnetic devices and more specially in high density magnetic read-heads.

## 11.2   Mathematical Treatment

The electronic noise we shall be discussing corresponds to random fluctuations in the current flowing through a material or in the voltage measured across the terminals of the sample. The random aspect of the fluctuations versus time means that we deal with a stochastic process: the instantaneous values of the fluctuating entity cannot be predicted. Besides, most of the electronic noise we observe in materials and devices are stationary functions of time. The statistical analysis is independent of the time at which the signal is recorded. In the following, the specific mathematical treatments related to physical processes which exhibit some drifts of average values over time will be ignored.

A fluctuating quantity $\nu(t)$ can be characterised by a double statistical approach, in time and frequency space. We present an overview of mathematical techniques usually applied in solid states physics (for a detailed description see [1,4]).

## 11.2.1  The Time Domain Analysis

The basic treatment of a fluctuating quantity $\nu(t)$ in time space consists in calculating the average value $\bar{\nu}$, the variance $\sigma^2$, the probability density function $P(\nu)$ and the auto-correlation function $\Psi_\nu(\tau)$ as follows:
The average value:

$$\bar{\nu} = \lim_{T\to\infty} \frac{1}{T} \int_{-T/2}^{T/2} \nu(t)dt. \qquad (11.1)$$

$T$ is the duration of the observation.
The variance:

$$\sigma^2 = \overline{(\nu(t) - \bar{\nu})^2} = \overline{\delta\nu^2} = \overline{\nu^2} - \bar{\nu}^2. \qquad (11.2)$$

The probability density function $P(\nu)$: corresponds to the normalised distribution function of the magnitude of the fluctuations. When the fluctuations result from the sum of a large number of independent and random events, the fluctuating quantity follows a Gaussian law:

$$P(\nu) = \frac{1}{\sqrt{2\pi\sigma^2}} \exp\left[-\frac{(\nu - \bar{\nu})^2}{2\sigma^2}\right]. \qquad (11.3)$$

For a Gaussian fluctuation, defined by an average value $\bar{\nu}$ and a variance $\sigma^2$, 68% of the voltage magnitude lies between $\bar{\nu} \pm \sigma$. It is worth mentioning that once the electrical noise is recorded, the first statistical analysis of the fluctuations consists in checking their Gaussian behaviour. If it is so, a second-order statistical treatment corresponding to the power spectral density calculation will contain all the characteristics of the signal. In case of non-Gaussian behaviour, a higher order treatment is required (Sect. 11.3.4).
The auto-correlation function:

$$\Psi_\nu(\tau) = \overline{\nu(t)\nu(t+\tau)} = \lim_{T\to\infty} \frac{1}{T} \int_{-T/2}^{T/2} \nu(t)\nu(t+\tau)dt. \qquad (11.4)$$

It provides a measure of the memory of the process. In other words, how long does a given fluctuation persist at later times? The auto-correlation function at zero-time is equal to the mean square value of the fluctuations for $\bar{\nu} = 0$, $\Psi_\nu(0) = \overline{\nu(t)^2}$.

## 11.2.2  The Fourier Analysis of the Fluctuating Quantity

A powerful tool to specify a stochastic process is the distribution of the power of the signal in the frequency domain. Let us suppose a noise process $\nu(t)$ observed between the times $[-T/2; T/2]$ and assumed zero outside. The total energy of the signal $E$ is defined by:

$$E = \int_{-\infty}^{\infty} [\nu(t)]^2 dt. \qquad (11.5)$$

It can be expressed as a function of the Fourier transform $V_T(\mathrm{i}\omega)$ of $\nu(t)$, using Plancherel's theorem (with $\omega = 2\pi f$ and $\mathrm{i}^2 = -1$):

$$E = \frac{1}{2\pi} \int_{-\infty}^{\infty} |V_T(\mathrm{i}\omega)|^2 \mathrm{d}\omega \qquad (11.6)$$

The average power of the fluctuations $P$ is therefore defined by the relationship:

$$P = \lim_{T\to\infty} \frac{1}{T} \int_{-\infty}^{\infty} [\nu(t)]^2 \, \mathrm{d}t = \lim_{T\to\infty} \frac{1}{2\pi} \int_0^{\infty} 2 \frac{|V_T(\mathrm{i}\omega)|^2}{T} \, \mathrm{d}\omega \,. \qquad (11.7)$$

From the above equation, we directly derive the expression of the power spectral density (PSD) of the fluctuating process:

$$S_\nu(\omega) = \lim_{T\to\infty} 2 \frac{|V_T(\mathrm{i}\omega)|^2}{T}\,. \qquad (11.8)$$

In case of a statistically stationary process, the auto-correlation function and the PSD are related through the Wiener–Khintchine theorem:

$$\Psi_\nu(\tau) = \frac{1}{2\pi} \int_0^{\infty} S_\nu(\omega) \cos(\omega\tau)\mathrm{d}\omega\,, \qquad (11.9)$$

which is a Fourier transform relationship; the inverse gives rise to:

$$S_\nu(\omega) = 4 \int_0^{\infty} \Psi_\nu(\tau) \cos(\omega\tau)\mathrm{d}\tau \qquad (11.10)$$

If the fluctuating entity is a voltage fluctuation, units of $S_\nu(f)$ are $\mathrm{V}^2/\mathrm{Hz}$. It is interesting to notice that the integration of the PSD over all frequencies is equal to the variance of the signal. It is also called the net magnitude of the noise:

$$\Psi_\nu(0) = \overline{\delta\nu(t)^2} = \int_0^{\infty} S_\nu(f)\mathrm{d}f\,, \quad \text{with} \quad \bar{\nu} = 0\,. \qquad (11.11)$$

A well-known example of direct application of the Wiener–Khintchine theorem in solid-state physics consists in considering a relaxation process defined by a relaxation time $\tau_1$. This mechanism is characterised by an exponentially decaying auto-correlation function $\Psi_\nu(\tau)$ defined by:

$$\Psi_\nu(\tau) = \Psi_\nu(0) \exp\left[-\frac{|\tau|}{\tau_1}\right]\,. \qquad (11.12)$$

Using the Wiener–Khintchine theorem, a straightforward integration yields the expression for the power spectral density of the fluctuating quantity coupled to the relaxation process:

$$S_\nu(\omega) = 4 \int \Psi_\nu(0) \exp\left[-\frac{|\tau|}{\tau_1}\right] \cos(\omega\tau)\mathrm{d}\tau = 4\Psi_\nu(0) \frac{\tau_1}{1 + \tau_1^2\omega^2} \qquad (11.13)$$

The PSD is therefore a Debye-Lorentzian spectrum with a corner frequency equal to $1/\tau_1$ and a $1/f^2$ frequency dependence, above $1/\tau_1$.

## 11.3   The Most Common Types of Noise

Electrical noise in solids may originate from various sources such as defect motion, magnetic domain or spin fluctuations, charge carriers crossing an energy barrier, electronic traps, current redistribution within inhomogeneous materials. All these potential microscopic sources behave like "fluctuators"; once they are activated and physically coupled to the charge carriers constituting the current, they induce specific resistance or current fluctuations giving rise to the electrical noise we measure. Despite the random aspect of the fluctuating variable, a clear classification of the different kinds of noise has been carried out. It is mainly referring to the frequency dependence of the PSD. We essentially distinguish four types of electrical noise. The two most frequently encountered are thermal noise (Sect. 11.3.1) and shot noise (Sect. 11.3.2). These both exhibit a flat PSD and are rather well understood. Thermal noise is basically used to calibrate the background noise of an experiment. The third type is "flicker noise" also called "$1/f$ noise"; it refers to noise with a PSD spectrum of the form $1/f\alpha$, with $\alpha$ close to one (Sect. 11.3.3). The PSD of the fluctuations diverges at low frequencies. Fluctuations with such spectra have been observed not only in a tremendous variety of physical phenomena but also in other fields like human biology or geology. This omnipresence has led to a vast amount of research activity since the fifties. The last type of noise is random telegraph noise (RTN), which is a very beautiful and particular case of non-Gaussian fluctuations (Sect. 11.3.4). The resistance switches back and forth between two or more well-defined states under the effect of a single electron or one fluctuator and the resistance value of each state remains constant over time. Compared to $1/f$ noise, the RTN is a more recently studied phenomenon. In most cases, its manifestation is a direct consequence of a reduction of the dimensionality of the material toward the micron scale. Statistical study of lifetimes in the different states provides extremely rich information on the dynamics and the energy scales of the switching events as well as the related electronic properties.

The challenge for the physicist is to distinguish the microscopic origin of the electrical fluctuations from an almost "standard" and macroscopic transport measurement [17]. Through several examples of noise studies, mainly focussed on magnetic materials, we shall see how noise measurement and analysis are unique tools to probe the electronic and magnetic ground state. Investigations aiming at the general understanding of the solid state also offer insight into current industrial preoccupations. On the way to noiseless transistors or giant magnetoresistive (GMR) read-heads, the first step is an academic study of the electrical noise, with the understanding of its magnitude, its origins and, why not, its eradication.

### 11.3.1   Thermal Noise

The first observations of thermal noise, also called the Johnson or Nyquist noise, are due to Johnson in 1927 [18]; these were followed by the theoretical analysis developed by Nyquist in 1928 [19]. The Nyquist approach was based on thermodynamic calculations of the exchange energy between resistive elements in equilibrium. Thermal noise appears in all resistors, resulting from a random thermally-activated motion of charge carriers in equilibrium with a thermal bath.

The phenomenological description of its origin is based on the thermal activation of a large number of independent and random "events". One event is related to a departure from the equilibrium state (like one random electron motion) followed by a relaxation of the system to compensate the local perturbation of the charge distribution. A Fourier transform analysis of the thermal noise in all resistors yields a power spectral density of the terminal voltage fluctuations expressed by: $S_\nu(f) = 4k_BTR$ and the mean square noise voltage is equal to:

$$\overline{V^2} = 4k_BT\Delta fR\,. \tag{11.14}$$

Here, $\Delta f$ is the frequency bandwidth in which the voltage is measured, $R$ is the resistance value of the resistor and $k_BT$ is the thermal energy. For example, the rms noise voltage measured at the terminals of a 1 k$\Omega$ resistor in a 1 MHz frequency bandwidth is $V_{rms} = 4$ µV. This means that 68% of the voltage magnitude fluctuations fall within $\pm 4$ µV. Matching the frequency bandwidth of the experiment with the frequency domain of the desired signal is an obvious and well-known way to improve the signal-to-noise ratio of the measurement.

We note that the PSD of the thermal noise is not frequency dependent, which is why it is also called "white" noise. Quantum corrections are expected at high frequencies (above the microwave regime) taking into account the lifetimes of the charge carriers. The $S_\nu(f)$ spectrum is then given by:

$$S_\nu(f) = 4k_BTR\left\{hf/k_BT\left[1/2 + 1/\left(\exp\left(hf/k_BT\right) - 1\right)\right]\right\}\,. \tag{11.15}$$

## 11.3.2   Shot Noise

First experimental evidence and discussions dealing with shot noise are due to Schottky in 1918 [20]. This noise is related to the passage of the current across an energy barrier; it is a non-equilibrium form of noise. The elementary processes giving rise to shot noise are usually described by considering the current fluctuations in a thermionic tube: electrons which are randomly emitted from the cathode and flow to the anode under the influence of the electric field effect generate a current which fluctuates around a mean level. The fluctuations are caused by the random and discrete nature of the electronic emission as characterised by the work function. It is a direct consequence of the quantum character of the charge carriers. The PSD of the shot noise is also a flat spectrum expressed by: $S_I(f) = 2eI$. The current noise corresponds to:

$$I_{sh} = \sqrt{2eI\Delta f}\,. \tag{11.16}$$

$I$ is the average current and $\Delta f$ the frequency bandwidth. Quantum corrections involving the lifetimes $\tau$ of the charge carrier emission affect the PSD above microwave frequencies and yield:

$$S_I(f) = 2eI\left[\frac{\sin(\omega\tau/2)}{\omega\tau/2}\right]^2\,. \tag{11.17}$$

It should be emphasised that correlations between current pulses induce a space charge smoothing and consequently, a reduction of the shot noise magnitude. For this reason, shot noise is not observable in a homogeneous resistor.

Much work has been devoted to the characterisation of shot noise in semiconductor devices (for a complete review, see [4,6,21]). Very recently, Novak *et al.* [22] have found experimental evidence for shot noise in ferromagnetic-insulator-ferromagnetic tunnel junctions. The tunnel magnetoresistive junctions (TMR) with a 35% signal at room temperature exhibit voltage fluctuations with a cross-over from Johnson noise to shot noise as a function of the bias voltage, see Fig. 11.1. In the absence of interaction between tunnelling electrons and the oxide layer, the tunnel junction may be viewed as a ballistic conductor with a very small transmission probability. Shot noise dominates the flat part of the PSD, once enough current is applied through the junction to overcome the thermal noise. For MHz and higher frequency applications like TMR read-heads, shot noise may set the optimum sensitivity of the sensor. The set-up point of the bias voltage should be a compromise between the absolute $V(H)$ voltage, the decreasing $\Delta V(H)/V(H = 0)$ MR value versus the applied electric field and the shot noise level defined by the current flowing through the device.

To summarise, Johnson noise and shot noise are respectively the equilibrium and non-equilibrium noise both with a flat PSD for frequencies into the microwave band, but with different microscopic origins. They are inherent and correspond to the lowest noise level one may expect in a material or a device. Johnson noise provides a direct calibration of an electronic transport experiment: the background noise, in zero current, must tend to the thermal noise of the sample. The noise level above the theoretical thermal noise is usually estimated in dB, originates from the electronic equipment, the impedance mismatch and an inefficient shielding and grounding of the circuit [17,23].

However, in most cases, once a current is applied through a material, voltage fluctuations appear at low frequency, which can be several orders of magnitude greater than the Johnson or the shot noise. These fluctuations are expected to constitute a limiting factor for low frequency applications. Their PSD spectrum follows a $1/f$ dependency.

### 11.3.3    $1/f$ Noise

The commonly called "$1/f$ noise" refers to fluctuations of a physical variable with a PSD following a $1/f\alpha$ law, where the exponent $\alpha$ is equal or close to 1. This noise is also called "flicker" or "excess" noise: "excess" actually means in excess compared with the thermal noise level. Its first characterisation goes back to Johnson's experiments on current fluctuations of the electronic emission in the thermionic tube (1925) [24]. In addition to the shot noise, Johnson measured current noise whose spectral density increases with decreasing frequency $f$. The striking aspect of the $1/f$ noise which motivates a vast amount of research activity (one third of all publications on noise problems deal with $1/f$ noise) is its ubiquitous nature. Over the last fifty years, it has been observed in a tremendous variety of systems, far beyond the borders of solid state physics: $1/f$ fluctuations

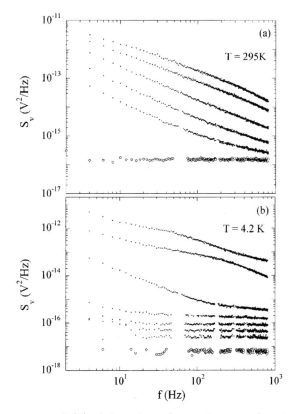

**Fig. 11.1.** Power spectra $S_\nu(f)$ of the voltage fluctuations as a function of dc current bias measured in a [3 nm $Co_{84}Fe_{16}$/2 nm $Al_2O_3$/8 nm $Co_{84}Fe_{16}$] tunnel junction [22]. For successive spectra from the bottom to the top, $I = 0$, 3, 6, 12, 24 and 36 µA in panel (a) and 0, 0.75, 1.5, 3, 6, 12, 24 and 36 µA in (b). Shot noise is the prevailing noise at 4.2 K and at low biases, but it is outweighed by the $1/f$-like noise (Sect. 11.3.3) at higher currents and higher temperatures.

have been reported in earthquakes studies [25], in the time dependence of the water level of the river Nile [26], in music [27], in biology [28]. Ionic current fluctuations through a neuro-membrane exhibit a $1/f$ spectrum [29], and so does the human heartbeat frequency which fluctuates with a PSD close to $1/f$ below 0.3 Hz [30].

Restricting the description of the $1/f$ noise to electrical noise measurements in condensed matter, we point out that $1/f$ fluctuations have been observed on the voltage probes in a vast number of different materials, like semiconductors, metallic and magnetic films, spin glasses, heterogeneous conductors, superconductors in the normal state, tunnel junctions, electronic devices, magnetic sensors [22,31–38] (based on AMR, CMR or TMR) and also in half metallic ferromagnets [39] and CMR perovskites [8,39–48] independently of the dimensionality of the material.

In all materials and devices mentioned above, the $1/f$ fluctuations are related to resistance fluctuations $\delta R(t)$. These are measured by applying a current $I$ and deducing the resistance fluctuations from the voltage fluctuations $\delta V(t) = I\delta R(t)$ at the terminal probes. The applied current does not create the fluctuations but just reveals them, above the white noise. This intrinsic nature of $1/f$ noise was clearly demonstrated in a major experiment performed by Voss and Clarke in 1976 [49]: in zero current, fluctuations of the variance of the Johnson noise exhibit a $1/f$ power spectrum. This rules out any contribution of the driving current to the resistance fluctuations.

In the following, we describe the relevant properties of the $1/f$ electrical noise. Comments will be offered on some open questions and paradoxes encountered in the $1/f$ noise studies. Finally, after a brief overview over the existing theories, we report on $1/f$ noise in various magnetic materials. Attention will be paid to the intrinsic electrical and magnetic properties which can be deduced from the $1/f$ noise measurements.

## Fundamental Aspects of Electrical $1/f$ Noise

The ubiquitous nature of this noise in various systems suggests the existence of an universal theory. Much work has been devoted to this ambitious approach. However, the available experimental evidence shows that the origins of the $1/f$ noise are quite different according to the material. It is now accepted that the $1/f$ spectrum is more a common mathematical feature, which can be phenomenologically modelled in a rather simple way.

**Resistance Fluctuations** As has been specified in the introduction, $1/f$ electrical noise on the voltage probes is measured by applying a current through the sample. The basic experimental configuration is shown in Fig. 11.2. Consequently, the noise level of the voltage is proportional to the square of the applied current $S_\nu(f) \propto V^2 \propto I^2$. This is of fundamental experimental interest as it allows a clear discrimination between the intrinsic $1/f$ noise coming from the sample and noise due to the preamplifier used for the measurement, which, in some case, may strongly dominate all other sources of noise.

**The Size Effect** In a rough approach, it is normally assumed that the noise level is inversely proportional to the volume of the sample between the probes [4,5]. The size effect is simply explained by the following argument: when the size of the sample is reduced, the size of the noise sources remains unchanged. Therefore, they induce stronger fluctuations on the overall electronic transport. The same argument holds for the number of charge carriers involved in the noisy volume: the charge carrier fluctuations due to electronic trapping and de-trapping affect the conductivity much more if the total number of charge carriers is small. Two straightforward consequences of the size effect on the effective noise level are the following: one way to measure the $1/f$ noise in a rather quiet material such as a well crystallised metal consists in working on thin films with reduced lateral sizes. This experimental approach holds as long as the noise originates from independent and local sources. And finally, the $1/f$ resistance fluctuations may cause a drastic degradation of the signal-to-noise ratio for small devices in low frequency applications (Sect. 11.4.3).

**A Major Paradox** The main feature of the excess noise is its frequency dependence proportional to $1/f^\alpha$. The $\alpha$ exponent has been determined in metals, semiconductors, oxides, electronic devices, nanocomposites.(see [1] and references therein). It ranges from 0.8 to 1.4. The frequency domain in which most of the experiments are performed is between $10^{-2}$ and $10^4$ Hz. However, experimental attempts have also been made to probe the very low frequency dependence of the noise. In 1974, Calayamides [50] has measured the PSD of resistance fluctuations in an operational amplifier down to $10^{-6.3}$ Hz, which requires times of averaging of the order of a month! No experimental evidence of any low frequency cut off in the $1/f$ dependency has been observed. This immediately raises a fundamental paradox in the magnitude of the noise level. We know that the auto-correlation function at zero-time, in other words, the variance of the fluctuations is defined by:

$$\overline{\delta V^2} \propto \int_0^\infty \frac{1}{f^\alpha}\,\mathrm{d}f\,. \tag{11.18}$$

This physical variable diverges at the upper (lower) limit if $\alpha < 1$ ($\alpha > 1$) and at both limits logarithmically for $\alpha = 1$. If no frequency limits are experimentally observed, it follows that the $1/f$ resistance-fluctuation variance is infinite! In practice, high frequency limits naturally exist and correspond to the finite

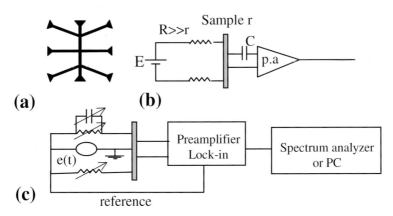

**Fig. 11.2.** Example of required sample geometry for electrical contacts outside the current lines. (b) Standard 4 probe noise measurement with a dc applied current (p.a = preamplifier). (c) Noise experiment in a 5 probe configuration with a ac applied current [17].

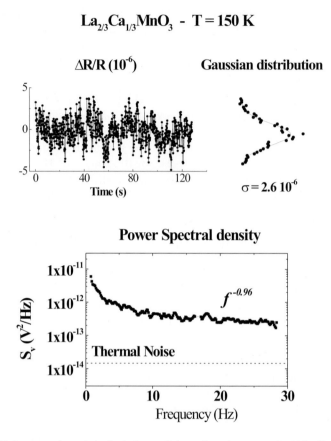

**Fig. 11.3.** Noise experiment analysis in a calcium doped manganite thin film at 150 K. Notice the Gaussian behaviour of the resistance fluctuations and the $1/f$ noise spectrum well above the thermal noise for an applied current density of $10^2$ A/cm$^2$.

scattering time of charge carriers or more basically to the frequency for which the $1/f$ noise sinks into the background noise. For the low frequency divergence, the problem remains open: a low frequency cut-off inherent to theories or to the lifetime duration of the system should be present. In practice, the stability of the experiment and the patience of the experimentalist usually dictate the existence of a low frequency limit which necessarily yields a finite experimental value of the fluctuation variance.

**A General Formula to Quantify the $1/f$ Noise** The $1/f$ resistance fluctuations are customarily quantified by a phenomenological equation given by Hooge in 1969 [51,52] which can be expressed as follows:

$$S_\nu(f) = \gamma_{\mathrm{H}} V^2 / N_c f . \tag{11.19}$$

Here, $V$ is the applied voltage between the probes, $N_c$ is the number of charge carriers in the noisy volume and $\gamma_H$ is a dimensionless constant for $\alpha = 1$, which is called Hooge's constant and refers to the noise level once the PSD is normalised by the volume and the applied voltage. Over the years, Hooge's constant was found to be equal to $2 \times 10^{-3}$ in very different materials [53], which was quite exciting in the search for an universal theory. Hooge's expression also implies that the noise level is independent of the temperature. However, since the eighties, strong experimental evidence has been found that Hooge's constant actually ranges from $10^{-6}$ to $10^7$ and is temperature dependent. The $2 \times 10^{-3}$ value seems to be more a contact noise characteristic related to the sample geometry than a real estimate of fluctuations in bulk materials [54]. Besides, the normalisation by the number of charge carriers is strongly questionable because it would imply that each mobile carrier individually carries the noise. This property is inconsistent with the $1/f$ law: we cannot assume fluctuations due to individual charge carriers which last longer than the duration of the mobile charges within the sample [2]. Despite the above arguments against a physical meaning of Hooge's equation, it remains a very convenient way to normalise the noise level between different systems and to provide an estimate of how noisy a device is at room temperature. The lowest $\gamma_H$ values have been obtained for bismuth and semiconductors with very clean surfaces. $10^{-3}$-$10^{-2}$ are the "standard" Hooge's constants for well crystallised metallic films and semiconductors. And the noise level is usually between 4 and 6 orders of magnitude higher in magnetic materials, oxides and nanocomposites.

**Other Aspects** Rather than going through some controversial aspects of the $1/f$ fluctuation properties in detail, we simply mention them.

The first aspect is related to the surface versus volume effects: at present the debate of this problem still continues without having reached a categorical conclusion [5]. The surface or volume locations of the noise sources are strongly dependent on the material. With an obvious risk of oversimplifying the physics, we may report experimental evidence which demonstrates that surface treatments and an applied electric field perpendicular to the surface of a semiconductor drastically affect its noise level. A significant similarity has been found between the gate voltage dependence of the resistance fluctuations and the density of surface states in a n-MOSFET [55]. Therefore, much evidence for surface mechanisms has been put forward in semiconductors, without excluding bulk effects. An opposite conclusion is reached for metallic films in which the noise level is inversely proportional to the film thickness [56].

The second dispute concerns the physical origin of the noise. In the Drude model, it is well known that the number of carriers $N$ and their mobility $\mu$ govern the resistivity; so it is of sense to expect that: $\delta R/R = +\delta N/N + \delta\mu/\mu$. Do resistance fluctuations originate from the carrier number fluctuation $\delta N$ or from the mobility fluctuation $\delta\mu$? Once again, the debate which has been going on for a long time has not reached a definite conclusion since more recently it has been confirmed that it naturally depends on the physics of the system.

In semiconductors, it is thought that most of the $1/f$ noise comes from charge trapping and de-trapping in localised states in the close vicinity of the surface. The trapping effect clearly affects both the mobility and the charge carrier number. In metals, if one only involves the number fluctuation, the noise level may only be reached with an unrealistic number of traps, comparable to the density of free carriers [4]. Experiments have been attempted to distinguish the contributions of $\delta N$ and $\delta\mu$ to the noise level. The idea consists in measuring both the $1/f$ voltage fluctuations on standard probes and those on the Hall voltage [54,57–59]. In some cases, the ratio between resistance noise and Hall voltage fluctuations plus the sign and the magnitude of their correlation give rise to pertinent information like the fraction of mobility fluctuations and the separate contribution of the majority and minority free carriers [54]. However, it seems that such experimental approaches do not lead to any definite conclusion and it is quite surprising that no recent results refer to the cross correlation between the Hall and the standard voltage fluctuations.

We conclude the survey of general aspects of $1/f$ resistance noise with comments on the microscopic properties of the fluctuations and the related experiments. It is of great interest to probe the anisotropy of the conductivity fluctuations. This provides evidence on the microscopic behaviour of the fluctuators responsible for the noise and their coupling to the resistivity. The relevant parameter is the asymmetrical factor S ranging from 1 (in case of scalar fluctuations) to $-1$ (for 2D traceless fluctuations) [1,60]. Noise experiments in metallic films have shown a strong asymmetry of the fluctuations. This strongly supports the concept of noise in metals originating from the hopping of defects on sites whose symmetry is lower than the point-group symmetry of the crystal [1]. Interesting attempts have also been performed to probe the correlation length of the conductivity fluctuations. Initially, the motivation was to give evidence for frequency-dependent spatial cross correlation in samples where the noise was supposed to be dominated by some heat-diffusion processes. However, measurements on a variety of metallic films and semiconductor devices show no correlation down to millihertz and into a distance scale of the order of few μm [61,62]. Therefore, within a micron scale, the $1/f$ noise is generated locally and not transported.

In the frame-work of noise experiments on a micron scale and below, we do expect that a new promising area of noise studies is now open with the achievement of nano-structured materials and devices. For example, nano-scale ferromagnets patterned in a cross-shape can be used to probe the resistance fluctuations asymmetry in case of magnetic domain fluctuations. We speculate that the asymmetry parameter may probe the anisotropic magnetoresistance or the magnetic domain anisotropy.

Besides, let us imagine a magnetic structure in which we locate a unique domain wall. The measurement of resistance fluctuations through the wall should be strongly influenced by its scattering processes. It may probe intrinsic properties like the time-scale of spin dynamics within the magnetic inhomogeneity. Furthermore, the achievement of magnetic materials with sizes comparable to the mean free path of the charge carriers or the spin-coherence length induces a crossover in the conductivity processes toward the ballistic regime. Noise measurements are thought to be an accurate tool to investigate the new conductivity regime [63]. The spatial correlation of the fluctuations within few tens of nanometers should also be revisited.

## Theoretical Approaches

The ubiquitous nature of the electrical $1/f$ noise in very distinct systems requires theoretical approaches which should be consistent with various physical origins of the noise sources. This task is extremely difficult. Many theories that have actually been proposed in an attempt to explain the $1/f$ frequency dependence of the PSD lack a reasonable physical basis.

The most customary concept on which most of the theoretical approaches are based is the concept of superposition of random and independent events coupled to the resistivity [64,65]. If one assumes the existence of a single two level process (TLP), also called a "fluctuator" with one relaxation time $\tau$, it has been demonstrated by the use of the Wiener–Khintchine theorem (Sect. 11.2.2) that the PSD of the fluctuations is a Lorentzian spectrum. If one considers now a physical system with a distribution of relaxation times $D(\tau)$, associated with independent TLPs, the corresponding PSD is defined by:

$$S_\nu(f) \propto \int \frac{\tau}{1+\omega^2\tau^2} D(\tau)\mathrm{d}\tau . \tag{11.20}$$

Considering a distribution of relaxation times equal to $D(\tau) \propto 1/\tau$, between two relaxation times $\tau_1$ and $\tau_2$, the integration of the Lorentzian yields:

$$S_\nu(f) \propto \frac{1}{f} \quad \text{for} \quad \tau_2^{-1} \le f \le \tau_1^{-1} . \tag{11.21}$$

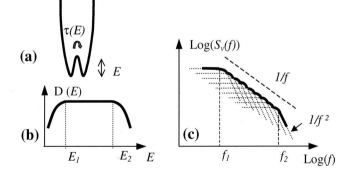

**Fig. 11.4.** Mathematical construction of $1/f$ noise: a superposition of a large number of Lorentzian spectra related to Two Level Processes (a) with a broad energy-barrier distribution (b) yields a $1/f$ spectrum of the fluctuating entity (c).

In case of thermally activated processes following an Arrhenius law [65], $\tau = \tau_0 \exp(E/k_B T)$, where $\tau_0$ is the attempt frequency, usually related to the phonon frequency in solids and $E$ is the activation energy in a symmetric TLP, see Fig. 11.4, the energy distribution required to provide $1/f$ noise between $[\tau_2^{-1}; \tau_1^{-1}]$ is then:

$$D(E) = \text{constant} \quad \text{for} \quad E_1 = k_B T \ln\left(\frac{\tau_1}{\tau_0}\right) \leq E \leq E_2 = k_B T \ln\left(\frac{\tau_2}{\tau_0}\right).$$
(11.22)

This concept exists since 1950 and shows how $1/f$ fluctuations are "mathematically" built from a superposition of independent and random TLPs with a wide distribution of relaxation times whose time-scale is coherent with low frequency processes. That is the key of the ubiquitous character of the $1/f$ noise, even if no light is cast on the physical origins of the fluctuators. In the following, we present a brief description of some well accepted models and the domains in which they have been successfully applied.

**The McWhorter Model** The McWhorther model is probably the most popular and recognised model of $1/f$ noise directly based on the concept developed above [66]. In this model, applicable to semiconductors, the $1/f$ fluctuations are related to fluctuations of the number of charge carriers at the semiconductor-oxide interface. The resistance fluctuations arise from tunnelling exchange of free charge carriers between the conduction channel near the surface and the traps lying in the oxide layer covering the surface. A distribution of traps inside the oxide layer induces a distribution of tunnelling distance $D(d)$ which has to be consistent with the $1/f$ noise frequency domain. For tunnelling events, $\tau = \tau_0 \exp(d/l_0)$. Let us fix $\tau_0 \simeq 10^{-12}$ s and $l_0 \simeq 0.1$ nm. Then, a constant distribution of the tunnelling distance between 1 and 4 nm yields an excess noise with a $1/f$ spectrum ranging from $10^{-8}$ to $10^6$ Hz, which is large enough to account for the $1/f$ noise measured in the experimental frequency window. The McWhorter model is widely used to interpret the $1/f$ noise in electronic devices; it also supports the strong noise difference between the JFET and the MOSFET.

**The Dutta–Dimon–Horn Model** Following the du Prè concept [65], in 1979 Dutta, Dimon and Horn (DDH) proposed a model based on a superposition of thermally activated random and independent processes with a broad distribution of transition energies [67]. Under major assumptions which will be explained later, the richness of their approach is to extract from the temperature dependence of the PSD slope $\alpha(T)$ the shape of the transition energy distribution $D(E)$ responsible for the resistance fluctuations. In some cases, the shape of the energy distribution provides a clear signature of the microscopic origins of the fluctuators.

In the framework of the DDH model, for a given temperature and within an experimental frequency range $[f_{\min}; f_{\max}]$ transition energies in the range $[k_B T \ln(f_0/f_{\max}); k_B T \ln(f_0/f_{\min})]$ are accessible. This domain is called the experimental energy window. Then, at a given temperature $T$, we probe the noise arising from fluctuators whose transition energies belong to this energy window. For a $10^{-2}$-$10^3$ Hz frequency domain, at 300 K and $f_0 \simeq 10^{12}$ Hz, the observable fluctuations necessarily involve transition energies between 0.45-0.9 eV. Sweeping the temperature from 4 to 300 K probes the energy scale between 6 meV and 0.9 eV. As a consequence, the PSD spectrum slope reflects the shape of the energy distribution within an energy window defined by $T$ and the experimental frequency domain.

Assuming a large distribution of thermally activated and symmetric two level fluctuations, the analytical calculation of the integral:

$$S_\nu(f) \propto \int \frac{\tau_0 \exp\left(E/k_B T\right)}{1 + \omega^2 \tau_0^2 \exp\left(2E/k_B T\right)} D(E) dE\,, \qquad (11.23)$$

permits the prediction of $\alpha(T)$ values from the experimental $S_\nu(f,T)$ data [67]:

$$\alpha(T) = -\frac{\partial \ln(S_\nu(f,T))}{\partial \ln(f)} = 1 - \frac{1}{\ln(\omega \tau_0)} \left[\frac{\partial \ln(S_\nu(f,T))}{\partial \ln(T)} - 1\right]\,, \qquad (11.24)$$

and an estimate of the energy distribution responsible for the fluctuations observed in the experimental frequency and temperature window:

$$D(E) \propto 2\pi f S_\nu(f,T)/k_B T\,. \qquad (11.25)$$

Let us notice that a flat $D(E)$ distribution gives rise to a pure $1/f$ spectrum and a linear temperature dependence of the noise level. However, departure from the linearity with $\alpha(T)$ values different from 1 implies a non-zero derivative of the energy distribution, $\partial D(E)/\partial E \neq 0$. Assuming $\alpha(T) > 1$ (respectively $\alpha(T) < 1$) implies an excess in the density of the low (high) energy fluctuators and a negative (positive) derivative $\partial D(E)/\partial E$ over the energy window.

The DDH model has been essentially applied to noise in metals [67,68]. In the case of Ag films for example, an energy distribution centred around 0.8 eV has been found. It is though that the rather high energies associated with the noise sources correspond to the energy necessary to create and induce atomic defect hops between equivalent sites in energy [61,69]. The coupling to the resistivity is due to different effective scattering cross-sections between accessible sites.

The DDH model is based on assumptions which considerably reduce its range of applicability [2]. Dutta *et al.* explicitly consider that $\delta R^2/R^2 \neq g(T)$, which means that the net magnitude of the noise is temperature independent. The number of fluctutators and/or their coupling strength to the conduction processes are assumed to be independent of temperature. Such a consideration may remain correct in degenerate systems with no particular electronic or magnetic transitions over a large temperature range, like in $3d$ transition metals.

Otherwise, the introduction of a temperature dependent function $g(T)$ is required in the $S_\nu(f,T)$ expression:

$$S_\nu(f,T) \propto \int g(T) \frac{\tau(E)}{1 + \tau(E)^2 4\pi^2 f^2} D(E) dE\,. \qquad (11.26)$$

The $g(T)$ function accounts for the temperature dependence of the net magnitude of the noise. It may include, for example, the activation of new fluctuators, some changes in their coupling to the conductivity process or modifications of the free carrier number. The final expression of the energy distribution is therefore:

$$D(E) \simeq \frac{2\pi f S_\nu(f,T)}{k_B T} \frac{1}{g(T)} , \qquad (11.27)$$

and the predicted $\alpha(T)$ values are:

$$\alpha(T) = 1 - \frac{1}{\ln(2\pi f \tau_0)} \left[ \frac{\partial \ln(S_\nu(f,T))}{\partial \ln(T)} - \frac{\partial \ln(g(T))}{\partial \ln(T)} - 1 \right] . \qquad (11.28)$$

Several authors have applied this improvement with some success [39,70,71]. The major objection is that even if the shape of $g(T)$ is inferred from a comparison between the experimental and the predicted $\alpha(T)$ values, only speculations can be made regarding its physical origin.

**Other Models** During the seventies, various studies suggested thermal fluctuations [49] as a possible cause of resistance fluctuations in materials and especially in metals. This concept considers energy fluctuations which give rise to enthalpy fluctuations at a constant pressure. The enthalpy fluctuations, usually referred to as temperature fluctuations, are coupled to the resistivity by the heat capacity of the specimen $C_p$ and the temperature coefficient of resistance $\beta = \partial R/\partial T$. A thermodynamic calculation yields $S_\nu(f,T) \propto V^2 \beta k_B T^2 / C_p f$, where $V$ is the applied voltage. Despite the initial success of this model in certain cases like in manganin, experimental evidence rules out the temperature fluctuation-contribution to the electrical $1/f$ noise. The strongest argument against this model is the lack of space-time correlation of the resistance fluctuations [71].

Among other theories, let us mention the quantum theory of $1/f$ noise [72], which assumes that the electrical $1/f$ noise is a fundamental feature of charge transport. The current fluctuations through a sample would originate from the interaction between charge carriers and low frequency photons emitted by the scattering of carriers by arbitrary potential barriers. Recently, in a search for a universal mechanism of $1/f$ noise, De los Rios *et al.* [73] proposed an extended version of the concept of self-organised critical systems [74] by introducing an activation-deactivation process and dissipation. The SOC model considers that systems are driven by their own dynamic which is critical in the sense that no characteristic time or length scale exists. Numerical calculation yields a $1/f$ power spectrum. The new theoretical development reproduces the hyper-universality and the apparent lack of a low frequency cut-off in the power spectral density.

### $1/f$ Electrical Noise in Magnetic Materials

Clear experimental evidence exists that resistance fluctuations in magnetic materials are strongly related to the magnetic ordering and its stability versus time.

Electrical excess noise is indeed a unique tool to probe the equilibrium magnetic noise (which differs from the Barkhausen noise) in small magnetic volumes or in magnetic materials with a weak magnetic signal like antiferromagnets. Such measurements provide an insight into the spin or domain fluctuations. In some cases, the switching relaxation time and the magnetic volume of the fluctuating entities can be inferred. Let us note that the frequency range $[10^{-2};10^4$ Hz] in which the electrical $1/f$ noise is usually observed is consistent with the relaxation time of a magnetic domain, which depends on its size, the local magnetic anisotropies and the temperature; these relaxation times usually fall in the range 1 μs-1 Ms. The coupling between resistance fluctuations and magnetic noise involves spin dependent scattering processes like the spontaneous resistive anisotropy (SRA), domain wall scattering, GMR effects.

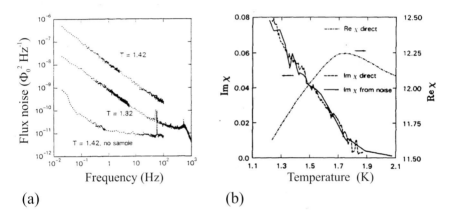

**(a)**                                                    **(b)**

**Fig. 11.5.** (a) Flux noise versus frequency for the detection system alone (miniature SQUID susceptometer) and for the spin-glass $Eu_{0.4}Sr_{0.6}S$ at two different temperatures above and below the freezing temperature, $T_f \sim 1.53$ K [75]. (b) Imaginary part of the ac susceptibility $\chi''(T, 50\,\mathrm{Hz})$ deduced from the magnetic noise measurement compared with the directly measured ac susceptibility at 50 Hz. The units for the susceptibility are arbitrary.

Only a few examples are reported in the literature of direct magnetic equilibrium noise measurements in small magnetic materials. W. Reim et al. [75] have used an integrated miniature SQUID susceptometer chip at low temperature to measure the magnetic fluctuations of a spin glass, in which $1/f$ magnetic noise was predicted due to the intrinsic broad spectrum of relaxation times [76]. They simultaneously measured the $1/f$ magnetic spectrum and the AC susceptibility of a $(15\ \mu\mathrm{m})^3$ $Eu_{0.4}Sr_{0.6}S$ sample, see Fig. 11.5. From the PSD of the magnetic noise and the imaginary part of the susceptibility, they demonstrated the validity of the fluctuation-dissipation theorem [77] below and above the freezing temperature. The F-D theorem establishes a direct relationship between a fluctuating quantity and a source of noise responsible for the fluctuations. In the present case, the applicability of the F-D theorem means that the equilibrium magnetic fluctuations are related to the dissipation that arises when the system is driven by an external magnetic field. As a consequence, noise measurements also offer an alternative way to estimate $\chi''$ in zero field.

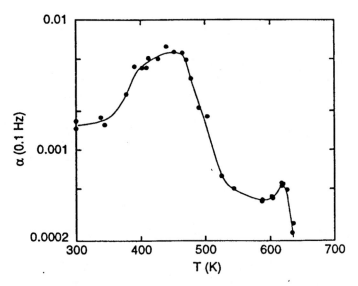

**Fig. 11.6.** Normalised noise power of the resistance fluctuations, $\alpha$, at 0.1 Hz as a function of temperature for a Ni thin film [80]. The smooth curve is a guide to the eye. The second and smaller noise peak corresponds to $T_C$.

We now come back to electrical noise. In case of antiferromagnets, the magnetic signal related to magnetic domain fluctuations is too small for a direct magnetic fluctuation measurement. However, several authors have measured the $1/f$ electrical noise in chromium [68,78,79]. A two order of magnitude increase has been observed when Cr is cooled through the Néel temperature ($T_N \sim 320$ K). The strong increase of the resistance fluctuations is related to fluctuations in the magnetic structure, i.e., the dynamics of the transverse spin-density wave (TSDW). At low temperatures, well below $T_N$, noise asymmetry studies (the $S$ parameter) reveal that the electrical noise originates from the rotation of a fluctuating entity. It has been concluded that the fluctuations come from the rotation of the polarisation $n$ of the TSDW between two states within volumes of about $10^{-16}$ cm$^3$. Near $T_N$, on small samples (less than $2 \times 10^{-13}$ cm$^3$), the time trace of the resistance fluctuations is highly non-Gaussian (Sect. 11.3.4). This noise is the signature of rotation of the $Q$ vectors of the TSDW domains. From the magnitude of the resistance steps, the minimum magnetic volume $V_D$ in which the $Q$ vector fluctuates is estimated to be of the order of $3 \times 10^{-16}$ cm$^3$.

In Ni thin films [80], the $1/f$ resistance fluctuations and the temperature dependence of the PSD have been correlated with spin cluster fluctuations with temperature dependent cluster volumes. Giordano [80] has measured the $1/f$

electrical noise in Ni from 300 K to above $T_C$ ($T_C \sim 625$ K). His study shows two maxima in the noise magnitude versus temperature; one around $T_C$ and the other one well below, around 450 K, see Fig. 11.6.

The maximum of resistance fluctuations near $T_C$ has been observed in various ferromagnets and clearly demonstrates the strong influence of domain dynamics on the low frequency electrical noise. In the framework of the DDH model, Giordano argued that the temperature dependence of the noise with the peak around 450 K necessarily implies microscopic changes in the fluctuating processes. Well below $T_C$, the magnetic correlation length is of the order of the lattice parameter; the spins fluctuate almost independently from the neighbourhood. The fluctuators responsible for the noise are therefore individual spins. Increasing the temperature induces an enhancement of the magnetic correlation length; the fluctuators change from an individual spin to spin clusters. According to Giordano, as $T_C$ is approached, the activation energies of the clusters, which are proportional to the volume, become too large to induce observable fluctuations. Only small clusters, much less numerous, contribute to the $1/f$ noise. Such a picture tends to explain the noise level decrease above 450 K and the change in the reduction of the number of efficient fluctuators predicted by the DDH model.

Curiously, apart from the above study, little has been reported in the literature on noise in $3d$ ferromagnetic epitaxial thin films. New studies on Co or Fe nano-structured epitaxial ultra-thin layers may give rise to unexpected results on the magnetic ordering in low dimensional systems.

### 11.3.4   Non-Gaussian Noise and Random Telegraph Noise (RTN)

The noise with which we have dealt up to now has been exclusively of the Gaussian type. From a statistical point of view, all the information is contained in the second-order correlation function. As it has been previously described, the common way to generate such a noise is by the superposition of many independent sources contributing individually and weakly to the variance of the fluctuating quantity. In case of a very small number of independent events or if one of them is much more strongly coupled to the resistivity, then singular events are noticeable on the resistance time trace. The lack of Gaussianity requires the use of high-order correlation functions. One technique consists in measuring the spectrum of fluctuations in the noise power within the frequency bandwidth of the ordinary spectrum. This spectrum is called the second spectrum [81].

Non-Gaussian (NG) effects on resistance fluctuations are usually observed in small samples (below few $\mu m^3$), once a decrease of the size of the material significantly reduces the number of fluctuators. Nevertheless, in rare cases, mostly related to strongly inhomogeneous materials, NG noise is measured in macroscopic volumes [40,41,46,82,83]. Fig. 11.7a shows the temperature dependence of the power spectral density of the resistance fluctuations in Fe-SiO$_2$ nano-composite films [83], very close to the percolation threshold, in the metallic regime. A drastic increase of the $1/f$ noise level is observed when the film is cooled down to 60 K. Hall effect measurements on the same sample reveal

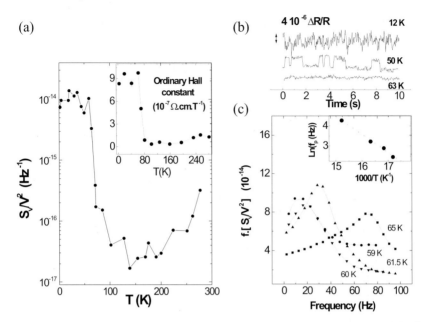

**Fig. 11.7.** (a) Temperature dependence of the normalised $1/f$ noise $S_\nu(f)/V^2$ measured in a $Fe_x$-$SiO_2$ ($x \sim 0.56$) nanocomposite thin film [83]. An abrupt increase of the noise level occurs when the sample is cooled down to 60-70 K. In the inset, one notices a one order of magnitude increase of the ordinary Hall voltage occurring at the same temperature in the same film. (b) Resistance fluctuations versus time at different temperatures. A strongly non-Gaussian noise is observed in the vicinity of 60 K. (c) The product $f \times S_\nu(f)$ as a function of frequency for several temperatures showing a maximum at a temperature dependent frequency. In the inset, the frequency at maximum is shown as a function of inverse temperature; the data follow an Arrhenius law, from which the activation energy of the fluctuating process responsible for the NG noise is inferred.

a one order of magnitude decrease of the effective number of carriers at the same temperature (see inset Fig. 11.7a). Both measurements provide evidence for a temperature induced electrical transition toward a more insulating regime in the infinite cluster. Besides, in the close vicinity of the electrical transition, resistance fluctuations exhibit a strong non-Gaussian behaviour, see Fig. 11.7b. The non-Gaussian noise around 60 K reveals that one or few "fluctuators" are active and are related to some change in the conduction process. This has been discussed in terms of current redistribution in the conducting paths, when the sample is cooled down to 60 K. The authors interpret the changes in the ordinary Hall constant as a cutting up of the metallic network. In the vicinity of the singularity, the product $f \times S_\nu(f)$ of the resistance fluctuations is not flat as it should be for pure $1/f$ noise. Rather, this quantity exhibits bumps with the frequency at the maximum shifting with temperature, see Fig. 11.7c. This is a clear signature of a thermally activated fluctuator responsible for the NG noise around 60 K. A 100 meV activation energy is inferred from the temperature dependence of the frequency peaks (see inset of Fig. 11.7c). This has to be related in some way to the thermally activated microscopic processes which induce a cut-off in the percolating network and current redistribution in the vicinity of 60 K. Quantum size effects were put forward as a possible origin of current redistribution. The authors speculate that these could be located in iron constrictions in the narrowest parts of the metallic paths where these isolate some portions of the backbone once the energy difference between discrete levels is larger than $k_BT$.

It is worth mentioning that the analysis of the $f \times S_\nu(f)$ plot which may exhibit particular shape shifts with temperature is a well known way to estimate the activation energies of the thermally activated fluctuation processes strongly coupled to the resistance and responsible for the NG noise [81].

An extreme case of NG noise is Random Telegraph Noise (for a review, see [84]). It occurs when a single fluctuator or a single electron is involved. The fluctuation process switches between two states which could be of electrical or magnetic origin. Each state has a different coupling to charge carriers thus contributing differently to the overall conductivity; hence the switching results in discrete jumps in the sample resistance. The time intervals between switching are random but the two values accessible by the resistance are time independent. In most cases, the RTN is described by a thermally activated two level process (TLP). If one assumes an asymmetrical two-well system with an energy barrier $E \pm \Delta E$ separating the two states, the average time $\tau_i(T, H)$ spent in the $i^{\text{th}}$ state is expressed by: $\tau_i(T, H) = \tau_{i,0} \exp(E_i/k_BT)$. A straightforward calculation of the PSD of the noise produced by the TLP yields [1]:

$$S_\nu(f) = \frac{S_\nu^0(0)}{\cosh\left[\frac{\Delta E}{k_BT}\right]\left[\cosh^2\left(\frac{\Delta E}{k_BT}\right) + \omega^2\tau^2\right]}, \qquad (11.29)$$

where $\tau^{-1} = \tau_1^{-1} + \tau_2^{-1}$ is the total transition rate and $S_\nu^0(0)$ is the zero-frequency spectral density at $\Delta E = 0$. The PSD follows a Lorentzian spectrum with a corner frequency equal to $\tau^{-1}$. We notice that the PSD of the noise decreases for an asymmetrical ($\Delta E \neq 0$) double-well as the low energy state is more stable. The same Lorentzian spectrum is obtained at low temperature when the switching process is governed by a tunnelling mechanism.

We point out that a statistical analysis of the occupancy lifetimes of the two states as a function of temperature or an applied electric or magnetic field provides an unique insight into the energies of the system and its dynamics. Fundamental microscopic variables like the energy differences between two states, the volume of the fluctuating quantity and its intrinsic nature (defect motion, magnetic domains.) can be inferred. From the experimental point of view, the delicate task is to stabilise a well-resolved RTN over a long time for statistical averages. A sporadic behaviour of the RTN has been experimentally found in various systems like hydrogenated amorphous silicon (a-Si:H) [82] or charge-ordered manganites [40] (Sect. 11.4.2). The "instabilities" in the RTN activation

have been theoretically predicted [86] and interpreted in terms of current redistribution in inhomogeneous samples. Such a sporadic feature prevents any
quantitative treatment.

The first observations of RTN began in the fifties in reversed-biased p-n junctions [84]. Lately, with the achievement of new structures with reduced lateral
sizes, resistance switching was discovered in various conductors and devices like
MOSFET, Metal-Insulator-Metal junctions, small semiconductors and metallic
samples.

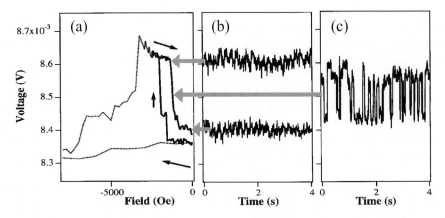

**Fig. 11.8.** (a) Magnetoresistance minor loop (full line) and main loop (dashed line)
at 80 K, in a bandwidth of 0.1 Hz for an ultra small Ni/NiO/Co junction. (b), (c)
Fluctuation at fields indicated by the arrows, measured in a bandwidth 0.1-12 Hz [37].
The authors conclude that the MR curve corresponds to a two level system defined by
the magnetisation state of the electronic trapping site in the NiO layer.

Outstanding noise measurements in metallic nanobridges close to the ballistic
regime (volume less than 8000 nm$^3$) have shown Random Telegraph switching
at low temperature [86]. It is originating from a single mobile defect, switching
back and forth between two locations with different scattering cross sections.
The effective cross-section is found to be of atomic dimension 0.01-0.1 nm$^2$ and
the required activation energy ranges from 30 to 300 meV. In some specimens
and at some temperatures, multi-level resistance switching has been observed.
The superposition of two correlated switching processes reveals the existence of
two distinct defects with the configuration of one influencing the lifetime of the
other. Above 150 K, the noise becomes Gaussian with a $1/f$ spectrum.

RTN may also be the dominant low frequency noise in magnetic and non-
magnetic tunnel junctions. As a general rule, only tunnel junctions with a small
active area (less than a few μm$^2$) exhibit RTN, whereas otherwise a $1/f$ PSD is
expected. The microscopic origin is related to trapping and de-trapping of electrons in defects located in the insulating layer. The Coulomb field induced by a
charged trap strongly affects the energy barrier. Resistance fluctuations as high
as 50% can be reached [37,87]. In the case of all-magnetic junctions, like elec-

trodeposited [Ni/NiO/Co] nanowires with cross-sections less than 0.01 μm² [37], the MR signal is dominated by resistance steps which correspond to two level fluctuations, see Fig. 11.8. The relevant parameter of the resistance fluctuations is the magnetisation state of the electronic trap in the oxide. The trapping probability is magnetic field dependent: under a given field, when the local magnetic state of the insulating magnetic oxide differs from the electrode magnetisation, the trapping site acts like a spin blockage. Fluctuations of the defects' magnetisation induce RTN in GMR curves which may occur well above the electrode saturation fields.

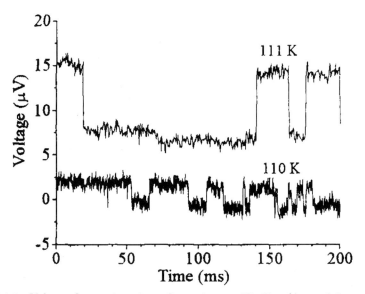

**Fig. 11.9.** Voltage fluctuations in a discontinuous $Ni_{82}Fe_{18}/Ag$ multilayer under an applied magnetic field of 6.24 mT [16]. The two curves reveal a discrete increase in the switching amplitude with a step of temperature from 110 K to 111 K. It is interpreted in terms of a change in the coupling strength between the fluctator to the charge carriers. One may involve a temperature induced growth of the fluctuating domain.

In tiny magnetic structures and in magnetic conductor devices like GMR metallic multilayers, RTN has also been reported as a predominant source of noise in the low frequency range. It is attributed to magnetic domain fluctuations [7,16,31–33]. A crossover from $1/f$ noise to RTN occurs when the size of the sample becomes comparable to the magnetic domain size. A statistical analysis of the RTN gives rise to an estimate of the magnetic domain volumes, their energy scale and the eventual domain-domain interaction. In a discontinuous GMR permalloy-silver multilayer with a lateral size reduced to a few μm, Kirschenbaum *et al.* [16] have presented first data of RTN with resistance switches around 1.2% due to thermally activated fluctuations of magnetic domains, see Fig. 11.9. From the analysis of the field dependence of the lifetimes of the "up" and "down" resistance states, they have inferred a magnetic volume of the switching entity equal to $10^6 \mu_B$. Based on a rough estimate of the permalloy grain size in the "pancake" structure of the multilayer, the authors have concluded that the discrete resistance jumps involve correlated switching of a multi-grain complex. This is of great importance for the understanding of the GMR magnitude. The activation energies of the "up" and "down" states, deduced from the temperature dependence of the RTN are of the order of 80 meV and 400 meV, respectively. The large energy difference refers to a strongly asymmetrical two energy well model. Local magnetic interactions may tend to favour one state in detriment to the other.

Finally, let us mention that telegraph noise spectroscopy also probes the magnetisation reversal of extremely small magnetic clusters. From the RTN study, Coppinger *et al.* investigated the single domain switching of nanometric ErAs clusters in a matrix of GaAs [7].

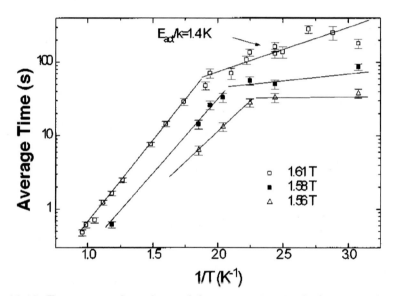

**Fig. 11.10.** Temperature dependence of the average time in the low state of two level fluctuations of nanometric ErAs antiferromagnetic clusters in various magnetic fields [7]. The saturation of the average dwell time is due to tunnelling of the magnetisation at very low temperature. At 1.56 T, the ground states of each well are aligned and the tunnelling process is temperature independent. Under a higher field, one state is favoured which leads to a thermally activated tunnelling regime.

The antiferromagnetic nature of the isolated ErAs cluster impedes a direct measurement of its magnetisation reversal. An angular study of the influence of the applied magnetic field on the resistance switching rates demonstrates the sixfold magnetic anisotropy of the ErAs nanometric clusters. From the temperature dependence of the lifetimes spent in both resistance states, the authors observed a strong deviation from the expected lifetime values in case of thermally activated processes below 300 mK, see Fig. 11.10. They interpreted their result as the manifestation of a macroscopic tunnelling reversal of the EuAs cluster magnetisation at very low temperatures.

It is worth mentioning that much care has to be taken to estimate the magnetic volume of the switching entity from the field dependence of the occupancy probabilities. For a thermally activated process, the general expression of the relaxation times is:

$$\tau_i(\boldsymbol{H}, \boldsymbol{E}, T) = \tau_{0,i} \exp\left[E_i(\boldsymbol{H}, \boldsymbol{E}, T)/\mathrm{k_B}T\right] , \qquad (11.30)$$

where $E_i(\boldsymbol{H}, \boldsymbol{E}, T)$ is the energy barrier for the $i^{\mathrm{th}}$ state which, in general, may depend on a magnetic and electrical field, and on the temperature. In the low magnetic field limit, we assume that the angular positions of the magnetic moments in each state are fixed. Therefore, the field effect on the energy barriers can be linearised and thus corresponds to the Zeeman energy. The Zeeman shift affects both the two stable states and the virtual state associated with the top of the barrier. The field dependence of the energy barrier is therefore [7]: $E_i(H) = E_i(H = 0) + \Delta\boldsymbol{m}_i \cdot \boldsymbol{H}$, where $\Delta\boldsymbol{m}_i = \boldsymbol{m}_i - \boldsymbol{m}_v$. $\boldsymbol{m}_i$ and $\boldsymbol{m}_v$ are the magnetisation vectors for the $i^{\mathrm{th}}$ state and the virtual state. The applied magnetic field probes the projection of the magnetic moment difference $\Delta\boldsymbol{m}_i$ onto its axis. From the statistical lifetime analysis, the $\Delta m_i$ value along $\boldsymbol{H}$ is inferred. This value can be orders of magnitude smaller than the magnetic moment of the $i^{\mathrm{th}}$ state. Such a situation occurs, for example, in case of fluctuations of the magnetisation between canted states. The estimate of the real magnetic volume requires either a strong assumption concerning the easy magnetisation axis within the magnetic entity or an accurate angular dependence study of the lifetimes versus the orientation of the magnetic field.

## 11.4   Electronic Noise Studies in Materials for Spin Electronic Applications

The achievement of electronic devices which are sensitive both to the charge of the carriers and their spin-quantum number initiates a revolution in future applications [9–11]. A new degree of freedom in the concept of electronics can be reached once the orientation, the transport and the lifetime of the spin magnetic moment of charge carriers are fully controlled by semiconductor technology. We may distinguish two kinds of magneto-electronic devices. The first one is directly issued by the discovery of GMR based on specific electron spin orientations across ferromagnetic layers [89]. The GMR effects and related phenomena like TMR are already exploited in the high technology market with a huge economic potential. GMR sensors in high density magnetic read-heads or non-volatile magnetic random access memory (MRAM) [15] are typical examples. In principle, the ultimate improvement of these devices rests in the use of ferromagnets with the strongest spin electronic density asymmetry at the Fermi level [89]. That is why, half metallic oxides [13] and half metallic ferromagnets like the Heusler compounds [90–92] which theoretically exhibit 100% spin polarisation are currently investigated in the light of potential applications. Right now, GMR based sensors use spin effects in metals, even though encouraging results have been obtained in other materials like all-oxide trilayers [93]. A fascinating improvement would be the implementation of the GMR effect in semiconductor technology, within conventional electronics. The magnetoelectronic devices based on spin-polarised transport correspond to the second type of spin electronic applications. New magnetic devices like spin transistors or spin-memory cells are at the earliest stage of development [94,95]. They are based on spin injection from a ferromagnet into a non-magnetic semiconductor. Spin injection is actually an extremely delicate task since spin dependent processes at the interface are thought to have a severe effect on the spin polarisation of the injected charges. Current work is devoted to spin injection and detection in all-semiconductor devices [96,97]. Results are promising although restricted for the moment to the low temperature range.

The development of spin electronics requires investigations on materials with unique properties like 100% spin polarisation or a high spin coherence length. The second step consists in the implementation of these materials in convenient devices adapted to the existing technology. In each case, electrical noise experiments provide an efficient tool for characterising the material and its eventual application. In the first stages of the development, noise studies in the material define its intrinsic noise level and in most cases, give rise to microscopic information which contribute to a better understanding of its electrical and magnetic properties. Recent noise studies have been performed on half metallic oxides [39] (Sect. 11.4.1) and CMR perovskites [8,39–48] (Sect. 11.4.2), both being potentially interesting for magnetic sensors. An intriguing high noise level in the low frequency range has been measured in these materials and the search of the physical origins permitted an insight into their electrical and magnetic ground states. The significance of this noise level as a limiting factor for low frequency applications might be assessed on the basis of these data.

The implementation of a device in an appropriate design also requires noise experiments to optimise the signal to noise ratio of the sensor in its final state. For example, the change from AMR to GMR high density magnetic read-heads is now well-advanced in several companies (Sect. 11.4.3). However, the earliest noise experiments on GMR multilayers showed electrical noise levels much higher than what is currently observed in commercial AMR based sensors [98]. Much work is therefore nowadays devoted to the optimisation of magnetic fluctuations in GMR devices.

## 11.4.1    Low Frequency Noise in Half-Metallic Oxides

A large $1/f$ noise, at least five or six orders of magnitude greater than normally found in metals has been measured in $CrO_2$, $Fe_3O_4$ and $La_{2/3}Sr_{1/3}MnO_3$ [39], which are currently under investigation because of their high spin polarisation. However, such low frequency noise is not at all unusual in oxides; a comparable noise magnitude has previously been observed in cuprates in the normal state

[99,100]. Nevertheless, it is of major interest whether the electrical noise properties in half metallic oxides are related in any manner to the half-metallicity of their electronic band structure.

In $CrO_2$ thin films [39], at room temperature, Hooge's constant is around 2000 and the magnetic field has little effect on the excess noise compared to its total magnitude. The temperature dependence of $S_\nu(f)$ reveals a significant increase of the noise above 200 K. The physical origin of the noise has been investigated in the framework of the Dutta–Dimon–Horn model. The form of the energy distribution of the excitation responsible for the $1/f$ noise exhibits two-well defined maxima, a first one around 0.4 eV and a larger one above 0.8 eV, see Fig. 11.11a.

The large activation energy is consistent with the activation energy of 0.93 eV for oxygen diffusion. Local oxygen motion and also oxygen deficiency create thermally activated vacancies and defects. These may act either as dynamic traps of free carriers and induce carrier fluctuations, or they may jump between atomic sites that are not equivalent in their contribution to the electrical resistivity.

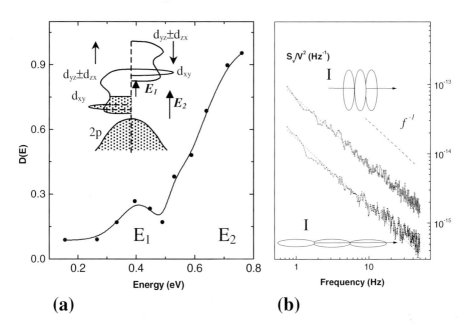

**(a)**                                                                **(b)**

**Fig. 11.11.** (a) Energy distribution of the fluctuators responsible for the $1/f$ resistance fluctuations within the experimental frequency window in $CrO_2$ thin film [39]. In the inset the schematic electronic band structure of $CrO_2$ is shown with the activation energies $E_1$, for $d \to d$ excitations and $E_2$ for $p \to d$ excitations. (b) Structure dependence of the noise in $CrO_2$ films for a low and high grain boundary density along the current path.

Such effects are usually more pronounced near grain boundaries which are one of the most probable places in a polycrystalline sample to find mobile atoms. The extrinsic noise sources have been studied in $CrO_2$ thin films with a needle-like structure with grains of approximately 30 μm length and 0.5-2 μm width. A significant decrease by a factor of 4 of the $1/f$ noise has been found at room temperature, when the current is applied along the path with the lowest density of grain boundaries, see Fig. 11.11b.

An intrinsic origin of the noise has also been put forward in relation to the half metallic character of $CrO_2$. The fluctuations providing the noise may be of electronic rather than structural origin. The band structure of $CrO_2$ is shown schematically in the inset of Fig. 11.11a. The activation energy for creation of down-spin carriers in the $t_{2g}$ band is $E_1$, for $d \to d$ excitations and $E_2$ for $p \to d$ excitations. Any reversed spin in the conduction band has a severe effect on the conductivity because it will be associated with a reversed $Cr^{4+}$ site moment, which not only withdraws that ion from the conduction process, but also reduces the bandwidth of the local density of states of neighbouring sites [101]. In this model, $E_1$ and $E_2$ are associated with the peaks at 0.4 and $\geq 0.8$ eV. No significant effect of an applied magnetic field is expected because the energy shift of the spin polarised bands (0.67 K/T) is much less than their separation ($\geq 1000$ K). Further work is needed to study the $1/f$ noise in epitaxial $CrO_2$ films and determine the extent to which structural and electronic excitations contribute to the noise. It should be noted that electronic transitions with spin reversal in the electronic band structure have also been regarded as a possible intrinsic source of noise in $La_{2/3}Sr_{1/3}MnO_3$ [39].

Concerning the Heusler compounds, various attempts have been made to use Heusler electrodes in TMR junctions, the goal being to inject highly spin polarised electrons through the junction. The TMR signal, at room temperature is around a few %, which is actually much weaker than the expected value for half metallic ferromagnet-insulator-ferromagnet junctions [90]. Spin scattering processes at the interface originating from the Heusler layer degradation (roughness and stochiometry) are thought to be the limiting factor. To our knowledge, no noise measurements have been performed on those systems.

## 11.4.2  Electrical Noise in CMR Perovkites

Mixed-valence magnetic perovskites are under intense study since the rediscovery of the colossal magnetoresistance in certain members of this group of materials. CMR near $T_C$ driven by the double exchange mechanism and the dynamic Jahn-Teller distortion [12] as well as the low field MR due to grain-boundary effects [102] provide a fertile ground for application as magnetic sensors [103].

In mixed-valence perovskite manganite, where the hopping conductivity is strongly dependent on the oxygen stoichiometry and the tolerance factor, the influence of oxygen dynamics as an extrinsic $1/f$ source of noise cannot be neglected, especially in the high temperature range [39]. Much more relevant are the temperature dependences of the $1/f$ resistance fluctuations observed in CMR manganite thin films in the ferromagnetic regime. In calcium and strontium doped manganites, a 3 or 4 order of magnitude increase of the noise level occurs as the metal-insulator transition is approached [8,39–48]. The noise peak temperature is slightly lower than that of the resistivity peak. By applying a magnetic field, a large reduction of the noise is observed in high field, comparable to the magnetoresistance. Therefore, the $1/f$ fluctuations are mostly of magnetic origin. The effect has been attributed to magnetic domain fluctuations near the transition in a time scale consistent with low frequency noise [48], or spin fluctuations coupled to the resistivity by electron-magnon scattering according to the double exchange theory [39].

In low-$T_C$ manganite thin films, several studies also reveal a huge $1/f$ noise which peaks well below $T_C$ [8,40–43,46–48] with, in some case, non-Gaussian resistance fluctuations [8,40,41,46]. Let us distinguish two classes of low-$T_C$ manganites, one which undergoes a first order metal-insulator transition and another one with a non-hysteretic phase transition as a function of temperature.

In the latter case, very recent work reports an astonishing random telegraph noise[1] in the resistance of $La_{2/3}Ca_{1/3}MnO_3$ in the ferromagnetic regime [8,41]. Well below $T_C$ ($T_C \sim 210$ K), between 4 K and 170 K, Raquet et al. [8] observed alternatively large non-Gaussian noise and giant RTN with resistance steps varying from 0.01% to 0.2% which is surprisingly high for an almost macroscopic sample (see Fig. 11.12a). From a statistical analysis of the lifetimes of the low and high resistance states versus the temperature and applied magnetic field, the authors have inferred, in the low temperature regime (below 30 K), a constant energy difference between the two states, around 100 K, with a magnetic volume of the switching entity of the order of $10^5$ Mn atoms. It is worth mentioning that the resistance steps have been observed for applied fields above the saturation field. The large RTN magnitude and its persistence in high fields rule out magnetic domain instabilities as noise sources of the non-Gaussian fluctuations and the RTN. At higher temperatures, an atypical temperature dependence of the occupancy lifetimes has been observed. An increase of the temperature over few hundred of mK drastically reduces the occupancy probability of the low resistive state in favour of the high resistive one, see Fig. 11.12b. The dynamics of RTN demonstrate that the energy barrier difference for the two states is strongly temperature dependent. The authors put forward a model of the energy configuration which reveals a strong analogy with the free energy functional defined for first order transitions in the framework of the Landau-Ginzburg treatment, see the inset to Fig. 11.12b, lower panel. They concluded that the large magnitude of the RTN, its temperature and magnetic field dependence give evidence for a dynamic coexistence of two phases below $T_C$: a ferromagnetic metallic phase and a phase with relatively depressed magnetic and electrical properties. The RTN occurs when a cluster, the fluctuator, switches back and forth between the two phases. Consequently, the noise measurements provide a direct proof of conduction in $La_{2/3}Ca_{1/3}MnO_3$ dominated by a mixed-phase percolation process

---

[1] Large resistance steps have also been observed in $La_{2/3}Sr_{1/3}MnO_3$ thin layers and $La_{2/3}Sr_{1/3}MnO_3/SrTiO_3/La_{2/3}Sr_{1/3}MnO_3$ tunnelling junctions by M. Viret et al. (unpublished).

below $T_C$. In this picture, the location of the noise-level peak well below $T_C$ and its surprising amplitude are a direct consequence of the mixed-phase: near the percolation threshold for the metallic state, the conduction is dominated by the narrowest current paths. Few switching clusters located in these critical bonds have dramatic effects on the overall connectivity of the metallic network, which results in a large increase of the noise level.

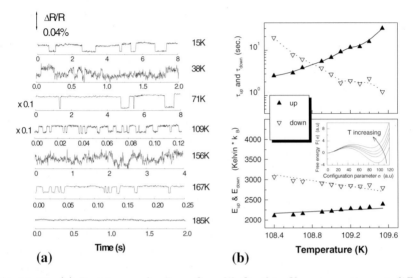

**Fig. 11.12.** (a) Resistance of a $La_{2/3}Ca_{1/3}MnO_3$ thin film versus time at different temperatures. The noise alternates from a strongly non-Gaussian fluctuation-type to RTN with $\Delta R/R$ ranging from 0.01% to 0.2% [8]. There is no evidence for RTN above 180 K ($T_C \sim 210$ K). (b) *Top panel:* temperature dependence of the average life-times for a fluctuator activated around 109 K. Note that by increasing the temperature the high-resistance state is stabilised and the low-resistance state becomes less probable. The lines are a guide to the eye. *Bottom panel:* temperature dependence of the deduced activation energies for both states. Inset: free energy functional $F(\sigma, T) = a(T-T_0)\sigma^2 + b\sigma^3 + c\sigma^4$ of the two level system versus a configuration parameter $\sigma$ at different temperatures ($\sigma = 0$ corresponding to the high-resistance state). $a$, $b$, $c$ and $T_0$ are fitting parameters chosen to describe simultaneously the temperature dependence of the activation energies $E_{up}$ and $E_{down}$ (solid and dashed lines in panel (b)).

Simultaneously, very similar results have been obtained by the Weissman group [41] on $La_{2/3}Ca_{1/3}MnO_3$ thin films and single crystals. They reached the same conclusion of a dramatic current inhomogeneity in the ferromagnetic regime, but with substantial differences in the energy analysis. A temperature independent energy difference between the two states with an enhancement of their entropy difference versus temperature was pointed out. They also deduced a large difference in the magnetic moment between the two states equivalent to $10^4$-$10^6$ unit cells. The high resistive state is thought to have a zero magnetic moment. Finally, these noise studies provide an insight into one of the most actively

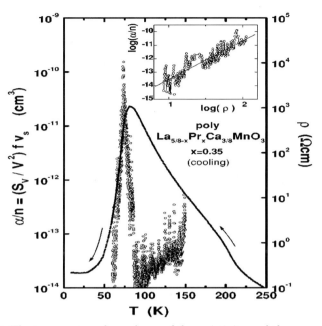

**Fig. 11.13.** The temperature dependence of the resistivity and the normalised spectral density at 10 Hz of the $1/f$ noise in a polycrystalline sample of $La_{5/8-x}Pr_xCa_{3/8}MnO_3$ ($x \sim 0.35$) [42]. Note that the noise peaks below the resistivity maximum, typical of a percolating two phase process. The inset shows the scaling dependence of the normalised magnitude of the $1/f$ noise versus the resistivity in an interval 61 K-73 K ($T_C \sim 73$ K). The solid line is a power fit: $\alpha/n \propto \rho^{2.9}$.

debated questions related to the spatial homogeneity of the ferromagnetic ground state in the manganites [104–106]. The last reports on noise experiments put forward a dynamic phase separation in $La_{2/3}Ca_{1/3}MnO_3$, while the intrinsic nature of the segregation (electronic or chemical inhomogeneity) is still under study.

In charge-ordered manganites, electrical noise measurements also exhibit a huge $1/f$ noise in the hysteretic region, between the ferromagnetic-metallic and the insulating-charge ordered phases [40,42]. Podzorov *et al.* [42] have observed an unprecedented magnitude of $1/f$ noise near $T_C$ in $La_{5/8-x}Pr_xCa_{3/8}MnO_3$ ($x \sim 0.35$), see Fig. 11.13. They have shown that the power spectrum of the resistance fluctuations near the transition scales with the resistivity as follows: $S_v(f)/V^2 \propto \rho^{\kappa/t}$.

The critical exponents $\kappa$ and $t$ obtained are fully consistent with theoretical noise predictions for a percolation metal-insulator transition [1,107,108]. It has to be mentioned that the existing theoretical approaches on noise behaviour in percolating systems refer to nanocomposite materials containing two physical phases, a metallic one and an insulating one. Podzorov's results clearly demonstrate the percolating nature of the CO-FM transition in polycrystalline samples of low-$T_C$ manganites.

However, the authors do not report on non-Gaussian resistance fluctuations in the vicinity of the percolation transition, which are normally expected for current-redistribution processes within an inhomogeneous sample. Giant non-Gaussian noise has actually been observed by A. Anane *et al.* in $Pr_{2/3}Ca_{1/3}MnO_3$, in the mixed region around $T_C$ [40]. The non-Gaussian behaviour has been discussed in terms of percolation behaviour of the conductivity. The dominant fluctuators are thought to be manganese clusters switching between the M-F and I-CO phases. Finally it is worth mentioning that, contrary to RTN in $La_{2/3}Ca_{1/3}MnO_3$, the non-Gaussian noise related to a percolation picture is not surprising in charge ordered systems as it undergoes a strongly hysteretic first order metal-insulator transition due to the formation of the CO state [109].

We conclude that noise measurements in magnetic oxides like CMR perovskites go further than a simple estimate of the magnitude of the noise level. Fundamental properties of the ferromagnetic ground state and the related energy scales have been inferred. That contributes, with other experiments, to the understanding of the electrical and magnetic properties of the CMR perovskites. In terms of applications, RTN is obviously unacceptable for low frequency sensors. Even in high frequency applications, the resistance steps induce high frequency components in the frequency spectrum of the desired signal. More work is needed to estimate the resistance fluctuation contributions to the minimum field detectable by a CMR perovskite prototype sensor.

### 11.4.3    Electrical Noise in GMR based sensors

In the following, we shall distinguish the electrical noise (mainly the $1/f$ noise) in GMR elements and in GMR sensors such as high density magnetic read-heads. Both originate from magnetic instabilities but the particular environment of the particular junction in the sensor design requires special comment.

The indisputable proof of the direct relationship between the electrical noise in GMR multilayers and magnetic fluctuations has been carried out by Hardner *et al.* [98,110]. Their experiment consisted in establishing the validity of the fluctuation-dissipation theorem, assuming that the dominant noise sources in $[Co/Cu]_n$ multilayers are magnetic instabilities. In the framework of the F-D theorem, it has been demonstrated that the power spectral density of the magnetic fluctuations is related to the imaginary part of the magnetic susceptibility (the magnetic loss) by the following expression:

$$S_M(f) = (2/\pi)\chi''(f)k_BT/\Omega , \qquad (11.31)$$

where $\chi''(f)$ is the out-of-phase magnetic susceptibility, $k_BT$ the thermal energy and $\Omega$, the sample volume. The authors proposed a new expression for the F-D theorem related to resistance fluctuations. The PSD of the magnetisation $S_M(f)$ is coupled to the PSD of the resistance $S_R(f)$ through the GMR effect and the

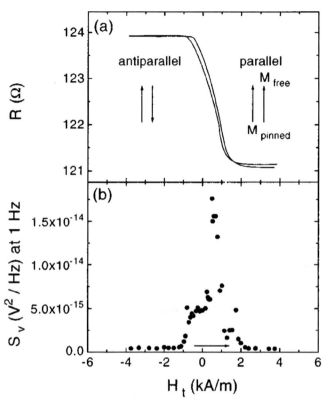

**Fig. 11.14.** Transverse magnetic field dependence of (a) resistance and (b) the voltage spectral density at 1 Hz for a GMR-based magnetoresistive element [36]. The data given in (b) are taken with increasing the applied field $H_t$, as indicated by the arrow.

authors have estimated $\chi''(f)$ from the experimental determination of the out-of-phase component of the resistivity for an ac applied magnetic field. They finally demonstrated that the noise level calculated from the F-D theorem is in perfect agreement with the measurement of Hooge's constant for various applied magnetic fields.

Besides, several studies [34,36,38,98,111] have shown that the maximum electrical $1/f$ noise level is reached for an applied magnetic field corresponding to the maximum sensitivity $\partial R/\partial H$ of the GMR effect, see Fig. 11.14. This means that the GMR multilayers are noisiest when the magnetic domains and/or the magnetisation orientation within a domain are in the least stable state, around the coercive field.

Van de Veerdonk *et al.* [36] developed a model which correlates the $1/f$ noise level to the magnetic instability within an AMR layer and a GMR multilayer. For an applied magnetic field, they calculated the magnetisation orientation in the free layer within the framework of a coherent rotation described by the well known Stoner-Wohlfarth model. Using Boltzmann statistics, they furthermore estimated the angular distribution of the magnetisation vector due to thermal activation. Finally, the angular distribution was coupled to the resistance fluctuations through the AMR and the GMR effects. The results illustrate the direct correlation between the calculated magnetisation instabilities within a single domain and the experimental electrical $1/f$ noise in the MR signal.

As a consequence, the noise level in a GMR multilayer drastically depends on its magnetic state. In a saturated state, the noise level is comparable to the one measured in non-magnetic granular materials. In a demagnetised state, or near $H_c$, with no bias field ensuring a monodomain configuration, Hooge's constant is around 1-20. For patterned GMR multilayers in a picture frame with a bias field to guarantee a single domain spin-valve, $\gamma_H$ is much smaller, around $2 \times 10^{-1}$, in the regime where $\partial R/\partial H$ is maximum.

From a practical point of view, it is of interest to get a rough estimate of the minimum detectable magnetic field $\Delta H_{\min}$ of a GMR device when the dominant sources of noise are supposed to be $1/f$ magnetic fluctuations $S_\nu(f)$ at low frequencies and thermal noise $4k_BTR$ at higher frequencies. The voltage fluctuations measured at the MR probes are simply expressed by:

$$\left[\frac{\delta V}{V}\right]^2 \simeq \left[\frac{\delta R}{R}\right]^2 \simeq \int_{\Delta f} \left[\frac{4k_BTR}{V^2} + \frac{S_\nu(f)}{V^2}\right] \, df. \tag{11.32}$$

With $\delta R \simeq (\partial R/\partial H)\Delta H$, we deduce:

$$(\Delta H_{\min})^2 \simeq \left[\frac{\partial H}{\partial R} R\right]^2 \int_{\Delta f} \left[\frac{4k_BTR}{V^2} + \frac{S_\nu(f)}{V^2}\right] \, df. \tag{11.33}$$

Here $\Delta f$ is the experimental frequency bandwidth and $\Delta H_{\min}$ corresponds to the minimal field above which the sensor detects a change in the external magnetic field. We note that the first term is linear in $\Delta f$, whereas, the $1/f$ noise contribution has a logarithmic frequency bandwidth dependence. Besides, $\Delta H_{\min}$ is inversely proportional to the MR sensitivity, which corresponds to the pre-factor of the above equation. At low frequencies, up to 10 kHz-1 MHz, the minimum detectable field is defined by the $1/f$ noise level.

In a commercial AMR sensor, $\Delta H_{\min} \sim 0.2$ mOe for a frequency cutoff of the $1/f$ noise around 100 kHz. For a $[Co/Cu]_n$ GMR multilayer, with an intrinsic complex multi-domain structure, $\Delta H_{\min} \simeq 1 - 5$ mOe. In the case of a polycrystalline $La_{2/3}Ca_{1/3}MnO_3$ thin film, at room temperature and with a small bias field, CMR and noise measurements lead to $\Delta H_{\min} \simeq 1$ Oe. Therefore, the gain in the magnetoresistance sensitivity may not induce an enhancement of the sensor sensitivity, if intrinsically noisier materials are used.

The significant difference in the noise level between AMR and GMR elements is thought to originate from the stronger magnetic sensitivity and hence, a stronger coupling between magnetic instabilities and the resistivity [36], a difficult control of the domain structure in a GMR process [34,111] and the use of thinner magnetic layers in GMR multilayers [36].

For sensors in a final design like GMR high density read-heads, much less relevant results have been published. We summarise the following points:

The dominant noise is still a low frequency $1/f$ noise originating from magnetic fluctuations. It is thought that, for 100 MHz-200 MHz applications (which is roughly the case for read-heads), this noise should not be a crucial problem [35].

The signal to noise ratio scales with the square root of the active sensor area. If $L$ is the strip length and $W$ the strip width, then [35]:

$$\frac{S}{N} = \sqrt{WL}\,\frac{\dfrac{\Delta H}{R}\dfrac{\partial R}{\partial H}}{L_c\dfrac{dR_c}{R_c}} \qquad (11.34)$$

This is easily verified if the sensor area is divided into small elements, each having statistically independent resistance fluctuations $dR_c$ and an associated coupling length $L_c$.

The $1/f$ noise level is mainly dependent on the probing current [31,32], the bias current (used to polarise the device in a constant field) [32,33] and the effective magnetic anisotropy of the free layer [112]. Xiao *et al.* [31] have shown that, in GMR recording heads, the square current dependence of the noise level, as predicted by Hooge's expression, is no longer observable: extra low frequency noise of purely magnetic origin appears. Under a high current density ($10^7$ A/cm$^2$), some heating effects probably activate new magnetic sources, i.e. new magnetic domains or orientation fluctuations.

Wallash [33] and Hardner *et al.* [32] provide clear evidence that much care has to be taken in the choice of the bias current; in some cases, the induced field may activate new fluctuators, giving rise to random telegraph noise. The RTN usually dominates the low frequency noise in small active areas, or after a strong electrical stress [31], see Fig. 11.15, and/or under particular bias fields [32,33]. The location of the magnetic instability can be inferred combining the effects of temperature, bias current and external magnetic field on the lifetime of the resistance switches [32].

We know that an accurate estimate of the magnetic size of the fluctuators based on the RTN analysis is rather delicate (see Sect. 11.3.4). For RTN in GMR sensors, the practical question is to know whether it arises from a large magnetic rotation within a small volume, or from a small angular instability in a much larger area. Xiao *et al.* have provided a simple argument based on the $\Delta R/R$ resistance switching magnitude [31]. It is well known that the MR signal is roughly proportional to $M^2 \propto \cos^2(\theta/2)$, where $\theta$ is the angle between the magnetisation vectors of the pinned layer and of the free layer. If one assumes that half of the sensor is affected by the magnetic instability (the exchange length is around 0.5 μm), then $5 \times 10^{-5}$ resistance steps are due to a magnetic instability with a 0.3° angular distribution into a magnetic domain of half a micron length. However, the resistance jumps persist under applied fields which rotate the free layer up to 30°. It is questionable that 0.3° fluctuations are still active over such a large free layer rotation. Therefore, Xiao *et al.* concluded that the RTN in GMR sensors probably originates from a large angle rotation in a small area due to a local non-uniformity of the anisotropy.

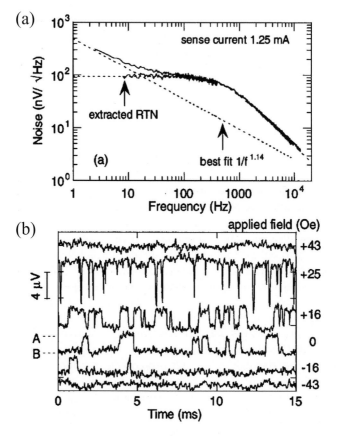

**Fig. 11.15.** Random telegraph noise of a GMR head after electrical stressing [31]. (a) Noise spectrum with the extracted $1/f^2$ dependence. (b) Time trace and its dependence on an external applied magnetic field.

Several authors [31–33] agree in the conclusion that $1/f$ noise is not a limiting factor for high frequency applications. However, it is thought that low frequency RTN may cause significant problems for head operation. Its sharp switching induces high frequency components in the data channel's passband and it is responsible for multiple metastable states in the sensor which leads to a popping baseline in the MR signal.

## 11.5   Concluding Remarks

We have shown that electrical noise studies in condensed matter are now an effective tool to probe dynamic microscopic processes coupled to the charge carriers. Resistance fluctuations may be fingerprints of defect generation and motion between equivalent atomic sites, electronic surface traps interacting with the conduction channel, current redistribution in inhomogeneous materials, slow exchange of electrons between the infinite cluster and small isolated donor clusters, charge carriers crossing an energy barrier, magnetic domain fluctuations, magnetisation instabilities within a magnetic cluster. All these microscopic events are of fundamental interest in the understanding of the electronic and magnetic ground states of the material.

In magnetic materials, we have stressed by various examples, the strong coupling between magnetic instabilities and electronic transport via spin dependent scattering processes. Depending on the balance between the sample size, the volume of the fluctuating magnetic cluster and the strength of the coupling, the resistance fluctuations range from $1/f$ Gaussian fluctuations to random telegraph noise between two distinct resistive states. Statistical analysis in frequency and time space provide an insight into the fluctuation regime (thermally activated or tunnel events), the energies and the volumes of the fluctuations. The noise results are also interpreted in terms of noise level above the thermal noise. They provide an estimate of the signal to noise ratio for materials and devices with a view on applications.

Finally, with the development of sub-micron magnetic structures, electronic noise measurement is a promising transport experiment to probe singular events on a nanometer scale. It is an unique tool able to identify the dynamics of the nanometric magnetic inhomogeneities. Future experiments which combine, for example, both electronic noise measurements and magnetic imaging may provide an ultimate insight into spin dependent scattering processes coupled to magnetic instabilities.

## Acknowledgements

Helpful discussions with A. Anane are gratefully acknowledged. One part of the work has been performed during a stay in the MARTECH laboratory (Florida). B. R. is very indebted to the MARTECH laboratory for financial support and to its director, Professor S. von Molnár, for illuminating discussions.

## References

1. Sh. Kogan, in "Electronic Noise and Fluctuations in Solids", Cambridge University Press, Cambridge (1996).
2. M. B. Weissman, Rev. Mod. Phys. **65**, 4829 (1993); **60**, 537 (1988).
3. D. A. Bell, in "Noise and the Solid State", Pentech Press, London (1985).
4. M. J. Buckingham, in "Noise in Electronic Devices and Systems", J. Wiley & Sons, New York (1983).
5. P. Dutta and P. M. Horn, Rev. Mod. Phys. **53**, 497 (1981).
6. A. van der Ziel, Adv. Electron. Phys. **49**, 225 (1979).
7. F. Coppinger, J. Genoe, D. K. Maude, U. Gennser, J. C. Portal, K. E. Singer, P. Rutter, T. Taskin, A. R. Peaker, and A. C. Wright, Phys. Rev. Lett. **75**, 3513 (1995); F. Coppinger, J. Genoe, D. K. Maude, X. Kleber, L. B. Rigal, U. Gennser, J. C. Portal, K. E. Singer, P. Rutter, T. Taskin, A. R. Peaker, and A. C. Wright, Phys. Rev. B **57**, 7182 (1998).

8. B. Raquet, A. Anane, S. Wirth, P. Xiong, and S. von Molnár, Phys. Rev. Lett. **84**, 4485 (2000).

9. D. D. Awschalom and J. M. Kikkawa, Phys. Today **52**, 33 (1999).

10. G. A. Prinz, Science **282**, 1660 (1998); Phys. Today **48**, 58 (1995).

11. J. L. Simmonds, Phys. Today **48**, 26 (1995).

12. J. M. D. Coey, M. Viret, and S. von Molnár, Adv. Phys. **48**, 167 (1999).

13. J. M. D. Coey, Phil. Trans. R. Soc. Lond. A **356**, 1519 (1998).

14. H. Ohno, Science **281**, 951 (1998).

15. S. S. P. Parkin, K. P. Roche, M. G. Samant, P. M. Rice, R. B. Beyers, R. E. Scheulerlein, E. J. O'Sullivan, S. L. Brown, J. Bucchigano, D. W. Abraham, Y. Lu, M. Rooks, P. L. Trouilloud, R. A. Warner, and J. G. Gallagher, J. Appl. Phys. **85**, 5828 (1999).

16. L. S. Kirschenbaum, C. T. Rogers, S. E. Russek, and S. C. Sanders, IEEE Trans. Magn. **31**, 3943 (1995).

17. J. H. Scofield, Rev. Sci. Instrum. **58**, 985 (1987).

18. J. B. Johnson, Nature **119**, 50 (1927); Phys. Rev. **29**, 367 (1927).

19. H. Nyquist, Phys. Rev. **32**, 110 (1928).

20. W. Schottky, Ann. Phys. (Leipzig) **57**, 541 (1918).

21. A. van der Ziel, Physica B **83**, 41 (1976).

22. E. R. Nowak, M. B. Weissman, and S. S. P. Parkin, Appl. Phys. Lett. **74**, 600 (1999).

23. S. Demolder, M. Vandendriessche, and A. van Calster, J. Phys. E **13**, 1323 (1980).

24. J. B. Johnson, Phys. Rev. **26**, 71 (1925).

25. S. Machlup, Proc. 6$^{th}$ Int. Conf. on Noise in Physical Systems, Gaithersburg, MD, USA, 157 (1981).

26. M. Gardner, Scientific American **238**, 16 (1978).

27. R. F. Voss and J. Clarke, J. Acoust. Soc. Am. **63**, 258 (1978).

28. T. Musha, Proc. 6$^{th}$ Int. Conf. on Noise in Physical Systems, Gaithersburg, MD, USA, 143 (1981).

29. D. E. Burgess, T. A. Zimmerman, M. T. Wise, S.-G. Li, D. C. Randall, and D. R. Brown, Am. J. Physiol. Regulatory. Integrative. Comp. Physiol. **277**, R894 (1999).

30. T. R. Wigton, R. E. Sabbagha, R. K. Tamura, L. Cohen, J. P. Minogue, and J. F. Strasburger, Obstet. Gynecol. **82**, 219 (1993).

31. M. Xiao, K. B. Klaassen, J. C. L. van Peppen, and M. H. Kryder, J. Appl. Phys. **85**, 5855 (1999).

32. H. T. Hardner, M. J. Hurben, and N. Tabat, IEEE Trans. Magn. **35**, 2592 (1999).

33. A. Wallash, IEEE Trans. Magn. **34**, 1450 (1998).

34. M. A. M. Gijs, J. B. Giesbers, P. Belien, J. W. van Est, J. Briaire, and L. K. J. Vandamme, J. Magn. Magn. Mater. **165**, 360 (1997).

35. N. Smith, A. M. Zeltser, and M. R. Parker, IEEE Trans. Magn. **32**, 135 (1996).

36. R. J. M. van de Veerdonk, P. J. L. Belien, K. M. Schep, J. C. S. Kools, M. C. de Nooijer, M. A. M. Gils, R. Coehoorn, and W. J. M. de Jonge, J. Appl. Phys. **82**, 6152 (1997).

37. B. Doudin, G. Redmond, S. E. Gilbert, and J.-Ph. Ansermet, Phys. Rev. Lett. **79**, 933 (1997).

38. M. A. M. Gijs, J. B. Giesbers, J. W. van Est, J. Briaire, L. K. J. Vandamme, and P. Belien, J. Appl. Phys. **80**, 2539 (1996).

39. B. Raquet, J. M. D. Coey, S. Wirth, and S. von Molnár, Phys. Rev. B **59**, 12435 (1999).

40. A. Anane, B. Raquet, S. von Molnár, L. Pinsard-Godart, and A. Revcolevschi, J. Appl. Phys. **87**, 5025 (2000).

41. R. D. Merithew, M. B. Weissman, F. M. Hess, P. Spradling, E. R. Nowak, J. O'Donnell, J. N. Eckstein, Y. Tokura, and Y. Tomioka, Phys. Rev. Lett. **84**, 3442 (2000).

42. V. Podzorov, M. Uehara, M. E. Gershenson, T. Y. Koo, and S-W. Cheong, Phys. Rev. B **61**, R3784 (2000).

43. M. Rajeswari, R. Shreekala, A. Goyala, S. E. Lofland, S. M. Bhagat, K. Ghosh, R. P. Sharma, R. L. Greene, R. Ramesh, T. Venkatesan, and T. Boettcher, Appl. Phys. Lett. **73**, 2670 (1998).

44. S. K. Arora, R. Kumar, D. Kanjilal, R. Bathe, S. I. Patil, S. B. Ogale, and G. K. Mehta, Solid Stat. Comm. **108**, 959 (1998).

45. H. Jianhua and H. Kangquan, Chinese Science Bulletin **42**, 163 (1997).

46. H. T. Hardner, M. B. Weissman, M. Jaime, R. E. Treece, P. C. Dorsey, J. S. Horwitz, and D. B. Chrisey, J. Appl. Phys. **81**, 272 (1997).

47. M. Rajeswari, A. Goyal, A. K. Raychaudhuri, M. C. Robson, G. C. Xiong, C. Kwon, R. Ramesh, R. L. Greene, T. Venkatesan, and S. Lakeou, Appl. Phys. Lett. **69**, 851 (1996).

48. G. B. Alers, A. P. Ramirez, and S. Jin, Appl. Phys. Lett. **68**, 3644 (1996).

49. R. F. Voss and J. Clarke, Phys. Rev. B **13**, 556 (1976).

50. A. Caloyannides, J. Appl. Phys. **45**, 307 (1974).

51. F. N. Hooge and A. Hoppenbrouwers, Physica (Amsterdam) **45**, 386 (1969); 42, 331 (1969).

52. F. N. Hooge, Physica B **83**, 14 (1976).

53. T. G. M. Kleinpenning, J. Phys. Chem. Solids **37**, 925 (1976).

54. R. D. Black, P. J. Restle, and M. B. Weissman, Phys. Rev. B **28**, 1935 (1983).

55. C. T. Sah and F. H. Hielscher, Phys. Rev. Lett. **17**, 956 (1966).

56. F. N. Hooge, Phys. Lett. **29**A, 139 (1969).

57. T. G. M. Kleinpenning, J. Appl. Phys. **51**, 3438 (1980).

58. M. B. Weissman, Physica B **100**, 157 (1980).

59. H. M. J. Vaes and T. G. M. Kleinpenning, J. Appl. Phys. **48**, 5131 (1977).

60. M. B. Weissman, J. Appl. Phys. **51**, 5872 (1980).

61. J. H. Scofield, J. V. Mantese, and W. W. Webb, Phys. Rev. B **32**, 736 (1985).

62. Z. Çelik-Butler and T. Y. Hsiang, Solid State Electron. **31**, 241 (1988).

63. M. Viret, *Private Communication*.

64. J. Bernamont, Ann. Phys. (Leipzig) **7**, 71 (1937).

65. F. K. du Prè, Phys. Rev. **78**, 615 (1950).

66. A. L. McWhorter, in "Semiconductor Surface Physics", R. H. Kingston (ed.), University of Pennsylvania Press, Philadelphia, p.207 (1957).

67. P. Dutta, P. Dimon, and P.M. Horn, Phys. Rev. Lett. **43**, 646 (1979).

68. J. H. Scofield, J. V. Mantese, and W. W. Webb, Phys. Rev. B **34**, 723 (1986).

69. J. Pelz and J. Clarke, Phys. Rev. Lett. **55**, 738 (1985).

70. D. M. Fleetwood, T. Postel, and N. Giordano, J. Appl. Phys. **56**, 3256 (1984).

71. J. H. Scofield, D. H. Darling, and W. W. Webb, Proc. 6th Int. Conf. on Noise in Physical Systems, Gaithersburg, MD, USA, 147 (1981).

72. P. H. Handel, Phys. Rev. Lett. **34**, 1492 (1975); **34**, 1495 (1975); Phys. Lett. **53**A, 438 (1975).

73. P. de Los Rios and Y. C. Zhang, Phys. Rev. Lett. **82**, 472 (1999).

74. P. Bak, C. Tang, and K. Wiesenfeld, Phys. Rev. Lett. **59**, 381 (1987); Phys. Rev. A **38**, 364 (1988).

75. W. Reim, R. H. Koch, A. P. Malozemoff, M. B. Ketchen, and H. Maletta, Phys. Rev. Lett. **57**, 905 (1986).
76. M. B. Weissman, Rev. Mod. Phys. **65**, 829 (1993).
77. H. B. Callen and T. A. Welton, Phys. Rev. **83**, 34 (1951).
78. N. E. Israeloff, M. B. Weissman, G. A. Garfunkel, D. J. Van Harlingen, J. H. Scofield, and A.J. Lucero, Phys. Rev. Lett. **60**, 152 (1988).
79. M. B. Weissman and N. E. Israeloff, J. Appl. Phys. **67**, 4884 (1990).
80. N. Giordano, Phys. Rev. B **53**, 14937 (1996).
81. P. J. Restle, R. J. Hamilton, M. B. Weissman, and M. S. Love, Phys. Rev. B **31**, 2254 (1985).
82. C. Parman, N. E. Israeloff, and J. Kakalios, Phys. Rev. B **44**, 8391 (1991).
83. B. Raquet, B. Aronzon, V. V. Rylkov, E. Z. Meilikhov, N. Negre, M. Goiran, and J. Leotin, *unpublished*.
84. M. J. Kirton and M. J. Uren, Adv. Phys. **38**, 367 (1989).
85. L. M. Lust and J. Kakalios, Phys. Rev. Lett. **75**, 2192 (1995).
86. K. S. Ralls and R. A. Buhrman, Phys. Rev. Lett. **60**, 2434 (1988).
87. K. R. Farmer, C. T. Rogers, and R.A. Buhrman, Phys. Rev. Lett. **58**, 2255 (1987).
88. M. N. Baibich, J. M. Broto, A. Fert, F. Nguyen van Dau, F. Petroff, P. Etienne, G. Creuset, A. Friedrich, and J. Chazeles, Phys. Rev. Lett. **61**, 2472 (1988).
89. A. M. Bratkovsky, JETP, **65**, 452 (1997); Phys. Rev. B **56**, 2344 (1997).
90. C. T. Tanaka, J. Nowak, and J. S. Moodera, J. Appl. Phys. **81**, 5515 (1997).
91. C. T. Tanaka and J. S. Moodera, J. Appl. Phys. **79**, 6265 (1996).
92. J. S. Moodera and D. M. Mootoo, J. Appl. Phys. **76**, 6101 (1994).
93. M. Viret, M. Drouet, J. Nassar, J. P. Contour, C. Fermon, and A. Fert, Europhys. Lett. **39**, 545 (1997).
94. S. Gardelis, C. G. Smith, C. H. W. Barnes, E. H. Linfield, and D. A. Ritchie, Phys. Rev. B **60** ,7764 (1999).
95. P. R. Hammar, B. R. Bennett, M. J. Yang, and M. Johnson, Phys. Rev. Lett. **83**, 203 (1999).
96. R. Fiederling, M. Keim, G. Reuscher, W. Ossau, G. Schmidt, A. Waag, and L. W. Molenkamp, Nature **402**, 787 (1999).
97. Y. Ohno, D. K. Young, B. Beschoten, F. Matsukura, H. Ohno, and D. D. Awschalom, Nature **402**, 790 (1999).
98. H. T. Hardner, Thesis, University of Illinois (1996).
99. S. Scouten, Y. Xu, B. H. Moeckly, and R. A. Buhrman, Phys. Rev. B. **50**, 16121 (1994).
100. L. Liu, K. Zhang, H. M. Jaeger, D. B. Buchholz, and R. P. H. Chang, Phys. Rev. B **49**, 3679 (1994).
101. A. Barry, J. M. D. Coey, L. Ranno, and K. Ounadjela, J. Appl. Phys. **83**, 7166 (1998).
102. K. Steenbeck, T. Eick, K. Kirsch, H.-G. Schmidt, and E. Steinbeiß, Appl. Phys. Lett. **73**, 2506 (1998); N. K. Todd, N. D. Mathur, S. P. Isaac, J. E. Evetts, and M. G. Blamire, J. Appl. Phys. **85** 7263 (1999).
103. Y. Xu, V. Dworak, A. Drechsler, and U. Hartmann, Appl. Phys. Lett. **74**, 2513 (1999).
104. E. L. Nagaev, Physics-Uspekhi **39**, 781 (1996).
105. A. Moreo, S. Yunoki, and E. Dagotto, Science **283**, 2034 (1999).
106. P. Schlottmann, Phys. Rev. B **59**, 11484 (1999).
107. R. Rammal, C. Tannous, P. Breton, and A-M. S. Tremblay, Phys. Rev. Lett. **54**, 1718 (1985).

108. Z. Rubin, S. A. Sunshine, M. B. Heaney, I. Bloom, and I. Balberg, Phys. Rev. B **59**, 12196 (1999).
109. A. Anane, *Private Communication.*
110. H. T. Hardner, M. B. Weissman, M. B. Salamon, and S. S. P. Parkin, Phys. Rev. B **48**, 16156 (1993).
111. H. T. Hardner, M. B. Weissman, B. Miller, R. Loloee, and S. S. P. Parkin, J. Appl. Phys. **79**, 7751 (1996).
112. J. X. Shen, C. Xie, J. Ding, A. Shultz, and S. H. Liao, IEEE Trans. Magn. **35**, 2595 (1999).

# Part III

# Materials, Techniques and Devices

# 12  Materials for Spin Electronics

J. M. D. Coey

Physics Department, Trinity College, Dublin 2, Ireland

**Abstract.** Materials which are currently used in spin electronic devices, and materials which may be useful in future are discussed. These include iron- cobalt- and nickel-based alloys for spin polarization and analysis, metallic and insulating antiferromagnets for exchange bias and oxides for tunnel barriers. The $3d$ alloys also serve as detection or sensor layers. Permanent magnet materials play a role in biasing some device structures. Novel materials are half-metallic oxides for all-oxide devices, and magnetic semiconductors which may allow the integration of spin electronics and optoelectronics.

## 12.1  Introduction

This chapter presents magnetic materials of interest for spin electronic devices. The focus is on crystal structure and intrinsic magnetic properties of the bulk materials, although it must be remembered that when incorporated into devices these materials frequently form part of a thin film stack with a layer thickness $< 10$ nm. The structure and magnetic properties of thin films can differ significantly from those of the bulk. To cite just two examples, the atomic magnetic moments in a free surface layer of a ferromagnetic film may be enhanced because of band narrowing, and surface anisotropy is present which is typically $\simeq 0.1$ mJ m$^{-2}$ with the anisotropy direction normal to the film surface.

The properties that are exploited in spin electronic devices are of several kinds, but they relate mainly to the hysteresis curve and magnetic-field-dependent transport properties. Most semiconductors and semimetals are nonmagnetic; they exhibit the normal Hall effect, and the classical $B^2$ magnetoresistance due to the Lorentz force $-e\boldsymbol{v} \times \boldsymbol{B}$ acting on the electrons. When the mean free path is long, as in single-crystal films of bismuth [1], film dimensions influence the magnetoresistive response. However, it is unnecessary to consider the spin of the electrons to explain these magnetoelectronic effects; conventional electronics has ignored the spin on the electron.

For ferromagnets, it is convenient to distinguish *intrinsic* magnetic properties, which are independent of microstructure or nanostructure in all but the thinnest films, from *extrinsic* properties which derive from the microstructure or nanostructure in an essential way. Besides the big three: Curie temperature $T_\mathrm{C}$, spontaneous magnetization $M_\mathrm{S}$, and magnetocrystalline anisotropy $K_1$, intrinsic properties include band structure, conductivity ratio $\alpha$ for $\uparrow$ and $\downarrow$ electrons, magnetostriction $\lambda_\mathrm{S}$, anisotropic magnetoresistance (AMR) and colossal magneto-resistance (CMR). The main two extrinsic properties are remanence

$M_r$, and coercivity $H_c$, but there is also induced anisotropy $K_u$, granular and powder magnetoresistance (PMR), giant magnetoresistance (GMR), and tunneling magnetoresistance in planar tunnel junctions (TMR). GMR and TMR are at the heart of spin electronics, as we know it at present.

The magnetic materials principally used in spin electronics are soft ferromagnetic alloys of the late $3d$ metals. These serve as sources and conduction channels for the spin-polarized electrons, as well as magnetic flux paths and shields. Most progress has been made with sensors, ranging from simple position sensors and elements for nondestructive testing of ferrous metals to sophisticated miniature sensor elements in read heads for digital tape and disc recording where requirements are very demanding; high permeability is required with a sharp low-field switching response that extends to frequencies in the GHz range. Magnetic memory and logic elements require square hysteresis loops. All AMR, GMR, TMR and magnetic random access memory (MRAM) devices developed so far are based on $3d$ ferromagnetic metals and alloys. So too are magnetic three-terminal devices such as spin transistors and spin injection switches, as well as the magnetic Schottky barriers for injecting spin-polarised hot electrons into semiconductors.

Antiferromagnets, which may be metals or insulators, find a use in exchange biasing of magnetic thin film structures. Hard magnetic materials in thin film form can be employed to generate a stray field to stabilize a particular domain structure in a contiguous soft layer. Ferromagnetic oxides are at the research stage, but it is hoped that in future their half-metallic character will be exploited in sources and analysers of completely spin-polarized electrons. Magnetic semiconductors are another class of potentially-interesting materials, but they suffer from the critical defect that their Curie temperatures are far below room temperature.

Here, each of the main groups of actually or potentially useful materials will be presented, and some alloys of interest for particular applications are highlighted.

## 12.2    Iron Group Alloys

First we review the ferromagnetic elements Fe, Co and Ni, and then discuss alloy systems based on these three elements. Each has a different crystal structure, body-centred cubic (bcc) for iron, hexagonal close-packed (hcp) for cobalt and face-centred cubic (fcc) for nickel. Their electronic densities of states are compared in Fig. 12.1. All three transition elements have a broad, almost unpolarised $sp$ band superposed on a spin-split $3d$ band. The unsplit density of states $D(E)$ exhibits a peak at the Fermi level $E_F$ so that the Stoner criterion for spontaneous ferromagnetism $D(E_F)I > 1$ is satisfied. The exchange interaction $I$ in the $3d$ band is $\simeq 1$ eV for all three ferromagnetic elements [2]. Iron, which has the largest atomic moment of 2.22 Bohr magnetons ($\mu_B$), is a weak ferromagnet in the sense that there are both $3d \uparrow$ and $3d \downarrow$ electrons at the Fermi level. Cobalt and nickel, which have smaller moments, are strong ferromagnets in the sense that the $3d \uparrow$ states lie entirely below the Fermi level. The electronic configuration of Ni, for example, is approximately $3d^{\uparrow 5.0}3d^{\downarrow 4.4}4s^{0.3}4p^{0.3}$, which gives a spin-only moment of $0.62\mu_B$. Cobalt has an orbital moment of $\simeq 0.15\mu_B$ and a total moment of $1.73\mu_B$. The atomic moments quoted are at zero temperature.

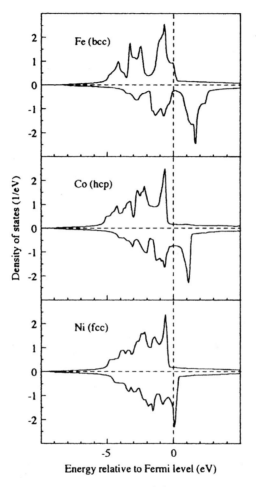

**Fig. 12.1.** The spin-split densities of states $D(E)$ calculated for iron, cobalt and nickel at zero temperature.

Strong ferromagnets were expected to show a higher value of spin polarization $P$ of emitted electrons and a larger resistivity ratio $\alpha$ for $\uparrow$ and $\downarrow$ carriers than weak ferromagnets because scattering of $sp$ electrons into the filled $3d\uparrow$ states is suppressed. In fact $P$ turns out to be almost the same in magnitude and, more significantly of the same sign for all three ferromagnetic elements. $P$ is easy to define, but difficult to measure. The definition is simply $(n^\uparrow - n^\downarrow)/(n^\uparrow + n^\downarrow)$ where $n^{\uparrow,\downarrow}$ is the number of conduction electrons of either spin in the unit cell, but in any experiment to measure $n^{\uparrow,\downarrow}$ a weighting factor is involved which depends on how the spin-polarized electrons are detected [3]. Methods for measuring $P$ include measuring the $I : V$ characteristic after applying a field to a tunnel junction between the ferromagnet and a thin film of superconducting aluminium, or measuring the $I : V$ characteristic of a point contact between a superconducting tip and the ferromagnet (Andreev reflection).

A summary of the main intrinsic properties of the ferromagnetic elements at room temperature is given in Table 12.1 [4]. Values refer to room temperature, except for the spin polarization, which was measured by Andreev reflection at 4.2 K [5].

**Table 12.1.** Intrinsic magnetic properties of Fe, Co and Ni.

| | structure /density $(kg\ m^{-3})$ | lattice parameters (pm) | $T_C$ (K) | $M_S$ $(MA\ m^{-1})$ | $K_1$ $(kJ\ m^{-3})$ | $\lambda_S$ $(10^{-6})$ | $\alpha$ | $P$ (%) |
|---|---|---|---|---|---|---|---|---|
| Fe | bcc 7874 | 287 | 1044 | 1.71 | 48 | $-7$ | 1.6 | 45 |
| Co | hcp 8836 | 251 407(fcc) | 1388 | 1.45 | 530 | $-62$ | 8.0 | 42 |
| Ni | fcc 8902 | 352 | 628 | 0.49 | $-5$ | $-34$ | | 44 |

A number of derived properties important for aspects of nanoscale magnetism are listed in Table 12.2. These are the exchange stiffness $A$, the exchange length $l_{ex} = \sqrt{A/\mu_0 M_S^2}$ and the Bloch domain wall width $\delta_W = \sqrt{A/K_1}$. The coherence radius $l_{coh} = \sqrt{24}l_{ex}$, the single-domain particle size $d_{sd} = 72\sqrt{(AK_1)}/\mu_0 M_S^2$ and the superparamagnetic blocking diameter at room temperature $(150k_BT/\pi K_1)^{1/3}$ refer respectively to the reversal mechanism, domain structure and stability of small particles. Analogous quantites can be defined for thin films. Other significant length scales are the mean free paths $\lambda$ for $\uparrow$ and $\downarrow$ electrons and spin-diffusion length $\lambda_{sd}$; the spin diffusion length is usually one or two orders of magnitude greater than the mean free path, because spin-flip scattering events are comparatively rare. The mean free path is relevant for in-plane conduction in multilayer stacks, the usual GMR configuration. For cobalt, $\lambda^{\uparrow} \simeq 5.0$ nm, $\lambda^{\downarrow} \simeq 0.6$ nm. The spin diffusion length is the appropriate scale for perpendicular-to-plane conduction. $\lambda_{sd}$ is about 50 nm for Co [6]. Some desirable properties sought in soft ferromagnetic 3d alloys are a high magnetization and high degree of spin polarization, low anisotropy and zero magnetostriction, since a stress $\sigma$ induces a uniaxial anisotropy $K_{stress} = (3/2)\lambda_S\sigma$. Often it is desirable to create a weak uniaxial anisotropy $K_u$ ($\simeq 1$ kJ m$^{-3}$) by processing a thin film of a disordered alloy in an applied magnetic field, which creates some slight texture or pair ordering on an atomic scale. The weak induced anisotropy increases the permeability in the longitudinal direction, giving a square hysteresis loop with little coercivity. In the transverse direction there is a straight anhysteretic magnetization curve saturating at $2K_u/\mu_0 M_S$ (Fig. 12.2).

**Table 12.2.** Derived properties for Fe, Co and Ni.

|  | $A$ (pJ m$^{-1}$) | $l_{\mathrm{ex}}$ (nm) | $\delta_B$ (nm) | $l_{\mathrm{coh}}$ (nm) | $d_{\mathrm{sd}}$ (nm) | $d_{\mathrm{sp}}$ (nm) |
|---|---|---|---|---|---|---|
| Fe | 8.3 | 1.5 | 41 | 7.4 | 12 | 16 |
| Co | 10.3 | 2.0 | 14 | 9.7 | 64 | 7 |
| Ni | 3.4 | 3.4 | 82 | 16 | 31 | 34 |

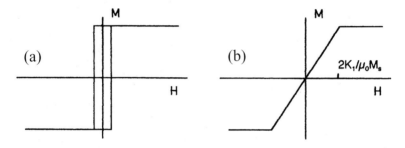

**Fig. 12.2.** Hysteresis loops for a soft magnetic material with weak induced uniaxial anisotropy, measured (a) in the longitudinal direction and (b) in the transverse direction.

Resistivity is also an issue in devices which switch at high frequency. Useful approaches are lamination of metallic and insulating films, or decorating grain boundaries to make them resistive, thereby minimizing eddy current losses. Amorphous alloys have the advantage of an intrinsically high resistivity, of order 1.5 μΩ m, which is the maximum possible for a homogeneous metal since the mean free path can be no shorter than the interatomic separation. The magnetocrystalline anisotropy of isotropic amorphous alloys is zero. As with disordered crystalline alloys, weak uniaxial anisotropy $K_u$ can be induced by annealing or depositing the material in a uniform magnetic field.

A famous summary of the magnetic moment per atom in binary $3d$ alloys is the Slater–Pauling curve, shown in Fig. 12.3. The main branch, with slope $-1$ accounts for the strong ferromagnets having a filled $3d^{5\uparrow}$ subband. Each electron removed comes essentially from the $3d^\downarrow$ subband and increases the spin moment by $1\mu_B$. Extrapolating the curve to hypothetical strongly-ferromagnetic iron gives a moment of $2.6\mu_B$. The branches with slope $\simeq 1$ represent the moments of alloys between early and late transition metals, where the $3d$ states of the early transition elements lie well above the Fermi level of the late transition

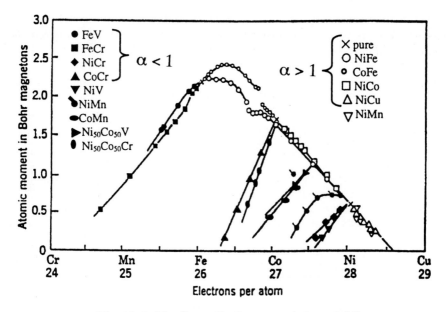

**Fig. 12.3.** The Slater–Pauling curve (after ref. [7]).

elements. The conductivity ratio $\alpha < 1$ for alloys on the branches with positive slope, whereas $\alpha > 1$ on the branch with negative slope [7]. Moments in strong ferromagnets are described quantitatively by the magnetic valence model [8], which is a generalization of these ideas. The chemical valence $Z$ of an atom is $n^\uparrow + n^\downarrow$, where $n^{\uparrow,\downarrow}$ are the number of valence electrons with either spin. The spin moment in units of $\mu_B$ is $n^\uparrow - n^\downarrow = 2n^\uparrow - Z$. Now $n_d^\uparrow$ is precisely 5 for strong ferromagnets, so the magnetic valence defined by $Z_m = 2n_d^\uparrow - Z$ is an integer. The moment $m$ is therefore $Z_m + 2n_{sp}^\uparrow$ where $n^\uparrow = n_d^\uparrow + n_{sp}^\uparrow$, and $2n_{sp}^\uparrow$ is the number of electrons in the $sp$ band, which is practically unpolarized. In an alloy, the average moment *per atom* is deduced by replacing $Z_m$ by its average over all the atoms present;

$$\langle m \rangle = \langle Z_m \rangle + 2n_{sp}^\uparrow . \tag{12.1}$$

Here $n_d^\uparrow$ is 5 for iron and atoms to the right, but zero for atoms at the beginning of the $3d$ series. $Z_m$ is $-3$ for Sc, Y, B ...; $-4$ for Ti, Zr, C ..., 2 for Fe, 1 for Co and 0 for Ni. $2n_{sp}^\uparrow$ is about 0.3.

### 12.2.1   Iron–based Alloys

Besides the fact that it is not a strong ferromagnet, the problems with iron are its anisotropy and magnetostriction (Table 12.1). $K_1$ is fairly large for a cubic material, and $\lambda_S$, which is an isotropic polycrystalline average, is the resultant of bigger values in the $\langle 100 \rangle$ and $\langle 111 \rangle$ directions, $21 \times 10^{-6}$ and $-20 \times 10^{-6}$, respectively.

Alloying iron with cobalt produces a strong ferromagnet at $Fe_{65}Co_{35}$ (*Permendur*) which holds the record room-temperature magnetization of 1.95 MA m$^{-1}$, corresponding to a ferromagnetic polarization $\mu_0 M_S = 2.45$ T. The magnetization and Curie temperature are almost constant in $Fe_x Co_{100-x}$ for $35 \leq x \leq 55$. The anisotropy of bcc Co is about $-60$ kJ m$^{-3}$, so zero anisotropy occurs at $x \simeq 55$. Unfortunately $\lambda_S$ is $60 \times 10^{-6}$ and the alloy usually has low permeability. The near-equiatomic compositions have a tendency to CsCl-type order (Fig. 12.4) and a unique axis may be induced by field annealing which can lead to $H_c \simeq 15$ A m$^{-1}$ and an initial permeability $\mu_I \simeq 800$.

Generally it is not possible to find a composition in a binary alloy system where $K_1$ and $\lambda_S$ go to zero simultaneously. By chance this almost happens in the Fe-Ni system (*Permalloy*) discussed below. A bcc iron-based ternary system which does have a $K_1 = 0$; $\lambda_S = 0$ point is Fe-Si-Al at the composition $Fe_{74}Si_{16}Al_{10}$ (*Sendust*). The alloy has a tendency to order in the $Fe_3Si$ superstructure, and atomic order and composition must be accurately controlled to achieve optimum properties. Polarization is 1.2 T. Sendust has been used in write heads for magnetic recording.

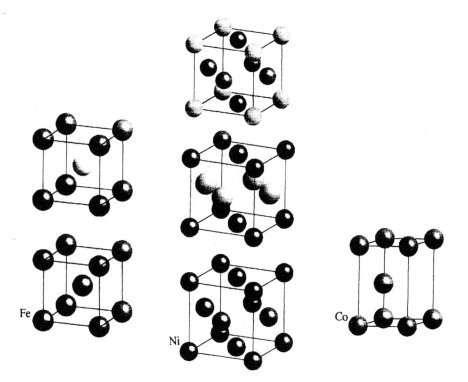

**Fig. 12.4.** Some simple crystal structures for metals; body-centred cubic (Fe) with the CsCl superstructure; face-centred cubic (Ni) with the $CuAu_3$ and tetragonal CuAu(I) superstructures; hexagonal close packed (Co).

Another approach is to prepare thin films with a concentration of dissolved nitrogen in excess of equilibrium by reactive sputtering. These have composition around $Fe_{97}N_3$ [9]. A few percent of an element such as Al, Ta, Ti or Rh serves to increase the solubility of nitrogen in iron or extend the stable $\alpha$-phase field. The saturation magnetostriction changes sign at about 3% N, and the use of additions inhibits grain growth and thereby helps to stabilize soft magnetic properties by *anisotropy averaging*. In soft ferromagnetic nanostructures, the characteristic length scale for anisotropy averaging is the domain wall width which sets the scale for the smallest possible domain size

$$\delta_B = \pi \sqrt{A/K_1} \,. \tag{12.2}$$

The number of crystallites of average diameter $D$ within a volume of $\delta_B^3$ is $N = (\delta_B/D)^3$. Anisotropy directions of the crystallites are random, so the effective anisotropy constant is $\langle K \rangle \simeq K_1/\sqrt{N}$. Hence

$$\langle K \rangle \simeq K_1(D/\delta_B)^{3/2} \,. \tag{12.3}$$

But it is this effective anisotropy constant which must be used to determine the domain wall width. Substituting from (12.2) with $K_1$ replaced by $\langle K \rangle$ yields

$$\langle K \rangle \simeq K_1^4 D^6/\pi^6 A^3 \,. \tag{12.4}$$

Taking $D = 20$ nm and the values for iron in Table 12.1 leads to $\langle K \rangle \simeq 0.6 \text{ kJ m}^{-3}$, a reduction of the anisotropy by two orders of magnitude. Anisotropy averaging in soft exchange-coupled nanostructures is a powerful way of making them very soft indeed [10]. An example here is *Finemet*, a two-phase nanostructure of crystalline $Fe_{80}Si_{20}$ regions in an amorphous Fe-B matrix. The composition is $Fe_{73.5}Si_{15}B_{7.5}Cu_1Nb_3$. Copper and niobium additions serve to promote nucleation of the Fe-Si crystallites and refine the grain structure, respectively. The anisotropy of the Fe-Si crystallites is exchange-averaged to zero and the contributions to $\lambda_S$ of the crystalline and amorphous regions are of opposite sign and cancel, yielding an iron-based nanocomposite with exceptionally high permeability. A variant on this is the Fe-Co-B system where nanometer-scale Fe-Co-rich regions are dispersed in a boron-rich amorphous matrix. A typical composition is $Fe_{62}Co_{21}B_{17}$, with polarization $\simeq 1.6$ T.

One other iron nitride that deserves a mention is the metastable $\alpha'$-$Fe_{16}N_2$. It has been reported to have a moment in thin film form as high as $3.5\mu_B/$Fe [11], but this result have not been independently confirmed, and are at variance with theoretical expectations. Recent surveys of the literature on this material place its likely room-temperature polarization at 2.3(1) T [9,12]. However, it is claimed that imperfectly-ordered thin films ($\simeq 40$ nm) with a large cell volume have a moment of $2.8\mu_B/$Fe [13], corresponding to a polarization of 2.5 T. The tetragonal $\alpha'$ phase has a large uniaxial anisotropy of order 1 MJ m$^{-3}$ [12], so anisotropy averaging here would need impracticably small crystallites, of diameter 2-3 nm.

## 12.2.2   Nickel–based Alloys

The fcc $Ni_xFe_{100-x}$ system includes the famous *Permalloy* composition range $78 \leq x \leq 81$. Permalloy is probably the best-studied soft magnetic material, as it is very suitable for thin film devices. Permalloy is a strong ferromagnet with a polarization $\mu_0 M_S \simeq 1.0$ T. The conductivity ratio for $\uparrow$ and $\downarrow$ electrons is $\alpha = 6$ and the degree of spin polarization for emitted electrons is $P = 0.37$ [5]. The degree of ordering of Fe and Ni in the $Cu_3Au$ structure (Fig. 12.4b) can be adjusted by heat treatment, and weak uniaxial anisotropy can be induced by field annealing. Cobalt is added to fcc Ni-Fe alloys around the permalloy composition range to increase their magnetization and improve their susceptibility to magnetic field treatment. It is then possible to induce the uniaxial anisotropy by depositing the film in an applied field of order 1 kA m$^{-1}$, which is preferable to field annealing for device structures as it avoids possible interdiffusion of the layers. In films thinner than 20 nm, the induced anisotropy $K_u$ is proportional to film thickness. A typical composition is $Ni_{65}Fe_{15}Co_{20}$.

The particular feature of permalloy is that $K_1$ and $\lambda_S$ change sign at nearly the same composition, making it possible to achieve an excellent soft magnetic response in a binary system (Fig. 12.5). Small additions of Mo or Cu are used to optimize the properties. Permalloy films have a good AMR response of 2% in a field of about 300 A m$^{-1}$. For this reason permalloy was used in AMR read heads. Thicker films ($\simeq 1$ µm) of permalloy are used in thin film write heads for hard discs and tapes. Produced by electrodeposition [14] from a single bath containing iron and nickel salts together with additives such as saccharine which serve to relieve the strain in the deposited film or increase its resistivity, these films are also employed for on-chip inductors. Uniaxial anisotropy is achieved by electrodeposition in a magnetic field of order 40 kA m$^{-1}$.

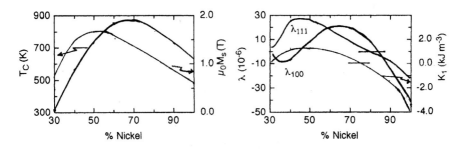

**Fig. 12.5.** Magnetic properties of Ni-Fe alloys.

The main drawback of permalloy is its relatively low polarization, which limits the field that can be generated and makes it unsuitable for ultrahigh-density write heads. The $Ni_{50}Fe_{50}$ composition is better in this respect, since $\mu_0 M_S = 1.6$ T. However the ultimate recording densities will need an excellent soft material with a polarization greater than 2.0 T, which cannot be achieved in the Ni-Fe system. A third region of interest in the Ni-Fe series is Invar around $Ni_{35}Fe_{65}$, which is at the limit of the fcc phase field. Here $T_C$ is low and the natural thermal expansion over a limited range of temperature around ambient is compensated by the temperature-dependent spontaneous volume magnetostriction $\omega_S$, which is independent of applied field for a strong ferromagnet.

Anisotropy, but not magnetostriction can be suppressed by preparing Ni-Fe alloys in an amorphous form, using boron as a glass former. *Metglas 2628* a-$Fe_{40}Ni_{40}B_{20}$) is one such alloy. It has a random-dense-packed Bernal structure, with boron occupying the large intersticies in the random packing. The polarization, 0.8 T, is much reduced compared to $Ni_{50}Fe_{50}$ because of the presence of boron, which has a magnetic valence of $-3$, and the lower density of the amorphous dense-packed structure.

### 12.2.3    Cobalt–based Alloys

Cobalt normally has an hexagonal close-packed structure with a fairly large uniaxial anisotropy of $K_1 = 530$ kJ m$^{-3}$, corresponding to an anisotropy field $H_a = 2K_1/\mu_0 M_S$ of 0.57 MA m$^{-1}$. Cobalt can be easily stabilized in an fcc structure and the quoted Curie temperature, which is the highest known for any material, actually refers to the fcc phase. Cobalt is used in alloys to increase $T_C$. The atomic moment is unusually robust and structure independent. A thin (0.4 nm) fcc film at the interface between magnetic and nonmagnetic layers serves to provide a magnetically-sharp interface which promotes spin-dependent scattering [15]. Iron and boron are sometimes added to the interfacial cobalt. A typical composition is $Co_{87}Fe_9B_2$. Co-Fe-B is also used for the free layer of spin valves, where the additives allow enhanced uniaxial anisotropy and improve the thermal stability.

The anisotropy of hcp cobalt is insufficient to make a true permanent magnet, for which the anisotropy field $H_a = 2K_1/\mu_0 M_S$ would have to be significantly greater than the magnetization. Nevertheless, thin films with in-plane c-axis orientation can exhibit useful coercivity. Magnetization is in-plane in most magnetic thin film device structures and the demagnetizing field is small. Thin film media for hard discs are based on hexagonal cobalt with Cr, Pt and B additions which help create a layer of magnetically-decoupled Co-rich crystallites about 20 nm thick and 10-20 nm in size. Coercivity is $\simeq 300$ kA m$^{-1}$. A typical composition is $Co_{67}Cr_{20}Pt_{11}B_6$.

Cobalt-based alloys with a uniaxial crystal structure easily develop a very large anisotropy field and exhibit permanent magnet properties. A good example is $YCo_5$ where yttrium occupies alternate planes in the hexagonal structure. The anisotropy field $H_a = 12.3$ MA m$^{-1}$. Another structure with uniaxial anisotropy is the face-centred tetragonal CuAu(I) structure adopted by CoPt, and illustrated in Fig. 12.4b. Alternate planes are composed of Co and Pt, and the anisotropy field is 9.8 MA m$^{-1}$. The cubic CuAu$_3$ type of order occurs in $CoPt_3$, which is a semihard material that has been used in demonstration MRAM devices. Polycrystalline films of Co-Cr-Pt and CoPt with the c-axis in-plane are used as permanent magnets to longitudinally bias and stabilize the domain structure in both AMR and GMR read heads.

Cobalt-based alloys with the fcc structure and amorphous cobalt-based alloys are much softer. In the Fe-Co-Ni system, nanocrystalline electrodeposited alloys at the border of the fcc and bcc phase fields have exchange-averaged anisotropy and near-zero magnetostriction. There are reports of polarization in excess of 2.0 T for $Co_{65}Ni_{12}Fe_{23}$ [16], $Co_{56}Ni_{13}Fe_{31}$ [17] and $Co_{52}Ni_{29}Fe_{19}$ [18]. Amorphous cobalt-rich alloys of composition $(CoFe)_{80}B_{20}$ can have zero magnetostriction, and are excellent high-permeability materials. The amorphous $Co_{90}Zr_{10}$ system also shows a zero magnetostriction point when a few percent of tantalum or rhenium is substituted for zirconium, or some nickel is substituted for cobalt. Polarization is about 1.4 T. The amorphous alloys are mechanically much harder than permalloy, and they are suitable for thin film write heads.

## 12.3   Antiferromagnets

It is common practice to pin the direction of magnetization of one of the ferromagnetic layers in a spin valve by exchange coupling to an antiferromagnet [19]. When the Curie temperature of the ferromagnet is greater than the Néel temperature of the antiferromagnet $T_C > T_N$ the direction of magnetization of the pinned layer may be set by cooling the exchange couple in a magnetic field. Otherwise it may be necessary to deposit or anneal the antiferromagnet in a large applied field. The direction of magnetization of the free layer in a spin valve can switch from antiparallel to parallel to the pinned layer under the influence of a small stray field which is sensed by the device (parallel anisotropies). Otherwise the direction of magnetization of the free layer may be set perpendicular to the direction of magnetization of the pinned layer with a small induced anisotropy $K_u$; the stray field then causes the magnetization of the free layer to rotate continuously (crossed anisotropies). These cases are illustrated schematically in Fig. 12.6.

Considering only the pinned layer of thickness $t_p$, its energy per unit area in the presence of an external field $H$ is

$$E/A = -\mu_0 M_f H t_p \cos(\theta) + K_u t_p \sin^2(\theta) - \sigma \cos(\theta), \qquad (12.5)$$

where $\sigma$ is the interface exchange coupling, which is of order 0.1 mJ m$^{-2}$. The origin of the exchange coupling at the interface is still a matter for discussion [19], but a common view is that the ferromagnetic and antiferromagnetic axes are perpendicular at the interface, and that a domain wall develops in the ferromagnetic layer, provided the antiferromagnetic layer exceeds the critical thickness $t_0$ needed to generate exchange bias. $t_0$ ranges from 7 nm for FeMn to 15 nm for a-TbCo$_3$ or $> 50$ nm for $\alpha$-Fe$_2$O$_3$ [20].

The exchange coupling in (12.5) is represented by a field $H_{ex} = \sigma/\mu_0 M_p t_p$ acting on the pinned layer, which depends on temperature and falls to zero at a blocking temperature $T_b < T_N$. The unblocking of exchange bias can reflect a thermally-excited relaxation mode impeded by weak anisotropy, such as rotation of the antiferromagnetic axis in the 111 plane of NiO, or else may reflect an atomic order-disorder transition. A typical room-temperature value of $H_{ex}$ for a 5 nm thick pinned layer with $\mu_0 M = 1$ T is 20 kA m$^{-1}$. Some representative multilayer structures including an antiferromagnetic layer are shown in Fig. 12.7. In dual spin valves, the stacks are mirrored about a central free layer, which is sandwiched between two pinned layers. The quality of a spin valve depends on the field needed to switch the free layer, and the quantity $\Delta\rho/\rho = (\rho^{\uparrow\downarrow} - \rho^{\uparrow\uparrow})/\rho^{\uparrow\uparrow}$ where the double superscripts refer to the antiparallel or parallel configurations for the pinned and free layers. In a simple two-current model, this is related to $\alpha$, the conductivity ratio for $\uparrow$ and $\downarrow$ electrons in the spin valve structure by the formula $\Delta\rho/\rho = (1 - \alpha^2)/4\alpha$ [20].

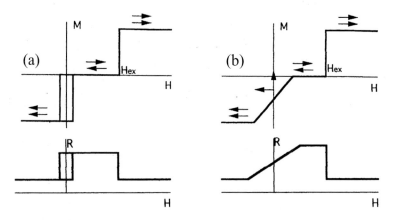

**Fig. 12.6.** Schematic response of a spin valve structure with (a) parallel and (b) crossed anisotropies.

Most of the antiferromagnets of interest for spin electronic devices are manganese alloys, whose properties are summarized in Table 12.3. Some oxides and amorphous materials are also useful. Manganese antiferromagnets close to the equiatomic composition may have a disordered fcc structure, or else adopt the face-centred tetragonal CuAu(I) structure illustrated in Fig. 12.4. The manganese alloys exhibit a great variety of collinear and noncollinear antiferromagnetic structures, yet all are able to provide exchange bias. For example, FeMn, which has been widely studied with permalloy as the adjacent ferromagnetic layer, has a disordered fcc crystal structure, and a magnetic structure with four sublattices oriented along the four $\langle 111 \rangle$ directions [4]. NiMn has a higher blocking temperature, and is chemically more inert than FeMn. It has the fct structure, and a magnetic structure of antiferromagnetic 002 planes, with $S \parallel a$. The quest for a high blocking temperature and the ability to conveniently set the antiferromagnetic axis has led to investigation of Ir-Mn, Rh-Mn and Pt-Mn alloys as well as pseudobinaries such as $(Pd_{1-x}Pt_x)Mn$ and Cr-Mn alloys. The bulk magnetic

**Table 12.3.** Antiferromagnetic materials for exchange bias [19–21]. $S$ denotes the spin direction. $^{\#}$Order-disorder transition. $^{*}$Sperimagnetic; $T_N$ is the Néel temperature, $T_b$ an irreversible transition.

| | | $T_N$ (K) | $T_b$ (K) | $\sigma$ (mJ m$^{-2}$) |
|---|---|---|---|---|
| FeMn | fcc; four noncollinear sublattices; $S \parallel 111$ | 510 | 440 | 0.10 |
| NiMn | fct; antiferromagnetic 002 planes, $S \parallel a$ | 1050$^{\#}$ | 700 | 0.27 |
| PtMn | fct; antiferromagnetic 002 planes, $S \parallel c$ | 975 | 500 | 0.30 |
| RhMn$_3$ | triangular spin structure | 850 | 520 | 0.19 |
| Ir$_{20}$Mn$_{80}$ | fct; parallel spins in 002 planes, $S \parallel c$ | 690 | 540 | 0.19 |
| Pd$_{52}$Pt$_{18}$Mn$_{50}$ | fct; antiferromagnetic 002 planes | 870 | 580 | 0.17 |
| a-Tb$_{25}$Co$_{75}$ $^{*}$ | $T_{comp} = 340$ K | 600 | > 520 | 0.33 |
| NiO | parallel spins in 111 planes, $S \perp \langle 111 \rangle$ | 525 | 460 | 0.05 |
| $\alpha$-Fe$_2$O$_3$ | canted antiferromagnet, $S \perp c$ | 950 | | |

structures of the antiferromagnets summarized in Table 12.3 are not necessarily those of the thin films used for exchange bias.

It is important that the processing conditions needed for the magnetic material are compatible with the other materials present in the device. If, for example, magnetic devices such as MRAM are to be integrated with silicon electronics, the exchange couple should be stable at temperatures used in silicon processing, typically $> 300°$C for one hour to reduce radiation damage, and $200°$C for up to six hours for packaging. Ir$_{20}$Mn$_{80}$ might be suitable in this respect [22,23].

Some antiferromagnetic oxides are also useful. NiO has the highest Néel temperature of the monoxides, but $T_b$ is rather low; the anisotropy can be enhanced by cobalt substitution. Nevertheless NiO has been used in commercial spin-valves. $\alpha$-Fe$_2$O$_3$ has a high Néel temperature, but a thick layer is needed because of the low anisotropy of the antiferromagnet due to the proximity to room temperature of the Morin transition, where the antiferromagnetic anisotropy constant $K_1$ changes sign. The orthoferrites RFeO$_3$, which have $T_N$ in the range 620-740 K, are also being investigated. Oxides have the bonus that they act as specularly reflecting layers, which enhance the efficiency of spin valve structures.

A more recent development is the *artificial antiferromagnet* (AAF) [24] (Fig. 12.7). This is a stack of two or more ferromagnetic layers separated by layers of a nonmagnetic metal whose thickness is chosen to provide an antiferromagnetic interlayer exchange. Best is cobalt separated by a very thin layer,

$\simeq 0.6$ nm, of ruthenium. Iron or iron and boron additions facilitate the creation of induced anisotropy in the cobalt. Copper can be used as a weaker coupling layer [25]. The upper cobalt layer of the AAF can serve as the pinned layer of the spin valve, and layer thicknesses adjusted to give no stray field on the free layer. One of the exchange bias antiferromagnets just discussed can then be used to pin the lower cobalt layer, (Fig. 12.7). Annealing in a rather large field ($\simeq 1$ MA m$^{-1}$) is needed to saturate the AAF and fix the antiferromagnetic axis, but spin valves with an AAF pinned layer (Fig. 12.7) show better thermal stability and larger exchange bias than the basic configuration (Fig. 12.7) [26,27]. Stacks for dual spin valves with artificial antiferromagnets become impressively complex, with up to 15 layers [28] composed of as many as seven different materials, four of which are magnetic (AF bias layer, AAF magnetic layers, free layer, interface layer).

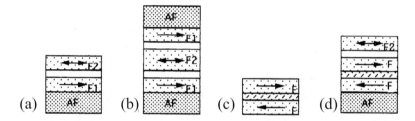

**Fig. 12.7.** Magnetic multilayers: (a) simple spin valve with an antiferromagnetic pin-ning layer, (b) double spin valve (c) an artificial antiferromagnet, (d) a spin valve based on an artificial antiferromagnet. The interfaces between the magnetic layers (F1, F2) and the spacer layer (unshaded) are often decorated with an ultrathin cobalt layer to improve $\Delta\rho/\rho$ for the devices.

## 12.4   Oxides and Half–metals

Thin oxide layers, usually 1-2 nm of nanocrystalline Al$_2$O$_3$, are used as barrier layers in planar tunnel junctions. These current-perpendicular-to-plane devices have at least twice the sensitivity ($\Delta\rho/\rho$) of GMR spin valves. Their high intrin-sic resistance and low power consumption makes them attractive for applications such as MRAM [29]. For read heads, a lower resistance is required, and the oxide barrier must then be very thin [30]. The most popular method for producing the Al$_2$O$_3$ barrier layer is by plasma oxidation of a layer of aluminum metal. Ther-mal oxidation in air is also used, but the best resistivities, of order 1 k$\Omega$ µm$^2$, may be obtained by oxidation assisted by ultraviolet light [31]. Other barrier oxides which have been investigated include SrTiO$_3$, TiO$_2$ and CeO$_2$. The mag-netoresistive response of the tunnel junction depends on the nature of the barrier layer [32]. The ferromagnetic electrodes in almost all the devices showing a useful effect at room temperature have been the $3d$ alloys discussed in Sect. 12.2.

Ferromagnetic metallic oxides and related compounds can act as sources and conduction channels for the spin-polarized electrons. The $3d$ metals, even those that are strong ferromagnets, suffer from incomplete spin polarization of the conduction electrons because of the presence of the $4s/4p$ bands, which are not spin-split. In principle, a more favourable situation can arise in oxides where hybridization of the outer metallic electron shells with the $2p(O)$ orbitals produces a gap of several eV between them. The $3d$ bands and the Fermi level tend to fall in this $s$-$p$ gap. When the Fermi level intersects only one of the spin-polarized $3d$ bands, and there is a gap in the density of states for the other spin direction we have a half-metallic ferromagnet (Fig. 12.8). A feature of a stoichiometric half-metallic ferromagnet is that the spin moment should be an integral number of Bohr magnetons. This follows since $n^\uparrow + n^\downarrow$ is an integer in a stoichiometric compound and $n^\downarrow$ is an integer on account of the gap. Hence $n^\uparrow - n^\downarrow$ is an integer.

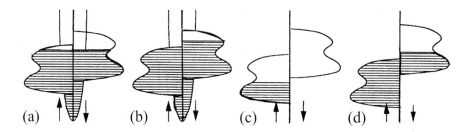

(a)          (b)          (c)          (d)

**Fig. 12.8.** Schematic densities of states for (a) a weak ferromagnet, (b) a strong ferromagnet and (c) and (d) half-metallic ferromagnets where a gap arises for minority or majority-spin electrons.

Other compounds containing a main group element such as Sb, Si which hybridizes with the outer metallic orbitals can also have half-metallic character. Examples are the Heusler alloys NiMnSb, PtMnSb and $Co_2MnSi$. These alloys

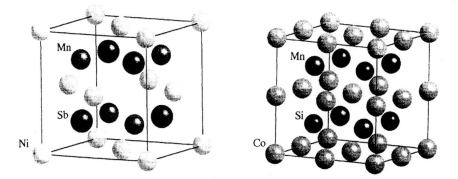

**Fig. 12.9.** Crystal structures of the Heusler alloys NiMnSb and $Co_2MnSi$.

have an    ordered fcc structure, with the atoms ordered on three or four of the
simple cubic sublattices (Fig. 12.9). Curie temperatures are 728, 572 and 985 K.
Tunnel junctions have been built using NiMnSb [33]. Data on some half-metals
is collected in Table 12.4.

We now consider a few oxides in turn. The first is $CrO_2$ which has the distinction of being the only simple oxide that is a ferromagnetic metal. The rutile
structure is illustrated in Fig. 12.10. There the $Cr^{4+}$ ion is surrounded by a
nearly-undistorted oxygen octahedron. The primary effect of the crystal field
due to the six oxygen anions is to split the $3d$ orbitals into a $t_{2g}$ triplet (xy,
yz, zx) and an $e_g$ doublet ($x^2$-$y^2$, $3z^2$-$r^2$), with a crystal-field splitting of about
1.5 eV. (Fig. 12.10) The $3d$ orbitals overlap to form bands; the overlap of the xy
orbitals in the rutile structure is slight, so they form an occupied nonbonding
level with a localized $S = 1/2$ core spin. The other $t_{2g}$ orbitals mix to form a
broader half-filled band with a dip in $D(E)$ at $E_F$. The exchange mechanism
in $CrO_2$ is a combination of ferromagnetic superexchange together with double
exchange due to hopping of the band electrons from site to site, where they are
coupled to the $S = 1/2$ cores by the on-site Hund's rule exchange. $CrO_2$ is a
black metal with a low resistivity ($\simeq 0.05$ $\mu\Omega$ m) in the liquid helium range.
There the mean free path is long enough for a classical $B^2$ magnetoresistance to
be observed [34]. However, $\rho$ increases rapidly as $T$ approaches the Curie point
$T_C = 398$ K, and the mean free path is reduced to the scale of the interatomic
spacing by strong spin-flip scattering.

**Table 12.4.** Half-metallic ferromagnets.

|  | Structure | Lattice parameter (pm) | $T_C$ (K) | $m_0$ ($\mu_B$/ formula) | $M_S$ (MA m$^{-1}$) |
|---|---|---|---|---|---|
| NiMbSb | Cubic | 592 | 728 | 4.0 | 0.71 |
| $CrO_2$ | Tetragonal | 442; 292 | 398 | 2.0 | 0.40 |
| $(La_{0.7}Sr_{0.3})MnO_3$ | Rhombohedral | 548; 60.4° | 380 | 3.6 | 0.31 |
| $Sr_2FeMoO_6$ | Tetragonal | 557; 791 | 426 | 3.5 | 0.15 |

Other metallic ferromagnetic oxide systems where the double exchange mechanism is important are the mixed-valence manganites $(La_{1-x}A_x)MnO_3$; A = Ca,
Sr or Ba, $x \simeq 0.3$ [35]. These oxides exhibit a metal-insulator transition at the
Curie point, which reaches a maximum value of 380 K in $(La_{0.7}Sr_{0.3})MnO_3$. This
is accompanied by colossal magnetoresistance, an intrinsic effect associated with
a field-induced increase of spontaneous magnetization near $T_C$. The oxides have
the perovskite structure, and the electronic structure is shown schematically in
Fig. 12.11. The half-filled $e_g$ band associated with $Mn^{3+}$ is split in $LaMnO_3$ by
the Jahn-Teller effect. The band splitting is sufficient to make the end-member a
narrow-gap antiferromagnetic semiconductor. Doping with $A^{2+}$ introduces holes

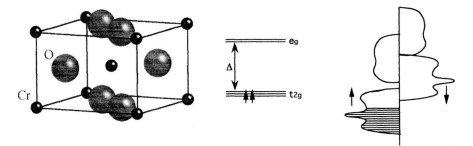

**Fig. 12.10.** Crystal structure of $CrO_2$. The effect of the crystal field on the one-electron $3d$ levels is shown, together with a schematic density of states.

into the lower spin-split $e_g$ band, and when these are sufficiently numerous, the holes can move freely among the ferromagnetically-aligned $t_{2g}^3$, $S = 3/2$ cores. Hopping between the core spins provides the double exchange. If the cores are misaligned by an angle $\Psi$, the hopping probability varies as $\cos(\Psi/2)$. Electron transport is therefore inhibited in the magnetically disordered state above $T_C$, where the carriers are polarons of some description.

Magnetite, $Fe_3O_4$, is a ferrimagnet crystallizing in the spinel structure with a single $3d^{\downarrow}$ electron hopping among the $3d^{5\uparrow}$ cores on octahedral sites. This corresponds to a half-metallic density of states, but there is a strong tendency to form polarons below the Curie temperature (860 K), and the conductivity shows a small activation energy.

The magnetoresistance effects of most interest in the manganites, $Fe_3O_4$ and $CrO_2$ are associated with transport of spin polarized electrons from one ferromagnetic region to another with a different direction of magnetization. These regions are not usually separated by a domain wall, but by a grain boundary, an interparticle contact or planar tunnel barrier which does not transmit exchange coupling. The effects are seen in low fields and in the liquid helium tempera-

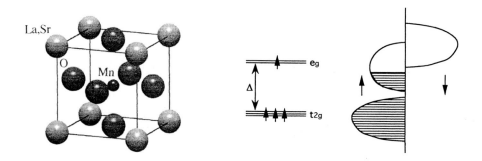

**Fig. 12.11.** Crystal structure of $(La_{0.7}Sr_{0.3})MnO_3$. The effect of the crystal field on the one-electron $3d$ levels is shown, together with a schematic density of states.

ture range they can reach 50% in $CrO_2$ pressed powder compacts, and several hundred % in planar manganite tunnel junctions [35]. Small effects have been observed in $CrO_2$ tunnel junctions [36,37]. The MR effects fall away fast on increasing temperature because of spin depolarization of the emitted electrons. Prospects of using $CrO_2$ or mixed-valence manganites in devices are dim, at least in a typical temperature range of $-40$ to $120°C$. Progress with oxide spin electronics will require half-metallic compounds with higher Curie temperatures.

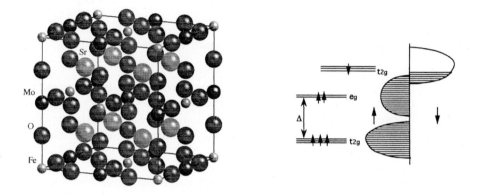

**Fig. 12.12.** Crystal structure of $Sr_2FeMoO_6$. The effect of the crystal field on the one-electron levels is shown, together with a schematic density of states.

Attention has recently turned to double perovskites with general formula $A_2BB'O_6$ where the B and B' cations occupy an NaCl-type superlattice (Fig. 12.12). The compound $Sr_2FeMoO_6$, for example, has a Curie temperature of 421 K and electronic structure calculations [38] indicate a half-metallic structure of the type shown in Fig. 12.6d. The majority spins are associated with the $Fe^{3+}$, $3d^5$ core spins, and the minority carriers are in a $\downarrow$ band of mainly $4d^1(Mo)$ character which is mixed with the empty iron $t_{2g}\downarrow$ orbitals. The ferromagnetic exchange is due to this electron hopping among the $3d^5\uparrow$ cores. There is no Fe-O-Fe superexchange on account of the NaCl-type order of Fe and Mo. Quite a large granular magneto-resistance has been reported at room temperature [39]. Other double perovskites such as $Sr_2FeReO_6$ have been reported to have a substantially higher Curie temperature (540 K).

Compared to the metallic multilayer structures which have undergone very rapid development in the ten years or so since the discovery of GMR, mainly in response to the urgent demands of the magnetic recording industry, research on optimizing oxide structures is in its infancy. Much has to be done to understand and prevent spin depolarization at the interfaces, and there is scope for new materials development focussing on increasing the Curie temperature. The oxides offer the prospect of very large magnetoresistance effects which could eliminate the need for associated electronics in MRAM, as well as providing streams of spin-polarized electrons which can advance the science of spin electronics in the

21st century. The oxides are robust and may lead to low cost sensors for a range of mass-market applications.

## 12.5   Ferromagnetic Semiconductors

Integration of spin electronics with conventional electronics would entail the manipulation of spin-polarized currents in silicon or gallium arsenide (see chapter 17 for further details on ferromagnetic semiconductors). There is evidence that the spin diffusion length in these semiconductors is long, with values of many microns being reported for Si or GaAs. The difficulty has been to find an effective way of injecting the spin-polarized electrons.

A fully-dopable ferromagnetic semiconductor would be a formidable advance for spin electronics. Some ferromagnetic semiconductors do exist [40], including EuX; X=O, S, $B_6$, and the chalcogenide spinels $CdCr_2X_4$; X = S, Se. The $MCr_2S_4$ spinels with M = Mn, Fe, Co are ferrimagnetic semiconductors. Mn-doped GaAs is a tetrahedrally bonded material which has been successfully used for spin injection into GaAs [41], opening the prospect of a marriage of spin electronics and opto-electronics. The outstanding problem with all this that the Curie temperatures of all these ferromagnetic semiconductors is far below room temperature. It is predicted that Mn-doped GaN or ZnO should have Curie temperatures in excess of 300 K [42]. If this is true, a new chapter in spin electronics may open.

## Problems

1. Use the magnetic valence model with $2n_{sp}^{\uparrow} = 0.6$ to deduce $m_0$, the magnetic moment in $\mu_B$/formula unit, for the following alloys: $Ni_{65}Fe_{15}Co_{20}$, $Fe_{40}Ni_{40}B_{20}$ and $Co_{88}Zr_8Ta_4$. Give the corresponding values of the polarization $\mu_0 M_S$ in tests assuming the first two alloys are fcc with a packing fraction of 0.74, and the $3d$ transition elements in the second two alloys are random close-packed with a packing fraction of 0.64. Why are your values of polarization overestimated?

2. How small would the cobalt crystallites have to be if a polycrystalline film of hcp cobalt was to have an effective anisotropy constant of 1 kJ m$^{-3}$? Explain why alloy additions are used to decouple the cobalt crystallites in thin film magnetic media.

3. Use (12.5) to deduce the values of the external magnetic field which must be applied along the anisotropy axis to switch the magnetization of a pinned layer. Evaluate these fields for the case of a 10 nm layer of permalloy pinned by NiMn if $K_u = 500$ J m$^{-3}$. Estimate how big a field would be needed to obtain a symmetric hysteresis loop.

4. From the value of the magnetic moment $M_0$ for $Sr_2FeMoO_6$ given in Table 12.4, deduce the fraction of Fe and Mo atoms that are on the wrong sites in the NaCl-type superlattice. Justify the assumptions you make regarding the directions of magnetization of the misplaced atoms.

5. You are looking for a new ferromagnetic material to be used as a source of polarized electrons for spin electronics. Make a list of properties, in order of importance, that it should possess.

## The Bibliography

R. C. O'Handley, "Modern Magnetic Materials", Wiley-Interscience, New York 1999, 740 pp.
U. Hartmann (ed.) "Magnetic Multilayers and Giant Magnetoresistance", Springer, Berlin 1999, 321 pp.
E. du Trémolet de Lacheisserie (ed.) "Magnétisme", 2 vols., Presses Universitaires de Grenoble 1999, 1006 pp.

## References

1. F. Y. Yang, K. Liu, C. L. Chien, and P. C. Searson, Phys. Rev. Lett. **82**, 3328 (1999).
2. E. P. Wohlfarth in "Handbook of Ferromagnetic Materials" (E. P. Wohlfarth, ed.) vol. 1 North Holland, Amsterdam p.1 (1980).
3. I. I. Mazin, Phys. Rev. Lett. **83**, 1427 (1999).
4. "Magnetic Properties of Metals" , H. P. Wijn (ed.) Springer, Berlin 1991.
5. R. J. Soulen, J. M. Byers, M. S. Osofsky, B. Nadgorny, T. Ambrose, S. F. Chong, P. R. Broussard, C. T. Tanaka, J. S. Moodera, A. Barry, and J. M. D. Coey, Science **282**, 88 (1998).
6. A. Fert and L. Piraux, J. Magn. Magn. Mater. **200**, 338 (1999).
7. A. Barthélémy, A. Fert, and F. Petroff in "Handbook of Ferromagnetic Materials" (K. H. J. Buschow, ed.) vol. 12 North Holland, Amsterdam p.1 (1999).
8. A. R Williams, V. L. Moruzzi, A. P. Malozemoff, and K. Tekura, IEEE Trans. Magn. **19**, 1983 (1983).
9. J. M. D. Coey and P. A. I. Smith, J. Magn. Magn. Mater. **200**, 405 (1999).
10. G. Herzer in "Handbook of Ferromagnetic Materials" (K. H. J. Buschow, ed.) vol 10, North Holland, Amsterdam p.415 (1997).
11. Y. Sugita, M. Mitsuoka, M. Komuta, H. Hoshiya, Y. Kozono, and M Hanazono, J. Appl. Phys. **79** 5576 (1996); H. Takahashi, M. Igarishi, A. Kaneko, H. Miyajima, and Y. Sugita, IEEE Trans. Magn. **35**, 794 (1999).
12. M. Takahashi and H. Shoji, J. Magn. Magn. Mater. **208**, 145 (2000).
13. S. Okamoto, O. Kitakami, and Y. Shimada, J. Magn. Magn. Mater. **208**, 102 (2000).
14. P. C. Andricarcos and N. Robertson, IBM J. Res. Dev. **42**, 671 (1998).
15. S. S. P. Parkin, Appl. Phys. Lett. **61**, 1358 (1992).
16. T. Osaka, M. Takai, K. Hayashi, Y. Sogawa, K. Ohashi, Y. Yasue, M. Saito, and K. Yamada, IEEE Trans. Magn. **34**, 1432 (1998).
17. T. Yokoshima, M. Kaseda, M. Yamada, T. Nakanishi, T. Momma, and T. Osaka, IEEE Trans. Magn. **35**, 2499 (1999).
18. X. Liu, G. Zangari and L. Shen, J. Appl. Phys. **87**, 5410 (2000).
19. A. E. Berkowitz and R. Takano, J. Magn. Magn. Mater. **200**, 552 (1999).
20. R. Coehoorn in "Magnetic Multilayers and Giant Magnetoresistance", (U. Hartmann, ed.) Springer, Berlin 1999, p 65.

21. M. Ledermann, IEEE Trans. Magn. **35**, 794 (1999).
22. D. Wang, M. Tondra, C. Nordman, and J. M. Daughton, IEEE Trans. Magn. **35**, 2886 (1999).
23. S. Tehrani, J. M. Slaughter, E. Chen, M. Durlam, J. Shi, and M. de Herrera, IEEE Trans. Magn. **35**, 2814 (1999).
24. H. A. M. van den Berg in "Magnetic Multilayers and Giant Magnetoresistance", (U. Hartmann, ed.) Springer, Berlin 1999, p 179.
25. H. A. M. van den Berg, J. Altman, L. Bär, G. Gieres, R. Kinder, R. Rupp, M. Veith, and J. Wecker, IEEE Trans. Magn. **35**, 2892 (1999).
26. H. Nagai, M. Ueno, F. Hikami, T. Sawasaki, and S. Tanoue, IEEE Trans. Magn. **35**, 2964 (1999).
27. Y. Sugita, Y. Kawawake, M. Satomi, and H. Sakakima, IEEE Trans. Magn. **35** 2961, (1999).
28. H. C. Tong, X. Shi, F. Liu, C. Qian, Z. W. Dong, X. Yan, R. Barr, L. Miloslavsky, S. Zhou, J. Perlas, S. Prabhu, M. Mao, S. Funada, M. Gibbons, Q. Leng, J. G. Zhu, and S. Dey, IEEE Trans. Magn. **35**, 2574 (1999).
29. J. M. Daughton, J. Appl. Phys. **81**, 3758 (1997).
30. R. Coehoorn, S. R. Cumpson, J. J. M. Ruijrok, and P. Hidding, IEEE Trans. Magn. **35**, 2586 (1999).
31. P. Rottländer, H. Kohlstedt, H. A. M. de Gronckel, E. Girgis, J. Schelten, and P. Grünberg, J. Magn. Magn. Mater. **210**, 251 (2000).
32. J. M. de Teresa, A. Barthélémy, A. Fert, J. P. Contour, F. Montaigne, and P. Seneor, Science **286**, 507 (1999).
33. C. T. Tanaka, J. Nowak, and J.S. Moodera, J. Appl. Phys. **86**, 6239 (1999).
34. S. M. Watts, S. Wirth, S. von Molnár, A. Barry, and J. M. D. Coey, Phys. Rev. B **61**, 149621 (2000).
35. J. M. D. Coey, M. Viret, and S. von Molnár, Adv. Phys. **48** 167 (1999).
36. A. Barry, J. M. D. Coey, and M. Viret, J. Phys.: Condens. Matter **12**, L173 (2000).
37. A. Gupta, X.W. Li, and Gang Xiao, J. Appl. Phys. **87**, 6073 (2000).
38. K. I. Kobayashi, T. Kimura, H. Sawada, K. Tekura, and Y. Tokura, Nature **385**, 677 (1998).
39. R. P. Borges, R. M. Thomas, C. Cullinan, J. M. D. Coey, R. Suryanarayanan, L. Ben-Dor, L. Pinsard-Gaudet, and A. Revcolevschi, J. Phys.: Condens. Matter **11**, L445 (1999).
40. S. Methfessel and D.C. Mattis, "Handbuch der Physik", **18** (Springer, Berlin 1966) p389.
41. Y. Ohno, D.K. Young, B. Beschoten, F. Matsukura, H. Ohno, and D. D. Awschalom, Nature **402**, 790 (1999).
42. T. Dietl, H. Ohno, F. Matsukura, J. Cibert, and D. Ferrand, Science **287**, 1019 (2000).

# 13　Thin Film Deposition Techniques (PVD)

E. Steinbeiss

Institut für Physikalische Hochtechnologie Jena e.V. ,
Winzerlaer Strasse 10, D 07745 Jena, Germany

## 13.1　Introduction

The most interesting materials for spin electronic devices are thin films of magnetic transition metals and magnetic perovskites, mainly the doped La-manganites [1] as well as several oxides and metals for passivating and contacting the magnetic films. The most suitable methods for the preparation of such films are the physical vapor deposition methods (PVD). Therefore this report will be restricted to these deposition methods.

PVD are vacuum deposition methods, where the thin film material will be transferred from solid, liquid or gas phase precursors into a vapor phase by different vaporization methods and condensed after the transport as a molecular beam in vacuum or by diffusion through a diluted background gas on a suitable substrate, the temperature of which must be below the melting point of the deposit. Chemical compounds are usually decomposed during their vaporization. Their deposition therefore requires a chemical reaction of the growing film surface with a suitable reactive gas.

The properties of the vaporized species can be strongly influenced with respect to the particle energy and distribution by the vaporization process, during the transport and on the surface of the growing film by different activation or de-activation processes allowing the control of the growth process, thereby changing the structure and properties of the films in a complicated manner.

Regarding the vaporization methods we can classify the PVD as thermal evaporation and sputtering methods.

The properties of the deposited films are determined essentially by the growth conditions during their deposition. There are many factors, which can influence the growth processes and modify the structure of the films. In general three stages are essential in the thin film deposition processes: nucleation and coalescence, followed by different growth processes as special cases of crystal growth (columnar, polycrystalline, epitaxial). The most important growth mode for the preparation of epitaxial films is the 2-dimensional layer–by–layer growth, which can be controlled by RHEED and other in situ characterization methods. Potential candidates for this deposition technique are the superconducting and magnetic perovskites investigated in the frame of the OXSEN program.

## 13.2   Thin Film Deposition Methods

### 13.2.1   Thermal Evaporation

The thermal evaporation of a solid precursor of the thin film material takes place by heating the source above the melting point, where the vaporization of the melt increases with increasing temperature in an exponential manner and a vapor beam will be spread into the vacuum chamber having a low residual pressure. The most simple evaporation sources are resistance heated sources [2]. The basic components of a deposition system with a thermal evaporation source is illustrated in Fig. 13.1.

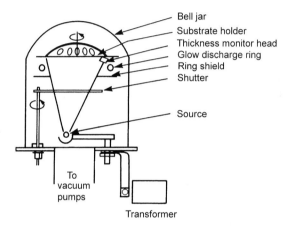

**Fig. 13.1.** Vacuum system for deposition from resistance heated sources

Because the distribution of the vapor flux has approximately a $\cos(\Theta)$ characteristic, the condensation on planar substrates provides films of inhomogeneous thickness. Therefore a large source-substrate distance is required or the substrates must be moved in a suitable manner. Resistance heated sources are restricted to materials of low or moderate melting points because of undesired reactions between the melt and the resistance carrier and his temperature limitations. These disadvantages can be avoided by the electron beam evaporation [3].

In the e-beam evaporator a focused electron beam with 10 keV and 1 A is used for the evaporation of the source material from a water cooled crucible. High melting metals and chemical compounds can be evaporated with high rates limited only by the development of the vapor pressure above the heated hearth (10 Pa).

The main disadvantage of thermal evaporators is the different evaporation speed for different materials at the same temperature leading to deviations of the film composition compared with the source material, when it consists of alloys or compounds.

A possible solution of this problem is the flash evaporation from a very hot crucible or the laser ablation [4] as a flash like evaporation of a target spot heated by a short laser pulse, suitable not only for the stoichiometric evaporation of complex compounds with high melting points but also for reactive evaporation at relatively high reactive gas pressures in the vacuum chamber. Figure 13.2 shows the typical arrangement of a laser ablation apparatus with the main components of a pulsed laser (Nd: YAG, Excimer) and a rotatable target carrier combined with a heated substrate holder and a RHEED system for epitaxial film preparation [5].

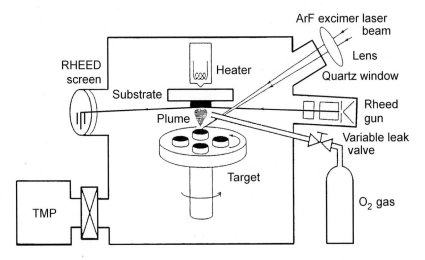

**Fig. 13.2.** Schematic diagram of a typical laser ablation apparatus for preparation of YBCO films [6]

The evaporation mechanism is somewhat complicated, consisting of several processes like thermal evaporation of a thin surface region followed by the explosion of the neighboring overheated target layer and radiation induced decomposition by pair breaking, resulting in emission of atoms, molecules, clusters and macroparticles with different velocities. The explosive part has a very sharp $\cos^{11}(\Theta)$ distribution.

Optical and electronic excitations of the emitted particles result in a luminous plasma plume affecting the film growth in a positive manner. Deposition in UHV for the growth of high quality epitaxial films as well as high rate deposition and reactive deposition at high reactive gas pressure are possible (suitable especially for $BiO_x$– and $PbO_x$–containing compounds). Disadvantages are the incorporation of droplets and the limitation to relative small substrate areas.

## 13.2.2   Ion Plating

A method to use the influence of energetic ions on the film growth process is the so called ion plating, a combination of thermal evaporation with a gas discharge, induced by a high voltage between the substrate/film-system and the grounded chamber walls. By the discharge the residual or reactive gas in the chamber and the evaporated material will be partly ionized and accelerated to the surface of the growing film. A typical set up is sketched in Fig. 13.3 [7].

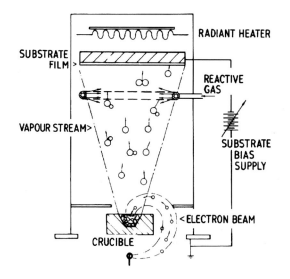

**Fig. 13.3.** Ion vapor deposition chamber with electron beam heated evaporation source [7]

### 13.2.3  Molecular Beam Epitaxy (MBE)

For the preparation of high quality epitaxial thin films the growth process must be carefully controlled layer by layer. Therefore low deposition rates (1 monolayer/s) are required. In order to avoid the disturbing influence of residual gas impurities UHV conditions are necessary allowing at the same time the installation of sensitive in situ characterization equipments like RHEED and LEED either in the deposition chamber or in a separate preparation chamber without vacuum breaking.

The construction of the evaporation sources must consider all the special conditions of a precise low rate deposition with controlled evaporation rates in a very pure UHV–system. In this sense MBE is a refined evaporation method for epitaxy of pure metals, defect-free semiconductors and superlattices with perfect interface morphology and even for the deposition of high quality epitaxial oxides like high-$T_c$- superconductors and magnetic perovskites. A schematic of typical MBE systems is shown in Fig. 13.4 [8].

As evaporation sources so called Knudsen cells will be applied, allowing the stabilization of a defined vapor pressure inside the cell by a precise temperature

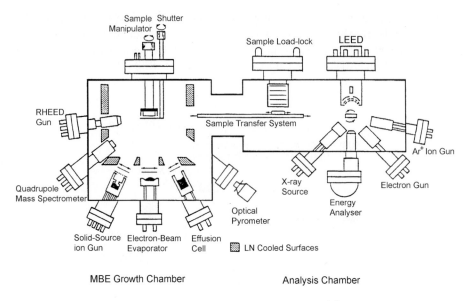

**Fig. 13.4.** Molecular Beam Epitaxy [9]

control of the heater as a presupposition for the realization of a constant flux vapor beam with a low flux rate. In the case of chemical compounds the Knudsen cell can be completed by a cracker section for the decomposition of the vaporized material [10]. The same cracker technique can be used for the production of a vapor beam from a gasphase precursor.

Instead of thermal cracking the modern effusion cells use plasma sources, either with rf-discharges or as ECR microwave discharges, which are useful also for the activation of reactive gases introduced into the UHV system for the epitaxial growth of oxides.

### 13.2.4  Sputtering Methods

In contrast to the evaporation methods the sputtering methods are characterized by a nonthermal mechanism for the transfer of the solid target material into the vapor phase, which is based on the impulse transfer from accelerated energetic particles to the surface and near surface atoms of the target [7].

The impact of energetic particles (with energies between a few eV and 1 keV) on the surface of a solid state results in a number of secondary processes summarized in Fig. 13.5.

Besides implantation, trapping, chemical effects, mixing and lattice destruction important essentially with respect to the growth of thin films under ion bombardment several types of particles can be emitted by single and multi-step impulse transfer and a part of the incoming ions are simply reflected. The sputtered particles consist mainly of single neutral atoms and a small part of molecules. The theoretical background was developed for the first time by Sieg-mund [11] and for the low energy range by Bohdansky [12]. From these theories the most important equations for the sputtering yield, the angle and energy distributions for the sputtered and reflected particles can be derived. For ion energies $E_0$ below 1 keV the sputtering yield is given by

$$Y(E) = 6.4 \times 10^{-3}\, M\, \gamma^{5/3}\, E_0^{1/4}\, (1 - E^{*\,-1})^{7/2}, \qquad (13.1)$$

where $\gamma = 4\, M_i\, M\,/\,(M_i + M)^2$, $E^* = E_0\,/\,E_{\text{th}}$ and $E_{\text{th}}$ is the threshold energy for sputtering. $M_i$ and $M$ are the masses of the incoming ions and the target atoms, respectively.

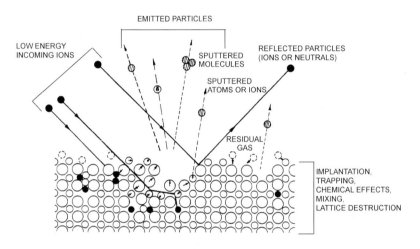

**Fig. 13.5.** General ion-surface interaction processes [7]

Essential for the impulse and energy transfer is the mass ratio between in-coming ions and the target atoms, which is optimal for equal masses. Target atoms of different masses are therefore initially sputtered with different yields until the surface density of the atoms is changed in such a degree, that the sput-tering yields of the different species correspond to the target bulk composition, a great advantage for the deposition of multicomponent alloys and compounds. A further feature is the dependence of the sputtering yield from the angle of incidence with a maximum of the yield near an incidence of 60°, important for surface planarization by ion etching or for optimization of the sputtering yield using ion sources under optimal incidence [13].

The sputtering yield shows a sharp threshold for ion energies near the binding energies of the target atoms. The energy distribution of the sputtered particles has a flat maximum just below the binding energy of the target atoms of a few eV, which can be shifted and strongly increased to lower energies by collisions with gas atoms between target and substrate resulting in a thermalization of the sputtered particles controlled by the total gas pressure in the vacuum chamber [14]. The same thermalization takes place for the reflected energetic particles.

The energetic particles for the bombardment of the sputter target can be produced by a separate ion source or by different discharge types inside the deposition chamber. Therefore the different sputtering methods are usually classified after the applied type of the gas discharge.

**Glow Discharge Diode Sputtering.** The simplest way to produce energetic particles in a deposition chamber is the ignition of a self sustaining dc glow discharge with a metallic target as a cathode and the substrate or the chamber walls as an anode.

The essential part of the discharge is the cathode sheath where a high electric field accelerates the positive ions of the discharge created by the electron impact around the negative glow region of the plasma. These ions bombard the cathode surface producing the sputtered particles and secondary electrons important for sustaining the discharge (Fig. 13.6).

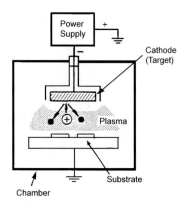

**Fig. 13.6.** Schematic of a dc glow discharge sputtering system [15]

The simple glow discharge sputtering system has many serious disadvantages connected with the relatively high pressure of the sputter gas, the high discharge voltage, the strong electron bombardment of the substrate and the limitation on conductive targets.

**Triode and RF–Sputtering Systems.** Some of these disadvantages can be avoided by the use of an rf-discharge or by a low pressure dc-discharge sustained by an auxiliary thermionic emitter.

An additional advantage of the rf-discharge is the possible use of isolating target material like oxides or nitrides. The application of an rf-discharge for sputtering is based on the different mobilities of electrons and ions leading at higher frequencies (typically 13.5 MHz) to a negatively biased cathode, when a capacitive coupling to the power supply is used [16]. In this case the ions are accelerated in a dc electric field, whereas the ionization is enhanced by the oscillating electrons, thus reducing the gas pressure necessary for the self-sustaining discharge.

**Ion Beam Sputtering (IBS).** More flexible is the ion beam sputtering method based on a separate ion source for the generation of energetic particles, where the ion flux and ion energy can be controlled independently and the angle of incidence on the target can be optimized with respect to a maximal sputter yield [7,17]. With dual beam ion sources the film growth can be influenced additionally by the bombardment with energetic particles of a defined energy and angle of incidence, frequently used for activated reactive sputtering or for the production of highly textured films by selective resputtering of misoriented crystallites (IBAD). The schematic diagram of the IBS-system, the ion source and their arrangements are illustrated in Fig. 13.7.

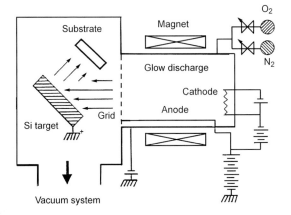

**Fig. 13.7.** Schematic diagram of the IBS system using a Si target and $O_2 - N_2$ discharges [18]

**Magnetron Sputtering.** One of the main disadvantages of dc glow discharges is the low ionization rate which can be strongly increased by introducing magnetic fields in suitable configurations. One of the most effective magnetic field configurations is a magnetic tunnel field above the target surfaces leading to a steered movement of the electrons and their confinement along a trace, where the magnetic field has a perpendicular component to the electric field.

   This so called magnetron configuration, illustrated in Fig. 13.8 [7] produces a region of high plasma density along the magnetic tunnel providing a high ion current at moderate voltages, increasing with $I \sim V^n$ with $n = 5 \ldots 10$ even at pressures below 1 Pa. Using an anode near the plasma track the substrate is protected against electron bombardment. In some cases the current and the sputtering rate can be increased to a degree, where the vaporized target material can sustain the discharge without an additional background gas. Therefore magnetron sputtering is frequently used for the high rate sputtering of metals like Al or Cu.

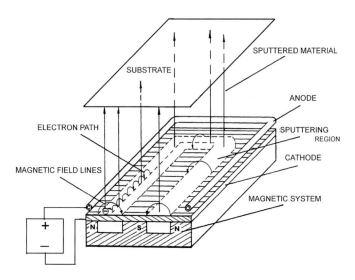

**Fig. 13.8.** Schematic of magnetron sputtering [7]

The magnetron system can be adapted to many special deposition problems simply by the change of the magnetic field configuration.

The so called unbalanced magnetron was proposed by Window and Savvides [19] and will be widely used for the preparation of hard films like TiN or diamondlike carbon. The substrate is biased by energetic electrons moving along field components perpendicular to the target and substrate surfaces.

More suitable and controllable is the bias effect using a magnetic field combination consisting of a central permanent magnet and a surrounding coil producing a magnetic field perpendicular to the target surface controlled by a dc current. Figure 13.9 shows the design of this unbalanced magnetron proposed by Orlinov [20] and the effect on the floating potential of the substrate as a function of current through the coil.

Whereas the bombardment of the growing film with energetic particles by bias methods is very useful for the deposition of simple metals, simple oxides and nitrides with special structural peculiarities, the bombardment has crucial consequences on the stoichiometric composition of complex alloys and chemical compounds. Because the bias voltage corresponds with the threshold voltage of several species a very selective resputtering takes place resulting in a large shift of the film composition compared with the target composition.

Extremely effective are negative ions generated at the surface of electronegative oxide targets like YBCO or Sr doped La-Manganites. These ions will be

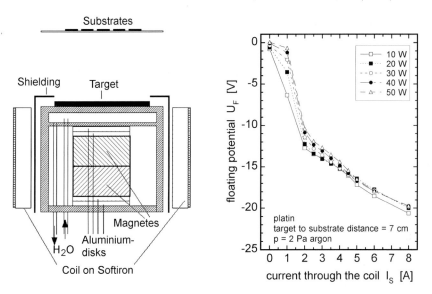

**Fig. 13.9.** (left) Design of an unbalanced magnetron and (right) floating potential of the substrate as a function of the current $I_S$ through the coil

accelerated by the cathode fall up to energies of 100 eV. In order to avoid the negative effects of ion bombardment so called off–axis arrangements are frequently used protecting the substrate and the growing film against resputtering effects [21].

Very suitable is the cylindrical magnetron design in combination with an anode, which can be used to control the extension of the plasma region, in this way controlling the bias voltage of the substrate (Fig. 13.10).

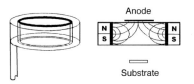

**Fig. 13.10.** Inverted cylinder magnetron with quasi off axis substrate arrangement

## 13.3   Thin Film Growth

The properties of thin films are determined essentially by the growth conditions during their deposition. There are many factors, which can influence the growth processes and modify the real structure of the film. Although many details of the thin film deposition can be explained already sufficiently, a complete theory of the thin film growth mechanisms does not yet exist, so that the optimization of the deposition parameters is dominated mainly by empirical points of view and experience. The reason is the depressingly large array of factors responsible for the growth processes and the developing of film structure and the difficulty to measure and control them in situ during the film deposition.

In general three steps are essential in the thin film deposition process: nucleation and coalescence, followed by different growth processes as special cases of crystal growth (columnar, polycrystalline, epitaxial).

### 13.3.1    Nucleation

The first steps of thin film deposition are determined by different interactions of the arriving atoms with the surface of the substrate, listed in Fig. 13.11 [15,22]:

- adsorption at special sites
- surface diffusion
- desorption
- creation of clusters (nuclei) by capture of adatoms.

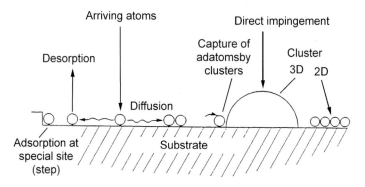

**Fig. 13.11.** Processes at the substrate surface [15]

A first theoretical description of these processes from first principles was carried out by Zinsmeister [22]. The creation rate of nuclei can be described by a set of differential equations as a function of the impact rate $q$, the desorption time $\tau_A$, the capture rate $\lambda_i$ and the decay rate of the clusters $\kappa_i$. In practice the solution of this system of differential equations is not possible. Many approximations are necessary resulting in a semi-empirical description of the nucleation process. The nucleation is finished by the coalescence of the nuclei followed by the growth of amorphous or polycrystalline films, if the nuclei have statistically distributed crystallographic orientations or by the growth of monocrystalline films, if the nuclei are epitaxially oriented by a monocrystalline substrate and a sufficient high substrate temperature.

### 13.3.2   Thornton Diagram

A phenomenological description of the film growth and structure as a function of
the surface mobility of the adatoms depending on the substrate temperature and
the residual gas pressure was given by Thornton [23,24]. The so–called Thornton
diagram is illustrated in Fig. 13.12.

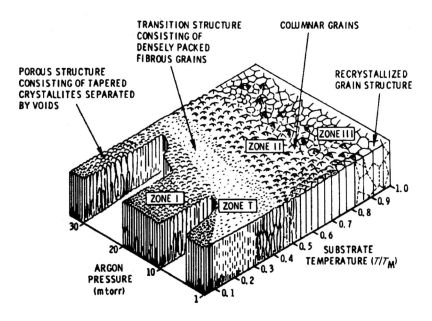

**Fig. 13.12.** Thornton Diagram [7]

Depending on substrate temperature and gas pressure four zones of the developing film structure can be distinguished. Zone I is characterized by a porous
structure caused by shadowing effects of the incoming atoms by the adatoms,
which can be demonstrated by computer simulations. In the transition zone T the
structure is characterized by densely packed fibrous grains developing to columnar grains in zone II. The columnar structure can be dissolved by increasing
surface mobility with increasing temperature (zone III) or by ion bombardment.

### 13.3.3   Epitaxial Growth

It is well known that the properties of thin films not only depend on their
chemical composition but mainly on their crystallographic structure. This is
true especially for semiconductors, high-$T_c$-superconductors and ferromagnetic
perovskites. The interesting unique properties of these thin film materials can be
achieved only with high quality monocrystalline films or film systems prepared
by epitaxial growth at sufficient high temperatures on monocrystalline substrates
with matched lattice constants.

The preparation of epitaxial films with atomic flat surfaces and interfaces important for the realization of superlattices, tunneling barriers and quantum wells requires very pure UHV conditions and low deposition rates necessary for the forming of equilibrium surface states under the condition of a limited surface mobility. The most important growth mode for the preparation of such films is the 2-dimensional layer–by–layer growth known as Frank van der Merwe mechanism. The necessary condition for this growth mode is [25]:

$$\gamma_{fv} + \gamma_{fs} \leq \gamma_{sv} , \qquad (13.2)$$

where $\gamma_{fv}$, $\gamma_{fs}$ and $\gamma_{sv}$ are the surface energies between film and vacuum, film and substrate and substrate to vacuum. Otherwise 3-dimensional growth of clusters takes place (Kossel–Stranski mechanism).

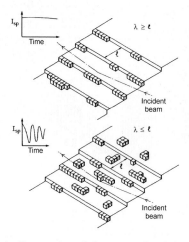

**Fig. 13.13.** Schematic illustration of the principle of the MBE growth method, showing the change in RHEED information as the growth mode changes from "step-flow" to 2-D nucleation. Steps lie along [100] [9]

Important for the layer–by–layer growth is further the surface roughness, especially the distance between step edges in relation to the surface mobility of adatoms [26]. These growth modes can be observed and in situ controlled by RHEED. Because of the grazing incidence of the focused electron beam the penetration depth is limited to a thin surface layer. The diffraction pattern provides therefore informations about the first monolayer, their lattice parameters, surface reconstruction and disorder effects.

MBE growth is usually characterized by two limiting cases: In one limit, the surface migration length is much smaller than the surface features so that the growth occurs by the nucleation of 2-dimensional islands. The other limit is the step-flow growth, in which the adatoms migrate to the next step edge, where they will be incorporated (Fig. 13.13).

The effect of an insufficient surface mobility can be compensated by a pulsed deposition mode allowing a rearrangement of the adatoms after the deposition pulse.

Sometimes a post deposition flattening of the film surface can be achieved by an annealing process, which is demonstrated in Fig. 13.14 for a sputtered epitaxial $La_{0.8}Sr_{0.2}MnO_{3-\delta}$ thin film on $SrTiO_3$.

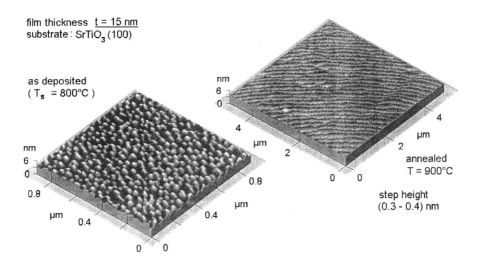

**Fig. 13.14.** AFM – scans of a sputtered epitaxial $La_{0.8}Sr_{0.2}MnO_{3-\delta}$ single crystal film before and after oxygen annealing at temperature T

These magnetic perovskites as well as the high-$T_c$-superconductors are characterized by a highly anisotropic layered lattice structure especially in the Tl- and Bi-based cuprates [27] and in the $(La,Sr)_3Mn_2O_7$-system [28] leading to intrinsic tunneling effects between the superconducting $CuO_2$ planes or the ferromagnetic $MnO_2$ planes of the lattices. These intrinsic tunneling structures are important for Josephson devices in cryoelectronics and spin dependent tunneling elements in spin electronics.

In these cases MBE is a very powerful deposition technique for the preparation of artificial tunneling structures by local modification of the lattice structure demonstrated for example by Bozovic and Eckstein [29].

### 13.3.4   Reactive Deposition of Compounds

The preparation of chemical compounds by PVD methods requires in general the film deposition in the presence of a reactive background gas. Because these compounds are usually decomposed during the evaporation or sputtering process, surface reactions during the film growth are necessary to form the stoichiometric compounds, for example oxides or nitrides [30]. The reactive sputtering of

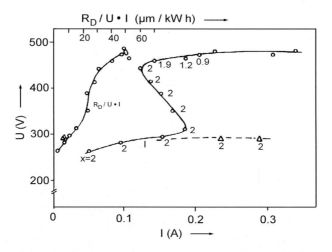

**Fig. 13.15.** Specific deposition rate $R_D/UI$ of $SiO_x$ films measured along the I–U characteristics for medium (○) and high (△) oxygen throughputs (substrate-to-target distance, 5 cm). The measured values of x are indicated at the corresponding U,I points [31]

elemental targets is complicated by additional effects caused by a strong feedback between the surface reactions and the reactive gas pressure and between the oxygen coverage and the discharge characteristic of the target. The oxygen adsorption on the growing films acts as a strong getter pump reducing the oxygen partial pressure $p$ at a constant oxygen flow rate $D_0$ in dependence of the sputtering rate $R$. On the other hand, the oxygen coverage $\theta$ of the target surface changes the exponential dependence of the discharge current $I$ from the discharge voltage $U$. In the case of the Si/O-system the current strongly increases with increasing $\theta$ leading to sudden jumps of the deposition parameters with increasing reactive gas pressure or discharge power, if the power supply is power or current controlled. Only a voltage controlled power supply allows a steady change of the deposition parameters including the possibility to achieve a high degree of film oxidation at a high deposition rate by sputtering a nearly pure elemental target. The essential features are summarized in Fig. 13.15 [31].

The I(U)–characteristic and the deposition parameters can be derived from a set of equations describing the equilibrium between the adsorption and desorption (sputtering) processes of oxygen on the target surface area $A$ and the surface area $F$ of the growing films with the degree of oxidation $\vartheta$ (getter surface):

$$\kappa p (1 - \theta) = \frac{\eta_0}{e} \frac{\theta I}{A}$$

$$\vartheta = \frac{\kappa p F (1 - \vartheta)}{2 R}$$

$$p = \frac{D_0}{S} - \frac{\gamma \vartheta R}{S} \tag{13.3}$$

$$R = \frac{\eta_{\mathrm{m}}}{\mathrm{e}} I\,(1-\theta)$$

$$I = C\,A\,\left(\frac{U}{U'}\right)^{\nu+\beta\theta}$$

$\eta_{\mathrm{m}}$ and $\eta_0$ are the sputtering yields of Si and oxygen on the target surface. $C$, $U'$, $\nu$ and $\beta$ are constants describing the special form of the magnetron discharge characteristic used for the measurements illustrated in Fig. 13.15, $\gamma \approx k_B T = 4.1 \times 10^{-18}$ Pa l and $\kappa = 3.81 \times 10^{18}$ cm$^{-2}$ s$^{-1}$ Pa$^{-1}$ (impact rate of O-atoms at 300 K). $S$ is the pumping speed of the vacuum pump.

### 13.3.5   Bias Effects

An effective tool for the control of the film structure during the deposition process is the bombardment with energetic ions using a bias voltage of the substrate to accelerate ions from the surrounding plasma of a gas discharge up to energies of a few eV to about 30 eV, comparable with the sputtering threshold for impurities and other weakly bonded species on the substrate and film surface. Therefore the bias effect can be used for substrate cleaning prior to the film deposition providing an improved adherence of the growing film. Furthermore the increased surface mobility results in a more perfect crystallographic structure with increased film density. Selective resputtering effects during the film growth can reduce the incorporation of weakly bonded species improving the long term stability of the films.

## Exercises

- Thermal evaporation systems with point like sources provide thin films on planar substrates with a characteristic thickness distribution.
  - What is the minimal distance between the evaporation source and a fixed planar substrate (3 inches in diameter), if the thickness of the deposited film is requested with a tolerance of 10% over the whole substrate?
- The sputtering yield of an oxygen covered Si-target strongly depends on the coverage parameter $\theta$, $(0 \le \theta \le 1)$, influencing the discharge characteristic and the deposition parameters for the deposition of SiO$_2$ thin films.
  - Calculate the discharge characteristic and
  - estimate the critical current $I$ for a given flow rate $D_0 = 3$ Pa l/s allowing a working point outside the negative resistance range of the I(U)-curve; use the set of equations 13.3 with the following parameters: $A = 10$ cm$^2$; $\eta_{\mathrm{m}} = 0.5$; $\eta_0 = 2.5$; $F = 10^3$ cm$^2$; $S = 6$ l/s; $C = 8 \times 10^{-4}$ A/cm$^2$; $U' = 200$ V; $\nu = 4.5$; $\beta = 9.5$.
- Bias effects have a strong influence on the composition of thin films sputtered from alloy or compound targets.
  - Estimate the composition shift between a Au$_{0.5}$Cu$_{0.5}$ alloy target and the growing film as a function of the rate $E_0/E_{\mathrm{th}}$ and $r/r_i$ in the parameter

range between 1 and 10. $r$ is the impact ratio of target atoms with a sticking coefficient of 1 and $r_i$ the impact rate of Ar-ions of energy $E_0$ on the film surface.

– Compare the composition shift in the same parameter range using targets of $Pt_{0.5}Rh_{0.5}$.

# References

1. J. M. D. Coey, M. Viret, S. von Molnár, Adv. Phys. **48**, 167 (1999).
2. E. B. Graper, "Resistance Evaporation", in "Handbook of Thin Film Process Technology", (eds.) D. A. Glocker and S. I. Shah (Inst. Phys. Publ., Bristol and Philadelphia 1995) A 1.1: pp. A1.1:1–A1.1:7.
3. E. B. Graper, J. Vac. Sci. Technol. **7**, 282 (1970).
4. S. Otsubo, T. Maeda, T. Minamikawa, Y. Yonezawa, A. Morimoto, and T. Shimizu, Jpn. J. Appl. Phys. **29**, L133 (1990).
5. M. Kawasaki, J. Gong, M. Nantoh, T. Hasegawa, K. Kitezawa, M. Kumagai, K. Hirai, K. Horiguchi, M. Yoshimoto, and H. Koinuma, Jpn. J. Appl. Phys.**32**, 1612 (1993).
6. A. Morimoto and T. Shimizu, "Laser Ablation" in "Handbook of Thin Film Process Technology", (eds.) D. A. Glocker and S. I. Shah (Inst. Phys. Publ., Bristol and Philadelphia 1995) A 1.5: pp. A1.5:1–A1.5:11.
7. P. J. Martin, J. Mater. Sci. **21**, 1 (1986).
8. M.-A. Hasan, J. Knall, S. A. Barnett, A. Rockett, J.-E. Sundgren, and J.-E. Greene, J. Vac. Sci. Technol. **B5**, 1332 (1987).
9. S. A. Barnett and I. T. Ferguson, "Molecular Beam Epitaxy, Intoduction and General Discussion" in "Handbook of Thin Film Process Technology", (eds.) D. A. Glocker and S. I. Shah (Inst. Phys. Publ., Bristol and Philadelphia 1995) A 2.0: pp. A2.0:1–A2.0:35.
10. E. H. Parker, "The Technology and Physics of Molecular Beam Epitaxy", (Plenum, New York 1985).
11. P. Siegmund, Phys. Rev. **184**, 383 (1969).
12. J. Bohdansky, R. Roth, and H.L. Bag, J. Appl. Phys. **51**, 2861 (1980).
13. B. Navinsek, Prog. Surf. Sci. **7**, 49 (1976).
14. F. Steinbeiss, W. Brodkorb, M. Manzel, J. Salm, and K. Steenbeck, "Reactive Thin Film Deposition by Planar Magnetron Discharges – Basic Problems and Experimental Procedures". in "Proc. 6. Int. Symp. High-Purity Mater. in Sci. and Technol., Dresden 1985, Pt. I, Plenary Papers/Preparation", ed. by Zentralinstitut für Festkörperphysik und Werkstofforschung, Dresden, Helmholtzstr. 20, Dresden 1985) pp. 245–258.
15. S.I. Shah, "Sputtering: Introduction and General Discussion". in "Handbook of Thin Film Process Technology", (eds.) D. A. Glocker and S. I. Shah (Inst. Phys. Publ., Bristol and Philadelphia 1995) A 3.0: pp. A3.0:1–A3.0:18.
16. H. S. Butler and G. S. Kino, Phys. Fluid **6**,1346 (1963).
17. J. M. E. Harper, J. J. Cuomo and H. Kaufman, J. Vac. Sci. Technol. **21**, 737 (1982).
18. T. Itoh, "Ion-Beam Sputtering", in "Handbook of Thin Film Process Technology", (eds.) D. A. Glocker and S. I. Shah (Inst.Phys.Publ., Bristol and Philadelphia 1995) A 3.3: pp. A3.3:1–A3.3:12.
19. N. Savvides and B. Window, J. Vac, Sci. Technol. **A4**, 504 (1986).

20. J. Kourtev, S. Grudeva-Zorota, I. Garnev, and V. Orlinov, Vacuum **47**, 1397 (1996).
21. T. Schüler, E. Steinbeiss, G. Bruchlos, and T. Eick, phys. stat. sol. (a) **134**, K 25 (1992).
22. G. Zinsmeister, Vakuum-Technik **22**, 85 (1973),
23. J. A. Thornton, J. Vac. Sci. Technol. **11**, 666 (1974).
24. J. A. Thornton, J. Vac. Sci. Technol. **12**, 830 (1975).
25. M. Ohring, "The Materials Science of Thin Films", (Academic, New York 1992) pp. 200.
26. J. H. Neave, B. A. Joyce, P. J. Dobson, and J. Zhang, Appl. Phys. Lett. **47**, 100 (1985).
27. E. Steinbeiss, M. Manzel, M. Veith, H. Bruchlos, T. Eick, S. Huber, K. Steenbeck, W. Brodkorb, W. Morgenroth, G. Bruchlos, H.G. Schmidt, S. Bornmann, T. Köhler, L. Redlich, H. J. Fuchs, K. Schlenga, G. Hechtfischer, and P. Müller, Vacuum **47**, 1117 (1996).
28. Y. Morimoto, A. Asamitsu, H. Kuwahara, and Y. Tokura, Nature **380**, 141 (1996).
29. I. Bozovic and J. N. Eckstein, Appl. Surf. Sci. **113/114**, 189 (1997).
30. W. Brodkorb, J. Salm, and E. Steinbeiss, phys. stat. sol. (a) **84**, 379 (1984).
31. K. Steenbeck, E. Steinbeiss, and K. D. Ufert, Thin Solid Films **92**, 371 (1982).

**Acknowledgement**

For the use of some figures I would like to thank the publisher and authors indicated in the references of the corresponding figure captions.

# 14 Magnetic Imaging

A. K. Petford–Long

Department of Materials, Oxford University, U.K.

## 14.1 Introduction

Spin-transport effects, such as giant magnetoresistance, rely on the fact that there is a difference in scattering between the spin-up and spin-down electrons in a ferromagnetic material. The degree to which each electron channel is scattered depends on the magnetisation direction within the material, and thus on the local magnetic domain structure. It is therefore of importance when analysing spin-transport devices to understand their magnetic domain structure, both as a bulk property and locally. The aim of this chapter is to review a number of the techniques currently used to image magnetic domain structure in materials. Although a considerable amount of information about the magnetic properties and behaviour of a piece of material, for example a thin ferromagnetic film, can be obtained from bulk magnetometry measurements, it is often extremely useful to image the magnetic domain structure of the film and thus gain information about its magnetic properties at a local level. The various magnetic imaging techniques yet to be described can be extended, by the application of in-situ magnetic fields which allow not only the magnetic domains but also the magnetisation reversal process to be followed in real-time.

The techniques detailed in this chapter can be loosely grouped into electron optical techniques, such as Lorentz microscopy, optical techniques which are based on the magneto-optical Kerr effect, and scanning probe techniques. There is an additional technique, the Bitter technique, which does not fall into any of these categories but can usefully be applied to the study of magnetic domain structures. The techniques to be described necessarily only comprise a sub-set of all those available. For a detailed description of the interpretation of domain images and of the origins and theory of magnetic domains, the reader is referred to the excellent book by Hubert and Schäfer [1].

## 14.2 Bitter Pattern Formation

The earliest technique used for imaging magnetic domains was the Bitter technique, first reported by Bitter in 1932 [2], which allows the domain structure at the surface of a bulk sample to be imaged with a spatial resolution down to $\sim 0.1$ μm. The imaging medium is a ferrofluid, which is a colloidal suspension of small magnetic particles in a liquid. When the surface of the sample to be analysed is coated in a thin layer of ferrofluid, the particles are attracted to the position of the stray fields above the magnetic domain walls. If the sample is then viewed using either an optical microscope or a scanning electron microscope (SEM) the position of the decorated domain walls can be seen. It should be noted that some uncertainty about the exact mechanism by which contrast is formed in Bitter patterns exists, because of the fact that a number of magnetic structures that would be expected to show contrast do not in fact do so. It is therefore usually advisable to use the Bitter technique in conjunction with another magnetic imaging technique such as magneto-optical Kerr microscopy. A big advantage of the technique is that it can be used to image surfaces without the need for extensive sample preparation, but the difficulty of making suitable colloid suspensions means that it is not widely used.

## 14.3   Electron Microscopy

There are several electron optical techniques that can be used to image magnetic domains, but only those applicable to the study of thin film materials will be discussed. These are the transmission electron microscope (TEM) techniques which include variations on Lorentz microscopy and electron holography, and scanning electron microscopy with polarisation analysis (SEMPA).

### 14.3.1   Transmission Electron Microscopy

#### Lorentz Microscopy

In a TEM a high energy (100–1000 keV) electron beam is incident on a thin specimen. The interaction of the electrons passing through the specimen results in magnetic contrast [3] which can be explained by considering the electrons either as waves or particles. If the electrons are considered as particles, then on passing through the magnetic induction in the specimen, they are deflected by the Lorentz force:

$$\boldsymbol{F} = |e| \left(\boldsymbol{\nu} \times \boldsymbol{B}\right) , \tag{14.1}$$

where e and $\boldsymbol{\nu}$ are the charge and velocity of the electrons, and $\boldsymbol{B}$ is the magnetic induction in the specimen. Note that: only components of the magnetic induction normal to the electron beam give rise to a deflection; the stray fields above and below the specimen also contribute to the image because Lorentz microscopy is a transmission technique; and the deflection direction depends on the magnetisation direction within the domain being imaged and is perpendicular to it. The magnitude of the deflection angle, $\beta$, is given by:

$$\beta = \left(e\lambda B t\right)/h , \tag{14.2}$$

where $\lambda$ is the electron wavelength, $t$ is the specimen thickness, and h is Planck's constant. The deflection is proportional to the product of the specimen thickness and magnetisation. Figure 14.1 shows ray diagrams indicating the way in which the electrons are deflected for a specimen containing two sets of domains magnetised in-plane and separated by 180° domain walls: the Lorentz deflection results in each spot in the electron diffraction pattern being split in two. There are then two methods by which the magnetic domain structure can be imaged: the Fresnel mode and the Foucault mode – these are discussed in more detail below. Reviews of Lorentz TEM techniques can be found in [4,5].

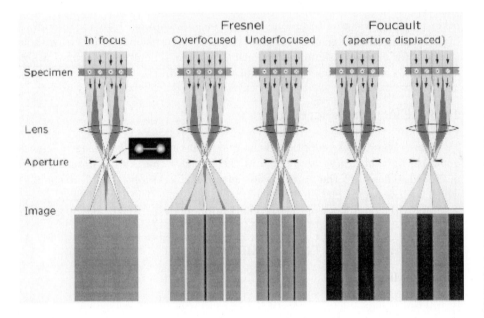

**Fig. 14.1.** Ray diagrams showing Fresnel and Foucault modes of Lorentz transmission electron microscopy.

**The Fresnel Imaging Mode** For the Fresnel imaging mode the objective lens is defocused so that an out–of–focus image of the specimen is formed (see Fig. 14.1). Under these conditions the magnetic domain walls are imaged as alternate bright (convergent) and dark (divergent) lines. The bright lines occur when the domain walls are positioned such that the magnetisation on either side deflects the electrons towards the wall. If a coherent electron source is used, the convergent wall images consist of sets of electron diffraction fringes running parallel to the wall. Detailed analysis and simulation of the fringe patterns can give information about the domain wall structure, but this is not easy to interpret. A Fresnel mode image of a Co thin film can be seen in Fig. 14.2a. Information about the magnetisation direction within the magnetic domains can also be obtained from the *magnetisation ripple* visible in Fresnel images of polycrystalline specimens as a result of small fluctuations in the magnetisation direction. The ripple is always oriented perpendicular to the magnetisation direction. A typical ripple image can be seen in Fig. 14.2b.

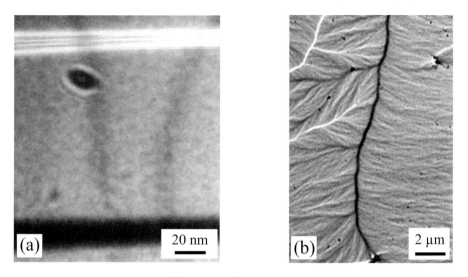

**Fig. 14.2.** Fresnel images of (a) Co film (note fringes in bright domain wall image) and (b) MnFe/NiFe bilayer film showing magnetisation ripple contrast.

**Foucault Imaging Mode** To image magnetic domains using the Foucault mode, the objective lens is kept in-focus but one of the split spots in the diffraction pattern is blocked by displacing the objective aperture. The contrast then results from the magnetisation within the domain, with the deflected electrons passing through the aperture. By knowing the relative direction of the aperture and image, the direction of magnetisation within the various domains can be qualitatively determined. To obtain good quality Foucault mode images the back-focal plane of the objective lens and the objective aperture must be as near co-planar as possible. A Foucault image of the magnetisation distribution in the sense layer of a NiFe/Cu/NiFe/MnFe spin-valve can be seen in Fig. 14.3.

**Foucault Differential Phase–Contrast Microscopy** The differential phase contrast (DPC) technique was first developed by Chapman *et al.* [6] and uses a scanning transmission electron microscope (STEM). The specimen is scanned with a small electron probe and the signal is detected on a circular detector split into four quadrants. The Lorentz deflection of the electrons results in the signal being displaced from the centre of the detector, and the component of magnetisation in two perpendicular directions can be calculated from the difference between the signals on opposite quadrants of the detector. Thus the DPC technique allows quantitative mapping of the magnetisation perpendicular to the electron beam direction, i.e. in the plane of the specimen. DPC contrast can be obtained in a conventional TEM by digitally combining series of Foucault images taken with small increments of electron beam tilt in two orthogonal directions [7]. Combining these series together again allows the in-plane magnetisation to be mapped.

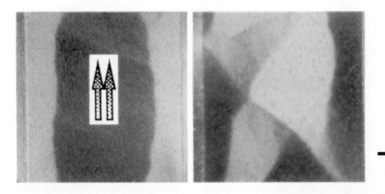

**Fig. 14.3.** Plan–view Foucault images of the magnetic domains in the sense layer of a 10 μm×10 μm spin-valve element at two different applied field values. The different contrast levels in the domains indicate different components of magnetisation in the vertical direction.

## Electron Holography

Electron holography is based on recording an interference pattern from which the amplitude and phase of an object can be reconstructed [8]. Magnetic thin films are strong phase objects and the phase shift of the electrons passing through the specimen is proportional to the magnetic flux enclosed by the electron paths. Provided that the phase shifts are caused only by the magnetic fields, adjacent interference fringes run parallel to the magnetisation direction and are separated by a flux quantum equal to h/e.

**Off–Axis Holography** A specimen is chosen that does not completely fill the image plane (for example a small magnetic element or the edge of an extended film) so that only part of the electron beam passes through the specimen. An electrostatic biprism is then used to recombine the specimen beam and the reference beam so that they interfere and form a hologram. This can be digitised and image processing techniques can be applied to reconstruct a quantitative image of the magnetic domain structure. An example of a plan-view electron hologram of a Co/Au/Ni/Al layered film is shown in Fig. 14.4.

**Coherent Foucault Mode Imaging** This technique, described by Chapman *et al.* [10], produces images which are similar to standard electron holography images. A thin semitransparent aperture containing a small hole is used, of a thickness that phase shifts electrons passing through the aperture film by $\pi$. The hole in the film is positioned so that half the central beam passes through the film and half through the hole. This results in interference fringes in the image as for holography.

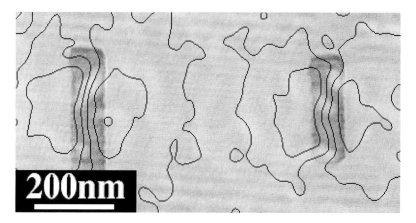

**Fig. 14.4.** Remanent state electron holographic phase image of two Co/Au/Ni/Al layered film elements (dark rectangular regions. Contour spacing of $0.064\pi$ radians is proportional to in-plane component of induction integrated in incident beam direction. The Co and Ni layers are coupled antiferromagnetically. (Courtesy of R. Dunin–Borkowski [9].)

### Extending Lorentz Microscopy Techniques

The Lorentz microscopy techniques described above can be made even more powerful if a magnetic field can be applied in-situ to the specimen so that rather than just observing a static domain structure, magnetisation reversal processes can be imaged. This can be done either by using magnetising coils built into the TEM specimen holder or by tilting the specimen into the vertical lens field (if a standard objective lens is used). Applying a field using magnetising coils in the holder removes the potential problem of the specimen experiencing a vertical component of magnetic field as well as an in-plane component. However, this technique can result in the need for an extra set of correction coils in the microscope column to realign the electron beam, which will be deflected by the applied magnetic field. Further developments are to heat or cool the specimen (with or without an external applied field) and to pass a current through the specimen (used to study active spin-valve devices [11]) to observe the effect on the magnetisation process.

The fact that Lorentz microscopy is a TEM technique means that it is limited to specimens with $t < \sim 100$ nm, which can lead to specimen preparation difficulties in the case of bulk material or thin films grown on bulk substrates. In addition, the fact that the sample needs to sit in a low magnetic field means that specially modified TEMs are needed. The spatial resolution of the Lorentz microscopy techniques described above is of the order of a few nm for the in-focus techniques and its somewhat worse for the defocused (Fresnel) technique.

### 14.3.2    Scanning Electron Microscopy (SEM) Techniques

Two standard SEM techniques exist for magnetic domain imaging on the surface of bulk specimens. Type I contrast images are obtained when low energy secondary electrons, sensitive to the stray fields above the specimen surface, are detected. Type II contrast images are obtained when higher energy back-scattered electrons, sensitive to the magnetisation within the specimen, are detected. A review of the various SEM techniques can be found in [12]. The spatial resolution of the techniques is limited to slightly better than 1 μm. A more recent development has been SEM with polarisation analysis (SEMPA), developed by Unguris *et al.* at NIST [13], in which the spin-polarisation of the secondary electrons emitted from the sample surface is measured using a Mott detector. A two-dimensional map of the spin-polarisation reveals the surface magnetisation distribution for ferro- (or ferri-) magnetic materials. The spatial resolution of this technique is much higher (of the order of 10 nm), and the depth probed is about 1 nm. A useful feature of this technique is that the magnetisation maps are independent of topography, but a topographic map can be collected simultaneously using standard SEM imaging techniques.

## 14.4    Scanning Force Microscopy

The atomic force microscope (AFM) was pioneered by Binnig, Quate and Gerber in 1986 [14] and involves scanning a fine tip on a flexible cantilever across a sample using piezoelectric scanners, as shown in Fig. 14.5. There is a force between the sample and the tip which can deflect the tip, and if the tip deflection can be measured at each point in the scan then a force image can be produced. Several techniques have been developed for detecting the cantilever deflection, for example optical interference between the tip and an optical fibre. There are a number of different interactions that can be detected to produce force images, namely: electrostatic (range, $d < 100$ nm), van der Waals ($0.2 < d < 10$ nm), magnetostatic ($d < 1$ μm).

### 14.4.1    Magnetic Force Microscopy

The magnetic force microscope (MFM) is a further development of the AFM, first reported in 1987 [15], in which the tip is either made of a magnetic material, or is coated in a magnetic layer. If the tip is then scanned over a magnetic sample, the tip can interact with the stray fields above the sample and an image of these stray fields, and thus of the magnetic structure of the sample, can be produced. The spatial resolution of the MFM is of the order of 10–100 nm, which is considerably worse than standard AFM (0.02–0.1 nm), but the force resolution (sensitivity) of an MFM is considerably higher ($10^{-13}$ N/m compared to $10^{-5}$ N/m for AFM)

In order to understand the basic principles of the tip/sample interaction in MFM, first consider the field $\boldsymbol{H}(\boldsymbol{r})$ from the sample to be the result of a point dipole, $m_1$, and consider the tip likewise to be modelled as a point dipole, $m_2$. The force between the tip and the sample is then given by:

$$F = \nabla \left[ \frac{3\left(m_1 \cdot r_u\right)\left(m_2 \cdot r_u\right) - \left(m_1 \cdot m_2\right)}{r^3} \right], \qquad (14.3)$$

where $r_u$ is the unit vector along $r$ (the tip/sample separation). The precise geometry of the tip, the cantilever and the sample then allows the various components of the force and also its gradient terms to be detected. In practice the forces must be integrated over the extended tip/sample volume and this represents a major challenge for MFM.

The fact that the MFM is a scanning system with the force at each point being collected sequentially, and that the tip-sample interaction only generates weak forces and force gradients, means that the MFM instrumentation requirements are very stringent. Thermal noise limits detection of the force gradients and stable detection conditions are needed to enable long scan times ($> 30$ min). Another complication is the need to have a micro-positioning system with sub-nm resolution and a wide range (50 μm or more). This is necessary because domain walls can show detail on the nm scale but domains extend over several μm.

MFM contrast depends in a complex way on stray fields and sample topography and so it is almost essential to use an additional imaging mode based on another interaction to acquire topographic data for in-situ correlation with the magnetic data. In addition, since MFM contrast can be complicated by the tip-sample interaction and potential artefacts, it is useful to have an alternative domain imaging technique to correlate in-situ with MFM (e.g. Kerr microscopy – see below). One of the main uses of MFM is to image written bits on magnetic storage media such as hard disks, as seen in Fig. 14.6. For this purpose the fact that the sequential collection of the data makes MFM a relatively slow technique is not a problem, and the relatively simple magnetic structure precludes the need for comparison with data from another technique.

In practice MFM resolution is ultimately limited by the specific tip/sample system, and different sample materials require specific tip/cantilever coatings, resulting in the need for a range of tips for specific applications (e.g. low stray

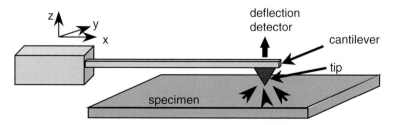

**Fig. 14.5.** Schematic diagram of a scanning force microscope.

**Fig. 14.6.** 20 × 20 micron force gradient image from a 2 μm bit length '1100' pattern on a commercial CoPt hard disk imaged by a 50 nm CoCr film on a Si nanoprobe cantilever scanned at 100 nm tip flyheight. Imaged in zero field. (Courtesy of Prof. S. Hoon, Manchester Metropolitan University).

field – soft materials) which can be fabricated in a reproducible manner. Care must be taken when choosing the tip material to ensure that the tip-sample interaction is not so strong that the magnetic domain structure of the sample is influenced by the tip i.e. the tip should be magnetically softer than the sample in most cases. Points that need to be considered are the micromagnetic state of the tip, the spatial frequencies of the magnetic domain structure in the sample, and the nucleation/pinning fields of the tip and sample. If quantitative image interpretation is required, the tips must be well characterised and calibrated and this can be carried out using magnetometry, DPC imaging, or *in situ* calibration. Micromagnetic modelling of the tip-specimen interaction is also necessary. Reviews of the MFM technique and some of its applications can be found in [16] and [17].

### Extending MFM Studies

Imaging in an external applied field is possible, but the micromagnetic state of the tip must be unperturbed by external applied field, or must change in a well-defined way to allow image interpretation. A recent application of MFM has been to use a commercial magnetoresistive recording head as the sensor in a scanning probe microscope [18]. This technique has been used very successfully to image the stray fields above written bits in magnetic recording media.

### 14.4.2  Atomic Force Microscopy

Although the AFM cannot be used for the imaging of magnetic domain structures, a brief description of some of the AFM techniques and their applications are included here because they are frequently applied to magnetic materials. In the **non-contact** mode of atomic force microscopy a tip is scanned over the sample at a height of 1–50 nm above the surface, allowing sample topography to be measured. At this separation, there is an attractive force between tip and sample (long-range van der Waals forces). The low total force between the tip and the sample results in a small signal, and the technique reduces sample contamination. The instrument is usually used in one of two modes: constant force, in which the height of the tip above the sample is adjusted to keep a constant cantilever deflection (the usual mode of operation), or constant height, which is used when measuring small changes in force. In the **contact-mode** the sample and tip are in contact, resulting in a repulsive force between the tip and the sample. The cantilever bends to follow the surface topography of the sample. The tip can be in constant contact with the sample surface, or the sample can be scanned in tapping mode, which is less likely to damage the sample surface and is good for imaging large scans with large variations in sample topography.

The AFM can also be used to measure surface friction – so-called friction force microscopy. Differences in the twisting of the cantilever at different positions across the specimen surface, as a result of friction at the sample surface produce image contrast. A further use of a scanning probe microscope is nanometric cartography [19] in which an atomic force microscope/scanning tunnelling microscope set-up is used to map the perpendicular tunnelling current in spin-tunnel junctions.

## 14.5   Polarised Light Microscopy

All polarised light techniques used for imaging magnetic domain structures rely on the fact that a piece of magnetic material rotates the plane of polarisation of linearly polarised light – the magneto-optical effect. If reflected light is used the effect is referred to as the magneto-optical Kerr effect [20,21] (MOKE), and this is the technique most widely used for domain imaging, with a spatial resolution of the order of 0.1 μm and a sampling depth of the order of 10 nm. If transmitted light is used the effect is referred to as the Faraday effect. This is not as widely used as the Kerr effect and will not be discussed further here. Both techniques image the magnetisation distribution in the sample rather than stray fields above the surface, and the images that are produced are thus directly comparable with Foucault mode Lorentz microscopy images.

### 14.5.1   Magneto–Optical Kerr Effect Microscopy

The rotation of the polarisation that is induced by the magneto-optical Kerr effect is usually very small ($\ll 1°$) although in some materials (e.g. Si-Fe [22]) the effect can be much larger. The small signal for most materials meant that Kerr effect imaging only really became widely used once digital imaging techniques were available which enabled background subtraction and contrast enhancement to be achieved. There are three configurations that can be considered, and which are illustrated in Fig. 14.7, namely:

1. The longitudinal Kerr effect, in which the MOKE signal is a maximum for incident angle $\simeq 60°$, which is used for study of in-plane magnetisation. See Fig. 14.8 for an example of an application of longitudinal Kerr microscopy.
2. The polar Kerr effect for which the signal is a maximum for normal incidence, used for the study of samples with a perpendicular magnetisation.
3. The transverse effect, which results in a change in reflected amplitude depending upon the magnetisation conditions of the sample.

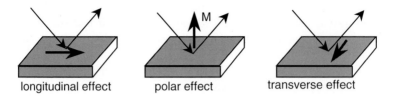

longitudinal effect      polar effect      transverse effect

**Fig. 14.7.** Schematic diagrams illustrating the various MOKE configurations.

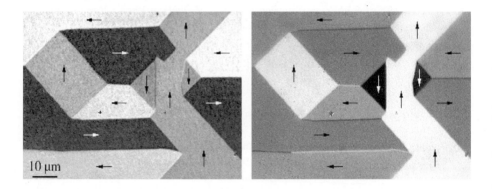

10 µm

**Fig. 14.8.** Longitudinal Kerr images of domains in a (100) oriented Si-Fe crystal. In the left image the sensitivity axis is horizontal and on the right it is vertical. (Courtesy of Dr. R. Schäfer, University of Dresden.)

The longitudinal and polar Kerr effects are usually modelled as inducing a Kerr component on reflection that is perpendicular to the incident polarisation direction. Since the Kerr component and the normal component are not usually in phase, the reflected light is elliptically polarised. Two parameters can then be measured, namely the Kerr rotation angle, $\theta_K$, and the ellipticity, $\eta_K$.

### 14.5.2   New Developments in Kerr Microscopy

#### Scanning Laser Microscope

In a scanning laser microscope (SLM), a diffraction-limited laser spot is scanned over the sample. The reflected light is detected, and the angle of the polarisation relative to the incident beam is detected by analysing the signal incident on two sets of quadrant photodiodes. The image is then built up from the pixels collected at each point using a camera or a CCD array. The advantages of this configuration are that the spatial resolution is improved relative to a parallel detection system, there is a large field of view (up to $\sim 25$ mm $\times$ 25 mm), there are many operating modes that can be used depending on the way in which the signals from the various quadrants are combined, and the images can easily be subjected to digital processing [23,24]. If used in the confocal mode, the depth discrimination becomes extremely good, allowing information to be obtained as a function of depth into the sample. An SLM image of bits written in a CoPt MO material is shown in Fig. 14.9.

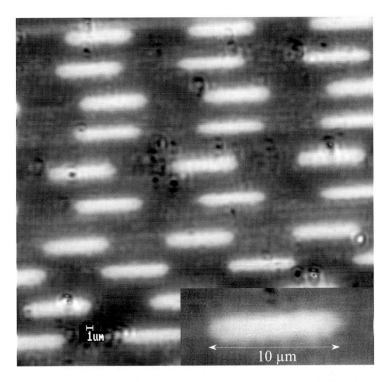

**Fig. 14.9.** SLM image, recorded using a 100 μm diameter confocal pinhole of 100 μm × 50 μm rectangular bits written into a Pt/Co multilayer sample (used for MO storage applications). The bit edges are clearly not perfect , with 'worm' like domains visible, which are a result of a relatively low coercivity of the Pt/Co layer. (Courtesy of Prof. C. D. Wright, University of Exeter.)

As well as being capable of imaging static domains, magnetic fields can be applied that enable real-time imaging of magnetisation reversals to be recorded, although the fact that the information is recorded serially does limit the time-resolution. Two further applications of the SLM have been: for obtaining local hysteresis loops from small regions of a sample, by keeping the laser spot fixed and following the change in Kerr rotation signal with applied field; and for mapping spatial variations in a magnetic property, for example permeability [23].

In order to improve the time-resolution of dynamic scanning Kerr microscopy, a number of groups have developed systems capable of nano- or pico-second time resolution, with a spatial resolution down to about 100 nm. This has been achieved either by stroboscopic imaging [25] in which a magnetic material is repeatedly cycled around a hysteresis loop and a stroboscopic technique (such as a fast pulsed laser) is used to record variations in magnetic domain structure with time at a fixed spatial position, or by scanning the spot over the sample, to build up an image that is fixed in time (relative to the hysteresis cycle) over a larger area. This technique has been used to great advantage to study rapid processes such as the magnetisation dynamics of write-heads, for which the whole magnetisation process occurs over a time-scale of 10–20 ns.

### Kerr Effect Near–Field Optical Microscope

Near-field optical microscopy gets around the problem of the diffraction limit by passing the light through a sub-microscopic aperture which is scanned over the sample. This results in an enhanced resolution of the order of 10–50 times greater than conventional optical imaging [26]. The technique can be used either in transmission or in reflection although, as with all optical techniques, the reflection technique is more widely applicable but more difficult [27]. However near-field optical microscopy is still under development and relatively few results have been published so far.

### 14.5.3    Polarised Light Microscopy: Advantages and Disadvantages

There are a number of advantages to the MOKE microscopy techniques described above, such as the fact that the Kerr contrast is directly related to the *magnetisation* in the sample, rather than to the stay fields above the sample surface. In addition, no specific sample preparation is required and the shape and size of the sample are not constrained. The technique can be made quantitative, although this is not necessarily easy to achieve, and dynamic effects such as magnetisation reversal processes can be observed, in addition to observing the effects of parameters such as temperature and applied stress.

Against this, there are a number of disadvantages that should be taken into account. Firstly, although no specific sample preparation is required, it is necessary for the sample to have a flat, reflective surface – surface irregularities can severely degrade the image quality. In addition, the spatial resolution is limited (except for near-field techniques) to about 0.15 μm.

## 14.6   Summary

In this chapter I have tried to present a brief overview of a number of the techniques that are used for magnetic domain imaging. The techniques all have their own particular merits and difficulties and it is often best to use a number of different techniques to analyse the same sample and thus gain the maximum information. For example, combining MOKE or Lorentz microscopy (which image the magnetisation within the sample) with MFM (which images the stray fields above the sample surface). I have not touched at all on the domain imaging techniques based on X-rays or neutrons and for this I would refer the reader to section 2.7 in Chapter I of Hubert and Schäfer [1]. These techniques differ from those discussed in this chapter in that they rely on access to synchrotrons and nuclear reactors and are therefore not 'lab-based' techniques.

## Acknowledgements

I am very grateful to the Royal Society for support for my research and to my colleagues who have allowed me to publish the domain images shown: Prof. S. Hoon, Manchester Metropolitan Univ., Dr. X. Portier and Dr. R. E. Dunin–Borkowski, Dept. of Materials, Oxford, Dr. R. Schäfer, University of Dresden and Prof. C. D. Wright, University of Exeter.

## 14.7   Problems

1. Electrons with energy 100 keV are incident normally on a Co foil of thickness 100 nm in an electron microscope. Calculate the angle through which the electrons can be deflected after passing through the specimen.
2. A spin-valve of width 10 μm has the following structure: Ta/NiFe/Cu/Co/-NiFe/MnNi (5/8/3/2/6/25 nm) where the numbers in brackets indicate the layer thicknesses in nm. The resistivities of the layers are: Cu – 2.7 μΩcm, NiFe – 20.0 μΩcm, Co – 7 μΩcm, Ta – 175 μΩcm and MnNi – 180 μΩcm. Use Ampere's law to estimate the field induced by the current in the NiFe sense layer for a 6 mA current.
3. (a) The saturation resistance of a spin-valve structure is 51.75 Ω and the resistance in the antiparallel configuration is 53.60 Ω. Calculate the GMR ratio of the system, stating the formula that you have used.
   (b) If the resistance of the system varies as $\cos(\theta)$, where $\theta$ is the angle between the magnetisation directions in the pinned and sense layers, calculate the resistance of the spin valve for $\theta = \pi/3$ and for $\theta = \pi/2$.

## 14.8   Solutions

**Answer to 1:**

$100 \text{ keV} = \frac{1}{2}mv^2$      $v^2 = (2 \times 100 \times 10^3 \times 1.6 \times 10^9)/9.11 \times 10^{-31}$
$v = 1.874 \times 10^8 \text{ ms}^{-1}$

Maximum force $\perp$ to $v$ is $F = -|e|vB = ma$
      $|F| = ev\,\mu_0\,M$          ($M$ is magnetisation)
$\Rightarrow |F_{\max}| = 1.6 \times 10^{-19} \times 1.874 \times 10^8 \times 4\pi \times 10^{-7} \times 1.43 \times 10^6$
                $= 5.39 \times 10^{-11} \text{ N}$

Time taken for electrons to travel through foil $= 100 \times 10^{-9}/1.874 \times 10^8$ s

Horizontal displacement $= \frac{1}{2}at^2 = \frac{1}{2}Ft^2/m = 8.44 \times 10^{-12}$ m
$\tan(\theta)$ ($\theta$ in radians) $= 8.44 \times 10^{-12}/100 \times 10^{-9} = 8.44 \times 10^{-5}$ rad

**Answer to 2:**

We assume that approximately 75% of the current contributes to the induced field, $H_I$, based on the resistivity and thickness of the layers that comprise the SV. We also assume that $H_I$ will only affect the sense layer magnetisation. Then:

$$H_I = 0.75I(4\pi \times 10^{-3})/2h$$

where $I$ is the sensing current, $4\pi \times 10^{-3}$ is the conversion factor from SI to CGS and $h$ is the height of the SV element. For a 6 mA current $H_I$ has a value of 2.8 Oe.

Note: This is a very oversimplified version of the truth, but it agreed very well with the experimental determined value for a spin-valve of this structure of $H_I = 2.5$ Oe.

Think about configurations of applied field direction, induced field direction and easy axis direction and how these parameters are related, and for which configurations the induced field will have an effect.

**Answer to 3:**

Assume that the thickness of the conducting layers corresponds only to the pinned + spacer + sense layers. The height of the spin-valve is the term used for its width.

$\Rightarrow$ Area $= 10 \times 10^{-6} \times (8 + 3 + 2 + 6) \times 10^{-9}$ m$^2$

Current densities are: 0.3 mA - $1.6 \times 10^5$ A cm$^{-2}$, 9.0 mA - $4.7 \times 10^6$ A cm$^{-2}$

# References

1. A. Hubert and R. Schäfer, in "Magnetic domains", Springer-Verlag, Berlin, (1998)
2. F. Bitter, Phys. Rev. **41**, 507 (1932).
3. M. E. Hale, H. W. Fuller, and H. Rubenstein, J. Appl. Phys. **30**, 789 (1959).
4. J. N. Chapman, J. Phys. D: Appl. Phys. **17**, 623 (1984).
5. J. P. Jakubovics, in "Handbook of Microscopy", Vol. 1, (eds.) S. Amelinckx *et al.*) VCH, Weinheim, New York (1997) p. 505.
6. J. N. Chapman and G. R Morrison, J. Magn. Mag. Mat. **35**, 254 (1983).
7. A. C. Daykin and A. K. Petford-Long, Ultramicroscopy **58**, 365 (1995).
8. A. Tonomura, T. Matsuda, J. Endo, T. Arii, and K. Mihama, Phys. Rev. Lett. **44**, 1430 (1980).
9. R. E. Dunin-Borkowski, B. Kardynal, M. R. McCartney, M. R. Scheinfein, and D. J. Smith, Mat. Res. Soc. Symp. Proc. (in press).
10. A. B. Johnston and J. N. Chapman, J. Microsc. **179**, 119-128 (1995).
11. X. Portier, A. K. Petford-Long, R. C. Doole, T. C. Anthony, and J. A. Brug, IEEE Trans. Magn. **33(5)** , 3574 (1997).
12. D. E. Newbury, D. C. Joy, P. Echlin *et al.*, in "Advanced Scanning Electron Microscopy and X-ray Microanalysis", Plenum, London, New York, (1986) p. 147.
13. J. Unguris, D. T. Pierce, R. J. Celotta, and J. A. Stroscio, in "Magnetism and structure in systems of reduced dimension", (eds.) R. F. C. Farrow *et al.*), Plenum, New York (1993).
14. G. Binnig, C. Quate, and C. Gerber, Phys. Rev. Lett. **56**, 930 (1986).

15. Y. Martin and H.K. Wickramasinghe, Appl. Phys. Lett. **50**, 1455 (1987).
16. P. Grütter, MSA Bulletin **24**, 416 (1994).
17. A. Wadas, in "Handbook of Microscopy", Vol. 2, (eds.) S. Amelinckx *et al.*) VCH, Weinheim, New York (1997) p. 845.
18. S. Y. Yamamoto, S. Schultz, Y. Zhang, and H.N. Bertram, IEEE Trans. Mag. **33**, 891 (1997).
19. V. Da Costa *et al.*, J. Appl. Phys. **83**, 6703 (1998).
20. J. Kerr, Phil. Mag. **3(5)**, 321 (1877).
21. J. Kranz and A. Hubert, Z. Angew. Phys. **15**, 220 (1963).
22. A. Hubert, Z. Angew. Phys. **18**, 474 (1965).
23. P. Kasiraj, R. M. Shelby, J. S. Best, and D. E. Horne, IEEE Trans. Mag. **22**, 837 (1986).
24. W. W. Clegg, N. A. E. Heyes, E. W. Hill, and C. D. Wright, J. Magn. Magn. Mat. **83**, 535 (1990).
25. M. R. Freeman and J. F. Smyth, J. Appl. Phys. **79**, 5898 (1996).
26. D. Courjon and M. Spajer, in "Handbook of Microscopy", Vol. 1, (eds.) S. Amelinckx *et al.*) VCH, Weinheim, New York (1997) p. 83.
27. C. Durkan, I.V. Shvets, and J.C. Lodder, Appl. Phys. Lett. **70**, 1323 (1997).

# 15 Observation of Micromagnetic Configurations in Mesoscopic Magnetic Elements

K. Ounadjela[1,2], I. L. Prejbeanu[1], L. D. Buda[1], U. Ebels[1], and M. Hehn[3]

[1] Institut de Physique et Chimie des Matériaux de Strasbourg, Unité Mixte CNRS-ULP-ECPM, 23, rue du Loess, 67037 Strasbourg Cedex, France
[2] VEECO instruments, 3100 Laurelview Court, Fremont, Ca 94538, USA
[3] LPM, Université Henri Poincaré BP 239, 54506 Vandoeuvre lès Nancy, France

**Abstract.** Advances in materials growth and characterization have, over the past ten years, made possible the investigation of basic physical processes in new "artificial" materials. These materials are artificial in the sense that the geometry and composition are controlled during growth on micrometer and nanometer length scales. This results in macroscopic behaviour that can be dramatically different from that of a material in its bulk form. Magnetic order and reversal processes, which have been extensively studied since the turn of the century, are now being re-examined for nanostructured materials.

The results presented here for the different magnetization configurations observed in submicron magnetic dots, rings and wires exemplify current state–of–the–art growth, lithography and imaging technologies. Using these geometries the potential for precise control of micromagnetic behaviour in patterned materials by control of shape and size is demonstrated. The boundaries between the different ground state configurations have been established experimentally as a function of the lateral width and height. Furthermore, metastable configurations can be induced following specific magnetization histories.

## 15.1 Introduction

During the last decade, much attention has been devoted to artificial layered magnetic materials which revealed a large variety of fascinating new phenomena such as the oscillatory interlayer exchange coupling in magnetic/non-magnetic multilayers, [1–6], surface and interface anisotropy [7–9], the giant magnetoresistance effect[1,5,10,11], quantum size effect in electronic properties [6,12] as well as in magneto-optical properties [12,13] of magnetic and metallic ultrathin films and related layered structures [14]. Those fundamental developments made such systems also of great interest from a technological point of view in the area of communication devices and storage media. Stimulated by this physics resulting from the layering and reduction of the system size in the vertical direction, a natural extension was the venture into a further reduction of the lateral sizes and quite general into low dimensional systems of nanometer extend [15]. Great interest has been developed for these mesoscopic magnetic structures. The term "mesoscopic" is used here to emphasize that the material dimensions are comparable to fundamental length scales associated with the transport and magnetic properties such as the conduction electron mean free path, the exchange lengths or the domain wall width.

The control of the unique micromagnetic properties at nanometer length-scales through a variation of the system dimensions made these low dimensional magnetic structures interesting not only from a fundamental, but also from a technological point of view. Examples for applications of high quality artificial low dimensional materials are well known for some time from the world of semiconductors, such as quantum wires and quantum dots [16–18]. In contrast a variety of tantalizing new possibilities for devices, structured from magnetic low dimensional systems have only been reported in the literature over the last years. Research and development of new magnetic structures has largely profited from these potential applications, in particular in high density data storage materials [19–21].

For the study of the static and dynamic properties of very small particles, say a few 10 nm to a few 100 nm, two approaches are possible. The first one consists in performing an ensemble average measurement on an assembly of many presumably identical (monodisperse, likely shaped) particles [22–24]. Due to the small particle volume, however, magnetization measurements are then limited to the study of a large number (millions) of small particles. The disadvantage of such an ensemble average is that it masks the intrinsic magnetic properties of the individual particle by the inevitable distribution of size or shape.

This can be overcome by state-of-the-art deep UV [25], X-ray [26,27] and e-beam ([28], see also Chap. 16) lithography techniques, making it possible to study one single particle at the time with a very local technique such as local near field probes [29,30], electrical measurements [31–34] or SQUID loop surrounding the particle to be studied [35–37]

Studies of the magnetic properties of individual particles have become possible with the development of the Magnetic Force microscopy (MFM) scanning probe technique. MFM has proven to be a well suited tool for imaging the stray fields of individual laterally confined elements [29,38–40] for which studies can be performed for example in the as-grown state, after applying different magnetic field histories or even as a function of an applied field following the hysteresis loop.

In this chapter, a comprehensive overview is presented on the current state of the art concerning the correlation between micromagnetic configurations and the shape and size of magnetic mesoscopic structures. Several examples from literature and our own work will be given. The various types of systems considered are submicronic dot, ring and wire structures, sufficiently far from each other, so that any interactions can be neglected. The effect of dipolar interaction in closely packed arrays is an interesting subject area on its own and goes beyond the scope of the presentation.

The chapter is organized as follows: In Sect. 15.2 several fabrication techniques for nanomagnets are described, while Sect. 15.3 gives a summary on the magnetic force microscopy imaging technique used. Section 15.4 summarizes the numerical micromagnetic calculations used to complement the MFM imaging. Section 15.5 gives a summary on the domain formation in thin films with particular emphasis on stripe domains developing in films with perpendicular magnetic anisotropy. Section 15.6 discusses the micromagnetic configurations in submicronic dots, in particular the transition from a vortex to a single domain state, driven by the balance between the exchange energy and demagnetization shape energies. In Sect. 15.7, similar aspects are described for circular rings, with more emphasis on stable and metastable magnetic structures. The last part of this chapter deals with epitaxial flat Co wires for two cases of orientation of a strong in-plane uniaxial anisotropy. The transformation between the stripe domains and single domain structures is discussed in one case, whereas for another head-to-head domain walls separate regions of uniform magnetization.

## 15.2 Fabrication Methods of Nanomagnets

Advanced lithographic techniques are currently employed to make regular periodic arrays of submicronic magnetic wires, dots and pillars [28]. The standard fabrication process used by the semiconductor industry involves electron beam lithography for the formation of designed patterns on a set of masks followed by optical lithography for the reproduction of the mask patterns at a high throughput level. A typical lithographic process consists of three successive steps: (1) coating a substrate with irradiation sensitive polymer layer resist, (2) exposing the resist with light, electron or ion beams depending on the lithography of choice, (3) developing the resist image with a suitable chemical. Exposures can be done either by scanning a focused beam pixel by pixel from a designed pattern (electron-beam or ion-beam lithography techniques), or exposing through a mask for parallel replication. We will discuss here only the lithography techniques used for making the constrained structures presented in this chapter.

### 15.2.1 E-beam Lithography

Electron beam (e-beam) lithography is used for primary patterning directly from a computer-designed pattern [28]. It is the essential basis for nanofabrication in addition to mask and prototype device manufacturing. This technique is however not suitable for mass production because of the limited writing speed. The resolution of e-beam lithography depends on the beam size and several factors related to the electron-solid interaction. Scanning electron beam lithography has demonstrated 10 nm resolution by using low energy electrons ($\sim$ 1 keV) to reduce the proximity effects. E-beam lithography has been found ideal for use in a research environment and has allowed us to study well defined wires as discussed in Sect. 15.8.2. These wires with lateral widths down to 100 nm, see Fig. 15.1d, were patterned from epitaxial cobalt films [41]. Furthermore, this technique was used to pattern circular rings, see Fig. 15.1c, discussed in Sect. 15.7.

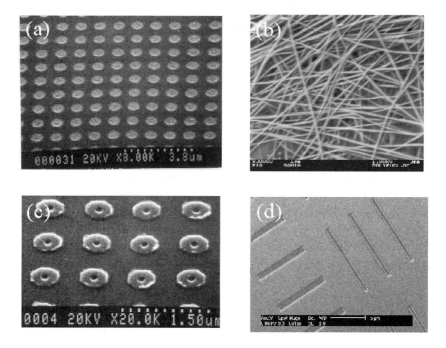

**Fig. 15.1.** High resolution scanning electron microscopy images of (a) an array of Co dots of $\sim 1\,\mu$m period and $\sim 0.5\,\mu$m basal plane, patterned by X-ray lithography from a continuous epitaxial Co film, (b) electrodeposited Co nanowires of 35 nm diameter, after dissolution of the polycarbonate membrane in dichloromethane, (c) polycrystalline Co rings of 500 nm outer diameter, fabricated by electron beam lithography and lift-off and (d) 100 nm wide flat Co wires, patterned by electron beam lithography from a continuous epitaxial Co($10\bar{1}0$) film. E-beam lithography allows to computer design the wires along any spatial direction. For instance, wires can be patterned along the anisotropy axis or perpendicular to the anisotropy axis.

## 15.2.2  X-ray Lithography

X-ray lithography is part of the next generation lithography, which has already demonstrated 20 nm resolution in a contact printing mode and can have a high throughput [26,27]. However, its mask technology using e-beam lithography and the exposure systems are currently rather complex and expensive. A typical X-ray mask consists of a 2 $\mu$m thick membrane of silicon carbide and absorber features of heavy metals such as Au, W, or Ta. Exposures can be done at a mask-to-wafer distance of $\sim 10\,\mu$m with synchrotron radiation or a laser-induced plasma source. The resolution of proximity X-ray lithography is defined by the Fresnel diffraction and the diffusion of photoelectrons in the resist. High resolution and high aspect ratio features can thus be obtained. Figure 15.1a shows an array of 1 $\mu$m-period and 0.5 $\mu$m wide epitaxial dots made on a total surface of 5 mm by 5 mm. The patterning process begins with the realisation of holes in a high sensitivity resist using X-ray lithography followed by an aluminium lift-off process [42]. The edges are straight with nearly vertical profile and the dots surface retain the smoothness already observed on the as-grown films. This quality of patterning is kept up to 150 nm thick cobalt films. The magnetic properties of these dots are discussed in Sect. 15.6.

### 15.2.3  Electrodeposition Into Porous Templates

As an alternative to these advanced lithographic techniques, electrochemical deposition of ferromagnetic metals into porous templates is performed to produce arrays of nanopillars (or nanowires) [43] of extremely small diameters, the most commonly used templates being polycarbonate membranes [44]. Although, at present, nanowires cannot be grown at prespecified locations in polycarbonate templates, this method has the attractive features of simplicity in operation and high cost-effectiveness. The membranes are first irradiated with $Ar^+$ ions accelerated at 120 MeV and subsequently etched chemically. The irradiation is performed at normal beam incidence with respect to the plane of the polycarbonate films and the dispersion in the direction of the ion tracks is usually less than 10 degrees. The etching conditions are adapted to produce regular cylindrical pores of varying diameters. The growth is generally performed by electrodeposition at room temperature from a sulfate bath containing the ions of the material which is suitable ($Co^{2+}$ in order to make arrays of Co nanowires as shown in Fig. 15.1b). The porosity of the membranes allows defining the average spacing between the pillars, making possible study of interacting or non interacting particles. The Co nanowires discussed in Sect. 15.8.3 were shown to be made of large crystal grains, extending transversally across the full wire diameter and longitudinally over several micrometers. Second and more important, electrodeposited Co proved to adopt a rather good quality hexagonal compact structure with a preferential texture along the wire axis for the 35 nm Co wires [45].

## 15.3  Magnetic Force Microscopy

Most of the images reported in this chapter have been taken using the very popular and widely used technique of Magnetic Force Microscopy. We used the commercial AFM/MFM scanning probe microscope from VEECO-Digital Instruments Dimension 3100, equipped with standard CoCr-coated tips magnetized along the tip axis, which applies the powerful TappingMode/LiftMode$^{TM}$ interleave technique developed by Digital Instruments (see Fig. 15.2b) [46]. This particular technique allows disentangling the topographical and magnetic data and to collect both kinds of information during the same image acquisition, hence making easy the correlation of the magnetic data with topographical features.

Magnetic Force Microscopy is a development of the technologies of Atomic Force Microscopy and Scanning Tunneling Microscopy [47] in order to image magnetization patterns with high resolution and minimal sample preparation. The technique relies on measuring the interaction of the stray magnetic field emerging from the sample with the magnetic moment of a sharp magnetic tip

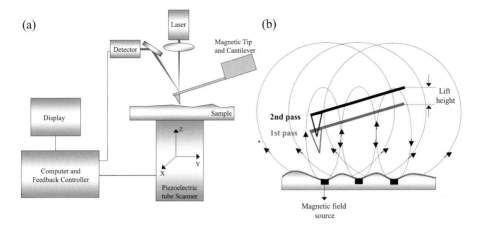

**Fig. 15.2.** (a) Tapping mode AFM operates by scanning a tip attached to the end of an oscillating cantilever across the sample surface. (b) Using a tip coated with a magnetic material, the system can be used in the lift mode, which allows to distinguish the long-range magnetic and the short-range topographic information during the same image acquisition.

attached to a flexible cantilever which is placed close to the surface (typically 0.5–500 nm).

The interaction is inferred by measuring the cantilever deflection using either the tunnel effect [48], optical interferometry [49] or optical deflection (see Fig. 15.2a). The image is formed by scanning the tip laterally with respect to the sample and recording the interaction strength as a function of position.

We introduce shortly the principle of this instrument, see also Chap. 14.

### 15.3.1   Principle of the Magnetic Force Microscope

**Frequency Shift:** A variety of different techniques can be used to probe the interaction between tip and sample. From those the non-contact ac-detection mode MFM (Fig. 15.2b) enjoys wide popularity in the literature. It operates by monitoring the vertical dynamics of the cantilever as it is scanned over the magnetised surface. A piezoelectric bimorph is used to oscillate the flexible cantilever that supports the probing tip transversely. This is generally done by driving the cantilever at a fixed frequency slightly higher than its mechanical resonance and observing changes in the cantilever deflection amplitude.

The mechanical resonance frequency of the cantilever is determined not only by its indigenous spring constant but also by the vertical component of any force gradient which it experiences since any such force which varies with displacement appears in the cantilever equation of motion as an additional effective spring [46]. This, in turn, shifts the resonance frequency [49] and hence changes the amplitude of the tip oscillation.

Force gradients can arise from several sources; magnetic force between tip and sample magnetizations, interatomic forces between tip and surface atoms and other forces such as electrostatic interaction [50]. The component of the force that actually contributes to the cantilever deflection is $F_{\mathrm{mag}} = \boldsymbol{n} \cdot \boldsymbol{F}$, where $\boldsymbol{n}$ is a unit vector normal to the cantilever, and the component of the force gradient that modifies the cantilever spring constant is the one normal to the cantilever. Therefore, such an ac detection technique yields a signal related to the force derivative $F'_n = \boldsymbol{n} \cdot \nabla(\boldsymbol{n} \cdot \boldsymbol{F})$.

**Separation of Topography and Magnetic Contrast:** The MFM image is constructed by extracting the magnetic force gradient component and suppressing the interatomic force gradients. This suppression can be done in several ways. The most straightforward is to operate in the large separation limit of the interatomic potential where the interatomic force is attractive and the third derivative of the potential with respect to tip height is small: thus the cantilever resonant frequency is insensitive to topographical changes as the tip is scanned.

In the TappingMode/LiftMode$^{\mathrm{TM}}$ interleave technique developed by Digital Instruments, a complete image acquisition is achieved by performing, at evenly spaced positions along the so-called slow translation axis, a set of two successive line scans along the fast translation axis (perpendicular to the slow translation axis). During the first of these two scans, the tip flies very near to the sample surface (10 nm at most). The tip-sample separation is constantly adjusted by a feedback loop so as to maintain constant the amplitude of the oscillation of the tip, the latter being mainly subjected to short range Van der Waals forces. This allows to generate a contour of constant force gradient that defines the sample topography to be stored. For the second of the two scans, the tip-sample distance is increased to some value (typically ranging from 50 to 200 nm in the case of the reported experiments) chosen so that the lifted tip be then predominantly subjected to the long range magnetic forces due to dipolar interactions between the tip magnetization and the stray field emanating from the specimen. The feedback control is turned off and the tip is driven along a trajectory that mimics precisely the sample topography measured during the first scan. The recorded signal then consists of the variation of the magnetic force gradient at constant height above the specimen surface, yielding a contrast mapping of the magnetic stray field above the sample.

The disadvantage of a too large tip-sample distance for the second (magnetic) scan is that spatial frequencies in a magnetic field pattern fall off rapidly with distance from the magnetization distribution. As a result the resolution decreases with which the magnetization distribution can be imaged. A compromise has therefore to be found for the correct tip-sample distance in each scan, to obtain good resolution of the magnetic contrast (reduce distance) and at the same time reduce any contributions from the interatomic interactions (increase distance).

## 15.3.2   Modelling of the MFM Response

Because of the proportionality of the recorded signal to the magnetic force derivative, the technique employed provides a very good signal to noise ratio. However, it concomitantly makes the recovering of the local magnetic configuration within the sample not straightforward. Indeed, modeling is most often required to ensure a correct interpretation of the observed MFM images [51–53]. To evaluate theoretically the magnetic force exerted on the tip and its gradient, various levels of approximation are possible, depending on the degree of complexity of the model used to describe the shape and magnetization state of both the tip and the sample and their interaction. One simple but very instructive approximation that we will use in the following is to assume that the probing tip consists of a point dipole with effective magnetic moment $\boldsymbol{m}$. In this case, the magnetic force acting on the tip follows the equation

$$\boldsymbol{F}_{\mathrm{mag}} = \nabla(\boldsymbol{m} \cdot \boldsymbol{h})\,, \qquad (15.1)$$

where $\boldsymbol{h}$ is the stray field originating from the sample. If, as is usual and was always the case for the reported MFM experiments, the cantilever is vibrated along the $z$ axis perpendicular to the $x - y$ plane in which lies the flat substrate supporting the sample and if the effective magnetic moment of the tip is oriented along the $z$ direction ($m_x = m_y = 0$, $m_z = \pm m$ ), the magnetic force derivative to which the MFM response is proportional will be then:

$$F_{\mathrm{n}}' = \pm m \, \frac{\partial^2 h_z}{\partial z^2}\,. \qquad (15.2)$$

Thus, the MFM response is simply proportional to the second derivative with respect to the $z$ coordinate of the vertical component of the stray field produced by the sample at the tip location. It should, however, be emphasized that this approach is only valid under the assumption that the stray field from the sample is not sufficient to alter the magnetization in the tip and that, conversely, the stray field from the tip has no significant effect on the magnetization distribution within the sample.

## 15.4   Micromagnetic Calculations

Numerical micromagnetic modeling is a vital tool to predict and understand the various magnetic configurations, which are strongly dependent on the material parameters but more importantly on the shape of the nanostructures. These calculations will then have to be compared to the results obtained from direct imaging. For this purpose we have developed a 3D micromagnetic code which has been used to obtain magnetization vector plots presented later on. The confidence of this code has been checked by treating several benchmark problems [54] such as the magnetization configuration in small cubic particles (Problem 3) as well as the reversal dynamics of a thin permalloy platelet (Problem 4) published by McMichael [54].

The magnetization configurations presented in this chapter were obtained by minimizing the total free energy of the system, which includes contributions from the magnetocrystalline anisotropy, the demagnetization, the exchange and the Zeeman energy [39]. For more details on the micromagnetic theory, refer to the chapter by R. Skomski, Chap. 10 in this book. The minimization is carried out with respect to $M = M_s m$ under the constraint $|m| = 1.0$. Furthermore the magnetization at the element surfaces satisfies the stationary boundary condition $\partial m / \partial n = 0$, where $n$ is the unit vector normal to surface pointing outwards. Starting from a given configuration, the system proceeds towards a local minimum by following the states according to the Landau–Lifshitz–Gilbert equation (LLG) [55,56]. The real system is discretized into $N_x \times N_y \times N_z$ cubic cells of constant magnetization. The cell size is chosen to be smaller than the characteristic magnetic lengths. This cell size is 2.5 nm in the case of Co as a material used for many studies presented in this chapter. The characteristic magnetic lengths using the material parameters of Co, saturation magnetization $M_s = 1400$ emu/cm$^3$, exchange constant $A_{ex} = 1.4 \times 10^{-6}$ erg/cm, magnetocrystalline anisotropy constant $K_u = 5 \times 10^6$ erg/cm$^3$, are: exchange length $l_{ex} = \sqrt{A_{ex}/(2\pi M_s^2)} \sim$ 3.37 nm and Bloch wall width parameter $\Delta_0 = \sqrt{A_{ex}/K_u} \sim 5.29$ nm. The magnetostatic energy is evaluated in the approximation of uniform magnetized cubic cells and the demagnetization field is substituted by its value averaged over the cell [57]. The fast Fourier method is implemented for the stray field evaluation. The numerical stability of the time integration of the LLG equation is assured by the use of an implicit forward difference method for the time discretization [58]. A constant time step of $dt = 0.1$ ps has been used and the damping parameter was set to $\alpha = 1.0$ since we are only interested in the static stable state. The convergence is reached when the residual torque $|m \times h_{eff}| < 10^{-6}$.

## 15.5   Domain Formation in Thin Films

### 15.5.1   Origin of Domains

Exchange due to Pauli exclusion tends to align magnetic moments parallel to each other in a ferromagnet whereas interaction with crystalline fields via spin orbit coupling leads to a preferential orientation of the magnetization along particular directions. This behaviour is often described in terms of effective exchange and anisotropy fields acting on a position dependent magnetization vector, see also Chap. 10.

The concept of domains was originally introduced by Weiss [59] to explain why ferromagnetic materials can have zero average magnetization while still having a non-zero local magnetization. The essential idea is that alternating the direction of the magnetization with respect to a surface can minimize the energy in the static magnetic fields associated with the magnetization in a finite material [39].

The transition from one direction of magnetization to another between adjacent domains involves a rotation of the magnetization vector. The rotation occurs over a finite distance whose width is determined by a competition between exchange and anisotropy. The resulting magnetic structure is called a domain wall. When a magnetic field is applied, domains with the magnetization oriented along the applied field direction grow by displacement of the walls at the expense of domains with the magnetization oriented opposite to the field direction.

The deviations of the magnetization from uniform inside the domain wall incorporates exchange, anisotropy and dipolar energies, so that the formation of the wall is energetically costly. The ground state of an infinite bulk material therefore would be the homogeneously magnetized single domain state. However, real materials have finite boundaries, which involve at some point or another a discontinuity in the magnetization and with this magnetic surface charges giving rise to shape demagnetization fields. It is the tendency to reduce these surface demagnetization fields (pole avoidance principle) which finally give rise to the formation of domains, where the reduction in demagnetization energy and the cost of wall energy are balanced against each other.

In the following only thin film systems will be considered. In this case the shape demagnetization field from the perpendicular surfaces prefers an alignment of the magnetization parallel to the film plane. However, the presence of magnetic anisotropies (others than shape, of magnetocrystalline origin for example) can re-orient the magnetization. One important case considers a uniaxial anisotropy which favors an alignment of the magnetization out of the film plane (called perpendicular films here). This case represents a model system for the discussion of the origin of domain formation as well as the confinement effects induced by lateral reduction. Those aspects are presented in detail in the following sections.

### 15.5.2   Stripe Domains in Thin Films with Perpendicular Anisotropy

**Magnetic Oxides:** A large amount of work was performed in the 70's on perpendicular films mainly in single crystals of orthoferrites, hexagonal ferrites and magnetic garnets with the idea of using the observed stable domain structures for storing and processing binary data in magnetic recording devices [60,65,66].

In such films the competition between the shape demagnetization fields and the out-of-plane magnetocrystalline anisotropy induces a periodic variation of the order parameter leading to typical domain patterns which, depending on the magnetic history can have the form of circular cylinders (bubble domains) or serpentines (stripe domains) [39,60–64]. Typical bubble diameters in these materials are on the order of 10 $\mu$m [60,65,66], and do not allow for sufficiently dense packing of information to be competitive. The bubble domain diameter or stripe domain width are given by the material parameters of exchange energy, saturation magnetization and uniaxial anisotropy strength and scale with the inverse of the magnetization [39,60,63,64].

**Metallic Thin Films:** The material parameters of metallic thin films (compare Chap. 10) are all at least one order of magnitude larger than those of the oxide materials, which give rise to stripe or bubble domains with dimensions between 10 nm and 100 nm [67–71], being thus 100–1000 smaller than in the oxide materials studied earlier. This is obviously exciting from the point of view of possible applications.

Recently, the domain configurations in thin hcp Co(0001) films grown on Ru buffers have been studied in our group as a function of film thickness [67]. The relatively strong perpendicular uniaxial magnetocrystalline anisotropy ($K_u = 6 \times 10^6$ erg/cm$^3$) of these films has to be compared to the shape demagnetization energy $2\pi M_s^2$ yielding a $Q$ factor $Q = K_u/2\pi M_s^2 = 0.4$ [39,67,68]. This means that the anisotropy is in principle not strong enough to overcome the demagnetization field. However, upon domain formation the effective demagnetization field is reduced such that a perpendicular orientation of the magnetization inside the domains can be stabilized.

Typical MFM images for stripe domains in Co films are shown in Fig. 15.3 from which the domain width $L$ as a function of film thickness was deduced as given in the graph of Fig. 15.3b. $L$ decreases with decreasing film thickness proportional to $\sim \sqrt{t}$ (line) in accordance with Kittel's law for stripe domains [61]. In contrast, the ratio $L/t$, which scales the demagnetization field inside a single domain, increases. This means that the effective demagnetization field inside a single stripe domain increases upon reducing $t$, causing the magnetization to rotate into the film plane for a ratio $L/t$ potentially larger than 1 [63,64,72]. This canting is seen in the MFM images (Fig. 15.3) by the decreasing contrast of the stripe domains for $t = 50$ nm and $t = 25$ nm and the loss of any stripe domain contrast in the thinnest film. The reorientation of the magnetization is a continuous rotation process as indicated by the thickness dependence of the in-plane remanence ratio $M_r/M_s$ shown in Fig. 15.3a. Above 60 nm (region I) the in-plane remanence is low, given only by the canted spins inside the domain wall, while the domain magnetization is perpendicular to the film plane. In the intermediate range II the remanence increases indicating a canting of the domain magnetization towards the film plane. Below 20 nm, region III the magnetization is fully in-plane.

## 15.6    Micromagnetic Configurations in Mesoscopic Dots

### 15.6.1    Reduction of Lateral Sizes

As outlined in Sect. 15.5.1, domains are formed to reduce demagnetization shape energies arising at the sample surfaces. For each system different domain configurations or magnetization distributions can be induced, depending on the magnetic history. These different configurations will generally be characterized by different energies. Besides the dependence on the particle shape and size, the various possible domain configurations will also be determined by the strength and symmetry of its magnetic anisotropy.

A reduction of the system size enhances quite generally the relative importance of the surface boundaries. For example, starting from an in-plane magne-

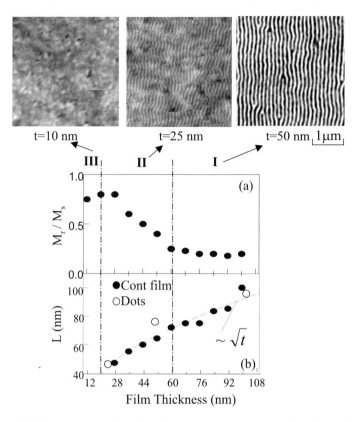

**Fig. 15.3.** MFM images of stripe domain structures developing in epitaxial Co(0001) thin films. Images of different thickness are shown for a strong stripe structure for $t = 50$ nm, a weak stripe structure for $t = 25$ nm and an in-plane magnetized film $t = 10$ nm. These three images correspond to the three regions I, II, III respectively given in (a) where the thickness dependence of the in-plane remanence ratio $M_r/M_s$ is shown. (b) The thickness dependence of the domain width $L$ for continuous Co(0001) films (closed circles) and 500 nm square Co(0001) dots (open circles). The dotted line is a fit to $\sqrt{t}$.

tized continuous film, the demagnetization fields arising at the edges are negligible compared to the bulk of the film and and the ground state of the film is the single domain state. Reducing the lateral sizes of the film, the demagnetization fields across the film increase and induce at some point a domain structure [39,73]. Interesting properties are expected when the geometrical dimensions are further reduced, such that they become comparable to either the domain sizes, the domain wall width (20 nm in Co hcp) or the exchange length (3.37 nm for Co) [29,39,74–77]. In particular, below a critical system size it was predicted that domain formation is suppressed and the magnetic particle is in a single domain state [39,78].

In the following the influence of the reduction of the lateral sizes in two dimensions will be discussed for Co(0001) dots with perpendicular uniaxial anisotropies for the three thickness regions of Fig. 15.3a with (i) perpendicular stripe domains, (ii) canted stripe domains and (iii) in-plane magnetization. The reduction of the lateral sizes of the continuous film in only one dimension, yielding wires is discussed in Sect. 15.8.

### 15.6.2 Preparation

The epitaxial Co(0001) dots with perpendicular uniaxial anisotropy presented in the following were prepared using X-ray lithography and ion beam etching from continuous epitaxial Co(0001) hcp films in arrays of $5 \times 5$ mm$^2$ square. The dots have a square basal plane with 0.5 $\mu$m lateral dimension, 1 $\mu$m array periodicity and a thickness varying from 10 to 150 nm [29,38].

### 15.6.3 Domains in Perpendicular Dots: Effect of Thickness and Shape

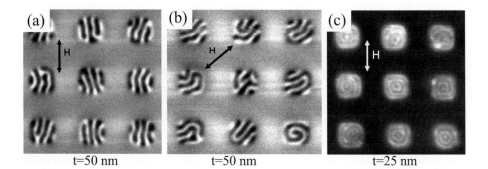

$t=50$ nm            $t=50$ nm            $t=25$ nm

**Fig. 15.4.** Stripe domains induced in $\sim 0.5$ $\mu$m wide epitaxial Co(0001) dots 50 nm thick and demagnetized (a) parallel to the edge of the square dot and (b) along the diagonal. (c) Circular stripe domains induced in $\sim 0.5$ $\mu$m wide epitaxial Co(0001) dots 25 nm thick and demagnetized parallel to the edge of the square dot.

MFM images for square Co dots (500 nm) are shown in Fig. 15.4 for two different in-plane demagnetization states for fields applied (a) parallel to the dot edge and (b) parallel to the dot diagonal. These dots have a thickness of 50 nm corresponding to the thickness region I, Fig. 15.3, of perpendicular domain magnetization. For these dots the influence of the thickness and the effect of shape are considered [29,38,79,80].

**Dot Height:** The number of domains inside the dots increases with decreasing dot thickness, corresponding to a decrease in domain size in much the same way as in the continuous thin films, see Fig. 15.3b. When the domain size is much smaller than the lateral dimension of the dot, one expects identical behavior for a continuous film and a magnetic dot. This is verified in Fig. 15.3b where the domain size as a function of thickness is compared for the continuous films ($\bullet$) and the dots ($\circ$) [80]. The dependence of the domain size on thickness is the same in both cases. This indicates that the magnetostatic energy at any point inside a domain is primarily determined by nearby domains.

**Shape Effects:**   While the domain sizes do not differ much in the square dots and the continuous films, effects of the shape and finite size can be clearly observed for the alignment and order of the stripe domain pattern. For the continuous films, the stripe domain structure shown for $t = 50$ nm in Fig. 15.3 is induced by an in-plane demagnetization procedure which results in well aligned stripes oriented along the applied field direction, independent of the orientation within the film plane. In the case of the square dots, shown in Fig. 15.4a,b, it is observed that the stripes are not very straight and can show a strong bending away from the field direction. Furthermore this missorientation and bending is more pronounced when applying the field along the diagonal, compare Fig. 15.4a,b. The bending in the case of Fig. 15.4a can be explained by the nucleation process of the stripe domains, as shown in [81]. More interesting is the dependence on the field orientation. It appears that upon demagnetization along the dot diagonal, domains are nucleated which have a tendency to align parallel to the edges rather than to the demagnetization field direction and which try to avoid the corners of the dots [79,80]. A possible explanation for this anisotropic order of the stripe pattern as a function of field orientation may be given in terms of the configurational anisotropy discussed further below in Sect. 15.6.3 for in-plane magnetized single domain square elements [77,82]. At saturation along the diagonal, the magnetization is in-plane yielding an inhomogeneous demagnetization field across the square. The demagnetization fields inside the corners are much stronger and induce an inhomogeneous alignment of the magnetization upon reducing the field similarly to the leaf state discussed below in Fig. 15.12. Since the domain walls orient parallel to the local field at nucleation, an inhomogeneous demagnetization field which superposes to the applied field can lead to strongly curved stripe domains avoiding corners.

### 15.6.4   Domains in Canted Dots

MFM images for square Co dots 500 nm wide and of 25 nm thick are shown in Fig. 15.4c [29,38,80]. This thickness corresponds to the thickness region II in Fig. 15.3, of strong canting of the domain magnetization towards the film plane. In the dots, the stripes form a complete concentric circular ring system of alternating black and white domains tending to keep the stripes parallel to the edges of the dots. This circular arrangement implies that a singularity occurs at the center of the dot where the in-plane magnetization reorients fully perpendicular to the plane to form a so-called vortex structure. This pattern is obtained independent of the orientation of the in-plane demagnetization field and must be analyzed in terms of the in-plane and out-of-plane components of the canted domain magnetization giving rise to in-plane and perpendicular demagnetization fields. The tendency to reduce the out-of-plane demagnetization field component induces the alternating stripes. In contrast, the presence of an in-plane magnetization component, giving rise to in-plane demagnetization fields induce the circular arrangement of the stripes in order to avoid surface charges at the dot edges. This indicates that strong effects are expected when the magnetization component is fully in-plane in dots patterned from Co(0001) films of region III of Fig. 15.3.

### 15.6.5    Domains in In-Plane Circular Dots

As discussed in the previous section, the presence of an in-plane magnetization component can have a drastic influence on the geometric arrangement of the stripe domains. In the following we will therefore consider only elements patterned from in-plane magnetized materials. These are materials corresponding either to those described in Fig. 15.3 in the thickness region III, where the magnetization in the continuous film is rotated fully in-plane, or materials without magnetic anisotropy.

For such in-plane films, the reduction of the lateral sizes induces first a multidomain structure in order to minimize the edge demagnetization fields. When the lateral dimensions of the element become comparable to the width of the domains only simple magnetization configurations can be realized as a direct consequence of the finite size effect [40,83]. Upon further size reduction comparable to the lengthscale of the domain-wall width the domain structure will finally turn into a single domain or near-single domain configuration [83,84]. It is this extreme limit which will be discussed in this section for in-plane circular dots and in Sect. 15.7 for in-plane circular rings. Focus will be given to (i) the stability of the different configurations and (ii) the transition as a function of thickness and lateral dimension between the single domain state and simple flux-closure configurations.

**Single Domain and Vortex Configurations:** MFM investigations of arrays of epitaxial circular Co(0001) dots of 200 nm diameter have shown that different magnetic states can be induced depending on the dot dimensions as well as on the magnetic history [74,85]. In Fig. 15.5, two examples are shown for $t = 10$ nm after (a) in-plane saturation and (b) out of plane demagnetization. The strong dipolar contrast in Fig. 15.5a is interpreted as a single domain state while the weaker contrast in Fig. 15.5b with a dark spot in the center is indicative of a vortex-like state for which the magnetization vectors remain parallel to the edges. The circular magnetization path results in a singularity in the center of the dot where the magnetization turns perpendicular out of plane, giving the name to this magnetization configuration. Similar observations have been made by several authors for polycrystalline circular dots which do not exhibit crystalline anisotropy [75,76,86,87] as well as on elliptical or elongated dots [88–91].

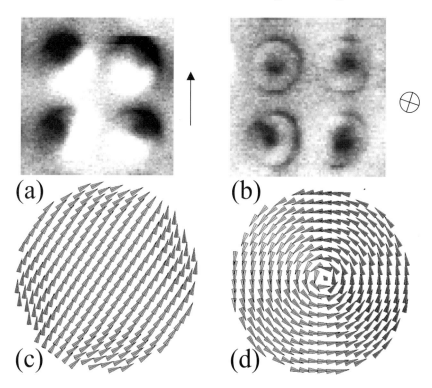

**Fig. 15.5.** MFM images of (a) a single domain structure and (b) a vortex type structure observed in 200 nm diameter epitaxial Co(0001) circular dots of $t = 10$ nm; (c) and (d) are the corresponding magnetization vector plots obtained from 3D micromagnetic modeling.

Particularly nice are the Lorentz transmission electron microscopy observations reported by Schneider *et al.* [76] and discussed further below in Fig. 15.8.

**Reversal of the Single Domain and Vortex Configurations:** Further experimental evidence of these two magnetization configurations was given by Cowburn *et al.* [84]. They have measured two classes of magnetization loops, see Fig. 15.6, probed by in-plane Kerr magnetometry for soft NiFeMo (Supemalloy) circular dots as a function of the lateral dot dimension and the dot height. These loops are reminiscent of either a flux closing type of magnetic structure with the formation of a central vortex (V) (Fig. 15.6a) or a single domain (SD) structure in which the magnetization is aligned along one specific direction within the dot (Fig. 15.6b).

**Single domain reversal in in-plane fields:** As shown on the magnetization curve, Fig. 15.6b, for a 100 nm wide and 10 nm thick NiFeMo dot array, the single domain loop retains a high remanence at zero field and switches abruptly at very low field because of the absence of anisotropy [84]. This reversal is classi-

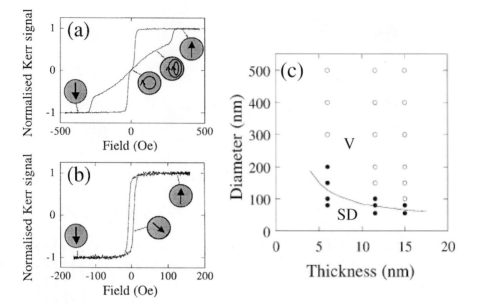

**Fig. 15.6.** Hysteresis loops measured by Kerr magnetometry from Supermalloy NiFeMo circular dots of $t = 10$ nm and (a) 300 nm and (b) 100 nm diameter. (c) Experimental phase diagram denoting the single domain (full dots) and vortex states (open circles) as a function of dot diameter and thickness. (From [84].)

cally described by the Stoner–Wolfarth model [39,92], which treats the reversal process as a coherent rotation of the magnetization. This reversal type is further confirmed by MFM imaging under a magnetic field performed on polycrystalline Co dots [93], which display a purely dipole character (Fig. 15.7). Varying the field shows that the magnetization rotates coherently and locks on the opposite direction.

**Vortex reversal in in-plane fields:** As seen from Fig. 15.6a for a 300 nm wide and 10 nm thick NiFeMo dot array [84], when the applied field is reduced from saturation, the dots retain full moment until a critical field slightly before crossing zero field, at which point nearly all magnetization is lost. The sudden loss of magnetization is characteristic of the formation of a flux-closing configuration.

This description of the macroscopic hysterisis loop taken from an array of dots is in agreement with the observation of the reversal of a flux-closure structure of soft NiFe dots, see Fig. 15.8a, reported recently by Schneider *et al.* [76] using Lorentz transmission electron microscopy. The observation started at an in-plane field of -200 Oe, which is sufficiently high to saturate the disks to the left. Increasing the magnetic field from saturation leads first to the formation of magnetization inhomogeneities at the edges of the dots. A vortex state appears for a field of $-60$ Oe, away from the center of the dot. This indicates that the

## D = 500 nm

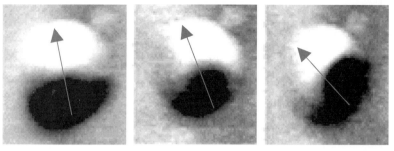

-200 Oe          -100 Oe          -50 Oe

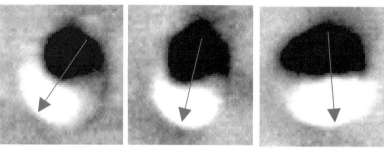

0 Oe          +100 Oe          +200 Oe

**Fig. 15.7.** MFM images as a function of the applied field for $t = 10$ nm polycrystalline circular Co dots of 500 nm diameter. The field values given correspond to those of the external applied field generated by two permanent magnets with variable separation. The actual field is the sum of the applied and the tip field.

magnetic rotation center moves perpendicular to the direction of the applied field to give rise to a net contribution of the magnetization along the direction of the applied field. Further increase of the external field results in the motion of the vortex (perpendicular to the applied field) until at zero field, the vortex center coincides with the geometric disk center. This then results in a zero remanent magnetization in agreement with the $M$–$H$ curves shown in Fig. 15.6a. Field reversal shifts the vortex center towards the particle border as shown in Fig. 15.8a.

This experimental observation is in agreement with 2D numerical micromagnetic calculations using the NIST OOMMF code [54]. As shown in Fig. 15.8b, at the early stage of the reversal process, a magnetization vortex is formed at the border of the element along the upper-edge. The polarity of the magnetization circulation is such that the bottom part of the magnetization vortex is in the same direction as the applied reversing field. At zero field the vortex is in the center and during the field reversal the vortex is expulsed at the lower-edge of the dot element.

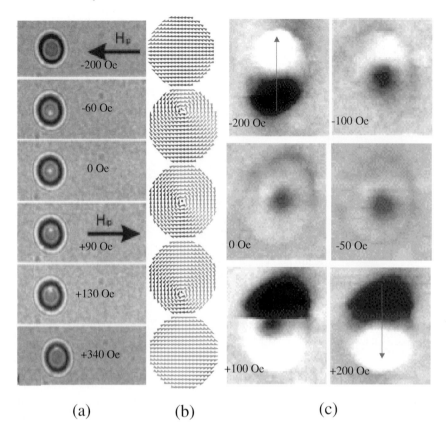

(a)                 (b)                 (c)

**Fig. 15.8.** (a) Lorentz microscopy images for circular Permalloy dots of $t = 15$ nm and 340 nm diameter showing the motion of a vortex as a function of an in-plane field (from [76]) (b) Magnetization vector plots obtained from 2D numerical micromagnetic calculations using the NIST OOMMF code [54] for circular Co dots of 400 nm diameter and 50 nm thickness. (c) MFM images showing the reversal for epitaxial Co dots 10 nm thick and 500 nm in diameter measured with the external field applied in-plane.

In Fig. 15.8c, corresponding MFM images are given for the reversal obtained on polycrystalline Co dots [93]. This field dependent sequence shows the transformation of the single domain state at negative saturation, into the vortex state at zero field and finally into the reversed single domain state at positive saturation. Similar MFM observation have been reported by Pokhil *et al.* [87] on NiFe discs of 800 nm diameter and varying in thickness (5 to 50 nm). Interesting to note is that in these experiments the deviations from the single domain state upon lowering the saturation field led for the thinner dots to the reversible formation of a double vortex structure (Fig. 1 in [87]) before an irreversible transition to the single vortex state occurs, with a vortex motion described in Fig. 15.8a. In the initial double vortex structure, the vortices are described as partially open with opposite chirality and appear at opposite positions along the edges.

**Phase Boundary and Energetics:** From the Kerr hysteresis loop measurements, Fig. 15.6a,b described by Cowburn [84] an experimental phase boundary between the vortex and the single domain state has been deduced as a function of dot thickness and diameter. This is given in Fig. 15.6c. For larger dot thickness and diameter the vortex state is stable, while a transition to the single domain state occurs upon decreasing thickness and diameter. This transition is explained in simple terms by the dominant energy contributions in the two configurations. The flux-closure vortex-structure minimizes the demagnetization field energy and is dominated by exchange energy. In contrast, the single domain configuration is dominated by demagnetization field energy with negligible exchange energy. Upon decreasing the diameter for example at constant thickness, the exchange energy of the vortex structure becomes more and more comparable to the magnetostatic energy of the single domain state. Consequently, below a critical diameter the circular magnetization mode can no longer be maintained and the single domain state becomes energetically more favorable. An example for this transition by calculating the total energy density of circular dots is shown in Fig. 15.10 and is described in more detail further below. Similar arguments apply for reducing the film thickness at constant diameter. It is interesting to note that Cowburn *et al.* [84] have compared the experimental phase diagram with the one calculated using micromagnetic simulations and found them in good agreement.

Another interesting aspect concerns the coexistence of magnetic states. In the MFM measurements performed on the Co(0001) dots [74], the single domain and vortex states are observed simultaneously with a larger probability of single domains to exist after in-plane saturation. However, those single domain states are found to be metastable in agreement with micromagnetic calculations and the phase diagram in Fig. 15.8c, which predict a vortex structure as the stable state. Indeed, as shown in Fig. 15.9 for a 200 nm diameter and 20 nm thick Co(0001) dot, small perturbations such as the stray field from the MFM tip, can induce a transition into the vortex-like state. This type of transition from one state to the other will be discussed in more details in Sect. 15.7 in context with the cobalt rings.

**Stability of the Vortex State in Circular Dots:** MFM or Lorentz microscopies have been performed by various authors on systems with perpendicular magnetic anisotropy [74] and zero anisotropy [75,76,86,87]. It is interesting to note that similar observations of the presence of the vortex state have been made. The recent experiments performed on soft NiFe dots [76] – Fig. 15.8a – or soft Co dots [93] – Fig. 15.8c – and epitaxial Co(0001) dots with large magnetic anisotropy perpendicular to the dot base [74] – Fig. 15.5 –, are a perfect illustration. This raises the question upon the role of the perpendicular uniaxial anisotropy in the epitaxial Co(0001) dots for the formation and stability of the vortex state, since one may suppose that such an anisotropy may favor the presence of vortices or stabilize the vortex state.

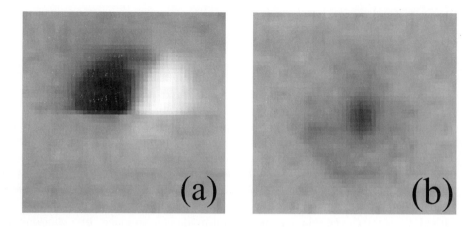

**Fig. 15.9.** The influence of the MFM tip on the single domain pattern of epitaxial Co(0001) dots. By decreasing the lift scan height from (a) 100 nm to (b) 50 nm, the metastable single domain state transforms into the vortex state, which is the ground state of the system (thickness 20 nm, diameter 200 nm).

In order to investigate the role of the Perpendicular Magnetic Anisotropy (PMA), 3D numerical micromagnetic calculations were performed [94]. Two cases were compared, Co dots with strong PMA and those having zero magnetocrystalline anisotropy. Using different starting configurations, the system relaxed either into a single domain state, compare Fig. 15.5c or a vortex-like state, compare Fig. 15.5d. In the thickness range between $t = 5$ nm to 20 nm and for dot diameters of 60 nm to 200 nm, the vortex-like state was found to be in both cases the energetically lowest state. The dependence of the total energy density on the diameter of the dots is shown in Fig. 15.10 for dots of $t = 5$ nm, yielding a very similar behaviour in both situations. The difference being that the total energy density of the single domain and vortex dots with PMA is shifted upwards by the amount of the magnetic anisotropy energy constant $K_u$. This shift is due to the fact that most spins are in plane and hence point into the hard direction of the magnetic anisotropy.

From Fig. 15.10, it can be seen that the presence of the PMA does not influence the ground-state configurations very much, nor the transition from the vortex state towards the single domain state. This transition takes place at a slightly lower critical diameter of 60 nm for Co(0001) dots with PMA compared to 67.5 nm for dots with no anisotropy. This weak dependence on the PMA is related to the fact that the thickness range investigated here, corresponds to the one for which the continuous epitaxial Co(0001) films are in-plane magnetized ($Q = 0.4$), see Fig. 15.3, region III. Thus, thin Co(0001) dots with a thickness below 20 nm behave like in-plane isotropic elements and the transition from the vortex to the single domain state is determined by the the same energetics as for dots with zero magnetic anisotropy (pure balance between demagnetization shape energy and exchange energy). However, the presence of the PMA lowers

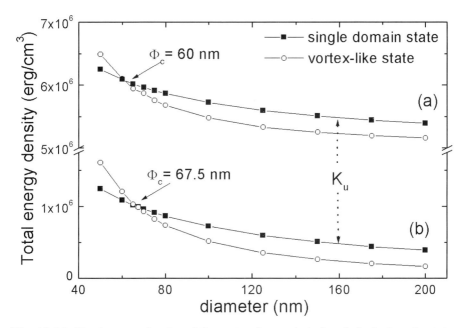

**Fig. 15.10.** Total energy density of the vortex (open circles) and single domain states (full squares) as a function of the dot diameter obtained from 3D numerical micromagnetic calculations for circular Co dots of $t = 5$ nm and having (a) perpendicular uniaxial anisotropy and (b) zero anisotropy.

the total energy density of the magnetization inside the vortex core slightly, since the spins inside the vortex point into the magnetocrystalline easy axis and thus lower their PMA energy. Close to the critical diameter of 60 nm ($t = 5$ nm), this gain in energy is most pronounced, since the relative volume fraction of the vortex is large (vortex diameter 30 nm [94,95]). This stabilizes the vortex state to smaller diameters and as a consequence the critical diameter for the transition into a single domain state is pushed to lower values as compared to dots with zero magnetic anisotropy, see Fig. 15.10a,b.

It is noted that the critical diameter $\Phi_c$ calculated here for Co dots ($\Phi_c = 60 - 67.5$ nm at $t = 5$ nm) when scaled with the magnetization is in good agreement with the values given in the phase diagram of Fig. 15.6c [84]. Due to the larger value of the saturation magnetization $M_s$ in Co (1400 emu/cm$^3$ to be compared to 800 emu/cm$^3$ for NiFeMo [84]), the critical diameter is pushed to lower values in the Co dots when compared to the corresponding critical diameters reported for NiFeMo (approx. 150 nm for $t = 5$ nm).

**Configurational Anisotropy in Mesoscopic Ferromagnets:** This section on dots will not be complete without discussing the effect of configurational anisotropy. This type of anisotropy, first theoretically proposed in 1988 by Schabes and Bertram [96] for the case of magnetic cubes, finds its origin in the fact that sharp edges in constrained nanostructures induce deviations from the uniform magnetization, deviations which are dependent on the direction of the magnetic moment with respect to the axis of the nanostructures. First experimental evidence has been reported by Cowburn *et al.* [82] for flat squares of permalloy. The case of single domain flat square structures is ideal because they do not exhibit any in-plane demagnetization shape anisotropy and thus make the observation of any other type of anisotropy easier. The results of Cowburn *et al.* [82] are illustrated in Fig. 15.11 for a 15 nm thick permalloy square with a sidelength of 150 nm. The polar plot of the effective field reveals a fourfold anisotropy with an abrupt minimum in the internal anisotropy field, whenever the magnetization is parallel to one of the edges of the square nanostructures. The strength of this configurational fourfold anisotropy field deduced from the polar plot has been found extremely strong, of the order of 360 Oe [82].

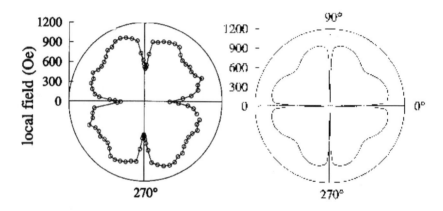

**Fig. 15.11.** Polar plot of the effective field (internal anisotropy field plus 660 Oe of an external field) of NiFeMo square elements as a function of the applied field direction (from [82]).

Although the energy of a perfect uniformly magnetized square is independent of the in-plane magnetization direction, when taking into account the non-uniformity of the magnetization distribution at the edges, energy differences do arise. The non-uniformities lead to magnetization patterns shown in Fig. 15.12 as derived from 3D micromagnetic calculations and depend on whether the sample is magnetized with the external field parallel to the edges (flower state) or at 45° from the edges (leaf state). These two magnetic configurations differ in energy [97]. The anisotropy energy measured is therefore related to the energy difference of the two magnetic configurations shown in Fig. 15.12, and thus to the geometric shape of the square structure. This is confirmed by numerical micromagnetic calculation as shown in Fig. 15.11b. Such anisotropy should be more difficult to evidence in any structures with small perturbations of magnetizations such as

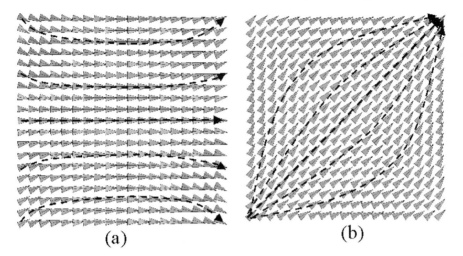

**Fig. 15.12.** Magnetization vector plot showing (a) the flower state and (b) the leaf state for a square Permalloy element of $t = 5$ nm and 95 nm side length. The dotted black lines indicate the demagnetization field lines. The vector plots were obtained from 3D numerical micromagnetic simulation after saturation (a) parallel to the edges and (b) at 45° from the edges.

circular or ellipsoidal type of magnets for which the magnetizations are more uniform.

## 15.7   Domains in Circular Rings

### 15.7.1   Linear versus Circular Magnetization Mode

In all reported MRAM designs up to date, the magnetization in the active device is oriented linearly, mainly controlled by the shape of the element, such as rectangles or long ellipses. In such type of memory elements, it is known that non-repeatable switching is caused by the presence of edge domains [83,98,99]. To overcome the irreproducibility in the switching, several groups have proposed to taper the end of the magnetic element [98–103]. The sharpness of the tapered ends becomes critically important for the repeatability of the switching field for an element with linear magnetization mode [20,83,99,103]. This means that geometric variation of the ends from element to element can yield significant variation of the magnetic switching, and effectively reduces the margin for memory element addressing [103].

Instead of having the magnetization linear in the memory element, the magnetization orientation can be circular, forming a flux-closure structure as discussed in the previous section. The problems associated with the end effects of the linear elements can thus be eliminated. In order to form a flux-closure or circular magnetization mode, the memory element can be either a ring or a circular dot. As discussed in Sect. 15.6.3, the circular magnetization configuration in circular dots, can only be maintained above a critical diameter (increasing with decreasing thickness) [84], see Fig. 15.6c. Below this critical diameter, the single domain state is the energetically lower state.

One way to stabilize the circular magnetization mode to smaller diameters, is to eliminate the central vortex which contains the dominant energy contribution of the flux-closure configuration in form of exchange energy. For such ring shaped elements, the flux-closure structure results in a significantly lower energy than the flux-closure structure of a circular dot with the same outer diameter. Hence it is clear, that upon reducing the outer diameter of the rings, the transition into a single domain state, which takes place in circular discs, is suppressed in the ring geometry to lower values. For example the 3D calculations show that for a circular dot of $t = 5$ nm the flux-closure state is the energetically lower state only above a critical diameter of 60 nm (see Sect. 15.6.3). In contrast, for the ring geometry the flux closure state is the energetically lower state well below 60 nm. The latter can only be considered correct as long as the inner diameter is larger compared to the diameter of the central vortex (30 nm for Co [94]), which forms in circular discs. Furthermore, the introduction of the inner edge reinforces the circular magnetization configuration. It is noted that recently a vertical magneto-resistive random access memory (VMRAM) design was proposed based on a ring-shaped magnetic multilayer stack [20,103].

### 15.7.2 Magnetization Configurations in Submicron Rings

The stability range of the flux-closure structure of polycrystalline Co rings as a function of film thickness and ring diameter using MFM images and micromagnetic simulations has been studied recently in our group [104]. The thickness range investigated lies between 10 nm and 50 nm. The outer ring diameter varies between 300 nm and 800 nm, while the inner ring diameter varies between 100 nm and 300 nm.

In contrast to the case of circular dots discussed in the previous section, for all dimensions investigated, the flux-closure state is the energetically lowest state. However, a metastable single domain state at zero field can be induced at remanence after saturation in an in-plane applied field. The probability to trap this single domain state increases for decreasing film thickness and increasing outer diameter. Figure 15.13 identifies the MFM images and the corresponding micromagnetic configurations of the single domain and flux-closure structures, respectively. The strong black and white contrast in Fig. 15.13a corresponds to the single domain structure whereas the weaker alternating contrast shown in Fig. 15.13b corresponds to the flux-closure state. Figure 15.13c,d gives the corresponding magnetization-vector plots obtained from 3D numerical calculations. It is noted that in order to make the flux-closure configuration visible in the MFM imaging, all rings were patterned in an octagonal rather than a circular shape.

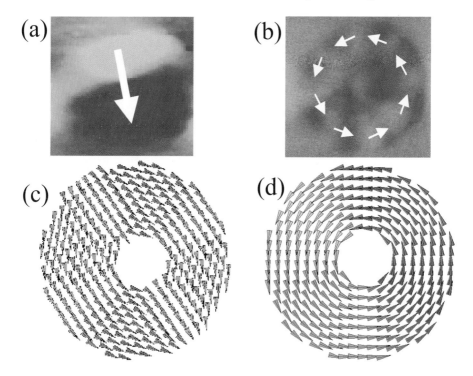

**Fig. 15.13.** MFM contrast of octagonal polycrystalline Co rings (outer diameter $d_o = 800$ nm, inner diameter $d_i = 400$ nm and $t = 20$ nm) denoting (a) a single domain state and (b) a flux-closure configuration. (c) and (d) are magnetization vector plots of circular ring elements (outer diameter 200nm, inner diameter 50 nm and thickness $t = 5$ nm) calculated using a 3D micromagnetic code

### 15.7.3  Metastable States Observed Using the MFM Tip Effect

The metastability of the in-plane remanent single domain state is demonstrated in Fig. 15.14, where repeated zero field MFM scans over the same area of rings are shown. The scan height is held at a large distance of 150 nm, which still allows one to visualize the strong dipolar contrast of the single domain state, but not the much weaker contrast of the flux-closure state. In Fig. 15.14a, almost all rings (96%) are in a single domain state before the MFM tip is scanned for the first time (from top to bottom). The tip field appears to be strong enough to switch the rings during the scan, as can be seen by the fact that for many rings only the upper black half is visible. Repeating the scan a second time (from bottom to top), only 25% of the rings are left in the single domain state, with more rings switching into the flux-closure state during the scan. After the fourth scan, only 14% of the rings are left in the single domain state, see Fig. 15.14b. It is noted that the reversal of the dots from the single domain into the flux-closure state is not a relaxation effect. Changing the scan area, after having 'erased' in

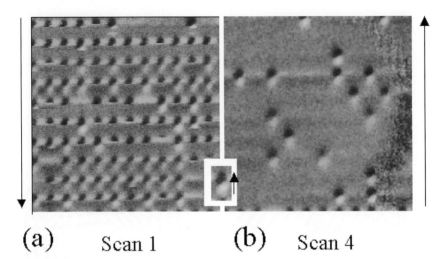

**(a)**     Scan 1          **(b)**     Scan 4

**Fig. 15.14.** Repeated MFM scans at remanence after in-plane saturation of the same area of $20 \times 20~\mu m^2$ for polycrystalline Co rings of $t = 20$ nm and $d_o = 800$ nm. The images were scanned (a) in a first scan from top to bottom and (b) in a fourth scan from bottom to top. For a certain number of rings only the upper black part of the single domain state is seen, revealing that the tip-sample interaction switches the rings into the flux-closure state when positioned above the center of the ring.

one area most of the single domain states by repeated scanning, the new area shows again the high remanence single domain state given in Fig. 15.14a.

From Kerr magnetometry performed on all ring arrays, it is found that the probability of the presence of the metastable single domain state at remanence depends on film thickness and outer ring diameter. This is summarized in the qualitative phase diagram of Fig. 15.15, where the experimental boundary between the low remanence loops at zero field (flux-closure state) and the high remanence loops (single domain state) is shown as a function of thickness and outer ring diameter. Micromagnetic calculations indicate that for increasing diameter the energy of the single domain and flux-closure state both decrease and approach each other. Hence, the probability increases for trapping the single domain state in a local energy minimum at in-plane remanence.

### 15.7.4   Reversal Processes in Rings

Although, as shown above, the in-plane remanent state can be the single domain state, the reversal for all rings with dimensions investigated takes place via the transformation of the saturation single domain state into the flux-closure structure and then into the reversed single domain state in a reversed bias field. This process is illustrated in Fig. 15.16 where the magnetization reversal has been studied by MFM imaging in an applied field for a large array of rings having a large remanence at zero field (as deduced from Kerr hysteresis-loops). The

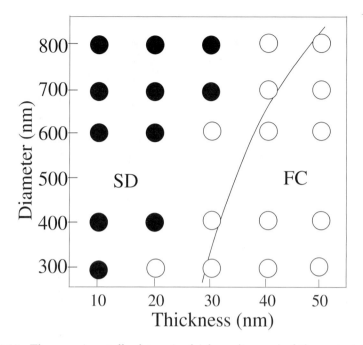

**Fig. 15.15.** The experimentally determined 'phase diagram' of the metastable single domain state (solid circles) and the flux-closure state (open circles) as a function of thickness and outer diameter for polycrystalline Co rings obtained from Kerr-effect hysteresis loop measurements. The solid line represents the calculated boundary [104].

MFM images shown in Fig. 15.16 were scanned in the reversed applied field at values indicated on top of each image, the positive saturation field pointing in the upward direction and the negative reversed field pointing in the downward direction. To minimize the tip-sample interaction, a lift scan height of 200 nm was chosen. The actual field value at each ring is hence the sum of the applied field and the tip field. Starting from the remanent state at 0 Oe (induced in the absence of the tip), a large number of rings switches from the single domain into the flux-closure state due to the tip field, compare Fig. 15.14. However, after several scans a number of rings are stable and a finite field of −100 Oe is required to switch all rings from the single domain into the flux-closure state as shown in Fig. 15.16. This means that the reversal occurs through the formation of a flux-closure configuration. Upon increasing the reversed field value to −450 Oe (maximum field of the magnet used) 70% of the rings are in the reversed single domain state. Reducing the field back to zero, the remanent state is lower (25%) than the initial remanent state (96%). This is because the initial remanent state was obtained in the absence of the tip field, while the remanent state at the end of the hysteresis cycle is obtained in the presence of the tip field, so a small additional field is sensed by the rings.

0 Oe                    -55 Oe                  -100 Oe

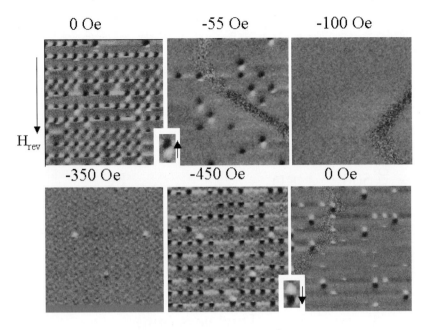

$H_{rev}$

-350 Oe                 -450 Oe                 0 Oe

**Fig. 15.16.** MFM images of an area of $20 \times 20 \ \mu m^2$ for polycrystalline Co rings of $t = 20$ nm and outer diameter of 800 nm as a function of an applied field. The arrow to the left indicates the reversed field direction $H_{rev}$ (negative).

## 15.8    Domain Configurations in Wires

In the last section of this chapter, we concentrate on the micromagnetic configurations of single magnetic wires. So far, we have addressed the reversal processes in dots and rings while most of the pioneering studies have been carried out on elongated particles [36,105–107] because they act as a model system for the nucleation-propagation reversal process. From these studies, basic micromagnetic information such as the nucleation volume, can be derived [36,106]. Moreover, due to the high aspect ratio of length to lateral cross-section, the magnetization direction can be well controlled by the shape of the nanostructures as well as by the magnetic anisotropy of the material. In the following, flat rectangular epitaxial cobalt wires are considered with in-plane uniaxial anisotropy oriented either perpendicular (Sect. 15.8.2) or parallel (Sect. 15.8.3) to the long wire axis.

### 15.8.1    Sample Preparation

The wires were prepared from epitaxial $(10\bar{1}0)$ Co thin films of thickness $t = 30, 40, 50, 60$ and 80 nm. The films were grown under ultrahigh vacuum conditions on (110) MgO substrates by molecular beam epitaxy using a Mo-Cr buffer layer. Structural investigations confirm the hcp structure and magnetic investigations confirm a strong in-plane uniaxial anisotropy [108]. The films were patterned using electron-beam lithography, lift-off techniques and ion beam etching. For each thickness, wire arrays were prepared for wire widths of $w = 100, 150, 200, 500, 800$ and $1000$ nm. The wires are 10 $\mu$m long and the separation between the wires is 5 $\mu$m, sufficient to neglect any dipolar interaction. For each set of $(t, w)$ – values, wires were patterned whose long wire axis is aligned either perpendicular or parallel to the magnetocrystalline anisotropy axis.

## 15.8.2   Wires with Crystal Anisotropy Field Perpendicular to the Wire Axis

**Single Domain to Stripe-Domain Transformation:** The flat rectangular Co wires are characterized by a strong uniaxial magnetocrystalline anisotropy ($K_u = 6 \times 10^6$ erg/cm$^3$) oriented in-plane and perpendicular to the wire axis. This magnetocrystalline anisotropy is in competition with the demagnetization shape anisotropy, which in turn favors an alignment of the magnetization parallel to the wire axis. A reduction of this demagnetization energy can be achieved by the formation of a periodic stripe domain pattern, in analogy to the stripe domains developing in continuous perpendicular Co(0001) films, see Sect. 15.5.2. In a further analogy, the stripe domain width can be controlled by the wire thickness and the wire width.

However, in contrast to the continuous films, which can be described by a thickness independent, constant Q-factor $Q_0 = K_u/4\pi M_s$ ($= 0.4$ for Co), for the wires an effective $Q_{\mathrm{eff}}$ factor can be defined, which varies as a function of system size $(t, w)$. This is because, in contrast to the continuous film (with $N_{\mathrm{eff}} = 1$), the effective demagnetization factor $N_{\mathrm{eff}}$ inside the wires is not constant upon variation of film thickness $t$ and wire width $w$. The corresponding effective demagnetization factor is given by $N_{\mathrm{eff}} = (2/\pi)\arctan(t/w)$ which has values between 0 (for $t \to 0$ or $w \to \infty$) and 1 (for $w \to 0$ or $t \to \infty$). This yields an effective Q-factor $Q_{\mathrm{eff}} = Q_0/N_{\mathrm{eff}}$, which can vary between the value of the continuous film ($Q_0 = 0.4$) and infinity.

For the continuous films different types of stripe-domain structures (open (strong) stripes $Q > 1$, flux closure (weak) stripes $Q < 1$) have been described depending on the Q-value [109]. Thus, one expects for the wires that different domain configurations can be stabilized upon variation of the wire dimensions $(t,w)$.

In Fig. 15.17 a phase diagram as a function of the lateral width $w$ and the thickness $t$ is shown [41], taking into account four possible magnetization configurations: (1) the transverse single domain (TSD) state ($\boldsymbol{M}$ perpendicular to the wire axis), (2) the longitudinal single domain (LSD) state ($\boldsymbol{M}$ parallel to the wire axis, (3) the open stripe structure (OS) and (4) the flux closure (FC) stripe structure. The boundaries between the different configurations (dotted lines in Fig. 15.17) were calculated from a domain theory model based on Kittel's formulation [61].

In order to be able to compare this predicted phase diagram to experimental data, one first has to define the procedure by which the magnetic system can

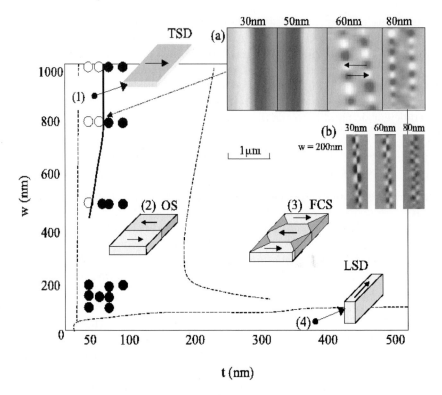

**Fig. 15.17.** Qualitative phase diagram $(t-w)$ of the ground state domain configurations of epitaxial Co($10\bar{1}0$) wires with uniaxial magnetocrystalline anisotropy aligned perpendicular to the long wire axis. The dotted lines are calculated boundaries while the bold line is the experimental boundary between the transverse single domain (open circles) and the open stripe-domain state (full dots). Inset (a) MFM images for wires of $w = 800$ nm showing the evolution of the ground state with $t$. Inset (b) MFM images for wires of $w = 200$ nm showing the thickness dependence of the domain period.

relax into its ground state. It was already mentioned in the case of dots and rings (Sects. 15.6,15.7), that depending on the magnetic history the induced configuration may correspond to a metastable state. For instance, as shown in Fig. 15.18, either a transverse single domain state is induced when magnetizing along the crystal anisotropy direction (in-plane and perpendicular to the wires). However, a stripe-domain pattern is induced when magnetizing along any other direction. For reasons explained below, the stable ground state is induced after hard axis demagnetization procedures (labelled D∥s and D⊥ in Fig. 15.18). Typical MFM images of stable single domain and stripe domain states are shown in Fig. 15.17 for different values of $t$ at constant $w$ with (a) $w = 800$ nm and (b) $w = 200$ nm. The stable single domain and stripe domain configurations are summarized in Fig. 15.17 by the open circles and full dots respectively.

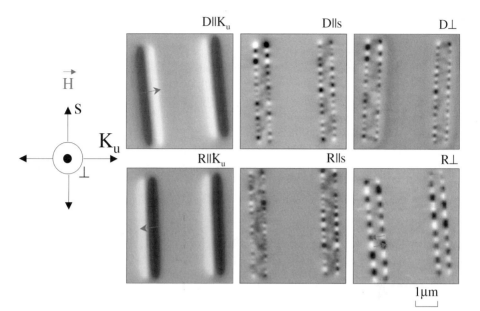

**Fig. 15.18.** MFM images for a Co(10$\bar{1}$0) wire with $w = 800$ nm and $t = 60$ nm obtained after different magnetization histories. D denotes a demagnetized state and R a remanent state. $K_u$ is the easy axis direction, s is the direction of the long wire axis and ⊥ denotes the application of the magnetic field perpendicular to the film plane

For decreasing film thickness and increasing wire width, the transverse single domain state is stabilized, because the gain in demagnetization energy by nucleating the stripe-domain state decreases. For some critical values the gain will not be sufficient to compensate the wall energy. Hence the system acquires a transverse single domain state. The bold line in the $(w, t)$–diagram of Fig. 15.17 summarizes the experimental boundary between the stable stripe-domain state (full dot) and the stable transverse single domain (open circle) state. The discrepancy between the calculation (dotted line) and the experiments is due to the fact that the Co bulk value for the domain wall energy (12 erg/cm$^2$) energy was used and taken as constant in the calculation. However, due to the vertical and lateral confinement, the wall energy in the wires should be much larger [108]. 3D micromagnetic calculations are currently in progress in order to determine the energy of the wall for such wires but more importantly the wall structure itself.

**Metastability of the Observed Magnetic Structures:** As already discussed, metastable states can be induced depending on the magnetic history. For instance, most of the wires investigated show a transverse single domain state as the remanent state, when saturated along the direction of the crystal anisotropy field (in-plane and perpendicular to the wire axis). This single domain state is the stable state only for those dimensions reported in the phase diagram as open circles. For all others, this state is metastable and is induced because the applied field parallel to the easy magnetocrystalline anisotropy axis gives a preferential orientation to the magnetization $M$, blocking $M$ in a local energy minimum, which is separated by an energy barrier from the multi-domain stripe state. This barrier decreases, the narrower and the thicker the wires are, due to the increase in the in-plane shape demagnetization fields (scaling with $t/w$), which favor the nucleation of reversed domains. This dependence is confirmed in Fig. 15.19a for which the remanent state after saturation along the direction of the crystal anisotropy direction (in-plane and perpendicular to the wire axis) is shown. For $w = 600$ nm a pure transverse single domain state is stabilized, while for $w = 150$ nm the stripe domain and transverse single domain state coexist and for $w = 100$ nm a complete stripe structure is induced.

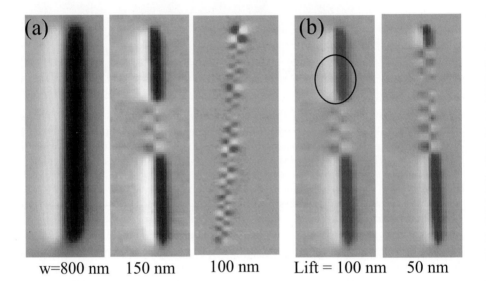

**Fig. 15.19.** MFM images taken on $Co(10\overline{1}0)$ wires after in plane easy axis saturation and at a lift scan height of 100 nm for $w = 800$ nm, 150 nm, 100 nm; $t = 60$ nm. (b) Upon decreasing the lift scan height from 100 nm to 50 nm (left image) the number of the stripe domains in a 150 nm wide wire is increased (right image), corresponding to an irreversible switch of the magnetization.

The stability or meta-stability of the single domain state described above can be experimentally verified by applying a small perturbation to the domain structure. This can be done using the stray field of the MFM tip. As shown in the left MFM image of Fig. 15.19b, the single domain and the stripe-domain state coexist at zero field for a narrow wire of $w = 150$ nm. Upon lowering the lift scan height from 100 nm to 50 nm when scanning the MFM tip over the wire, the enhanced stray field from the MFM tip produces in some regions an irreversible transition from the single domain into the stripe-domain state.

### 15.8.3  Crystal Anisotropy Field Parallel to the Wire Axis

In contrast to the previous section, in the following cobalt wires are discussed which are characterized by a magneotcrystalline uniaxial anisotropy that co-operates with the shape anisotropy thus maintaining the magnetization parallel to the wire axis. Two cases are considered, (i) wires having a flat rectangular cross-section of 100 nm wire width and prepared from epitaxial Co ($10\bar{1}0$) thin films by electron beam lithography and (ii) cylindrical wires with much smaller circular cross-sections ranging from 30 to 50 nm. The latter are prepared by electrodepositing Co into the pores of high quality track etched polycarbonate membranes [43–45], see Sect. 15.2. These wires proved to adopt a rather good quality hexagonal compact structure with a preferential texture along the wire axis [45].

In both cases, the magnetic ground state is the single domain structure. The magnetization reversal in a parallel field of such Co wires can, to a certain extent, be simulated by a coherent (or in unison) rotation of spins [36,39,92,110]. In this model, a single giant magnetic moment is subjected only, besides the external field, to a first order uniaxial anisotropy of constant value $K_u$, the direction of which is parallel to the wire axis. Micromagnetic theory predicts that in such type of nanostructures, two other types of nucleation (and reversal) mechanisms beside the coherent rotation can occur: buckling and curling [39,110–112]. All these reversal modes should result in a square hysteresis loop when the field is applied parallel to the axis of the wire as observed experimentally [36,106]. Hence, only two magnetization states, parallel and antiparallel to the wire axis, can be realized in this type of wires. This is confirmed by MFM imaging, as shown in Fig. 15.20a, for a cylindrical Co wire of 35 nm diameter, after application of a large field parallel to the wire axis [33]. The dark and bright contrasts at the wire extremities correspond to magnetic charge distributions at the end faces, which arise when the magnetization is in a single domain state and aligned parallel to the wire axis.

In contrast to the saturation along the wire axis, a multidomain state with head-to-head domain walls can be induced by saturation in a field perpendicular to the wire axis, as shown in Fig. 15.20b. This multidomain structure arises since upon reduction of the field from perpendicular saturation to zero, the magnetization may rotate clockwise or counter-clockwise towards the wire axis.

**Wall Structures:** The dark and bright contrasts visible along the wire axis, Fig. 15.20b, arise from the magnetic volume charges located at the domain walls. This corresponds to the strong accumulation of magnetic charges at the boundary between two longitudinally magnetized domains of opposite magneti-zation directions. In bulk materials and continuous films with in-plane uniaxial anisotropy, domains of opposite magnetization orientation are usually separated by 180° domain walls which are parallel to the magnetization inside the do-mains. This orientation arises essentially to avoid net magnetic charges of the walls, by virtue of the pole avoidance principle. In contrast, the lateral con-finement of the magnetization in small diameter nanowires with parallel-to-wire

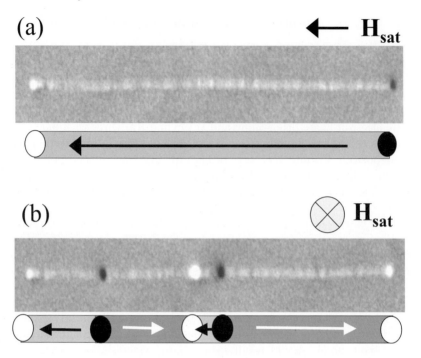

**Fig. 15.20.** Zero field MFM images of a 35 nm cylindricral electrodeposited Co wire after saturation in a field (a) parallel to the wire axis and (b) perpendicular to the wire axis.

anisotropy forces the domains to meet head-on and the separating walls to orient perpendicular to the wire axis, hence to the domain magnetization. This results in heavily charged walls having both volume and surface-pole distributions, exhibiting therefore strong MFM contrasts, as shown in Fig. 15.20b.

As sketched in Fig. 15.21, a simple one dimensional model for the domain-wall structure can be assumed in which the spins within the wall rotate perpendicular to the wire axis. One question to address is the precise configuration of the magnetization inside the wall. 3D numerical micromagnetic calculations carried out in our group have revealed two different wall magnetization modes for square wires, as sketched in the x-y cross-sections of Fig. 15.21. These wall magnetization modes are very similar to the single domain state and vortex state of circular dots shown in Fig. 15.5, Sect. 15.6. In correspondence they are called the transverse and the vortex-type wall. In the transverse wall the magnetization is almost uniform across the diameter and all spins point everywhere into the same direction perpendicular to the wire axis, giving rise to a transverse demagnetization field. In contrast, in the vortex wall the magnetization forms a circular magnetization path, where the wall spins try to stay parallel to the wire surface thus reducing the magnetostatic energy. Similarly to the circular dots, the total energy density of the two wall-magnetization modes depends on the

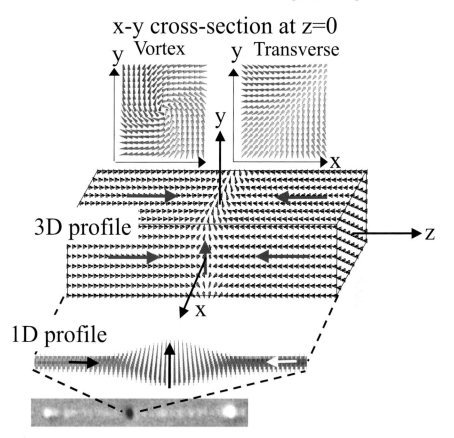

**Fig. 15.21.** Bottom: MFM image of the domain wall contrast along an electrodeposited Co nanowire. Above the MFM image, a 1D wall profile is shown where the magnetization is averaged over the wire cross-section. In this 1D model the wall spins point perpendicular to the wire axis. Above the 1D profile, the magnetization-vector distribution is shown for a wire with square cross-section, as obtained from 3D numerical micromagnetic simulations. Top: Magnetization vector plot showing the magnetization distribution across the x-y square cross-section at the wall center ($z = 0$). Two modes can be stabilized, a vortex type and a transverse mode.

diameter of the wires and a critical diameter exists below which the transverse wall has a lower energy, compare the corresponding Fig. 15.10 for circular dots. Using the parameters of Co this critical diameter is found to be approximately 20 nm. It is interesting to note that recently, numerical studies based on time quantified Monte Carlo methods yielded similar wall-magnetization modes which are nucleated in nanowires during the magnetization reversal.

Furthermore, it should be remembered that these configurations represent a theoretical solution, which do not take into account the defect structure of the real wires (variations of diameter, grain boundaries) at which the walls are pinned and which can modify the configuration shown above. Lastly, it will be difficult to differentiate by MFM these two configurations, since the contrast will be dominated by the strong magnetic volume charges which should be very similar in both cases.

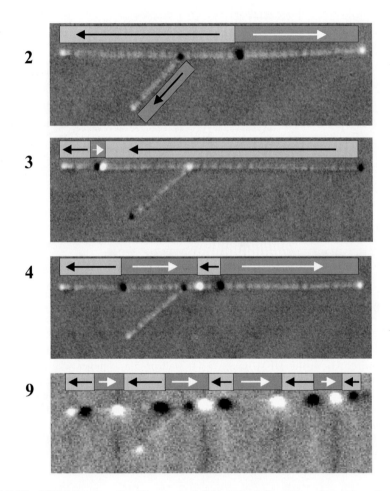

**Fig. 15.22.** MFM images of a 35 nm electrodeposited Co wire, after saturation in a field perpendicular to the wire axis. Repeated saturation may lead to a different number of domains. Attached to the horizontally aligned wire, a second smaller wire inclined at 45° is seen.

**Pinning Sites and Reversal Procedure:** It is observed that for the circular electrodeposited Co nanowires the number of domain walls induced in a single wire after perpendicular saturation (or demagnetization) can vary from one experiment to the next. For instance, three, four, five or even nine longitudinally magnetized domains of alternating magnetization directions were induced on a single wire as shown in Fig. 15.22. This variation in number probably arises from the fact that the exact field orientation varies slightly from one experiment to the next, with the field being slightly more or less inclined towards the wire axis. However, it is also observed that some walls occur at the same position along the wire, suggesting that they are stabilized at pinning sites. Consequently, in order to move the walls along the wire axis, a finite depinning field (applied parallel to the wire axis) value is required. This is demonstrated in the image sequence of Fig. 15.23 where it can be seen that weak and strong pinning sites are present.

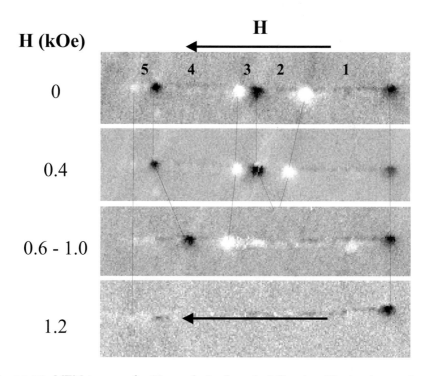

**Fig. 15.23.** MFM images of a 35 nm electrodeposited Co wire. The top image shows the wire in zero field after out of plane saturation. For the remaining images, a field was applied parallel to the wire axis, with field values noted to the left.

The number of pinning sites and their respective pinning strengths will depend on the crystalline quality of the wires. From TEM investigations of the electrodeposited wires it is known that the wires contain many stacking faults [44] and grains of micrometer sizes. For those wires a fairly large number of walls can be induced (up to ten along a wire of 10 $\mu$m). In contrast, for the flat rectangular 100 nm wide epitaxial Co(10$\bar{1}$0) wires it is found very difficult to induce a large number of walls in a perpendicular remanent state, unless a local constriction is artificially induced, see Fig. 15.24.

# Pinning centers

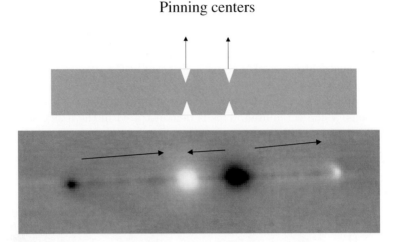

**Fig. 15.24.** MFM image taken from a flat rectangular epitaxial Co(10$\bar{1}$0) wire, patterned by electron beam lithography. In order to induce pinning sites, constrictions were defined along the wire axis as shown in the schematic on top.

**Resistance Associated to the Presence of a Domain Wall:** The giant magnetoresistance effect occurs in heterogeneous magnetic systems such as magnetic trilayers and multilayer stacks consisting of alternating magnetic and non-magnetic thin films [11,14]. It describes the change in resistance when the magnetization of adjacent layers is switched from a parallel to an antiparallel alignment. Recently, an analogy was drawn between such GMR multilayers and homogeneous magnetic media, in which a magnetic domain wall replaces the non-magnetic spacer layer [113,114]. Several experiments indicate, that the presence of a domain wall enhances the resistance [113–117], although the deduced effect has been small. This positive contribution to the domain-wall magnetoresistance (DW–MR) has been discussed theoretically [118] in terms of an enhanced domain-wall resistivity $\rho_w$ resulting from spin dependent scattering and spin channel mixing localized inside the domain wall. For thin enough domain walls, spin channel mixing arises from a mistracking effect [113,114,118], where the transport electron spins lag behind in orientation with respect to the local magnetization orientation inside the domain wall.

Many experiments on the DW–MR used a system of magnetic stripe domains [113,115–117], taking advantage of the high density of domain walls and assuming that the domain resistance and the DW resistance form a network of series resistors. However, the presence of other MR effects such as the anisotropic magnetoresistance (AMR) [119] or the Lorentz-MR [120], as well as a complex domain-wall structure often required a manipulation of the measured MR data in different magnetization configurations in order to extract the contribution from the domain wall [116]. These 'parasitic' effects can be avoided using the parallel wire configuration described in Figs. 15.21–15.24, in which a discrete

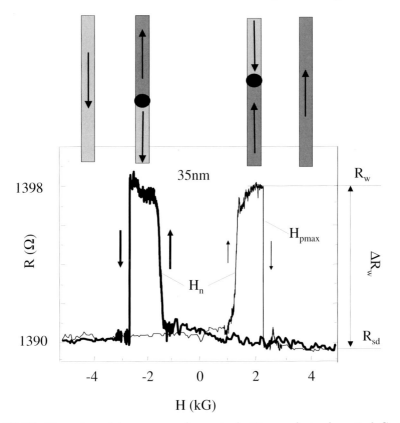

**Fig. 15.25.** Magneto-resistance curve for a single 35 nm electrodeposited Co wire, measured as a function of a field applied parallel to the wire axis. $R_{sd}$ denotes the background signal level for a single domain wire. $R_W$ denotes the signal level when one or two domain walls have been nucleated at $H_n$ during the reversal and are stabilized at pinning sites.

number of isolated head-to-head domain walls can be stabilized separating opposite domains. The advantage of such a configuration is that the geometry is well defined during the reversal process, avoiding other MR contributions which may mask the DW–MR effect. Indeed, the magnetization $M$ remains parallel or antiparallel to the current $J$ during the magnetization reversal process, whereas the domain walls are aligned perpendicular to the current and thus to the magnetization. The MR hysteresis loop of a 35 nm Co single wire shown in Fig. 15.25 illustrates the simplicity of this geometry [33]. There are two striking features in this loop: (i) the background resistance level is almost flat over the field range measured and (ii) two sharp jumps are visible, one upward at 1.3 kOe followed by a downward jump at 2.8 kOe. The flat background level indicates as expected that the magnetization remains parallel to the current $J$ during the whole reversal process, confirming the results of the MFM (Fig. 15.20a) observations that the effective easy axis is aligned very close to the wire axis. Since the upward jump appears around the nucleation field value of $1\,\text{kOe} < H_n < 2\,\text{kOe}$ as found also from MFM experiments, the upward and downward jumps are, respectively, identified as the nucleation field $H_n$ and the maximum depinning field $H_{pmax}$ at which all domain walls are expelled from the wire. From this it follows that the enhanced signal level in the field range $H_n < H < H_{pmax}$ can be attributed to the presence of a finite number of domain walls which nucleate during the reversal and propagate along the wire. Consequently, it is identified as a domain-wall magnetoresistance (DW–MR).

For all wires measured, the MR ratio for a single wall has a value of $\Delta R_w/nR_{sd} = 0.1$ to $0.3\%$ ($n$ = number of walls). In terms of the resistivity ratio this would correspond to a huge enhancement of the resistivity $\rho_w$ of Co inside the domain wall compared to its resistivity $\rho_{sd}$ in the single domain state. Indeed, taking as the relevant length scale the domain-wall width $\delta_w = 10$ to $15\,\text{nm}$ and $l_{Co} = 20\,\mu\text{m}$ for the length of the Co wire, one obtains a resistivity ratio of $\Delta\rho_w/\rho_{sd} = \Delta R_w/nR_{sd}(l_{Co}/\delta_w) = 100\%$ to $600\%$. Such a large resistance enhancement in the presence of a domain wall has initiated further studies of spin-dependent transport in chemically homogeneous structures supporting an in-homogeneous magnetic structure. In the current development of spin electronic devices, this effect may find potential applications with the domain wall as the parameter of control. This is a nice example for the importance of the prediction and understanding of the local micromagnetic structure in confined systems.

## 15.9   Summary

This chapter has reviewed an extensive study performed on the micromagnetic configurations in mesoscopic system and the influence of shape and size. After a short overview and summary on nanofabrication techniques [28] and magnetic force microscopy imaging [46,47,51–53], the first part of the chapter was dedicated to the stripe-domain formation with domain sizes below $100\,\text{nm}$ in Co(0001) thin films of large perpendicular anisotropy [67,68]. For the square magnetic Co(0001) dots fabricated from such films the influence of shape and size was discussed [29,38]. Domain configurations such as concentric rings and spirals have been stabilized, which drastically influence the magnetization reversal process. For dots fabricated in the thickness range below $20\,\text{nm}$ and diameters below $500\,\text{nm}$ where the magnetization lies in-plane [74,94], the magnetic configurations to be stabilized are identical to those in soft materials of the same sizes but zero magnetic anisotropy.

For such in-plane isotropic dots of nanometer extend two configurations exist, the single domain and the flux-closure state [84]. Their stability is determined by the balance between the quantum mechanical exchange energy and the classical magnetostatic energy. Equally important is the key role played by small deviations from uniformity in the demagnetization field which in some cases causes unexpected anisotropy to appear which can dominate the magnetic properties [82]. Similar configurations as in the dots are found in circular ring systems [104], however, the elimination of the central vortex pushes the transition to the single domain state to lower diameter values.

It is noted that furthermore quite similar properties were treated in the literature for elliptical dots [88–91], where the in-plane shape anisotropy is an additional parameter that influences the presence of different configurations.

As an example of 1D systems, magnetic wires fabricated from epitaxial Co($10\bar{1}0$) films with strong in-plane uniaxial anisotropy have been considered in this chapter [41]. The lowest energy states of these wires depend on the strength and character of its anisotropy. For the orientation of the uniaxial anisotropy perpendicular to the wire axis, regular stripe-domain patterns become energetically favorable, in analogy to the stripe domain-structure reported for the out-of-plane perpendicular Co(0001) films. A qualitative phase diagram as a function of thickness and wire width for the transition from the stripe domain into the single domain state has been deduced experimentally and compared to simple calculations using Kittel's domain theory [61]. This diagram will serve as a guide, since a rigorous treatment for the different configurations of this geometry requires intensive numerical micromagnetic modeling. As a last example at the end of the chapter, the case of crystal anisotropy oriented along the wire axis is discussed. We show that uniformly magnetized states prevail [33]. In certain conditions, a discrete number of walls can be induced in the nanostructures serving as a model system for the study of the resistance associated to the presence of a domain wall. These experiments underline the importance of the prediction and understanding of the local micromagnetic structure in confined systems.

## 15.10   Conclusion

Progress in nanofabrication, imaging techniques and micromagnetic modelling give new insight into the physics of nanostructured magnetic material. For the first time, we are able to compare experiments on the magnetization reversal, stable and metastable states, confinement of domains in laterally constrained structures to those predicted from numerical micromagnetic modeling. A large number of groups have contributed recently in this new field of mesoscopic systems. This combination of individual experience has facilitated and increased the possibility of creating a knowledge base for these novel magnetic structures never studied before. In order to achieve this, emphasis has been put on the fabrication of nanostructures of magnetic materials and the analysis of their respective magnetic and electrical properties. New preparation techniques have been developed and magnetic characterization techniques have been refined. Furthermore, the various types of systems are investigated with respect to their potential for controllable nanoscale design. This has helped to develop a better understanding of the different energetic contributions involved in systems with reduced dimensions. The prospective future applications of magnetic elements such as spin electronic devices, which control the resistance through a precise control of magnetic configurations, gives an additional drive to this field. For instance, integrated magnetoelectronic architecture is a moving target for the next dynamic

RAM generation. In order to stay in the ballpark it must anticipate an achievement of magnetic cells with a minimum feature size of 0.1 $\mu$m and an area of 0.06 $\mu$m$^2$ incorporating a reliable and reproducible magnetization switching on a time scale a few years to challenge the existing paradigm for main memory.

## Acknowledgments

The work presented in this chapter is part of the Ph.D thesis of several students performed in our group. We would like to thank our colleagues in Strasbourg, Marc Demand, Yves Henry, Victor Da Costa, Stephane Padovani, Jean Pierre Bucher, Ricardo Ferré and Jacek Arabski. The authors are grateful for the very fruitful collaborations with Luc Piraux and Aurel Radulescu in Louvain la Neuve, Yong Chen, Shunpu Li, Marco Natali and Françoise Rousseaux in Bagneux, Michel Viret and Claude Fermon in Saclay, Jean Marie George in Orsay, Phil Wigen in Columbus and John Gregg in Oxford. The authors wish to thank for technical support 'the Institut du Développement et des Ressources en Informatique Scientifique' (IDRIS). This work has been granted through several EC programmes: the EC-TMR program 'Dynaspin' n° FMRX-CT97-0124, the EC program 'Magnoise' n° IST-1999-11433, the EC-Brite-Euram program BE95-1761, and the EC program 'NanoPTT' n° G5RD-CT1999-00135.

## References

1. P. Grünberg, R. Schreiber, Y. Pang, M. Brodsky and H. Sowers, Phys. Rev. Lett. **57**, 2442 (1986).
2. K. Ounadjela, D. Muller, A. Dinia, A. Arbaoui, P. Panissod, and G. Suran, Phys. Rev. B **45**, 7768 (1992).
3. P. J. H. Bloemen, H. W. van Kerstern, H. J. M. Swagten, and W. J. M. de Jonge, Phys. Rev. B **50**, 13505 (1994).
4. P. Bruno and C. Chappert, Phys. Rev. Lett. **67**, 1602 (1991).
5. S. S. P. Parkin, R. Bhadra, and K. P. Roche, Phys. Rev. Lett. **66**, 2152 (1991).
6. P. Bruno, Phys. Rev. B **52**, 411 (1995).
7. P. Carcia, A. Meinhaldt, and A. Suna, Appl. Phys. Lett. **47**, 178 (1985).
8. Z. Q. Qiu, J. Pearson, and S. D. Bader, Phys. Rev. Lett. **70**, 1006 (1993).
9. A. Berger and H. Hopster, Phys. Rev. Lett. **76**, 519 (1996).
10. M. N. Baibich, J. M. Broto, A. Fert, Nguyen Van Dau, F. Petroff, P. Etienne, G. Creuzet, A. Friederich, and J. Chazelas, Phys. Rev. Lett. **61**, 2472 (1988).
11. For a review on GMR, see A. Barthelemy, A. Fert, F. Petroff, "Giant Magnetoresistance in Magnetic Multilayers", Handbook of Magnetic Materials, Vol. 12, Edited by K. Buschow, 1999, Elsevier Science.
12. P. Bruno in "Synchrotron Radiation and Magnetism" edited by E. Beaurepaire, B. Carrière, and J.-P. Kappler, Les Editions de Physique, Les Ulis (1997).
13. W. Geerts, Y. Suzuki, T. Katayama, K. Tanaka, K. Ando, and S. Yoshida, Phys. Rev. B **50**, 12581 (1994).
14. For a review on the properties of ultrathin films and multilayers see "Ultrathin Magnetic Structures" I, II, edited by B. Heinrich and J. A. C. Bland, Springer-Verlag, Berlin (1994).

15. K. Ounadjela and R. L. Stamps, "Mesoscopic Magnetism in Metals" in the Handbook of Nanostructured Materials and Nanotechnology, Vol. 2: Spectroscopy and Theory, p. 429, Editor Hari Singh Nalwa, Academic Press (1999).

16. M. A. Kastner, Rev. Mod. Phys., **64**, 849 (1992)

17. M. A. Kastner, Physics Today, **24** (1993).

18. Quantum Transport in Semiconductor Submicron Structures, edited by B. Kramer, NATO ASI Series, Kluwer Academic Publishers, Dordrecht/Boston/London (1996).

19. M. Johnson, "Magnetoelectronic memories last and last...", IEEE Spectrum, **33**, (February 2000).

20. J. Zhu and G. A. Prinz, "VMRAM memory holds both promise and challenge", Data Storage, **40**, (September 2000).

21. G.A. Prinz and K. Hathaway, "Magnetoelectronics", Physics Today, Vol. **48**, 24 (1995).

22. G. A. Gibson and S. Schultz, J. Appl. Phys. **73**, 4516 (1993).

23. A. D. Kent, S. van Molnar, S. Gider, D. D. Awschalom, J. Appl. Phys. **76**, 6656 (1994).

24. S. Manalis, K. Babcock, J. Massie, V. Elings, M. Dugas, Appl. Phys. Lett. **66**, 2585 (1995).

25. S. Okazaki, "Lithography for VLSI", Proc. SPIE 2440 (1995).

26. D. L. Spears and H. I. Smith, Solid State Technol. **12**, 21 (1972).

27. G. Simon, A. M. Haghiri-Gosnet, J. Bourneix, D. Decanini, Y. Chen, F. Rousseaux, H. Launois, and B. Vidal, J. Vac. Sci. Technol. B **15**, 2489 (1997).

28. For a review on different lithography techniques see: Y. Chen and A. Pepin, "Nanofabrication: Conventional and non-conventional methods", Electrophoresis, to appear in January 2001.

29. M. Hehn, K. Ounadjela, J.-P. Bucher, F. Rousseaux, D. Decanini, B. Bartenlian, and C. Chappert, Science **272**, 1782 (1996).

30. S. Chou, IEEE Trans. Mag., **85**, 664 (1997).

31. K. Hong and N. Giordano, J. Phys.: Condens. Matter **10**, L401 (1998).

32. U. Ruediger, J. Yu, S. Zhang, A. Kent, and S. S. P. Parkin, Phys. Rev. Lett. **80**, 5639 (1998).

33. U. Ebels, A. Radulescu, Y. Henry, L. Piraux, and K. Ounadjela, Phys. Rev. Lett. **84**, 983 (2000).

34. M. Viret, Y. Samson, P. Warin, A. Marty, F. Ott, E. Sondergard, O. Klein, and C. Fermon, Phys. Rev. Lett. **85**, 3962 (2000).

35. W. Wernsdorfer, K. Hasselbach, A. Suplice, A. Benoit, J. -E. Wergrowe, L. Thomas, B. Barbara, and D. Mailly, Phys. Rev. B **53** , 3341 (1996).

36. W. Wernsdorfer, B. Doudin, D. Mailly, K. Hasselbach, A. Benoit, J. Meier, J. -Ph. Ansermet, and B. Barbara, Phys. Rev. Lett. **77**, 1873 (1996).

37. W. Wernsdorfer, K. Hasselbach, B. Barbara, L. Thomas, D. Mailly, and G. Suran, J. Magn. Magn. Mater. **145**, 33-39 (1995).

38. K. Ounadjela, M. Hehn and R. Ferré, "Domain confinement in mesoscopic epitixial cobalt patches", in "Magnetic hysteresis in novel magnetic materials", edited by G. Hadjipanayis, Kluwer Academic Publishers, 485 (1997).

39. A. Hubert and R. Schäfer, "Magnetic domains", Springer, Berlin, (1998).

40. R. D. Gomez, T. V. Luu, A. O. Pak, K. J. Kirk, and J. N. Chapman, J. Appl. Phys. **85**, 6163 (1999).

41. I. L. Prejbeanu, L. D. Buda, U. Ebels, M. Viret, C. Fermon, and K. Ounadjela, IEEE Trans. Magn., June (2001).

42. F. Rousseaux, D. Decanini, F. Carcenac, E. Cambril, M.F. Ravet, C. Chappert, N. Bardou, B. Bartenlian, and P. Veillet, J. Vac. Sci. Technol. B **13**, 2787 (1995).
43. L. Piraux, S. Dubois, E. Ferain, R. Legras, K. Ounadjela, J.-M. George, J.-L. Maurice, and A. Fert, J. Magn. Magn. Mater. **165**, 352 (1997).
44. E. Ferain and R. Legras, Nucl. Instrum. Methods B **131**, 97 (1997) and references therein.
45. J.-L. Maurice, D. Imhoff, P. Etienne, O. Durand, S. Dubois, L. Piraux, J.-M. George, P. Galtier, and A. Fert, J. Magn. Magn. Mater. **184**, 1 (1998).
46. K. L. Babcock, V. B. Elings, J. Shi, D. D. Awschalom, and M. Dugas, Appl. Phys. Lett. **69**, 705 (1996).
47. R. Wiesendanger "Scanning Probe microscopy and spectroscopy: methods and applications", Cambridge University Press, 1995.
48. J. J. Saentz, N. Garcia, P. Grutter, E. Meyer, H. Heizelmann, R. Wiesendanger, L. Rosenthaler, H. R. Hidber, and H. J. Güntherodt, J. Vac. Sci. Technol. **A6**, 279 (1988).
49. Y. Martin, C. C. Williams, and H.K. Wickramasinghe, J. Appl .Phys. **61**, 4723 (1987).
50. D. Rugar, H. J. Mamin, P. Guethner, S. E. Lambert, J. E. Stern, J. McFadyen, and T. Yogi, J. Appl. Phys **68**, 1169 (1990)
51. H. J. Hug, B. Stiefel, P. J. A. van Schendel, A. Moser, R. Hofer, S. Martin, H.-J. Güntherodt, S. Porthun, L. Abelmann, J. C. Lodder, G. Bochi, and R. C. O'Handley, J. Appl. Phys. **83**, 5609 (1998).
52. R. Proksch, G. D. Skidmore, E. D. Dahlberg, S. Foss, J. J. Schmidt, C. Merton, B. Walsh, and M. Dugas, Appl. Phys. Lett. **69**, 2599 (1996).
53. S. Porthun, L. Abelmann, C. Lodder, J. Magn. Magn. Mater. **182**, 238 (1998).
54. http://www.ctcms.nist.gov/ rdm/mumag.org.html.
55. L. D. Landau and E. Lifshitz, Phys. Z. Sowjetunion **8**, 153 (1935).
56. T. J. Gilbert, Phys. Rev. **100**, 1243 (1955).
57. A. J. Newell, W. Williams, and D. J. Dunlop, J. Geophys. Res. **98**, 9551 (1993).
58. Y. Nakatani, N. Uesaka, and N. Hayashi, J. Appl. Phys. **28**, 2485 (1989).
59. P. Weiss, J. de Phys. Rad. **6**, 661 (1907).
60. For a review on magnetic bubbles in oxides see: A. P. Malozemoff and J. C. Slonczewski, "Magnetic Domain Walls in Bubble Materials", Academic Press, New York, (1979).
61. C. Kittel, Phys. Rev. **70**, 965 (1946).
62. C. Kooy and U. Enz, Philips Res. Rep. **15**, 7 (1960).
63. S. Chikazumi, "Physics of Ferromagnetism", chapter 17, Oxford University Press, 1997.
64. D. Craik, "Magnetism principles and applications", chapter 4.1, Wiley 1995.
65. A. H. Bobeck and E. Della Torre, "Magnetic Bubbles", North Holland, Amsterdam (1975).
66. A. H. Eschenfelder, "Magnetic Bubble Technology", Springer, Berlin, Heidelberg, New York (1981).
67. M. Hehn, S. Padovani, K. Ounadjela, and J. -P. Bucher, Phys. Rev. B. **54**, 3428 (1996).
68. U. Ebels, L. Buda, P. E. Wigen, and K. Ounadjela, Chapter 6 in "Spin Dynamics in Confined Magnetic Structures", edited by B. Hillebrands and K. Ounadjela, Springer, Spring 2001.
69. V. Gehanno, A. Marty, B. Gilles, and Y. Samson, Phys. Rev. B **55**, 12552 (1997); V. Gehanno, Y. Samson, A. Marty, B. Gilles, and A. Chamberod, J. Magn. Magn. Mater. **172**, 26 (1997).

70. A. Asenjo, J. M. Garcia, D. Garcia, A. Hernando, M. Vazquez, P. A. Caro, D. Ravelosona, A. Cebollada, and F. Briones, J. Magn. Magn. Mater. **196**, 23 (1999).

71. L. Folks, U. Ebels, R. Sooryakumar, D. Weller, and R. F. C. Farrow, J. Magn. Soc. Jpn. **23** No **S1**, 85 (1999).

72. R. Allenspach and M. Stampanoni, in "Magnetic Surfaces, Thin Films and Multi-layers", edited by S. S. P. Parkin et al., MRS Symposia Proceedings no. 231 , p.17, Material Research Society Pittsburg (1992).

73. This is illustrated for the case of eptiaxial Fe films in: E. Gu, E. Ahmad, S. J. Gray, C. Daboo, J. A. C. Bland, L. M. Brown, M. Rührig, A. J. McGibbon, and J. N. Chapman, Phys. Rev. Lett. **78**, 1158 (1997).

74. M. Demand, M. Hehn, K. Ounadjela, R. L. Stamps, E. Cambril, A. Cornette, and F. Rousseaux, J. Appl. Phys. **87** , 5111 (2000).

75. T. Shinjo, T. Okuna, R. Hassdorf, K. Shigeto, and T. Ono, Science **289**, 930 (2000).

76. M. Schneider, H. Hoffmann, and J. Zweck, Appl. Phys. Lett. **77**, 2909 (2000).

77. R. P. Cowburn, J. Phys. D **33**, R1 (2000).

78. W. F. Brown, Jr., J. Appl. Phys. **39**, 993 (1968).

79. M. Hehn, PhD thesis, University Louis Pasteur, Strasbourg (1997).

80. M. Hehn, R. Ferré, K. Ounadjela, J.-P. Bucher, and F. Rousseaux, J. Magn. Magn. Mater. **165**, 5 (1997).

81. R. Ferré, M. Hehn, and K. Ounadjela, J. Magn. Magn. Mater. **165**, 9 (1997).

82. R. P. Cowburn, A. O. Adeyeye, and M. E. Welland, Phys. Rev. Lett. **81**, 5414 (1998).

83. Chapter 5.7 in [39].

84. R. P. Cowburn, D. K. Koltsov, A. O. Adeyeye, M. E. Welland, and D. M. Tricker, Phys. Rev. Lett. **83**, 1042 (1999).

85. M. Demand, PhD thesis, University Louis Pasteur, Strasbourg (1998).

86. J. Raabe, R. Pulwey, R. Sattler, T. Zweck, and D. Weiss, J. Appl. Phys. **88**, 4437 (2000).

87. T. Pokhil, D. Song, and J. Nowak, J. Appl. Phys. **87**, 6319 (2000).

88. A. Fernadez and C. J. Cerjan, J. Appl. Phys. **87**, 1395 (2000).

89. A. Fernandez, M. R. Gibbsons, M. A. Wall, and C. J. Cerjan, J. Magn. Magn. Mater. **190**, 71 (1998).

90. E. Girgis, J. Schelten, J. Shi, J. Janeski, S. Tehrani, and H. Goronkin, Appl. Phys. Lett. **76**, 3780 (2000).

91. M. Kleiber, F. Kümmerlein, M. Löhndorf, A. Wadas, D. Weiss, and R. Wiesendanger, Phys. Rev. B **58**, 5563 (1998).

92. E. C. Stoner and E. P. Wohlfarth, Philos. Trans. London, Ser. A **240**, 599 (1948); L. Néel, Acad. Sci. Paris, **224**, 1550 (1947).

93. These experiments have been performed by our group in collaboration with M. Natali and Y. Chen, from L2M Bagneux.

94. L. D. Buda, I. L. Prejbeanu, M. Demand, U. Ebels, and K. Ounadjela, IEEE Trans. Magn., June (2001).

95. J. Miltat, "Applied Magnetism", NATO ASI Series, edited by R. Gerber, C. D. Wright, G. Asti, Kluwer Dordrecht, 221 (1994).

96. M. E. Schabes and H. N. Bertram, J. Appl. Phys. **64**, 1347 (1988).

97. R. P. Cowburn and M. E. Welland, Phys. Rev. B **58**, 9217 (1998).

98. Y. Zheng and J. Zhu, J. Appl. Phys. **81**, 5471 (1997).

99. K. J. Kirk, J. N. Chapman, and C. D. W. Wilkinson, Appl. Phys. Lett. **71**, 539 (1997).

100. J. Shi et al., IEEE Trans. Magn. **34**, 997 (1998).

101. J. Gadbois, J.-G. Zhu, and W. Vavra, A. Hurst, IEEE Trans. Magn. **34**, 1066 (1998).
102. M. Rührig, B. Khamesehpour, K. J. Kirk, J. N. Chapman, P. Aitchison, S. McVitie, and C. D. W. Wilkinson, IEEE Trans. Magn. **32**, 4452 (1996).
103. J.G. Zhu, Y. Zheng, and G. A. Prinz, J. Appl. Phys. **87**, 6668 (2000).
104. S. P. Li, A. Peyrade, M. Natali, A. Lebib, Y. Chen, U. Ebels, L.D. Buda, and K. Ounadjela, Phys. Rev. Lett. **86**, (2001).
105. M. Ledermann, R. O'Barr, and S. Schultz, IEEE Trasn. Magn. **31** 3793 (1997); R. O'Barr and S. Schultz, J. Appl. Phys. **81**, 5458 (1997).
106. R. Ferre, K. Ounadjela, J. M. George, L. Piraux, and S. Dubois, Phys. Rev. B **56**, 14066 (1997).
107. J.-E. Wegrowe, D. Kelly, A. Franck, S. E. Gilbert, and J.-Ph. Ansermet, Phys. Rev. Lett. **82**, 3681 (1999).
108. I. L. Prejbeanu, L. D. Buda, U. Ebels, and K. Ounadjela, Appl. Phys. Lett. **77**, 3066 (2000).
109. A. Hubert, IEEE Trans. Magn. **21**, 1604 (1985).
110. A. Aharoni, "Introduction to the theory of ferromagnetism", Clarendon Press, Oxford, (1996).
111. D. Hinzke and U. Nowak, J. Magn. Magn. Mater. **221**, 365 (2000).
112. R. C. O'Handley, "Modern Magnetic Materials, Principles and Applications", Chapter 9.2, Wiley, New York, (2000).
113. J. F. Gregg, W. Allen, K. Ounadjela, M. Viret, M. Hehn, S. M. Thomson, and J. M. D. Coey, Phys. Rev. Lett **77**, 1580 (1996)
114. M. Viret, D. Vignoles, D. Cole, J. M. D. Coey, W. Allen, D. S. Daniel, and J. F. Gregg, Phys. Rev. B **53**, 8464 (1996).
115. D. Ravelosona, A. Cebollada, F. Briones, C. Diaz-Paniagua, M. A. Hidalgo, and F. Batallan, Phys. Rev. B, **59**, 4322 (1999).
116. A. D. Kent, U. Ruediger, J. Yu, L. Thomas, and S. S. P. Parkin, J. Appl. Phys. **85**, 5243 (1999).
117. M. Viret, Y. Samson, P. Warin, A. Marty, F. Ott, E. Sondergard, O. Klein, and C. Fermon, Phys. Rev. Lett. **85**, 3962 (2000).
118. P. M. Levy and S. Zhang, Phys. Rev. Lett. **79**, 5110 (1997).
119. T. R. McGuire and R. I. Potter, IEEE Trans. Magn. **11**, 1018 (1975).
120. F. C. Schwerer and J. Silcox, Phys. Rev. Lett **20**, 101 (1968).

# 16  Micro– and Nanofabrication Techniques

C. Fermon

Service de Physique l' État Condensé, CEA-Saclay, 91191 Gif/Yvette, France

## 16.1  How Can We Go from Magnetic Layers to Submicron Scale Devices?

This chapter is intended to give readers a brief overview of the numerous techniques involved in the fabrication of small magnetic devices.

Until recently there has been a wide distinction between semi-conductor engineering and metallic and magnetic devices fabrication, the main reason being due to the huge investments in terms of money and manpower devoted to semiconductors rather than real technical limitations. With the advent of spin electronics, the number of metal and magnetic devices are increasing and in some instances, semiconductor and magnetic device fabrication have started to merge and are currently the topic of intensive research in some areas (e.g. in MRAMs). In the future it is anticipated, that metal and magnetic devices will be further employed at an accelerated pace in the electronics and computing sectors due to their inherent advantages, e.g. smaller, faster, more powerful non-volatile memories.

The construction of a magnetic device, like the fabrication of a semiconductor device consists of a succession of different steps, the deposition of layers, lithography (pattern generation), etching, oxidation and planarisation. However in semiconductors, there's an additional step, the doping of the semiconductor. Conventionally it is not used in metallic and magnetic devices, even though it is now possible to use ion beams to directly modify the magnetic properties [1]. For example a complete device like a read head requires more than 100 different steps and consequently the size of the wafers used for fabrication has tended to increase, so conventionally six inches wafers are used for fabricating read heads.

In the laboratory, prototype devices are usually made with typically ten or less steps and have smaller dimensions, the less the number of steps, the greater the probability a device will work!! Furthermore, in research laboratories the preferred tool for the pattern generation is mainly e-beam lithography which is rarely used in industry as its considered to be too expensive and too slow. This type of lithography has the advantage over orthodox techniques as its very versatile but its limited to small sample sizes.

Thus in this chapter, I will focus on a "laboratory" approach to the building of small magnetic devices.

## 16.2    Basic Processes

In this section I will describe the "standard" processes necessary to build a simple device and give a brief description of each fabrication step. There are a number of excellent text books [2,5] available and for a more detailed description of the fabrication procedures the reader is referred to these texts for further details.

The basic processes are widely used to fabricate a patterned magnetic layer (or magnetic multilayer) such as an array of dots or lines [6–11] or more sophisticated shapes. This can be, for example the patterning a hard disk surface creating dot arrays for magnetic storage.

### 16.2.1    Standard Patterning

Standard patterning consists of six basic steps:

This process is used when it is possible or necessary to deposit the magnetic layer first, e.g. for epitaxially grown magnetic films or when the growth of the films can only be done under external constraints (high temperature for example).

The standard patterning technique has a number of drawbacks: since it etches the magnetic layer, it is usually difficult to characterize the magnetic properties of small magnetic objects in addition the etching process may create defects on the surface or a heating of the layer.

Thus if the magnetic layer can be deposited at a rather low temperature then a lift off technique may be a much better solution.

### 16.2.2    Lift–Off Patterning

The lift off technique is very simple providing the deposition does not require a high temperature. It is also important to note that the surface of the wafer cannot be very clean (due to water and oxygen absorption etc.), hence if thick magnetic layers are being deposited ($> 1$ µm for photolithography and 200 nm for e-beam lithography), they may be difficult to lift off from the resist.

Also if the material is deposition by sputtering, the resist may be damaged by the plasma irradiation, additionally in high pressure deposition, the magnetic metal is deposited everywhere, even on the edges of the resist which makes the lift off extremely difficult. Evaporation or electro-deposition are the best deposition techniques when using a lift off process.

## 16.3    Deposition Techniques

Deposition techniques can be divided into three main groups:

– Physical Vapor Deposition (PVD) including evaporation and sputtering.

– Chemical Vapor Deposition (CVD) and Plasma Assisted Chemical Vapor Deposition (PACVD).

and

**1) Preparation of the wafer**.

**2) Layer deposition.** The magnetic layer or multilayer is deposited of the surface of a clean wafer. An overview of the different deposition techniques is given later on in Sect. 16.3.

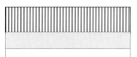

**3) Resist deposition.** The choice of the resist depends on the kind of lithography used (see Sect. 16.4).

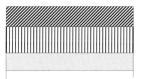

**4) Lithography** : resist exposure and development. The resist is patterned.

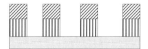

**5) Etching film**. This is usually the critical step in the fabrication : the transfer of the pattern to the metallic layer

**6) Lift off of the resist.**

**Fig. 16.1.** The stages involved in "standard" patterning.

**1) Clean wafer.**

**2) Resist deposition.**

**3) Lithography** : resist exposure and development.

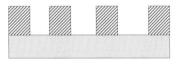

**4) Metal deposition.**

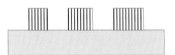

**5) Lift off resist.**

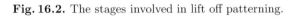

**Fig. 16.2.** The stages involved in lift off patterning.

– Electro-deposition.

It would be impossible to give here an extensive description of each technique, so once again the reader is referred to the text books [3,4] for further details.

Since the effectiveness of the resulting lithography, is incontrovertibly linked to the depositing technique used, I will use shape anisotropy, as the criterion to separate and define the deposition techniques.

Some deposition techniques like evaporation at very low pressures are strongly anisotropic and produces a good pattern transfer in the lift off process. Sputtering at high pressures is rarely used for a lift off process due to the greater isotropic deposition of the metal, thus it is then very difficult for the solvent to dissolve the resist. In addition as already mentioned above, splutter depositioning creates a high level of radiation. This radiation burns the resist, which likewise makes the removal of the resist extremely difficult, as usually plasma deposition hardens the resist. Since the dissolution of the resist is nearly impossible, it's necessary to plasma clean the sample using $O_2$.

Electro-deposition is a nice alternative deposition technique, since the deposition only occurs where it is necessary (by masking the electrodes). Electro-deposition does not cause damage or adhere strongly to the resist, but the operation of the electro-depositing baths is a very complicated technique. The skilled usage of the baths is very much a black art (for example for some materials adding saccharine to the bath aids the electro-depositing) and it may takes a user a while to obtain good electro-deposited layers. For devices, like read heads this technique is becoming increasingly popular due to the corresponding low cost in investment and production. Electro-deposition can be also used to produce very thin magnetic wires by deposition through membranes [12].

## 16.4  Resist Deposition

### 16.4.1  The Resists

Resists are made with polymers sensitive to radiation of some kind and they fall into two categories:

i) Positive resists: where there is a removal of the resist exposed to radiation and

ii) Negative resists: where there is a removal of the unexposed resist.

The irradiation breaks up the polymer network which then dissolves with an appropriate solvent. According to the type of lithography performed different resists are required so:

*For UV lithography*: A photoresist is composed of a base resist+photoactive compound+solvent. Positive and negative photoresists with their developers are available commercially.

*For electron and X-ray lithography*: A PMMA (polymethyl-meta-acrylate) or MAA (meta-acrylate acid) polymers dissolved in a solvent can be used for the resist. These resists can be home made and easily made in large quantities.

### 16.4.2   Resist Deposition

The most important step before the resist deposition is the cleaning of the substrate (+ layers), since the cleanliness of the substrate is so important resist deposition should be done in a clean room.

The resist is liquid and deposited on the sample, the sample is spun with typical speeds of 3000 to 6000 RPM with the resultant resist film thickness of course depending on the speed and on the viscosity of the resist. For best results the substrate is first spun at low speed and then given a final spin at high speed.

The shape of the substrate also determines the homogeneity of the resist film, so for example on a rectangular sample, the resist film is not perfectly homogeneous due to strain deformation at the edges of the substrate. Thus using a circular substrate gives the best homogenous resist film.

After deposition, the resist is baked to evaporate the solvent, with the baking time and temperature strongly dependent on the resist type.

## 16.5   Pattern Generation

Lithography can be divided in two different types, lithography using a mask and direct write lithography. Industrial processes use masks and mainly UV lithography and in the laboratories, e-beam lithography is often used due to its versatility and its resolution.

### 16.5.1   Lithography Through a Mask

Mask lithography is usually done with coherent light sources. The wavelength of the radiation used is between 0.436 μm (mercury G-line) and 0.157 μm ($F_2$ laser) for UV light. In the case of X-ray lithography, the wavelength is about 1 nm [13,14]. Of course the radiation wavelength determines the resolution of the technique, creating a pattern smaller than the wavelength is usually very difficult because of diffraction effects.

Lithography can be done either with the mask as close as possible to the substrate (contact lithography) or through an optical system, which allows a reduction of the mask pattern onto the substrate (projection lithography), see Fig. 16.3.

Contact lithography allows us to go below the wavelength limit by using the fact that light creates an evanescent wave through a hole smaller than the wavelength, but the distance between the sample and the mask must be very precise.

**UV Photolithography**   UV lithography is the easiest and cheapest way to produce large quantities of submicron magnetic systems. The best photolithography systems give resolutions of 0.25 μm and recently 0.18 μm has been obtained, the ultimate resolution of UV lithography is considered to be 0.1 μm. The resolution

(a)                          (b)

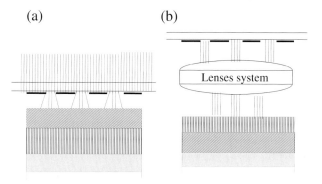

**Fig. 16.3.** The two types optical lithography (a) contact lithography and (b) projection lithography.

is limited by diffraction effects due to the distance between the mask and the sample. On cheap systems in a laboratory the resolution limit is about 0.5 μm.

The masks for UV lithography are made of chromium on glass and built using e-beam lithography and can be easily obtained commercially in any size.

**Electronic and Ion Beam Projection Lithography** Using large electron or ion sources, it is possible to pattern layers down to 100 nm, for example the SCALPEL, (SCattering with Angular Limitation Projection E-beam Lithography), approach combine high resolution with the throughput of a parallel projection system [15]. With ion projection lithography, 50 nm wide lines have been produced [16].

**X-ray Lithography** Soft X-rays with a wavelength of about 1 nm can be obtained from a synchrotron source, for example the synchrotron at Lure, Orsay has a beamline devoted to this application [14]. Due to the absence of efficient optics in this light regime, X-ray lithography is always a contact lithography. The difficulty of mask fabrication is the main problem, with X-ray lithography since most materials absorb at these energies. This problem is circumvented by making the mask from tungsten or gold on a thin (100 nm) carbon, silicon nitride or capton films. Access to cheap, high flux X-ray sources and the difficulties associated with masks fabrication limits X-ray usage.

### 16.5.2   Direct Writing

For direct exposure of the resist, a collimated beam can be used. This technique is much slower than lithography through a mask but it gives the allows mask fabrication and produces original patterns at a reduced price. The direct writing radiation can be a laser at a rather low wavelength or an e-beam.

**Table 16.1.** Correction factors for calculating the resistance per square for a circle and a rectangular sample

| Resist | minimum linewidth | aspect ratio | dose at 100keV$(C/cm^2)$ | Mechanism |
|--------|-------------------|--------------|--------------------------|-----------|
| PMMA   | 8–10 nm           | 45           | $5 \times 10^{-4}$       | Bond breaking |
| NaCl   | 1.5 nm            | > 40         | $10^2$–$10^3$            | Dissociation |
| $AlF_2$ | 1.5 nm           | > 40         | 1–10                     | Diffusion of Al, Dissociation of $AlF_2$ |

## Laser Photolithography

- *Direct laser writing.* This is used for low resolution masks e.g. masks for printed circuits for electronics.
- *Holographic lithography.* Two laser beams interfere and produce standing waves, grating periods approaching 50% of the laser wavelength can be obtained [17–19]. This is a convenient way to create gratings or large arrays of magnetic lines and dots.

**E–Beam Lithography** A scanning electron beam controlled by a computer is used to directly write the pattern. The e-beam main advantage is the high radiation damage it creates and its very short wavelength (less than 0.1 nm). A large variety of "resists" can be used, for example standard resists like PMMA can be deposited. The effect of the radiation breaks polymer bonds and thus enhances the dissolution effect by an appropriate solvent. The resolution is mainly given by the size of the polymer. It is also possible to use layers which can be destroyed by the e-beam. Table 16.1 gives some examples.

With PMMA, the resolution in a standard process is about 20–50 nm, depending on the magnetic material. The resolution limit is due to secondary electron creation, electrons are absorbed or scattered in the material and when they back scatter, then they irradiate the resist and the effective resolution is lowered.

There are several ways to partially solve this problem.

The first technique is to deposit a very thin resist layer like a Langmuir–Blodgett monolayer onto the sample, thus the resolution limit is determined by the resulting etching after the lithography. With this technique the contrast is very poor and so its only used in special circumstances.

A second technique uses a crystal modified by dissociation or diffusion for example a salt. Lithography in the nanometer range can be performed, but this type of lithography requires a high energy and long exposition time. Furthermore, the size of the grains must be smaller than about 150–200 nm.

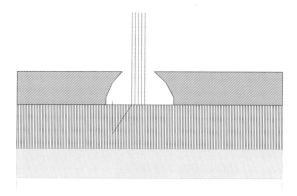

**Fig. 16.4.** Schematic diagram of e-beam lithography.

In Saclay, we currently use a third method the trilayer process which I will now briefly detail, the advantage of this technique is that it only uses standard processes.

### 16.5.3  Trilayer Technique

The idea is to use a thin germanium or titanium film as a suspended mask on the sample [20,21].

Thus a well defined metallic mask is obtained. This technique of tri-layer deposition allows us to carry out angular deposition (or shadow evaporation), see Fig. 16.6. Angular deposition has now become common place and its a convenient way to make nanostructures, such as junctions between two different materials (as used in squids) or devices. By evaporating two different metals at different angles through the same suspended mask, the same pattern at different locations can be obtained. By playing with the shape of the suspended mask an overlap of two kind of material can be created.

## 16.6  Etching of the Layers

The etching of the layers is a critical point in building a device. The etching must correctly transfer the designed pattern. In this section I will give a very brief summary of the various techniques used to transfer a pattern.

### 16.6.1  Wet Etching

Wet etching is material dependent and its the simplest way to etch a device. There are a large number of solutions for selective etching commercially available. Table 16.2 gives some examples.

Wet etching has a low final resolution and a rather poor etching homogeneity. The etching speed strongly depends on the layer crystalline structure, so with

**1) Deposition of MAA, PMMA Germanium and PMMA layers.**

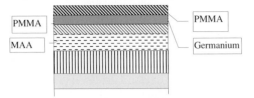

**2) Lithography of the first PMMA layer by e-beam and development. Etching of the Germanium layer by reactive ion etching.**

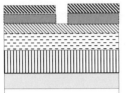

**3) Reactive ion etching of the resists underlayers by reactive ion etching.**

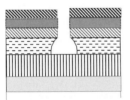

**4) Deposition of a metallic mask through the germanium suspended mask.**

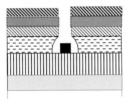

**5) Lift off resists.**

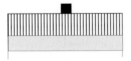

**Fig. 16.5.** Stages involved in the tri-layer technique.

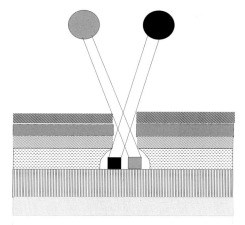

**Fig. 16.6.** Schematic diagram of the shadow evaporation technique.

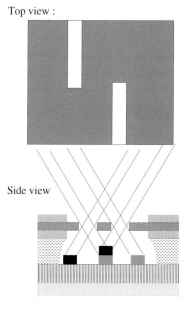

**Fig. 16.7.** Schematic diagram of the shadow evaporation technique viewed from the top and bottom .

**Table 16.2.** A list of materials and their etchent chemicals.

| Material | etched by |
|---|---|
| Au | KI–I$_2$ |
| SiO$_2$ | HF |
| Cu | FeCl$_3$ |
| Si | KOH(80°C) |

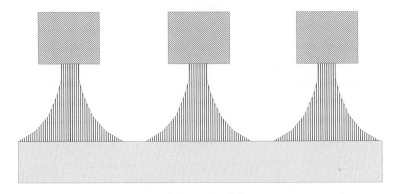

**Fig. 16.8.** Example of the isotropic nature of wet etching

textured samples, a rather homogeneous etch can be obtained but in polycrystalline films, a large difference can be observed. Wet etching is usually isotropic and its not possible to transfer a pattern with a high aspect ratio.

However, in some cases special aspects can be obtained. For example, the etching of silicon by KOH depends on the crystalline orientation. The (111) face is etched at a slower rate than the (100) or (110) planes, so very flat edges can be produced by choosing an appropriate crystal orientation [22].

### 16.6.2    Ion Beam Etching–Ion Milling [3,23]

Ion beam etching (IBE) produces ions in a cavity and then accelerate them to create a relatively intense and homogeneous beam. The ions are usually Ar$^+$ but oxygen or additional gas may be used. Typically the size of commercial sources vary from 2.5cm to 20cm and the ion energy is about a few hundreds of eV. The etching is anisotropic and hence can be used for a good transfer with high aspect ratios, with the speed of etching depends on the material type.

There is only a small dependence on the crystallinity of the material but the incidence of the beam is very important. The etching is maximum perpendicular to the sample and decreases to zero at grazing angle. For this reason IBE can be used to planarisation a sample.

**Table 16.3.** A list of materials and their reactive ion etchent gases

| gas | material it etches |
| --- | --- |
| $SF_6$ | Si, $SiO_2$ |
| $CF_4$ | Si, $SiO_2$, $Si_3N_4$, Ti, Mo, Ta, W |
| $CClF_3$ | Au, Ti |
| $CBrF_3$ | Ti, Pt |
| $CCl_4$ | Al, Cr |

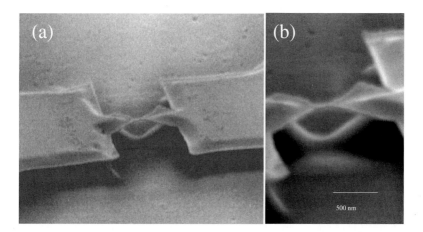

**Fig. 16.9.** (a) Tunnel junction fabricated by shadow evaporation, e-beam lithography and reactive ion etching. (b) enhanced view of the junction with typical dimensions 200 nm long ×15 nm wide. (courtesy of Ivan Petej, Clarendon Laboratory, Oxford University)

### 16.6.3  Reactive Ion Etching

Reactive ion etching (RIE) consists of creating a plasma with different types of gas in order to combine ion beam milling and chemical reaction, this type of etching can then be made more selective than a classical IBE.

If the sample is very close to the plasma, the etching is rapid and isotropic otherwise it becomes anisotropic and slow. The speed of etching varies from 0–1000 nm/minute depending of the material, gas pressure and crystalline structure. Table 16.3 list several gases commonly used in reactive ion etchers.

### 16.6.4  Focused Ion Beam (FIB) Etching

Since the focused ion beam (FIB) is a relatively new tool to the research community I will detail how it works and then expand on how it may be used to fabricate nanodevices.

The focused ion beam as its name implies, is basically a scanning electron microscope (SEM) which uses ions instead of electrons. The FIB has been available commercially for the last decade and has become a very popular tool for nanofabrication research because of its very versatile nature.

The ion source is generally a liquid metal source such as Ga (alternatively Be, Ar or He may be used) and the beam is extracted and ionised by a strong electric field applied to the tungsten injector needle. The ionised beam is then focused by electrostatic lenses and collimated with the aid of an aperture (which selects the beam current), the ionised beam then strikes the substrate emitting secondary electrons and ions, which can then be used to generate an image of the surface. Generally a FIB uses the secondary electrons to image the substrate as they are emitted in greater quantities and hence provided a greater image contrast. The contrast depends on material and topography, since an insulator emits less secondary electrons than a conductor the insulator regions appear bright and the conductive regions appear dark. Beam energy is typically 30 or 50 keV with the beam current in the range of 1pA–20 nA the best image resolution is about 10 nm.

Ion milling (and thus lithography see Fig. 16.10) can be additionally performed by the FIB with supplemental gases, injected by needles near the area of interest. During the gas assisted process the gas is adsorbed onto the surface, where it is broken down by the ion beam to produce volatile compounds that are pumped away, thus the local etching can be material dependent so only a metal, insulator or carbon based compound is etched. In a similar process the FIB can deposit metals such as platinum and tungsten and silicon oxide layers. To do this the beam parameters and the gas flow is optimised for the most efficient equilibrium between the breaking of the precursor gas and the milling action of the beam [27–29].

A FIB has conventionally been used in semiconductor failure analysis by making TEM cross-sections in the region of interest and for modifying tracks on prototype chips. However it may be used to do chemical analysis in a local region if used in combination with a secondary ion mass spectrometer. If a FIB has a secondary electron column – a dual beam FIB, then it has the colossal advantage of observing the device during etching (observing the etch while milling can also be done by the singe beam FIB, but there will be a small level of implantation [24–26]). Like the standard FIB, a dual beam FIB can also perform chemical analysis locally on the device by energy dispersive x-ray spectroscopy.

In some nanodevices the ion implantation by the ionised beam may cause problems, for example, in high-$T_c$ films, ion implantation due to migration is present on distances of 300–400 nm and then the superconducting properties are modified on that scale.

## 16.7   Additional Techniques

Finally I will conclude this chapter with a brief description of unconventional techniques used to patterning magnetic thin films.

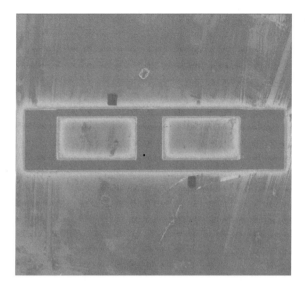

**Fig. 16.10.** Lithography performed with FIB which has defined the base and collector contacts of a magnetic field dependent transistor (Courtesy of Ivan Petej, Clarendon Laboratory, Oxford University).

### 16.7.1   AFM–STM Lithography

**Mechanical Lithography [30]** The principle is to etch mechanically a thin layer of resist by the tip of an AFM (Atomic Force Microscope) on a three layer deposited system. The tip is used to mechanically design the pattern.

First an image is stored and after, the tip pushed in the layer at the right place and moved. This technique can be used to fabricate very thin wires at the end of a lithographic process.

The advantage of this approach is the high achievable resolution, about 20 nm, and the possibility of visualization before the patterning. But it is convenient only for a small amount of very small objects like lines and the life of the tip is rather short. It is possible to avoid this last problem by using a tip made with a very small diamond crystal but with a loss in resolution.

**Electrical Lithography** A tip can also be used to apply locally an electrical pulse to destroy a resist on the surface. The advantage of that technique is a very precise lithography. But the resist must be very thin and then the transfer to the metallic layer is not obvious.

In presence of gas, an electrical pulse can help to deposit atoms on the surface. Then a metallic mask can be created.

**Arrays of Scanning Probes** Arrays using 50 SPMs have been fabricated [31] in order to perform patterning of large surfaces.

### 16.7.2   Chemical Transfer, Nanoimprint

There are two main techniques in chemical transfer and nanoimprinting they are: imprinting and inking. In imprinting, a reusable mold is stamped at high pressure into a polymer film on a substrate and removed. In inking, molecules which have an affinity for one metal and not the surface will wet only the corresponding metal. A "stamp" is prepared and immersed in a "inking" solution. Then, we have just to press the corresponding stamp on a surface where the liquid react to transfer the pattern of the stamp. The transfer of the pattern is surprisingly good and resolution better than 100 nm can be achieved [32]. With imprinting, 6 nm hole and 15 nm trenches have been obtained in PMMA [33]. This is a very cheap alternative to e-beam lithography.

# References

1. T. Devolder and C. Chappert, J. Appl. Phys. **87**, 8671 (2000).
2. I. Brodie and J. Muray, "The physics of micro/nano Fabrication", Plenum Press, New York and London 1992.
3. J. Cuomo, S. Rossnagel, and H. Kaufman (eds.), "Handbook of Ion beam Processign technology", Noyes publications, 1989.
4. R. F. Bunshah (ed.), "Handbook of deposition technologies for films and coatings", Noyes publications 1994.
5. J. L. Vossen and W. Kern (eds.), "Thin film processes", Academic press, 1978.
6. C. Miramond, C. Fermon, F. Rousseaux, D. Decanini, and F. Carcenac, J. Magn. Magn. Mater. **165**, 500 (1997).
7. P. Vavassori, O. Donzelli, V. Metlushko, M. Grimsditch, B. Ilic, P. Neuzil, and R. Kumar, J. Appl. Phys. **88**, 999 (2000).
8. M. Demand, M. Hehn, K. Ounadjela, R. L. Stamps, E. Cambril, A. Cornette, and F. Rousseaux, J. Appl. Phys. **87**, 5111 (2000).
9. A. Hirohata, H. T. Leung, Y. B. Xu, C. C. Yao, W. Y. Lee, J. A. C. Bland, and S. N. Holmes, IEEE Trans. Magn. **35**, 3886 (1999).
10. O. Fruchart, J. P. Nozieres, W. Wernsdorfer, D. Givord, F. Rousseaux, and D. Decanini, Phys. Rev. Lett. **82**, 1305 (1999).
11. J. P. Jamet, S. Lemerle, P. Meyer, J. Ferre, B. Bartenlian, N. Bardou, C. Chappert, P. Veillet, F. Rousseaux, D. Decaniniand, and H. Launois, Phys. Rev. B **57**, 14320 (1998).
12. J. L. Duvail, S. Dubois, L. Piraux, A. Vaures, A. Fert, D. Adam, M. Champagne, F. Rosseaux, and D. Dacanini, J. Appl. Phys. **84**, 6359 (1998).
13. J. P. Silverman, J. Vac. Sci. Technol. B **16**, 3137 (1998).
14. G. Simon et al., J. Vac Sci. Technol. B **15**, 2489 (1997).
15. L. H. Harriot, J. Vac. Sci. Technol. B **15**, 2130 (1997).
16. W. H. Brünger et al., MNE 98, Microel. Eng. **53**, 605 (2000).
17. P. Visconti, C. Turco, R. Rinaldi and R. Cingolani, Microel. Eng. **53**, 391 (2000).
18. M. Schneider and H. Hoffmann, J. Appl. Phys **86**, 4539 (1999).
19. S. Kreuzer, K. Prugl, G. Bayreuther, and D. Weiss, Thin Solid Films **318**, 219 (1998).
20. G. J. Dolan and J. H. Dunsmuir, Physica B **152**, 7 (1988).
21. J. Romijn and E. Van des Drift, Physica B **152**, 14 (1988).

22. E. Bassous, IEEE Trans. Electron Devices **10**, 1178 (1978).
23. S. Somekh, J. Vac. Sci. Technol **13**, 1003 (1976).
24. N. I. Kato, Y. Kohno and H. Saka, Technol A **17**, 1201 (1999).
25. T. Ishitani, H.Koike, T.Yaguchi, and T.Kamino, J. Vac. Sci. Technol B **16**, 1907 (1998).
26. R. M. Langford and A. K. Petford–Long, J. Vac. Sci. (in press).
27. K. Yamakawa, K. Taguchi, N. Honda, K. Ouchi and S. Iwasaki, J. Appl. Phys. **87**, 5422 (2000).
28. Y.Q. Fu, N. Kok, A. Bryan, N. P. Hung and S. N. Ong, Rev. Sci. Instrum. **71**, 1006 (2000).
29. H. C. Tong, X. Shi, F. Liu, C. Qian, Z. W. Dong, X. Yan, R. Barr, L. Miloslavsky, S. Zhou, J. Perlas, P. Prabhu, M. Mao, S. Funada, M. Gibbons, Q. Leng, J. G. Zhu, and S. Dey, IEEE Trans. Magn. **35**, 2574 (1999).
30. V. Bouchiat and D. Esteve, Appl. Phys. Lett. **69**, 3098 (1996).
31. S. C. Minne et al, *Phasdom 97, Aachen 10-13 March (1997)*.
32. S. Palacin et al, Chem. Matter. **8**, 1316 (1996).
33. S.Y. Chou and P. R. Krauss, *Phasdom 97, Aachen 10-13 March 1997.*

# 17   Spin Transport in Semiconductors

M. Ziese

Department of Superconductivity and Magnetism, University of Leipzig,
Linnéstrasse 5, 04103 Leipzig, Germany

## 17.1   Introduction

Recent months have seen a startling activity and a rapid development in the
area of spin-coherent transport in semiconductors. This research is driven by the
need for integration of spin-electronic devices into conventional semiconductor
technology. Devices based on spin-dependent effects such as read heads for hard
disks or non-volatile magnetic random access memory (MRAM) had (and are
supposed to have in the near future) a strong impact on storage technology
[1,2]. These areas, however, are well separated from standard Si technology and,
although these are highly market relevant, a new approach has to be developed
in order to transfer the benefits of spin-electronics into MOSFET technology.

Historically, the idea of spin-injection into semiconductors was first proposed
during the early seventies in the context of tunnelling studies using ferromagnetic
electrodes [3]. A theoretical account of spin-injection from a ferromagnet into
a semiconductor was given by Aronov and Pikus by calculating the spatially
decaying polarization of majority and minority carriers [4]. Alvarado and Renaud
successfully observed vacuum tunnelling of spin-polarized electrons from a Ni
tip into GaAs [5]. The spin-polarization was determined from the polarization
of the emitted radiation and a negative value of about $-30\%$ was found at
small injection energies. This early experiment proved that spin injection into
semiconductors is possible; the negative spin-polarization indicates that minority
$3d$ electrons are preferentially emitted from the Ni tip. Further tunnelling studies
with semiconducting barriers were conducted by Prins et al. [6].

The problem of spin-coherent transport in semiconductors can be conveniently split into three issues: the investigation of spin-coherence effects within
the semiconductor, spin-injection into the semiconductor from the outside world
and detection of a spin-polarized current. The first problem is mainly addressed
with ultra-fast optical spectroscopy techniques and will be discussed later. Currently two spin-injection methods are investigated, namely injection of spin-
polarized carriers from a metallic ferromagnetic electrode or through a ferromagnetic semiconductor. Whereas the first method seems to be the obvious choice,
results so far have proven to be poor and the second technique might actually
lead to a breakthrough. Some information on magnetic semiconductors can be
found in this volume in M. Coey's chapter on 'Materials for Spin Electronics'.
Recent work on ferromagnetic semiconductors has been reviewed by Ohno [7];
an account of early work was given by Methfessel and Mattis [8]; see also various
review articles on diluted magnetic semiconductors in [9]. The methods in use

to study spin-injection from ferromagnetic electrodes have been pioneered by Johnson in his investigations of spin-injection and -detection in metals [10,11] and superconductors [12]. Johnson's spin-transistor experiment was analyzed in detail by Fert and Lee [13].

This chapter concludes with a discussion of devices. A great deal of the present research activity was initiated by the proposal of Datta and Das for a spin-electronic transistor that is analogous to an electro-optic modulator [14]. In this transistor, spin-polarized electrons are injected into a two-dimensional electron gas (2DEG) via a ferromagnetic electrode and are analyzed by a ferromagnetic detector electrode. The spin direction can be controlled by a gate voltage through the Rashba effect. This device promised a straightforward route to a spin-transistor; the experiments, however, are so far disappointing.

## 17.2   Basics

In this section a simple idealization of spin-injection into semiconductors is discussed in order to introduce the relevant length scales. The discussion follows the treatment by Aronov and Pikus [4] slightly modified by recent ideas of Flatté and Byers [15] on spin diffusion.

The basic length scale relevant for spin-coherent transport processes is the spin-diffusion length given by

$$\lambda_s = \sqrt{D_s \tau_s} \,, \tag{17.1}$$

where $\tau_s$ denotes the spin-relaxation time and $D_s$ the spin-diffusion constant. $\lambda_s$ is the average distance a spin can diffuse without losing its spin memory. The spin-diffusion constant $D_s$ is not necessarily equal to the charge-diffusion constant $D$. This is addressed in more detail at the end of this section. There is experimental evidence, see Sect. 17.3, that the spin-diffusion constant $D_s$ is considerably larger than $D$.

Consider the following simple model: a doped semiconductor fills the half-space $x > 0$ and spin-polarized carriers are injected into this semiconductor by some kind of process. The spin-polarization of the injected carriers is given by $\boldsymbol{P}$ and the current density is denoted by $\boldsymbol{j}$. The evolution of the spin-density $\boldsymbol{S}$ in the semiconductor can be calculated from the Bloch equation

$$\frac{\partial \boldsymbol{S}}{\partial t} = \boldsymbol{S} \times \boldsymbol{\Omega} - \frac{\boldsymbol{S}}{\tau_s} - \nabla \cdot \mathbf{J}_s \,. \tag{17.2}$$

$\boldsymbol{\Omega} = g\mu_B \boldsymbol{B}/\hbar$ is the precession frequency of carriers with gyromagnetic ratio $g$ around a magnetic field $\boldsymbol{B}$. $\mu_B$ denotes the Bohr magneton. The second rank tensor $\mathbf{J}_s$ denotes the spin-current density. The boundary condition at $x = 0$ is given by

$$\mathbf{J}_s(x = 0) = q^{-1} \boldsymbol{j} \otimes \boldsymbol{P} \,. \tag{17.3}$$

Here spin-relaxation effects at the boundary are neglected and $\boldsymbol{P}$ has to be interpreted as some effective spin-polarization. $q$ is the charge of the carrier.

Let us first consider majority carriers; in this case recombination processes can be neglected and the carrier concentration $n$ is independent of the spatial coordinates and the current. The spin-current density is a sum of drift and diffusion terms:

$$\mathbf{J}_s = \boldsymbol{v} \otimes \boldsymbol{S} - D_s \nabla \otimes \boldsymbol{S} . \qquad (17.4)$$

The drift velocity is related to the current density by $\boldsymbol{v} = \boldsymbol{j}/qn$. In the absence of a magnetic field and for an injected current along the x-axis, the spin-polarization in the steady state is easily obtained by solving (17.2):

$$\boldsymbol{P}_s(x) = \frac{\boldsymbol{S}}{n} = \frac{2\lambda_d \boldsymbol{P}}{\lambda_d + \sqrt{\lambda_d^2 + 4\lambda_s^2}} \exp\left[ -\frac{x}{2\lambda_s^2} \left( \sqrt{\lambda_d^2 + 4\lambda_s^2} - \lambda_d \right) \right] . \qquad (17.5)$$

$\lambda_d = v\tau_s$ is the drift length. In the case of a non-degenerate superconductor the ratio of drift and spin-diffusion length is given by $\lambda_d/\lambda_s = qE\lambda_s/k_B T$ and can be high for typical electric fields $E$. In this regime the spin-polarization in the semiconductor is simply given by $\boldsymbol{P}_s(x) = \boldsymbol{P}\exp[-x/\lambda_d]$ with the exponential decay determined by the drift length $\lambda_d$. In the opposite limit, $\lambda_s \gg \lambda_d$, the exponential decay length is determined by the diffusion length $\lambda_s$; in this case however, the spin-polarization at the interface is reduced by a factor $\lambda_d/\lambda_s$.

In the case of a magnetic field $\boldsymbol{B}$ applied perpendicular to the spin-polarization $\boldsymbol{P}$ at the surface, the spin-polarization in the semiconductor $\boldsymbol{P}_s$ in the limit $\lambda_d \gg \lambda_s$ is given by

$$\boldsymbol{P}_s = P \exp\left[-x/\lambda_d\right] \left[ \hat{\boldsymbol{P}} \cos(\Omega x/v) + \hat{\boldsymbol{P}} \times \hat{\boldsymbol{\Omega}} \sin(\Omega x/v) \right] . \qquad (17.6)$$

$\hat{\boldsymbol{P}}$ and $\hat{\boldsymbol{\Omega}}$ denote unit vectors. The polarization is seen to rotate in a plane perpendicular to the applied magnetic field.

In the case of minority carriers the diffusion component is greater than the drift component as long as the minority-carrier density is much smaller than the majority-carrier density. The Bloch equation (17.2) is valid, when the spin-relaxation time is replaced by the combined spin-flip and recombination time $\tau^{-1} = \tau_s^{-1} + \tau_r^{-1}$. The minority-carrier density decays exponentially with distance from the injection point according to

$$n(x) = \lambda_r J/D \exp\left[-x/\lambda_r\right] , \qquad (17.7)$$

where $\lambda_r = \sqrt{D\tau_r}$ is the minority carrier diffusion length. The spin-polarization in the semiconductor is then given by

$$\boldsymbol{P}_s = \frac{\lambda \boldsymbol{P}}{\lambda_r} \exp\left[ -x \left( \frac{1}{\lambda} - \frac{1}{\lambda_r} \right) \right] \qquad (17.8)$$

with the minority carrier spin-diffusion length $\lambda = \sqrt{D\tau}$.

Flatté and Byers [15] qualitatively studied the motion of charge and spin packets in undoped and doped semiconductors. Due to the ineffective screening, a local spin/charge disturbance will in general be a multi-band perturbation. This is illustrated in Fig. 17.1 for a n-doped semiconductor.

charge packet              spin packet

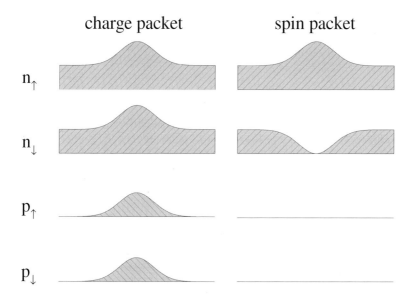

**Fig. 17.1.** Schematic picture of the carrier-density distribution in an n-doped semi-conductor for a local charge (left) or spin (right) disturbance. The ineffective screening in the semiconductor leads to the formation of hole packets accompanying the electron packets in a charge disturbance in order to preserve charge neutrality (left). A spin packet, however, can be entirely formed by disturbances in the conduction band (right panel). $n_{\uparrow(\downarrow)}$ and $p_{\uparrow(\downarrow)}$ denote the electron and hole densities with the indicated spin directions, respectively.

In the case of a n-doped semiconductor a charge packet consists of a localized increase of both electron and hole density. The resulting ambipolar diffusion constant is given by

$$D = \frac{n\mu_e D_h + p\mu_h D_e}{n\mu_e + p\mu_h} \tag{17.9}$$

where $\mu_{e(h)}$, $D_{e(h)}$ are the electron (hole) mobility and diffusion constant, respectively, and $n(p)$ denote the electron (hole) density. Since $n \gg p$, the ambipolar diffusion constant is dominated by hole motion, $D \sim D_h$. The spin packet, on the other hand, consists only of localized electron packets and therefore, the spin-diffusion constant is given by the electron-diffusion constant, $D_s \sim D_e$. Accordingly, the spin-relaxation time is given by the electron spin-relaxation time $\tau_s \sim \tau_{s,e} \gg \tau_r \gg \tau_{s,h}$. This in turn is determined by the Elliot–Yafet [16,17], D'yakonov–Perel' [18], Bir–Aronov–Pikus [19] or hot spot [20] mechanism depending on temperature regime and doping. A detailed discussion of these relaxation mechanisms is beyond the scope of this review. Since the electron mobility is generally much larger than the hole mobility, one has $D_s \gg D$.

In p-doped semiconductors the situation is reversed. In an undoped semiconductor both spin and charge packets consist of electron and hole excitations and the spin and charge diffusivities are of the same order of magnitude; in this case $\tau_{\mathrm{s}} \simeq \tau_{\mathrm{r}}$. This simple argument is in qualitative agreement with the doping dependence of the spin-relaxation time and the strongly enhanced spin-diffusion constant observed in n-doped GaAs, see Sect. 17.3.

## 17.3    Spin–Coherent Transport

Spin transport in semiconductors has been investigated with ultra-fast spectroscopy techniques. In this method a non-equilibrium spin-population is created in the conduction band after illumination of the semiconductor with circularly polarized light, thus bypassing spin-injection via solid-state techniques. The temporal evolution of the induced magnetization can be measured by the detection of "quantum beats" in the intensity of the photoluminescence with positive and negative helicities. This, however, is only possible for undoped systems with a large electron-hole recombination rate. In doped semiconductors the spin precession is monitored with time-resolved Faraday or Kerr rotation spectroscopy. When an electric field is applied to the sample the locally created spin density drifts through the sample. Scanning the probe beam over the sample surface allows the study of the spatio-temporal evolution of the spin packet; thus decoherence effects can be investigated as a function of drift distance and velocity. In this section a brief introduction to ultra-fast spectroscopy techniques is given and some recent results on spin-coherence times and lengths are discussed. A good introductory article on this topic was written by Awschalom and Kikkawa [21].

In a typical ultrafast spectroscopy experiment the system under study is excited by a coherent pump beam and then tested by a probe beam applied to the sample after some time delay $t$. In the specific case of spin-polarized transport in semiconductors, a circularly polarized pump beam is directed onto the sample that excites electrons with well defined spin direction into the conduction band. These electrons precess around a magnetic field $B$ applied perpendicular to the path of the probe beam. The time evolution of the magnetization is studied with linearly polarized radiation in transmission using the Faraday effect or (in the case of opaque samples) in reflexion using the Kerr effect. The plane of polarization is rotated by an angle $\Theta$ proportional to the projection of the magnetization onto the path of the probe beam. This is illustrated in Fig. 17.2. Thus the time dependence of the Faraday (or Kerr) rotation angle directly reflects the behaviour of the magnetization.

These experiments are performed with mode-locked Ti:sapphire lasers that typically have pump and probe pulses of about 100 fs duration. The pump beam creates a coherent wave packet in the conduction band. The electron spins precess around the magnetic field with the Larmor frequency $g\mu_{\mathrm{B}}B/\hbar$ and gradually lose the spin coherence under the influence of spin-spin relaxation processes characterized by an intrinsic relaxation time $\tau_2$ as well as extrinsic processes such as field inhomogeneities leading to a smaller relaxation time conventionally called $\tau_2^*$. These extrinsic properties are often called dephasing. The magnetization

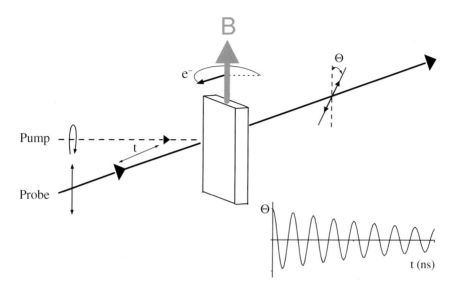

**Fig. 17.2.** This figure shows a schematic ultra-fast-spectroscopy setup. A circularly polarized pump beam excites electrons in the semiconductor with defined spin direction that precess around a magnetic field applied in the plane of the sample. The Faraday effect leads to the rotation of the linear polarization of the probe beam by an angle $\Theta$ proportional to the magnetization component parallel to the probe-beam path. The time evolution of the Faraday rotation angle is measured by varying the delay time between pump and probe; a typical variation is shown.

therefore performs an exponentially decaying oscillation:

$$M \propto \Theta \propto \exp\left[-t/\tau_2^*\right] \cos\left[g\mu_{\mathrm{B}}Bt/\hbar\right] . \tag{17.10}$$

Kikkawa *et al.* [22] and Kikkawa and Awschalom [23] investigated the spin relaxation in semiconductor quantum wells and bulk crystals using Kerr and Faraday rotation techniques, respectively. The quantum well was fabricated from a $Zn_{1-x}Cd_xSe$ II-VI semiconductor sandwiched between ZnSe layers grown on GaAs substrates. Electron-doped samples with typical sheet densities of $n_\square \sim 5 \times 10^{11}$ cm$^{-2}$, mobilities $\mu \sim 2700 - 7900$ cm$^2$/Vs and g factors $g \sim 1.1$ as well as an undoped sample were investigated. In this system the hole spins are fixed along the quantum well growth direction by the spin-orbit coupling; therefore, only the electron spins are precessing about the magnetic field. Kerr microscopy at 5 K and 4 T revealed the decay of the net magnetization in the doped quantum wells with two different lifetimes: at times shorter than 100 ps a fast decay was observed followed by a much slower decay over several nanoseconds. In contrast, the insulating undoped sample showed a rapid decay of the spin density within about 15 ps. The hole lifetime was determined to 50 ps by photoluminescence. The astonishingly large relaxation times in the doped samples were attributed to the absence of electron-hole recombination for times longer than 100 ps. The first rapid decay is due to this relaxation channel, but electrons from the conduction band Fermi sea neutralize most of the holes leaving a slowly dephasing spin density at longer times. This long time decay was found to be only weakly affected by temperature and even at room temperature clear "quantum beats" could be seen.

The spin relaxation measured in Si doped GaAs single crystals as a function of doping shows a similar behaviour. Kikkawa and Awschalom [23] studied an undoped crystal and crystals with dopings $n = 10^{16}$, $10^{18}$, $5 \times 10^{18}$ cm$^{-3}$. Faraday rotation spectroscopy was performed at low excitation densities of about $n_{ex} = 2 \times 10^{14}$, $1.4 \times 10^{15}$, $3 \times 10^{15}$ cm$^{-3}$ in case of the doped crystals and $n_{ex} = 10^{14}$ cm$^{-3}$ in case of the undoped crystal. The relaxation times were found to rise sharply at low dopings in agreement with the results on quantum wells; at higher dopings, however, a gradual decrease was detected. Kikkawa and Awschalom used "resonant spin amplification" in order to measure relaxation times of several nanoseconds. In this technique the time delay is fixed at some value and the spin precession is measured as a function of magnetic field. The repeated pulsing produces a spin-density amplification, if the magnetic field fulfils a resonance condition, thus producing a series of narrow resonance maxima in a magnetic field sweep. From the half-width the spin-relaxation time $\tau_2^*$ can be determined. At a doping of $10^{16}$ cm$^{-3}$ this is found to decrease with magnetic field, temperature and excitation density $n_{ex}$. At low temperatures and zero field, values of 100 ns were obtained.

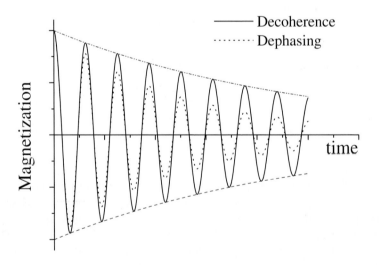

**Fig. 17.3.** Ultrafast spectroscopy measurements of the Faraday rotation angle in semiconductors reveal the Larmor precession of electrons around the applied magnetic field. The net magnetization decays due to spin-spin relaxation processes leading to a decoherence of the spin density as shown by the solid line. Other processes such as field inhomogeneities cause an additional dephasing indicated by the dotted line that is experimentally measured.

spin-packet amplitude

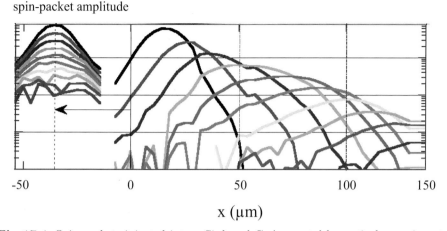

x (μm)

**Fig. 17.4.** Spin packets injected into a Si doped GaAs crystal by optical pumping at regular intervals of about 13 ns as a function of distance from the injection point. The spin density is subjected to an applied electric field of −37 V/m leading to a lateral drift. The temporal exponential decay causes a spatial decay along the drift direction. The spin transport over distances exceeding 100 μm, however, can clearly be observed. For comparison the zero bias spin packets are shown on the left, displaced by -36 μm for clarity. After Kikkawa and Awschalom [24].

These extraordinary long spin-relaxation times in electron-doped semiconductors form the basis for the investigation of spin-coherent transport. This was addressed in recent work by Kikkawa and Awschalom [24] as well as Hägele *et al.* [25]. In these experiments, the spin-coherent wave packets were dragged through the semiconductor by the application of an electric field and the spin-polarization was measured as a function of distance from the injection point. Kikkawa and Awschalom used a Si doped GaAs crystal with a carrier density of $10^{16}$ cm$^{-3}$ and applied an electric field parallel to the magnetic field. Scanning the probe beam over the sample with a resolution of 18 μm yielded data on the spatio-temporal evolution of the spin density. The spin-polarization at 5 K of ten wave packets dragged by an electric field of −37 V/m through the sample is shown as a function of distance from the injection point in Fig. 17.4. This illustration clearly shows that spin transport is observable over distances exceeding 100 μm. The temporal decay of the spin-polarization leads to a spatial decay along the drift direction. An analysis of the broadening of the spin packets yielded a spin-diffusion constant $D_s$ exceeding the charge-diffusion constant $D$ by more than one order of magnitude. This suggests that spin transport involves both electron diffusion as well as pure spin diffusion. The results of Hägele *et al.* [25] indicated no significant spin dephasing in GaAs at low temperatures for transport over 4 μm in electric fields up to 6 kV/m in qualitative agreement with Kikkawa's and Awschalom's results.

## 17.4    Spin–Injection

### 17.4.1    Ferromagnetic Metallic Electrodes

The studies of spin-injection using ferromagnetic metallic electrodes fall into two broad classes: investigations of a single ferromagnet/semiconductor interface or studies of the Datta and Das spin-transistor. The latter will be discussed in Sect. 17.6.1; here the results of Hirohata *et al.* [26] on a permalloy/GaAs interface and of Hammar *et al.* [27] on the interface resistance between permalloy and an InAs 2DEG are discussed.

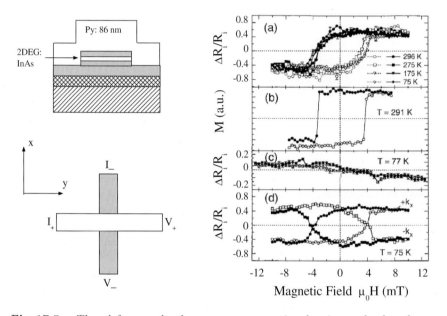

**Fig. 17.5.** The left panel shows a cross-sectional view of the ferromagnet/semiconductor device. An electron-doped InAs 2DEG is grown on a GaAs substrate buffered with a AlSb layer. After defining a mesa structure, the 2DEG is contacted with an e-beam deposited permalloy (Py: $Ni_{80}Fe_{20}$) layer. The interface resistance between the permalloy strip and the 2DEG is measured in the standard cross-strip geometry as indicated on the bottom left. In the right panel the response to a magnetic field is presented. (a) The interface resistance $R_i$ is shown as a function of magnetic field along $\hat{y}$ for a current along $\hat{x}$. $R_i$ is independent of temperature in the range between 75 K and 296 K. Panel (b) shows the magnetization in arbitrary units at room temperature as a function of magnetic field. The coercive field of 3.5 mT corresponds well with the magnetic field dependent change in the interface resistance. (c) The interface resistance proves independent of field when both current and field are applied along $\hat{x}$. (d) If the direction of the current is reversed from $\hat{x}$ to $-\hat{x}$, $R_i$ changes polarity. The magnetic field is again applied along $\pm\hat{y}$. After Hammar *et al.* [27].

Hammar *et al.* [27] investigated the interface resistance between a permalloy electrode and an InAs 2DEG. In general, spin-coherent transport between materials with different spin-polarization is thought to lead to a spin-coupled interface resistance induced by the spin-dependent chemical potentials [28]; therefore, the study of the interface resistance as a function of relative spin-polarization might provide clues to spin-injection. The spin-degeneracy in a 2DEG is lifted due to the Rashba effect [29]. The perpendidular electric field created by the band-gap mismatch in the quantum well transforms into a magnetic field acting on the spin in the rest frame of the charge carrier. The resultant spin-orbit Hamiltonian is given by

$$H_{\mathrm{so}} = \alpha \left[ \boldsymbol{\sigma} \times \boldsymbol{k} \right] \cdot \hat{z} \qquad (17.11)$$

where $\alpha$ denotes the spin-orbit coupling parameter, $\boldsymbol{\sigma}$ the Pauli spin-operator, $\boldsymbol{k}$ the wave vector and $\hat{z}$ a unit vector perpendicular to the 2DEG. This spin-orbit effect leads to a spin-polarization of the 2DEG that can be inferred from studies of Shubnikov-de Haas oscillations; for InAs/GaSb quantum wells a zero field splitting of about 3.5 meV was reported [30].

Hammar *et al.* [27] investigated InAs/GaSb quantum wells grown on a GaAs substrate by molecular beam epitaxy. The carrier density and mobility at 296 K (77 K) were determined from Hall effect measurements to $1.5 \times 10^{12}$ cm$^{-2}$ ($9.0 \times 10^{11}$ cm$^{-2}$) and 23500 cm$^2$/Vs (63500 cm$^2$/Vs), respectively. A mesa structure was patterned into the 2DEG using optical lithography and ion-beam milling. Subsequently a permalloy (Py: Ni$_{80}$Fe$_{20}$) layer was e-beam deposited. A schematic cross-sectional view of the structure is shown in Fig. 17.5 (left panel). The interface resistance between the 2DEG and the permalloy electrode was measured in a conventional cross-strip geometry as indicated in Fig. 17.5, bottom left. The interface resistance was in the range 20-110 $\Omega$ for temperatures between 75 K and 295 K. The response to a magnetic field is shown in Fig. 17.5, right panel; these measurements were performed in a regime of linear current-voltage characteristics. Panel (b) shows a magnetization hysteresis loop with the magnetic field applied along $\pm \hat{y}$ recorded at room temperature yielding a coercive field of 3.5 mT. The field dependent interface resistance $\Delta R_{\mathrm{i}}/R_{\mathrm{i}} = (R_{\mathrm{i}} - \langle R_{\mathrm{i}} \rangle)/\langle R_{\mathrm{i}} \rangle$ is shown in (a) with the current along the $+\hat{x}$ direction and the magnetic field along $\pm \hat{y}$. A distinctive, temperature independent change of the interface resistance is seen at the coercive field. In this field orientation, the spin-polarization within both the permalloy electrode and the 2DEG are along $\pm \hat{y}$. If the magnetic field is applied along $\pm \hat{x}$ leading to a spin-orientation in the ferromagnetic electrode along $\pm \hat{x}$, a null effect is seen (panel (c)). If the current direction is reversed, the spin-polarization in the 2DEG changes sign according to (17.11) and, correspondingly, the interface-resistance change is reversed as shown in panel (d). In conclusion, all these features are consistent with a spin-coupled interface resistance due to spin-injection from the permalloy into the 2DEG. The magnitude of the effect, however, is small about 0.9%. The null result shown in panel (c) seems to rule out spurious effects due to a Hall voltage induced by the stray fields of the electrodes.

The interpretation of the interface resistance is not straightforward, since it depends on a variety of parameters such as the spin-diffusion lengths in the 2DEG and the ferromagnet as well as the barrier transparency. The data reproduced here were interpreted by Hammar *et al.* within a model developed by Johnson [32]. This, however, might be flawed, since certain assumptions are in contradiction with transport theory, see Campbell and Fert [31]. If one adopts the general result that the interfacial magnetoresistance is proportional to the product of the spin-polarizations, $\Delta R_i/R_i \sim P_{Py}P_{2DEG}$, and using the known spin-polarization of permalloy ($P_{Py} = 0.4$) and the InAs 2DEG ($P_{2DEG} = 0.1$), a magnetoresistance $\Delta R_i/R_i \sim 0.04$ might be expected. The experimental value is clearly smaller indicating some spin-polarization loss at the interface.

Hirohata *et al.* [26] measured the photocurrent induced in a permalloy/GaAs (110) contact with respect to the current without illumination and as a function of the electrode magnetization [26]. Without illumination the current-voltage characteristics indicate Schottky barrier formation with a barrier height of about 0.7 eV. Subsequently, the contact was irradiated with circularly polarized light creating spin-polarized charge carriers with the spin direction along the contact normal. The photocurrent was measured as a function of bias voltage with the electrode magnetization parallel, $\Delta I_p$, and antiparallel, $\Delta I_{ap}$, to the carrier spin. For a forward bias exceeding the Schottky barrier, an increase in $\Delta I_p$ and a decrease in $\Delta I_{ap}$ was detected that might be related to spin injection. This corresponded to a relative change in the current of 1%.

In conclusion, spin-injection from ferromagnetic metallic electrodes into semiconductors has proven difficult. At present, effects due to spin-injection are small of the order of 1% at best. In some geometries this might be due to an impedance mismatch. If a ferromagnet/semiconductor contact is analyzed within the two-current model, the polarization of the majority ($I_\uparrow$) and minority ($I_\downarrow$) currents is given by

$$P_{semi} = \frac{I_\uparrow - I_\downarrow}{I_\uparrow + I_\downarrow} = \frac{R_\downarrow - R_\uparrow}{R_\downarrow + R_\uparrow + 4R_s}$$

$$= \frac{P}{1 + \frac{R_s}{R}(1 - P^2)} . \tag{17.12}$$

Here $R_\uparrow$ ($R_\downarrow$) denotes the majority (minority) resistance and $R$ the total resistance of the ferromagnet, $P$ the spin-polarization of the ferromagnet, $P_{semi}$ the spin-polarization of the ferromagnet-semiconductor structure and $R_s$ the resistance of the semiconductor that is assumed to be unpolarized. In typical geometries $R_s \gg R$, such that an impedance mismatch leads to a significant reduction of the spin-polarization in the whole circuit as compared to the ferromagnetic electrode. A more detailed analysis was performed by Schmidt *et al.* [33].

## 17.4.2   Magnetic Semiconductors

An alternative approach to spin-injection was suggested by Egues [34] and was experimentally studied by Oestreich *et al.* [35], Fiederling *et al.* [36] and Ohno *et al.* [37]. In this approach spin-injection is facilitated in a band-gap-matched semi-conductor structure containing a strongly paramagnetic component, a so-called spin aligner. At present three systems have been investigated, namely electron-doped $Cd_{0.98}Mn_{0.02}Te/CdTe$ [35], electron-doped $Be_xMn_yZn_{1-x-y}Se/AlGaAs$ [36] and hole-doped $Ga_{1-x}Mn_xAs/GaAs$ with $x = 0.045$ [37]. The Mn doped semiconductors are strongly paramagnetic such that electrons or holes passing through these can be easily aligned by an applied magnetic field. The spin-polarization of the injected charge carriers was detected by a polarization analysis of the photoluminescence. Since the spin-relaxation of holes is much faster than that of electrons due to valence band mixing [23,38], it is much more favourable to investigate electron-doped systems. Accordingly, the spin-polarization of the electron-doped systems introduced above is near 100% [35,36], whereas in the the GaMnAs system only a spin-polarization of about 1% was found [37]. In the following the results of Fiederling *et al.* are discussed in more detail, since this experiment was particularly clear and successful.

The heterostructures investigated by Fiederling *et al.* [36] are schematically shown in the inset to Fig. 17.6. A light-emitting diode (LED) consisting of an undoped GaAs layer sandwiched between n- and p-doped AlGaAs layers, was grown on a p-doped GaAs substrate. On this LED a spin-aligner consisting of a non-magnetic $Be_xMg_yZn_{(1-x-y)}Se$ capping layer and a strongly paramagnetic $Be_xMn_yZn_{(1-x-y)}Se$ layer were grown. BeMnZnSe is ideally suited for the growth on GaAs heterostructures, since it allows for high quality interfaces with a small conduction-band offset of about 100 meV. The n-doped BeMnZnSe layer orders antiferromagnetically at high doping values and is paramagnetic at low doping with a strongly enhanced g factor. This results in a large Zeeman splitting $\Delta E = g\mu_B B$ in an applied magnetic field; therefore electrons passing from the n-contact on the top through the BeMnZnSe layer are spin-aligned by this Zeeman splitting. The degree of spin-polarization depends on the magnetic layer thickness. Subsequently, spin-polarized electrons drift through the 100 nm thick n-AlGaAs layer and recombine with unpolarized holes supplied by the bottom p-contact in the undoped GaAs quantum well. Fiederling *et al.* [36] investigated four structures with varying thickness of the magnetic layer and found a high spin-polarization only in a device with a layer thickness of 300 nm. The degree of electron spin-polarization is inferred from the optical polarization of the photoluminescence. Let us assume that the injected electrons are fully polarized in the up-spin ($+1/2$) state. According to the selection rule for the magnetic quantum number, $\Delta m = \pm 1$, only two transitions are possible: a heavy hole transition (from $+1/2$ to $+3/2$) and a light hole transition (from $+1/2$ to $-1/2$). The radiation emitted in these transitions has opposite circular polarization. Since the matrix element for the heavy hole transition is by a factor of three larger than for the light hole transition, the resultant circular polarization of the emitted light can be related to the electron occupation numbers $n_\uparrow$, $n_\downarrow$ and the electronic

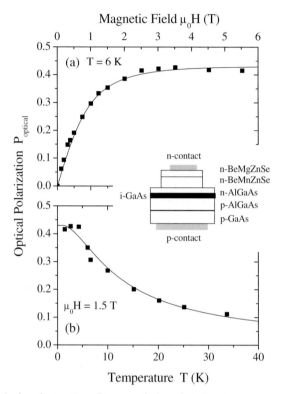

**Fig. 17.6.** Optical polarization $P_{\text{optical}}$ of the photoluminescence of an GaAs LED under injection of a spin-polarized current. The electrons injected into the LED pass through a paramagnetic quaternary II-VI semiconductor, $\text{Be}_x\text{Mn}_y\text{Zn}_{1-x-y}\text{Se}$ and are aligned with an applied magnetic field by the large Zeeman splitting. (a) Optical polarization at 6 K as a function of the applied magnetic field. (b) Optical polarization at 1.5 T as a function of the temperature. The solid lines are explained in the text. The inset shows a sketch of the structure. Electrons are injected from a n-contact through a non-magnetic BeMgZnSe layer and a magnetic BeMnZnSe layer into a GaAs LED. Through the p-contact at the bottom an unpolarized hole-current is supplied. After Fiederling *et al.* [36].

spin-polarization $P$ by

$$P_{\text{optical}} = \frac{(3n_\uparrow + n_\downarrow) - (3n_\downarrow + n_\uparrow)}{(3n_\uparrow + n_\downarrow) + (3n_\downarrow + n_\uparrow)} = \frac{1}{2}P. \tag{17.13}$$

Thus an optical polarization $P_{\text{optical}} = 0.5$ corresponds to fully spin-polarized electrons.

The results of Fiederling *et al.* [36] for the optical polarization are shown in Fig. 17.6. The top panel shows $P_{\text{optical}}$ as a function of the applied magnetic field at $T = 6$ K. Since $P_{\text{optical}}$ is proportional to the spin-polarization, the optical polarization might be expected to follow a Brillouin function

$$P_{\text{optical}} \propto B_J(x) = \frac{2J+1}{2J} \coth\left[\frac{2J+1}{2J} x\right] - \frac{1}{2J} \coth\left[\frac{x}{2J}\right] \qquad (17.14)$$

with $x = gJ\mu_B B/k_B T$. With a total spin $J = 5/2$ for the Mn ion, a satisfactory fit of (17.14) to the data can be made with a g factor $g = 6$, see solid line in Fig. 17.6a. The optical polarization decreases rapidly with increasing temperature as shown in the bottom panel. This decay is consistent with the temperature dependence of the Brillouin function as indicated by the solid line. A maximal value of $P_{\text{optical}} = 0.43$ is found at low temperatures and magnetic fields in excess of 3 T; this corresponds to a spin-polarization of the injected electrons of nearly 90% and shows the high efficiency of this heterostructure.

This demonstrator device works well at low temperatures far below room temperature and in large magnetic fields. This is due to the paramagnetic nature of the spin-aligner. An obvious remedy of this problem is the use of a ferromagnetic semiconductor as a spin-aligner. However, since Mn ions in Be-MnZnSe tend to couple antiferromagnetically at higher doping concentrations, it is not straightforward to fabricate a ferromagnet in this system. $Ga_{1-x}Mn_xAs$ shows a maximal Curie temperature of 110 K at a doping $x \sim 0.053$; it is intrinsically hole doped and therefore not a good choice, since the spin-relaxation time is quite small due to the large spin-orbit interaction. Magnetic semiconductors have been the object of many studies, see the review by Methfessel [8] and a brief account of recent trends by Ohno [7]. Now the race is on for an electron doped ferromagnetic semiconductor with a reasonable Curie temperature.

## 17.5   Spin–Detection

Spin-detection is the reverse process of spin-injection and therefore the techniques already discussed in this chapter might by employed. In the case of ferromagnetic metallic electrodes, at present both injection and detection efficiency are low. In the case of spin-injection using magnetic semiconductors, the spin-polarization was measured by optical techniques. This is infeasible for actual device performance; spin-detection using magnetic semiconductors has not been yet reported. Here a technique based on a ferromagnetic metal/semiconductor diode as recently discussed by Filipe et al. [39] is briefly reviewed.

Filipe et al. [39] fabricated an iron/GaAs diode by growing electron doped GaAs on a GaAs substrate and subsequently oxidizing this layer such that an oxide barrier of thickness 2 nm was formed. On this native oxide an iron film of thickness 3.5 nm was deposited and protected with a 5 nm Pd layer against oxidation. The magnetization of the Fe layer was found to be in-plane with a remanence of 90% of the saturation magnetization and a coercive field of 2 mT. In an ultra-high vacuum chamber this diode was irradiated with a spin-polarized electron beam emitted due to the photo-effect on GaAs using circular polarized light. The collector current through the GaAs was monitored as a function of the magnetization direction of the Fe layer and the energy of the incident electrons.

Electron injection was possible for energies above the work function of Pd that was found to be 4.8 eV. At an injection energy of 5.2 eV a considerable spin-filter effect of 20% was seen that decreases at higher energies. This hot electron effect is significantly higher than the detection efficiencies obtained by direct growth of metallic ferromagnets on the semiconductor.

## 17.6   Devices

### 17.6.1   The Datta and Das Transistor

A spin transistor based on a high electron mobility transistor (HEMT) structure that is analogous to an electro-optic modulator was proposed by Datta and Das [14]. Consider a 2DEG with two ferromagnetic electrodes for spin injection and detection. If the current direction is along $\hat{x}$ then let us assume that charge carriers with spin direction along $\hat{x}$ are injected into the 2DEG. While moving towards the collector electrode with wave vector $\boldsymbol{k}$, the carrier spins precess around a pseudomagnetic field $\boldsymbol{B}_{\mathrm{R}} = \alpha(\boldsymbol{k} \times \hat{z})$ due to the Rashba effect, see (17.11); $\hat{z}$ is directed along the surface normal. Since the Rashba precession arises from the electric field perpendicular to the 2DEG, it can be modulated by a gate electrode. Indeed, the spin-orbit coupling parameter is given by $\alpha = 2a_{46}E_z/\hbar g$, where $E_z$ denotes the electric field perpendicular to the 2DEG and $a_{46}$ is related to the band structure. Thus, in principle, a HEMT with control of the spin direction via the charging of a gate electrode and equipped with ferromagnetic electrodes for spin injection and detection can be constructed.

In the case of an ideal one-dimensional transistor, the pseudomagnetic field is along the $\hat{y}$ direction and the spin-polarization rotates in the xz-plane. The projection along the channel is given by $P(x) = P_0 \cos(E_z x/V_{\mathrm{R}})$, where $P_0$ stands for the spin-polarization at the injection point. $V_{\mathrm{R}}$ is a material constant, $V_{\mathrm{R}} = \hbar^2/(2m^* a_{46})$ of the order of a few volt. In the more realistic two-dimensional case, however, the wave vector component along $\hat{y}$ does not vanish and the pseudomagnetic field is oriented somewhere in the xy-plane. Since the carrier-wave vector is stochastically distributed after a scattering process, scattering events tend to randomize the precession vector and, consequently, lead to a vanishing spin-polarization for distances exceeding a mean free path. This issue was theoretically addressed by Bournel et al. [40] within a Monte-Carlo simulation of carrier motion in 2DEG channels with different widths. For a 2DEG channel of infinite width the spin-polarization was found to be strongly decreased. However, if the lateral size is restricted to values significantly below the channel length, the one-dimensional case is recovered. This simulation indicates that the fabrication of a Datta and Das spin-transistor should be possible in a suitably engineered design.

The dependence of the Rashba parameter $\alpha$ on a gate voltage was investigated by Nitta et al. [41] in an inverted $In_{0.53}Ga_{0.47}As/In_{0.52}Al_{0.48}As$ heterostructure and by Schäpers et al. [42] in $In_xGa_{1-x}As/InP$ quantum wells. Both groups report a dependence of $\alpha$ on the gate voltage; Nitta et al. [41] observed a variation of $\alpha$ by a factor of two for gate voltages in the range between $-1.5$ V and $+1.5$ V.

A first attempt to fabricate a demonstrator device based on the ideas of Datta and Das was reported by Cabbibo *et al.* [43]. Here a 2DEG formed in GaAs/AlGaAs and InGaAs/AlInAs heterostructures was contacted with sputter-deposited Fe contacts. Both heterostructures were n-doped with sheet densities of about $3 \times 10^{15}$ m$^{-2}$ and had mobilities at 77 K in excess of $10^5$ cm$^2$/Vs. Although the authors documented each fabrication step well using atomic force and scanning electron microscopy, the results were disappointing: the HEMT fabricated in GaAs proved insulating for voltages below 1 V and the device in InGaAs showed strongly non-linear current-voltage characteristics. These negative results might be related to Schottky-barrier formation and carrier trapping at defects introduced by the Fe sputtering. Meier and Matsuyama [44] reported the study of the micromagnetic properties of permalloy electrodes deposited on p-doped InAs crystals. This is the first step towards a spin-transistor fabricated in InAs; this might be a good choice of material, since InAs is known to form ohmic contacts with metals. Gardelis *et al.* [45] fabricated a spin-valve by contacting a 2DEG formed in an InAs quantum well with permalloy contacts; the 2DEG was n-doped with a sheet density of $6 \times 10^{15}$ m$^{-2}$ and a low temperature mobility of $5 \times 10^4$ cm$^2$/Vs. Different coercive fields as evidenced by the anisotropic magnetoresistance of the permalloy electrodes were achieved by shape anisotropy. Magnetoresistance measurements on the HEMT in magnetic fields applied parallel to the 2DEG showed small features of about 0.1% in magnitude near the coercive fields of the electrodes at a temperature of 0.3 K. These disappear at 10 K. Gardelis *et al.* [45] interpreted this small magnetoresistance as arising from two spin-valve effects, namely: an interfacial effect near the permalloy contacts and a second effect due to carrier motion between the contacts. Due to the smallness of the magnetoresistance, however, an interpretation in terms of spurious effects such as induced Hall voltages caused by the ferromagnetic electrodes cannot be ruled out.

In conclusion, a concept for a spin-transistor with gate control of the spin direction has been proposed. An experimental realization of this device has not been achieved so far, but many groups world-wide are currently working on HEMTs with magnetic electrodes.

### 17.6.2   The SPICE Transistor

A hybrid semiconductor/ferromagnetic metal structure, the SPICE (Spin Polarized Injection Current Emitter) Transistor, was designed and fabricated by Gregg *et al.* [46]. It behaves like a conventional transistor but its electrical parameters are tuneable by an external magnetic field. The device consists of a ferromagnetic material of low coercivity, between the base and collector, and a harder ferromagnetic material between the base and emitter. The hard layer acts to spin-polarize the current injected from emitter to base. These electrons diffuse through the base region and impinge on the back-biased collector barrier. The fraction which gets "collected" then depends on the magnetization of the soft ferromagnetic analyser layer which acts as a spin-dependent guard-rail on the base-collector potential drop. The common emitter current gain is thus magnetically variable and the overall device is a transistor with reasonable power gain and magnetically variable parameters. Initial prototypes show encouraging behaviour but their spin-dependence is small owing to interface scattering and the difficulties of spin injection between materials of very different conductivities [47].

In early versions of the spin transistor, which was first described in 1995, the ferromagnetic layers were laid down directly onto the doped silicon base layer, which resulted in the formation of metal silicides at the interface and these degraded the performance of the devices. This drawback has been removed by modifying the design of the transistor and interposing a thin tunnelling barrier between the emitter and base and also between base and collector in some cases. This reduces the formation of silicides and other contaminants at the interface, and has the added advantage that spin injection into the silicon is no longer subject to the constraints typical of direct contacts and may be targeted at particular parts of the silicon band structure chosen to optimize device sensitivity and gain.

Preliminary results on the tunnel variant of the SPICE transistor are promising, but further work is required to fully optimize and understand the behaviour of this device.

### 17.6.3   The Hot–Electron Spin–Valve

A hot-electron transistor as a magnetic field sensor was fabricated and investigated by Monsma et al. [48,49]. This device is based on a giant magnetoresistance (GMR) mulilayer used as metallic base electrode sandwiched between silicon emitter and collector, respectively. In some respect, this spin-valve transistor is the reverse of the heterostructures discussed so far, since the spin-dependent transport properties are manipulated in the metallic base, whereas the semiconducing electrodes are used for injection and collection. This spin-valve transistor shows a large magnetic field dependence of the collector current, much larger than the magnetoresistance of the GMR multilayer used as base. However, there is no current gain and this feature makes the spin-valve transistor less attractive for applications.

The spin-valve transistor was fabricated by sputtering a [Cu(2 nm)-/Co(1.5 nm)]$_4$ GMR multilayer on a freshly cleaned, n-doped Si substrate used as collector. The emitter is made from a n-doped Si wafer directly bonded by spontaneous adhesion to the multilayer. At the semiconductor-metal interfaces Schottky barriers form with typical barrier heights of 0.6-0.7 eV. The operation principle might be visualized as follows. If the transistor is operated in forward bias, electrons are accelerated in the emitter and injected as hot electrons with an energy about 1 eV above the Fermi energy into the base. The probability of electrons reaching the collector is limited by various scattering processes in the GMR multilayer that effectively cool the injected electrons. Electrons passing the base more or less ballistically reach the collector; typical values of the collector current for an emitter current of 100 mA are less than 1 μA, leading to a current gain below $10^{-5}$. The collector current, however, depends strongly

on spin-dependent scattering in the base: at 77 K a collector current of 0.1 μA
has been measured at the coercive field of the multilayer compared to a value of
0.5 μA for parallel magnetization orientation yielding a magnetocurrent change
of about 400%. This has to be compared with the modest current-in-plane
magnetoresistance of the Co/Cu-multilayer of about 3% [48]. The spin-valve
transistor might prove to be a valuable tool for fundamental research probing
spin-dependent transport of electrons with energies 0.2 to 3 eV above the Fermi
level.

## 17.7 Conclusions

The research area of spin dependent transport in semiconductors is very active and this review can only give a snapshot of the rapid development at this
particular moment in time. However, the basic ideas of spin-injection and transport in semiconductors have been reviewed and the relevant parameters and
length-scales were introduced. This review is focused on transport in magnetic
hybrid semiconductor-structures; superconductor/semiconductor interfaces are
discussed by Das Sarma *et al.* [50]. The first symposium on spin-electronics was
held in Halle, Germany, from 3. to 6. July 2000; this meeting showed the richness
of physical ideas involved in spin-dependent transport studies and demonstrated
the large potential of spin-electronic devices.

## Acknowledgement

The author is grateful to Dr. Martin Thornton, University of Oxford, for contributing the section on the SPICE transistor.

## References

1. G. A. Prinz, Science **282**, 1660 (1998).
2. J. de Boeck and G. Borghs, Physics World **12**, 27 (1999).
3. D. R. Scifres, B. A. Huberman, R. M. White, and R. S. Bauer, Solid State. Com-
mun. **13**, 1615 (1973).
4. A. G. Aronov and G. E. Pikus, Fiz. Tekh. Poluprovodn. **10**, 1177 (1976) [Sov.
Phys. Semicond. **10**, 698 (1976)].
5. S. F. Alvarado and P. Renaud, Phys. Rev. Lett. **9**, 1387 (1992).
6. M. W. J. Prins, H. van Kempen, H. van Leuken, R. A. de Groot, W. van Roy, and
J. de Boeck, J. Phys.: Condens. Matter **7** 9447 (1995).
7. H. Ohno, Science **281**, 951 (1998); J. Magn. Magn. Mater. **200**, 110 (1999).
8. S. Methfessel and D. C. Mattis, "Magnetic Semiconductors", in "Handbuch der
Physik", Vol. XVIII/1, (Eds.) S. Flügge and H. P. J. Wijn, (Springer-Verlag, Berlin,
1968) p. 389.
9. "Diluted Magnetic Semiconductors", in "Semiconductors and Semimetals", Vol. 25,
(Eds.) J. K. Furydna and J. Kossut, (Academic Press, New York, 1988).
10. M. Johnson and R. H. Silsbee, Phys. Rev. Lett. **55**, 1790 (1985).

11. M. Johnson, Phys. Rev. Lett. **70**, 2142 (1993).
12. M. Johnson, Appl. Phys. Lett. **65**, 1460 (1994).
13. A. Fert and S-F. Lee, Phys. Rev. B **53**, 6554 (1996).
14. S. Datta and B. Das, Appl. Phys. Lett. **56**, 665 (1990).
15. M. E. Flatté and J. M. Byers, Phys. Rev. Lett. **84**, 4220 (2000).
16. R. J. Elliot, Phys. Rev. **96**, 266 (1954).
17. Y. Yafet, "g Factors and Spin-Lattice Relaxation of Conduction Electrons", in "Solid State Physics", Vol. 14, (Eds.) F. Seitz and D. Turnbull, (Academic Press, New York, 1963), p. 1.
18. M. I. D'yakonov and V. I. Perel', Zh. Eksp. Teor. Fiz. **60**, 1954 (1971) [Sov. Phys. JETP **33**, 1053 (1971)].
19. G. L. Bir, A. G. Aronov, and G. E. Pikus, Zh. Eksp. Teor. Fiz. **69**, 1382 (1975) [Sov. Phys. JETP **42**, 705 (1976)].
20. J. Fabian and S. Das Sarma, Phys. Rev. Lett. **81**, 5624 (1998).
21. D. D. Awschalom and J. M. Kikkawa, Physics Today **June**, 33 (1999).
22. J. M. Kikkawa, I. P. Smorchkova, N. Samarth, and D. D. Awschalom, Science **277**, 1284 (1997).
23. J. M. Kikkawa and D. D. Awschalom, Phys. Rev. Lett. **80**, 4313 (1998).
24. J. M. Kikkawa and D. D. Awschalom, Nature (London) **397**, 139 (1999).
25. D. Hägele, M. Oestreich, W. W. Rühle, N. Nestle, and K. Eberl, Appl. Phys. Lett. **73**, 1580 (1998).
26. A. Hirohata, Y. B. Xu, C. M. Guertler, and J. A. C. Bland, J. Appl. Phys. **85**, 5804 (1999).
27. P. R. Hammar, B. R. Bennet, M. J. Yang, and M. Johnson, Phys. Rev. Lett. **83**, 203 (1999).
28. P. C. van Son, H. van Kempen, and P. Wyder, Phys. Rev. Lett. **58**, 2271 (1987).
29. Yu. A. Bychkov and E. I. Rashba, J. Phys. C: Solid State Phys. **17**, 6039 (1984).
30. J. Luo, H. Munekata, F. F. Fang, and P. J. Stiles, Phys. Rev. B **38**, 10142 (1988).
31. I. A. Campbell and A. Fert, "Transport Properties of Ferromagnets", in "Ferromagnetic Materials", Vol. 3, (Ed.) E. P. Wohlfarth, (1982, North-Holland Publishing Company, Amsterdam) p. 751.
32. M. Johnson, Phys. Rev. B **58**, 9635 (1998).
33. G. Schmidt, D. Ferrand, L. W. Molenkamp, A. T. Filip, and B. J. van Wees, Phys. Rev. B **62**, R4790 (2000).
34. J. C. Egues, Phys. Rev. Lett. **80**, 4578 (1998).
35. M. Oestreich, J. Hübner, D. Hägele, P. J. Klar, W. Heimbrodt, W. W. Rühle, D. E. Ashenford, and B. Lunn, Appl. Phys. Lett. **74**, 1251 (1999).
36. R. Fiederling, M. Keim, G. Reuscher, W. Ossau, G. Schmidt, A. Waag, and L. W. Molenkamp, Nature (London) **402**, 787 (1999).
37. Y. Ohno, D. K. Young, B. Beschoten, F. Matsukara, H. Ohno, and D. D. Awschalom, Nature (London) **402**, 790 (1999).
38. M. Oestreich, S. Hallstein, A. P. Heberle, K. Eberl, E. Bauser, and W. W. Rühle, Phys. Rev. B **53**, 7911 (1996).
39. A. Filipe, H-J. Drouhin, G. Lampel, Y. Lassailly, J. Nagle, J. Peretti, V. I. Safarov, and A. Schuhl, Phys. Rev. Lett. **80**, 2425 (1998).
40. A. Bournel, P. Dollfus, P. Bruno, and P. Hesto, Eur. Phys. J. AP **4**, 1 (1998).
41. J. Nitta, T. Akazaki, H. Takayanagi, and T. Enoki, Phys. Rev. Lett. **78**, 1335 (1997).
42. Th. Schäpers, G. Engels, J. Lange, Th. Klocke, M. Hollfelder, and H. Lüth J. Appl. Phys. **83** 4324 (1998).

43. A. Cabbibo, J. R. Childress, S. J. Pearton, F. Ren, and J. M. Kuo, J. Vac. Sci. Technol. A **15**, 1215 (1997).
44. G. Meier and T. Matsuyama, Appl. Phys. Lett. **76**, 1315 (2000).
45. S. Gardelis, C. G. Smith, C. H. W. Barnes, E. H. Linfield, and D. A. Ritchie, Phys. Rev. B **60** 7764 (1999).
46. J. F. Gregg, W. Allen, N. Viart, R. Kirschman, C. Sirisathitkul, J-P. Schille, M. Gester, S. Thompson, P. Sparks, V. Da Costa, K. Ounadjela, and M. Skvarla, J. Magn. Magn. Mater. **175** 1 (1997).
47. D. R. Loraine, D. I. Pugh, H. Jenniches, R. Kirschman, S. M. Thompson, W. Allen, C. Sirisathikul, and J. F. Gregg, J. Appl. Phys. **87**, 5161 (2000).
48. D. J. Monsma, J. C. Lodder, Th. J. A. Popma, and B. Dieny, Phys. Rev. Lett. **74**, 5260 (1995).
49. J. C. Lodder, D. J. Monsma, R. Vlutters, and T. Shimatsu, J. Magn. Magn. Mater. **198-199**, 119 (1999).
50. I. Žutić and S. Das Sarma, Phys. Rev. B **60**, R16322 (1999).

# 18 Circuit Theory for the Electrically Declined

J. F. Gregg and M. J. Thornton

Clarendon Laboratory, Oxford University, Parks Road, Oxford OX1 3PU, U.K.

## 18.1 The Soldering Iron and the Spin Electronician

Such is the sophistication of many contemporary University Physics Courses that their followers are at ease with the finer details of the Dirac equation and have no difficulty in thinking in a many-dimensioned Hilbert-space: however they are often less confident when faced with knowing which end of a soldering iron gets hot. Spin Electronics is above all a practical science which ultimately promises to implement a new and revolutionary technology in a form which will ultimately impact everyday existence. Card-carrying theoretical physicists doubtless have their part to play in this new and exciting field, but for the rapid and successful development of this science, the importance of practical knowledge and experimental dexterity is paramount. Those who would claim proficiency as Spin Electronicians must, above all, be capable of the simple, basic skills with which every TV repair engineer is acquainted. To those devotees of Spin Electronics whose degree courses have left you electrically deprived, this chapter is dedicated to you. Evidently, in the few pages available, only the surface of this topic may be scratched, but at least the basics can be laid, topics of major confusion like transistors and transformers can be treated and signposts pointed to further study.

## 18.2 Ohm's Law and Simple DC Circuits

Ohm's Law underpins all of electrical and electronic theory. On reflection it seems bizarre that, in a physical world where non-linearity seems to be the rule, not the exception, it is experimentally so easy to find systems which exhibit such a perfect linear relationship between current and voltage. The reason is simple. Components such as resistors are homogeneous: voltages applied to them are dropped uniformly over the entire structure and the local electric fields are small. Even if phenomena exist which invoke higher powers of electric field than the linear response, the smallness of the field implies that these non-linear effects are unobserved since higher power terms are vanishingly small. When inhomogeneous electrical devices are made, such as junction diodes where most of the voltage is dropped on a small region of the device, the non-linearities return with a vengeance – which is why we can use diodes for rectification purposes.

Ohm's Law, for all its simplicity, is a very capable tool. We will show later in this chapter that even quite complicated electronic circuits may be analyzed accurately using little more than Ohm's Law and a bit of common sense.

## 18.2.1   The Potential Divider

Ohm's Law affords us the tools to analyze a simple but useful circuit – the potential divider. Two resistors of values R and R* are connected as in Fig. 18.1. The question is what voltage appears at the output? The resistors pass current $I = V_{in}/(R + R^*)$. This current generates a voltage across $R^*$ given by $V_{out} = R^* I = V_{in} R^*/(R + R^*)$, so the input voltage has been divided in the ratio $V_{out}/V_{in} = R^*/(R + R^*)$.

   This circuit element is widely used in electronics wherever it is necessary to define a potential, for example for biasing devices. As discussed below, it has the disadvantage that its source impedance is high, i.e. the voltage drops if significant current is drawn from it.

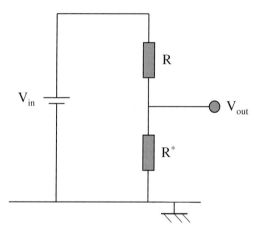

**Fig. 18.1.** Schematic diagram of a potential divider.

## 18.2.2   Voltage Sources

An ideal voltage source is one which maintains a given voltage between its two output terminal irrespective of the current which is drawn from it. For real voltage sources such as batteries, the voltage drops when a load current is drawn and this is modelled by treating such a source as an ideal voltage source in series with a source resistance, $R_s$, as shown in Fig. 18.2.

   Figures 18.3, 18.4 & 18.5 show examples of real voltage sources:

   Fig. 18.3 is a 12 volt car battery with a 0.01 Ω source impedance. When a starter current of 60 Amps is drawn, the battery output voltage thus drops to 11.4 Volts.

   Figure 18.4 is a potential divider consisting of a 9 Volt battery and divider resistors of value 6 kΩ and 3 kΩ respectively. The open-circuit (i.e. no-load) output voltage is 6 Volts. Having read the section below on Norton–Thevenin transforms, the reader will be able to deduce that the source impedance is 3 kΩ

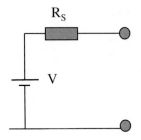

**Fig. 18.2.** Ideal voltage source in series with a source resistance R$_S$.

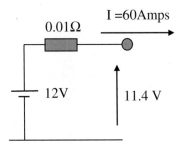

**Fig. 18.3.** Car battery.

in parallel with 6 kΩ (i.e. 2 kΩ) so if a load current of 1 milliamp is drawn, the output drops to 4 Volts.

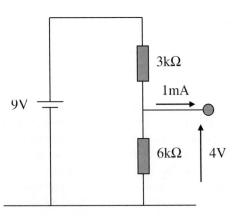

**Fig. 18.4.** Potential divider.

Figure 18.5 shows the use of a Zener diode to make a stable, low impedance voltage source. The characteristics of a typical Zener diode are shown in Fig. 18.6. The diode shown breaks down at approximately 6 Volts and is biased with a

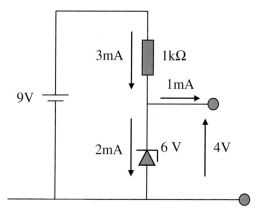

**Fig. 18.5.** Zener diode used to make a stable, low impedance voltage source.

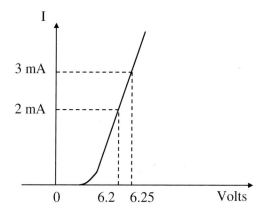

**Fig. 18.6.** Characteristics of a typical Zener diode.

current of about 3 mA from the 9 Volt battery and the 1 kΩ resistor. As shown from the characteristics, if a load current of 1 mA is drawn (leaving 2 mA for the diode) the voltage drops by only 0.05 Volts so the source impedance associated with the diode is only 50 Ω. The circuit may be modelled by treating the real Zener as an ideal diode (with the characteristics as shown in Fig. 18.7) in series with a resistance as in Fig. 18.8. The overall source impedance is now 50 Ω in parallel with the 1 kΩ bias resistor, i.e. 48 Ω which is a considerable improvement over the potential divider discussed above. Zeners are however more expensive than resistors and can generate more electrical noise, so are usually used only where voltage stability is crucial. Temperature compensated variants may be purchased for particularly critical applications.

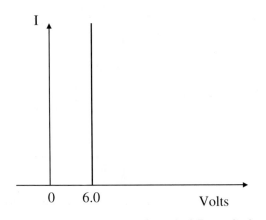

**Fig. 18.7.** Characteristics of an ideal Zener diode.

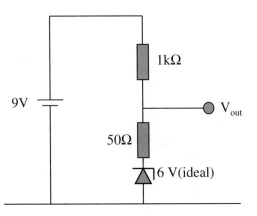

**Fig. 18.8.** Modelling a real Zener diode in the circuit of Fig. 18.5 in terms of the ideal Zener diode shown in Fig. 18.7.

### 18.2.3    Current Sources

An ideal current source is a more difficult concept to grasp, since, unlike voltage sources which are readily imitated by batteries, there is no simple electrical component which performs this function. An ideal current source outputs a constant current irrespective of the size of the resistor which is connected across its terminals. Clearly if the current source is open-circuited (i.e. the load resistor is disconnected), this is a tough call. To obey its job-description, the source must now increase the voltage across its output terminals until electrical breakdown of the intervening air occurs and the requisite current can be passed in the form of an electrical gas discharge! In practice, real current sources try their best by increasing the terminal voltage up to a certain limit, and then they give up. Two examples of circuits which function as current sources are shown in Fig. 18.9 together with their electrical characteristics. It may be seen that the current is

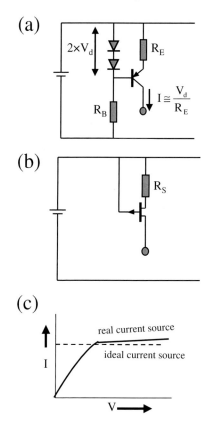

Fig. 18.9. Two examples of circuits (a) and (b) which function as an idealised current source and (c) their electrical characteristics.

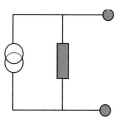

Fig. 18.10. A real current source may be modelled by as an idealised source with an admittance in parallel.

not wholly independent of voltage across the source. This may be modelled by representing the real source as an ideal source in parallel with an admittance as shown in Fig. 18.10.

## 18.3    Norton–Thevenin Transforms

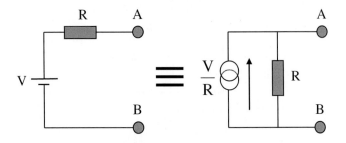

**Fig. 18.11.** Thevenin and Norton representations of the same circuit.

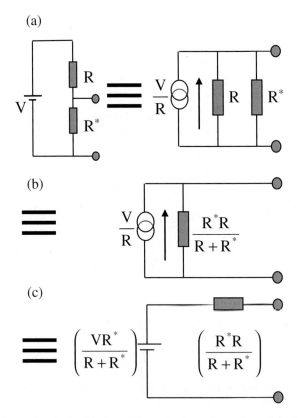

**Fig. 18.12.** Example of the Norton–Thevenin transformation. (a) Potential divider with its equivalent form reduced to its Norton equivalent (b) and (c).

There are two theorems ascribed to Norton and Thevenin whose upshot is that the two circuits shown in Fig. 18.11 involving respectively a voltage source and a current source are indistinguishable from the viewpoint of someone who has access only to the two terminals A and B. Clearly the open-circuit voltages of both circuits are identical – both are equal to $V$. If the terminals are short-circuited, the current that flows is in each case $V/R$. Indeed, no matter what experiment is conducted at the terminals A and B, it is impossible to distinguish between these two configurations. The Norton and Thevenin Theorems further imply that any combination of voltage sources, current sources and resistors which are connected by two wires to the rest of the world may be represented by either circuit with appropriate values. Moreover, such circuits may be analyzed rapidly and effectively by toggling between the Norton and Thevenin representations for selected fractions of the circuitry.

A simple example of the use of these theorems is to reduce the potential divider circuit of Fig. 18.12a to its Norton equivalent, as illustrated in Fig. 18.12b and c. It may easily be seen from the result that the source impedance of the potential divider is indeed $R_1$ and $R_2$ in parallel.

A more sophisticated example of the use of the Norton–Thevenin equivalents for circuit analysis is shown in the example below:

**Problem:** Use the method of repeated Norton–Thevenin transformations to determine the current $I$ in the following circuit.

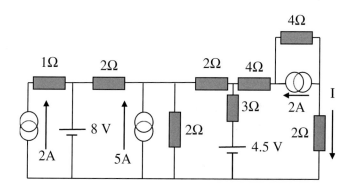

**Fig. 18.13.** Norton–Thevenin transformation problem.

**Solution:** The 4 V battery clamps the rails to 4 Volts so components to the left of it have no effect on $I$. The circuit transforms as follows:

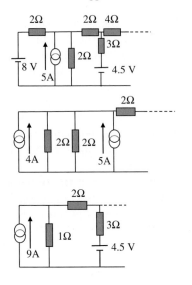

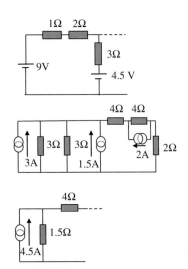

(a)                                                        (b)

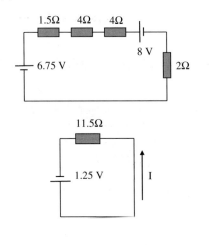

(c)

**Fig. 18.14.** Norton–Thevenin transformation solution.

Answer: $I = -\frac{1.25}{11.5}$ Amps

## 18.4    AC Circuit Theory

To DC currents and voltages, capacitors behave like open circuits and inductors look like lengths of wire with a bit of series resistance. For alternating currents however, the behaviour of a capacitor is described by (18.1):

$$CdV/dt = I \qquad (18.1)$$

Inductors obey the relation

$$V = LdI/dt \qquad (18.2)$$

For sinusiodal signals, the currents and voltages are thus an assortment of cos and sin functions with appropriate phases. This is messy. An elegant solution to this confusion lies in imagining that a sinusoidal signal (voltage or current) is the displacement viewed from the side of a mark on a rotating disk. In other words, there is a second hidden dimension which the eye does not see on the oscilloscope and the displacement in this dimension is also harmonic but displaced by 90 degrees. Equating this hidden dimension with the imaginary number axis and equating the real observed voltages and currents with the real number axis allows us to represent electrical signals in the form

$$V = V_0 \exp\left[j\omega t\right] \qquad (18.3)$$

and

$$I = I_0 \exp\left[j\omega t\right] \qquad (18.4)$$

Substituting into (18.1) and (18.2) produces the highly satisfactory result that inductors and capacitors can be shoe-horned into an Ohm's Law lookalike where complex impedances $Z_L = j\omega L$ and $Z_C = 1/j\omega C$ replace real resistance.

This is the basis of AC theory. The complex impedances of these components add in series and their corresponding admittances add in parallel. They are generally susceptible to all the same manipulations that we have seen performed with simple resistors.

### 18.4.1   Transfer Functions

The transfer function of a circuit is simply $V_{\text{out}}/V_{\text{in}}$ in complex notation such that the amplitude and phase information about the relationship between input and output are preserved. As a simple example, consider the following circuit:

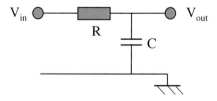
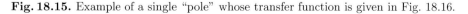

**Fig. 18.15.** Example of a single "pole" whose transfer function is given in Fig. 18.16.

$$\frac{V_{\text{out}}}{V_{\text{in}}} = \frac{1/j\omega C}{(R + 1/j\omega C)} = \frac{1}{1 + j\omega CR} \tag{18.5}$$

On a polar plot as a function of increasing frequency, the transfer function looks like this:

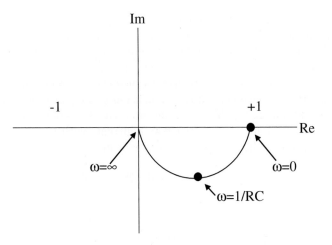

**Fig. 18.16.** Polar plot of the transfer function of Fig. 18.15.

As frequency increases, the amplitude gets smaller and the phase angle asymptotically approaches $-90°$. At a phase angle of $-45°$, the frequency is $\omega = 1/RC$ and the amplitude is $1/\sqrt{2}$ of the DC value.

A function of this form is know as a "pole" since it has a singularity on the complex frequency plane at $\omega = j/CR$. Such poles are mainly of interest to mathematicians since they may be used in contour integration to evaluate

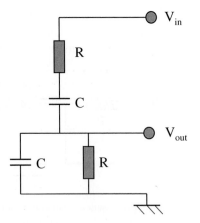

**Fig. 18.17.** Wien bridge circuit.

interesting and inaccessible definite integrals. For the purposes of electronics they are significant because they determine the good behaviour (or otherwise!) of amplifiers with feedback as we shall discuss below.

Another interesting circuit is the Wien bridge shown in Fig. 18.17. It is left to the reader to show (by calculating the transfer function $V_{out}/V_{in}$) that the output of this circuit is in phase with the input at a frequency $\omega = 1/RC$ and that the modulus of the transfer function is then $1/3$.

## 18.4.2   Norton–Thevenin Transforms Applied to AC Theory

Everything we discussed previously about Norton–Thevenin transformations applies equally to AC circuits. AC voltage sources replace batteries and complex impedances replace resistances. The following example shows the procedure for analyzing AC circuits using this method.

**Problem:** Use the method of repeated Norton–Thevenin transformations to determine the current $I$ in the following circuit.

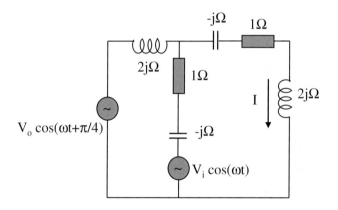

$V_o = V_i = 50$ Volts

**Fig. 18.18.** Norton–Thevenin AC circuit problem.

**Solution:** The circuit transforms as follows:

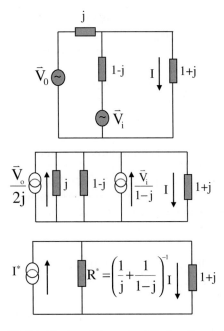

**Fig. 18.19.** Norton–Thevenin AC circuit solution.

$$I^* = \frac{\vec{V_o}}{j} + \frac{\vec{V_i}}{1-j} = \frac{50\,(1+j)}{j\sqrt{2}} + \frac{50}{1-j}$$

$$= \frac{100\,(1-j)}{2\sqrt{2}} + \frac{50\,(1+j)}{2}$$

$$\simeq 35\,(1-j) + 25\,(1+j) = (60 - 10j)\ \text{Amps}$$

also

$$R^* = \left(-j + \frac{1+j}{2}\right)^{-1} = \frac{2}{1-j} = 1 + j$$

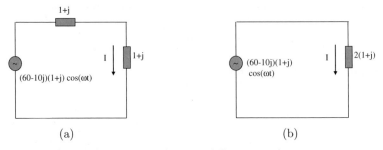

(a)                                                              (b)

**Fig. 18.20.** Norton–Thevenin AC circuit solution.

$$I = \frac{(60 - 10\mathrm{j})(1 + \mathrm{j})}{1 + \mathrm{j}} = (30 - 5\mathrm{j})\cos(\omega t)\ \text{Amps}$$

$$= \sqrt{925}\cos(\omega t + \phi)\ \text{Amps where}\ \phi = -\arctan\left(\frac{1}{6}\right)$$

## 18.5   Impedance Transformation

Impedance transformers are used in various ways: stepping down alternating voltages for power supplies, matching source and load impedances to maximize power transfer, and modifying effective source impedance to obtain the best noise performance from an amplifier. Transformers are also used for electrical isolation to improve personnel safety.

Two impedance transformation methods are in common use. The most usual is the two-coil transformer which is easy to use but bulky and expensive. Other methods include the use of "pi" and "tee" transformation networks or transmission lines with stub-matching.

### 18.5.1   The Transformer and its Uses

To understand the transformer, consider two coils situated close together such that the same magnetic flux goes through each (Fig. 18.21). The coils have $N_1$ and $N_2$ turns respectively and cross-sectional area $A$. The magnetic field seen by the coils is due to the currents $I_1$ and $I_2$ in each and is given by $B = C(N_1 I_1 + N_2 I_2)$ where $C$ is a geometrical constant common to both coils. Hence the flux through the coils is respectively:

$$\Phi_1 = N_1 A B = C A (N_1^2 I_1 + N_1 N_2 I_2) = L_1 I_1 + M I_2 \tag{18.6}$$

and

$$\Phi_2 = N_2 A B = C A (N_1 N_2 I_1 + N_2^2 I_2) = L_2 I_2 + M I_1 \tag{18.7}$$

where $L_1$, $L_2$ and $M$ are respectively the self and mutual inductances. Two things are apparent: firstly the mutual inductance between coil 1 and coil 2 is the same as between coil 2 and coil 1; secondly (and this arises because we assumed perfect coupling between the coils, i.e. all flux lines pass through both of them) $L_1 L_2 = M^2$. For imperfectly coupled coils $M^2 = k^2 L_1 L_2$ where $k < 1$. Differentiating (18.6) and (18.7) gives,

$$V_1 = L_1 \frac{dI_1}{dt} + M \frac{dI_2}{dt} = \mathrm{j}\omega L_1 I_1 + \mathrm{j}\omega M I_2 \tag{18.8}$$

$$V_2 = L_2 \frac{dI_2}{dt} + M \frac{dI_1}{dt} = \mathrm{j}\omega L_2 I_2 + \mathrm{j}\omega M I_1 \tag{18.9}$$

These equations are generally valid, even if the transformer coils are imperfectly coupled. A number of interesting results are apparent from (18.8) and (18.9).

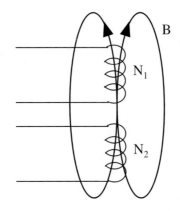

**Fig. 18.21.** Schematic diagram of a transformer.

For example, if the coils are perfectly coupled then $L_1 L_2 = M^2$ and by dividing (18.8) by (18.9) we have $V_2/V_1 = N_2/N_1$; in other words the voltage is transformed by a numerical factor equal to the turns ratio.

Likewise we may rewrite (18.8) as

$$I_1 = V_1/\mathrm{j}\omega L_1 - (M/L_1)I_2 . \tag{18.10}$$

Hence we can see that the primary current consists of two parts: a parasitic current (called the excitation current or magnetizing current) $V_1/\mathrm{j}\omega L_1$ which is present even when there is no load on the secondary and which corresponds to connecting an inductor of size $L_1$ across voltage source $V_1$; and a current proportional to $I_2$ but smaller by the ratio $M/L_2$, which in the case of perfect coupling is just $N_2/N_1$.

So leaving aside the excitation current, a perfectly coupled transformer steps down the voltage and steps up the current by the factor $N_2/N_1$; a load impedance viewed from the primary side then "looks" $(N_2/N_1)$ times larger than it really is. However, since the current/voltage are stepped up/down by the same factor, the powers entering the primary and leaving the secondary are the same (ignoring transformer losses) and energy is conserved – which is why a transformer cannot be used as a replacement for a power amplifier.

A numerical example is shown in Fig. 18.22a. The transformer has 1000 primary turns and 100 secondary turns and is fed with 240 Volts. A 24 $\Omega$ load is connected to the secondary. The secondary voltage is then 24 Volts (i.e. a step-down factor of $1000/100 = 10$). The secondary current is 1 amp and so the primary current is 100 mA (again stepped down by a factor 10) plus the excitation current. Suppose the transformer and its load are in a black box and the observer has access only to the primary terminals. When the load is connected or disconnected the primary current changes by 100 mA, and given a driving voltage of 240 Volts, this appears to the observer to represent a load of 2400 Ohms. Thus it is said that the load "reflected to the primary" has a value of 2400 Ohms and the equivalent circuit may be redrawn as in Fig. 18.22b.

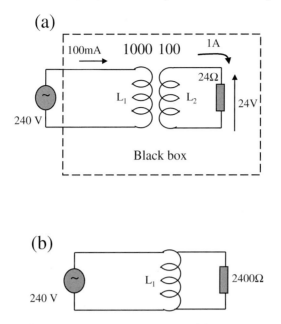

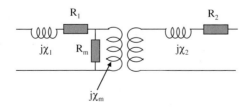

**Fig. 18.22.** (a) Example of transformer with (b) its equivalent circuit.

### 18.5.2   Real (i.e. Imperfect) Transformers

Real transformers are a bit more complex. For a start they are imperfectly coupled so the voltage and current ratios are not as clean as described above. Moreover there are power losses owing to the resistance in the copper windings of the primary and secondary and also due to magnetic hysteresis in the iron core and to eddy current heating (which may be reduced by laminating the core). For a real transformer the equivalent circuit therefore looks as in Fig. 18.23 where $j\chi_m$ is the primary reactance, $R_m$ represents the core losses, $R_1$ and $R_2$ represent the copper losses, and $j\chi_1$ and $j\chi_2$ are the stray reactances due to imperfect coupling.

**Fig. 18.23.** Example of a real transformer.

## 18.6   The Ideal Operational Amplifier

Operational amplifiers are black boxes with two inputs, $V^+$ and $V^-$, and an output, $V_{\text{out}}$. The output voltage is the voltage difference between the two inputs multiplied by a very large number like $10^8$. This large number is the Open Loop Gain of the OPA, usually denoted by A. The OPA is supplied with power rails usually of +/-15 Volts, outside of which the output cannot go. It follows that if the output is to be a sensible voltage and not squashed against one or other power rail, then the voltage difference between the inputs must be minute. For $A = 10^8$ and $V_{\text{out}} = 1$ Volt, $V^+ - V^- = 10$ nanovolts, which to most multimeters is unmeasurable. This suggests a very simple way to analyse OPA circuits to high accuracy: if the OPA is working properly, the voltages $V^+$ and $V^-$ are to all intents and purposes identical.

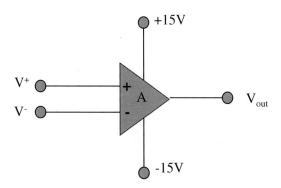

**Fig. 18.24.** Schematic diagram of an operational amplifier.

We now examine the two main configurations is which an OPA is used to amplify signal; these are illustrated in Fig. 18.25.

In Fig. 18.25a, the signal is fed to the inverting input $V^-$ and the noninverting input $V^+$ is connected to earth. It follows that the OPA will adjust its output to make sure that $V^-$ also is held at earth potential and any current which flows from $V_{\text{in}}$ to $V^-$ through $R_1$ gets sucked through $R_2$ by virtue of $V_{\text{out}}$ being held at just the right negative potential to remove the charge from $V^-$ as fast as it pours in. $R_1$ and $R_2$ pass the same current and Ohm's law then gives,

$$I = V_{\text{in}}/R_1 = -V_{\text{out}}/R_2 \tag{18.11}$$

and hence the gain of the circuit is

$$V_{\text{out}}/V_{\text{in}} = -R_2/R_1 \tag{18.12}$$

This circuit introduces the concept of a virtual earth – a point in the circuit which, although it is not actually connected to ground is nonetheless held at ground potential by the action of the circuitry.

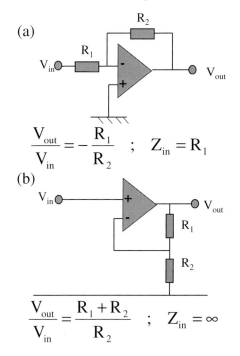

(a)

$$\frac{V_{out}}{V_{in}} = -\frac{R_1}{R_2} \quad ; \quad Z_{in} = R_1$$

(b)

$$\frac{V_{out}}{V_{in}} = \frac{R_1 + R_2}{R_2} \quad ; \quad Z_{in} = \infty$$

**Fig. 18.25.** Schematic diagrams of operational amplifiers with (a) parallel and (b) series feedback.

In passing we should note that the input impedance of this circuit is $R_1$ since $V^-$ is a virtual earth.

In the circuit in Fig. 18.25b, the signal is fed to the noninverting input and the inverting input is held at a fraction $R_2/(R_1 + R_2)$ of the output voltage determined by a potential divider with resistor values $R_1$ and $R_2$. As before, the OPA sets its output voltage to make sure that

$$V^- = V_{out} R_2/(R_1 + R_2) = V^+ = V_{in} \tag{18.13}$$

from which the voltage gain is found to be

$$V_{out}/V_{in} = (R_1 + R_2)/R_2 \tag{18.14}$$

Not only is the gain of this circuit positive, but the input impedance is that of the OPA itself and can be very large indeed – hundreds of megaohms in the case of an FET OPA. Unlike the first circuit the gain cannot be less than unity.

These circuits introduce the concept of negative feedback and serve to illustrate the difference between the two main types.

The first circuit is an example of parallel feedback where a current proportional to the output voltage is fed back via a path (i.e. through $R_2$) parallel to the amplifier to cancel the effect of the input current arriving via $R_1$.

In the second case, part of the output voltage is fed back to cancel the effect of the input voltage. This is achieved by arranging that the OPA and the fed back voltage are in series between the input voltage (at $V^+$) and ground. This is called series feedback.

In some instances it is desired to amplify the difference between two signals. An example might be the output of a tape recorder head where both the live and the earth of the leads coming from the tape head are contaminated with the same noise pickup. It is desired to reject the contamination which is common to both leads (the so-called common-mode signal) and amplify just the difference between the leads (the differential mode signal). The circuit shown in Fig. 18.26 performs this task. The example shows how to calculate the values of the various resistors. Note that since $V^+$ and $V^-$ for the OPA are always at the same potential, making $R_1$ and $R_3$ the same value will ensure that the positive and negative inputs, $V_{noni}$ and $V_i$, of the whole circuit have the same input impedance.

**Problem** Calculate the output $V_{out}$ of the opamp illustrated in terms of its inputs $V_i$ and $V_{noni}$. Assume its open-loop gain is large. Find the condition for which the common-mode gain is zero.

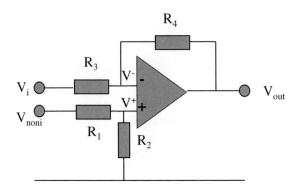

Fig. 18.26. Differential OPA circuit.

**Solution** If the open loop gain is large then $V^+$ and $V^-$ are effectively at the same potential at all times. So

$$V^+ = \frac{R_2}{R_1 + R_2} V_{noni} = V^-$$

$$V^- = V_{out} + \frac{R_4}{R_3 + R_4}(V_i - V_{out})$$

$$= \frac{R_3}{R_3 + R_4}V_{out} + \frac{R_4}{R_3 + R_4}V_i$$

So

$$\frac{R_3}{R_3 + R_4}V_{out} = \frac{R_2}{R_1 + R_2}V_{noni} - \frac{R_4}{R_3 + R_4}V_i$$

which we rewrite as

$$\alpha V_{\text{out}} = \beta V_{noni} - \gamma V_i$$

Defining

$$V_{cm} = \frac{V_{noni} + V_i}{2}$$

and

$$V_{dm} = V_{noni} - V_i$$

implies that

$$V_{noni} = V_{dm} + 2V_{cm}/2 \quad \text{and} \quad V_i = 2V_{cm} - V_{dm}/2$$

where $V_{cm}$ is the common mode voltage and $V_{dm}$ is the differential voltage.

$$\text{Thus} \quad \alpha V_{\text{out}} = \beta \left( \frac{V_{dm}}{2} + V_{cm} \right) - \gamma \left( V_{cm} - \frac{V_{dm}}{2} \right)$$

$$= \frac{\beta + \gamma}{2} V_{dm} + (\beta - \gamma) V_{cm}$$

so if

$$\beta = \gamma \quad \text{,i.e.} \quad \frac{R_2}{R_1} = \frac{R_4}{R_3}$$

the common mode gain is zero and the differential mode gain is

$$\frac{\beta + \gamma}{2\alpha} = \frac{\beta}{\alpha}$$

$$= \frac{R_2}{R_1 + R_2} \times \frac{R_3 + R_4}{R_3} = \frac{R_2}{R_1} = \frac{R_4}{R_3} .$$

### 18.6.1   Closed Loop Gain vs Open Loop Gain

Recall that the Open Loop gain of a typical OPA is of order $10^8$ while a typical closed loop gain is of order 100. The factor of $10^6$ between these two is called the Gain Reserve – in other words it's how much gain is waiting in the wings unused. Actually to say that it is unused is misleading - it is actually working hard behind the scenes suppressing the inherent nonlinearities of the OPA. If the OPA were for some reason to be used to give its full gain of $10^8$, the output would be found to distort badly. The reserve gain of $10^6$ is working to reduce any such distortions by approximately $10^6$ so that the observed output with a closed loop gain of 100 is very presentable. This is an example of one of the various used to which feedback may be put.

**Differential Input Amplifier with High Input Impedance** The differential amplifier circuit discussed above has one Achilles heel – its finite input resistance which is determined by the input resistors $R_1$ and $R_3$. The circuit in Fig. 18.27 overcomes this. It is left to the reader to show that the gain of this circuit is $(2R + R^*)/R^*$.

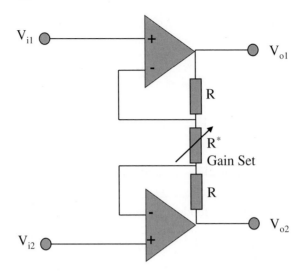

**Fig. 18.27.** Differential input amplifier circuit.

## 18.7    Transistors – How to Choose a Good One.

Transistors come in two basic flavours, bipolar transistors and field-effect transistors (of which we shall discuss just one variant – the JFET).

A bipolar transistor is a current amplifier. Its feature of merit is its current gain $\beta = I_C/I_B$. There are three terminals – emitter, base and collector – and in the NPN version illustrated in Fig. 18.28a the device is biased by holding the collector several volts positive relative to the emitter and injecting a current into the base. In practice, injecting $I_B$ requires the base to be biased about 0.6-0.7 Volts positive relative to the emitter since the base-emitter junction is just a junction diode. When this base current is injected, a collector current is pulled into the transistor which is $\beta$ times larger. Typical values of $\beta$ for a

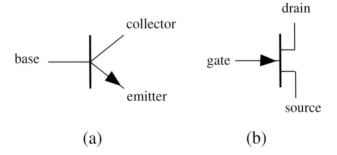

(a)                                 (b)

**Fig. 18.28.** Transistor symbols for: (a) bipolar NPN transistor and (b) N-channel JFET.

discrete transistor are about 100 or bigger (perhaps a bit less – say 20 – for power devices).

Figures 18.29 and 18.30 outline the most important behavioural characteristics of a typical NPN bipolar transistor.

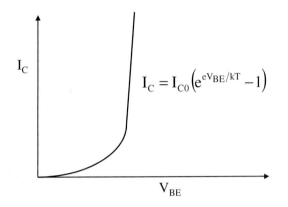

$$I_C = I_{C0}\left(e^{eV_{BE}/kT} - 1\right)$$

**Fig. 18.29.** Characteristics of a typical NPN bipolar transistor.

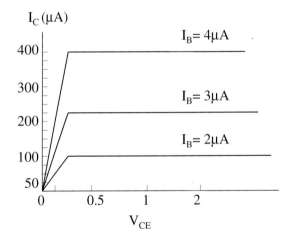

**Fig. 18.30.** Characteristics of a typical NPN bipolar transistor.

A FET has three terminals: a "drain" and a "source" between which is the "channel" in which a current flows ($I_D$) and a "gate" (see Fig. 18.28b) whose voltage relative to the source, $V_{GS}$, controls $I_D$.

To understand the operation of a FET, consider a capacitor consisting of two metal plates separated by an air gap. The resistance of one plate is measured between 2 opposite edges and its proportional to $1/\sigma$, where $\sigma = ne^2\tau/m^*$ is

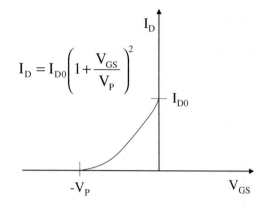

$$I_D = I_{D0}\left(1 + \frac{V_{GS}}{V_P}\right)^2$$

**Fig. 18.31.** Phase space of a FET plotting $V_{GS}$ against $I_D$.

the conductivity of the material which is proportional to $n$, the number of free electrons. Now charge the capacitor so that the plate whose resistance is being measured carries the positive charge. The electron density in this plate has decreased and hence the resistance of the plate (the "channel") has increased as a function of the voltage at which the other plate (the "gate") is held. This is the principle of the FET, the difference being that real FETs are made from semiconductor (not metal) channels. Since $n$ is so high in a metal, the metal channel can never be completely depleted of electrons (pinched-off) by the modest applied gate voltages in typical electronic circuits.

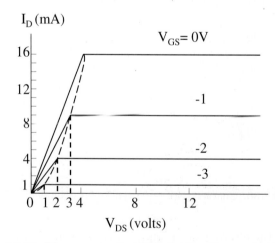

**Fig. 18.32.** Phase space of a FET plotting $V_{DS}$ against $I_D$.

The "phase space" of a FET is 3-dimensional with axes $V_{GS}$, $I_D$ and $V_{DS}$. Figures 18.31 and 18.32 show projections of this surface. Figure 18.31 is described by (18.15):

$$I_D = I_{D0}(1 + V_{GS}/V_P)^2 \tag{18.15}$$

where $I_{D0}$ and $V_P$ are parameters characteristic of a particular device.

## 18.8   Small Signal Analysis Using Differential Calculus – the Physicist's Approach

The amplifiers which we discuss in this chapter are all for use with small signals, i.e. AC amplitudes which are much smaller than the DC values of voltage and current which bias the circuit components. The physicists approach to analyzing these systems is to treat the signals as differential voltages and currents and use the principles of calculus. The method is best explained by illustration and there follows an example of this technique used to extract the input and output impedances of the common collector transistor configurations.

### 18.8.1   Common Collector

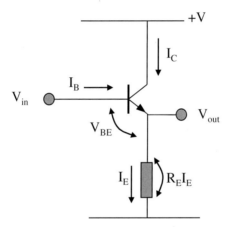

**Fig. 18.33.** Common collector circuit.

$$Z_{\text{in}} = \left(\frac{\partial V_{\text{in}}}{\partial I_B}\right)_{V_{CE}} \tag{18.16}$$

$$= \frac{\partial\,(V_{BE} + R_E I_E)}{\partial\left(\frac{I_C}{\beta}\right)} \tag{18.17}$$

$$= \beta\left\{\frac{\partial V_{BE}}{\partial I_C} + R_E\frac{\partial I_E}{\partial I_C}\right\} \tag{18.18}$$

$$\simeq \beta\left\{\frac{1}{g_m} + R_E\right\} \tag{18.19}$$

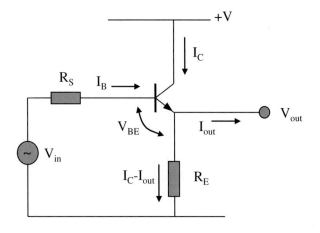

**Fig. 18.34.**

$$Z_{\text{out}} = -\left(\frac{\partial V_{\text{out}}}{\partial I_{\text{out}}}\right)_{V_{\text{in}}} = \frac{1}{Y_{\text{out}}} \tag{18.20}$$

$$V_{\text{out}} = R_E(I_C - I_{\text{out}}) \tag{18.21}$$

$$\Rightarrow I_{\text{out}} = I_C - \frac{V_{\text{out}}}{R_E} \tag{18.22}$$

$$\Rightarrow Y_{\text{out}} = -\left(\frac{\partial I_{\text{out}}}{\partial V_{\text{out}}}\right)_{V_{\text{in}}} = -\left(\frac{\partial I_C}{\partial V_{\text{out}}}\right)_{V_{\text{in}}} + \frac{I}{R_E} \tag{18.23}$$

but

$$V_{\text{out}} = V_{\text{in}} - R_S I_B - V_{BE} \tag{18.24}$$

$$\Rightarrow -\left(\frac{\partial V_{\text{out}}}{\partial I_C}\right)_{V_{\text{in}}} = -R_S\left(\frac{\partial I_B}{\partial I_C}\right)_{V_{\text{in}}} - \left(\frac{\partial V_{BE}}{\partial I_C}\right)_{V_{\text{in}}} \tag{18.25}$$

$$= -\left(\frac{R_S}{\beta} + \frac{1}{g_m}\right) \tag{18.26}$$

substituting for $\left(\frac{\partial I_C}{\partial V_{\text{out}}}\right)_{V_{\text{in}}}$ from (18.25) into (18.23) gives

$$\Rightarrow Y_{\text{out}} = \frac{1}{R_S/\beta + 1/g_m} + \frac{1}{R_E} \tag{18.27}$$

which means $Z_{\text{out}} = 1/Y_{\text{out}}$ appears as in Fig. 18.35.

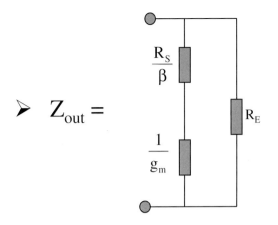

**Fig. 18.35.** Output impedance of the common collector transistor.

## 18.9   Equivalent Circuits – the Engineer's Approach

The engineering approach to analyzing transistor circuits involves constructing the AC equivalent circuit. To AC currents, power rails look like earth points (provided they are properly decoupled) since they no more wave up and down in voltage when a current is injected than the earth contact. In equivalent circuits resistors and capacitors feature unaltered, while transistors and FETs are represented by networks of impedances and current sources characterized by sets of interlocking parameters. The main substitution needed to create an equivalent circuit is that for the npn transistor. As shown in Fig. 18.36 the transistor symbol is excised and replaced by the network illustrated. Figure 18.37 shows the common emitter circuit and its equivalent circuit arrived at by this method

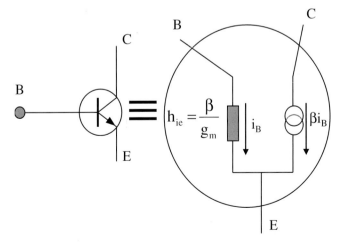

**Fig. 18.36.** FET and its equivalent circuit.

(see Fig. 18.38), together with the input and output impedances and the gain arrived at by this method of analysis. Figures 18.39 and 18.40 shows the equivalent circuit technique applied to analyzing the common collector configuration.

### 18.9.1   Common Emitter Equivalent Circuit

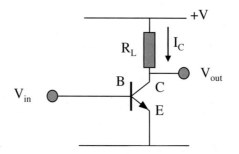

**Fig. 18.37.** Common emitter circuit.

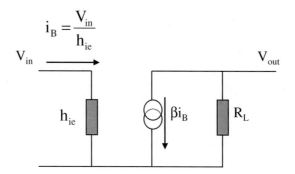

**Fig. 18.38.** Equivalent circuit of the common emitter.

$$Z_{\text{in}} = h_{ie} = \frac{\beta}{g_m} \tag{18.28}$$

$$\frac{V_{\text{out}}}{V_{\text{in}}} = -\beta i_B R_L = -\frac{V_{\text{in}}}{\beta/g_m} R_L = -g_m R_L \tag{18.29}$$

so

$$Z_{\text{out}} = R_L \tag{18.30}$$

### 18.9.2   Common Collector Equivalent Circuit

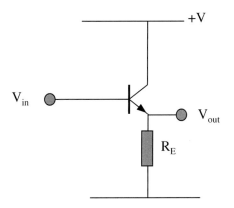

**Fig. 18.39.** Common collector circuit.

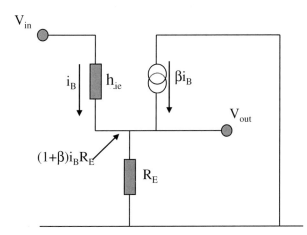

**Fig. 18.40.** Equivalent circuit of the common collector.

$$V_{\text{out}} = (1 + \beta)\, i_B R_E \tag{18.31}$$

since

$$i_B = \frac{V_{\text{in}} - V_{\text{out}}}{h_{ie}} \tag{18.32}$$

then

$$V_{\text{out}} = \frac{(1 + \beta)\, R_E (V_{\text{in}} - V_{\text{out}})}{h_{ie}} \tag{18.33}$$

$$\Rightarrow \left(1 + \frac{h_{ie}}{(1+\beta) R_E}\right) V_{\text{out}} = V_{\text{in}} \tag{18.34}$$

$$\Rightarrow G_V = \frac{V_{\text{out}}}{V_{\text{in}}} = \frac{1}{1 + h_{ie}/(1+\beta) R_E} = \frac{1}{1 + \beta/(1+\beta) g_m R_E} \sim 1 \tag{18.35}$$

## 18.10   The Loadline and its Uses

The principle of operation of loadlines rests on a very simple fact – that if a device and its load are in series across the power rails then the sum of the voltage dropped on the load and the voltage dropped by the device must equal the supply voltage. Since the load is ohmic, if no current flows then the load voltage is zero and the full power rail voltage is across the device. Thus on a current voltage diagram the device is sitting at a point on the current $= 0$ axis at a voltage equal to the rail voltage. For any other value of current, Ohm's law for the load tells us that the device must sit on a straight line projected back from this point with a slope of $-1/R_L$ where $R_L$ is the load resistance. This line is called a loadline. If we know that the device is passing a particular current then the circuit operating point must lie at the intersection of the device characteristic for that current and the loadline for the resistance for the simple reason that the device and the load must agree on what current they are passing! A loadline example is given below.

**Example** The transistor in the circuit in Fig. 18.41a has $\beta = 100$ and zero output admittance for $V_{CE} > 1$ V. Sketch its characteristics and superimpose the loadline appropriate to the circuit shown.

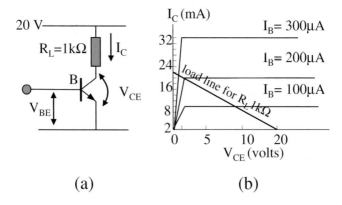

(a)                                    (b)

**Fig. 18.41.** Simple transistor circuit (a) and (b) the transistor characteristics with its appropriate loadline.

**Problem** The FET in Fig. 18.42 has $I_{DO}$ of 16 mA and a $V_P$ of 4 Volts. Draw its $I_D$ vs $V_{DS}$ characteristics for a selection of different gate voltages. Draw the line bounding the pinch-off region. Add a load-line corresponding to the 1 k$\Omega$ load resistor and determine for what range of gate voltages the amplifier operates in the pinched off zone. When below the pinched off zone, what value of resistance does this JFET offer when $V_G$ is 1 Volt?

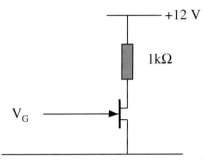

Fig. 18.42. Loadline problem.

**Solution**

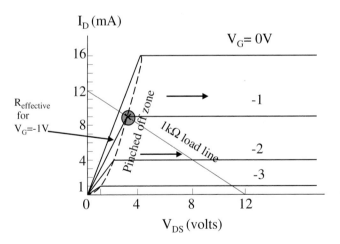

Fig. 18.43. Loadline solution.

The amplifier is pinched off when $V_G < -1$ Volt.
When $V_G$ is -1 Volt, the FET becomes like a resistor for $V_{DS} < 1$ Volt.

$$R = \frac{V}{I} = \frac{3V}{9mA} = 330\,\Omega$$

From the formula

$$R_{eff} = \frac{V_P^2}{I_{DO}(V_P + V_{GS})}$$

$$= \frac{4^2}{16(4-1)} = \frac{1}{3}k\Omega = 330\,\Omega$$

the same result is obtained.

A different type of loadline may be constructed for the particular case of a FET whose operating current is set by the insertion of a source resistor as shown in Fig. 18.44a. The rationale behind this is that the gate-source voltage $V_{GS} = RI_D$. This relationship appears as a straight-line on the $V_{GS}/I_D$ characteristics for the device, as shown in Fig. 18.44b.

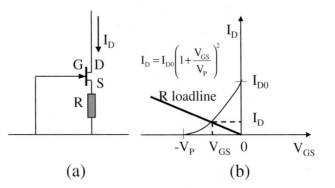

Fig. 18.44. (a) Common gate FET and (b) the loadline imposed on the device characteristics.

## 18.11   Miller Effect

The Miller effect is one of the main limitations on the high frequency performance of transistor amplifers and not only decreases the amplitude gain but also introduces large phase shifts. It is essentially a phenomena which magnifies the stray capacitance of the circuit by the voltage gain and thereby lowers the frequency of the corresponding pole by the same factor. To see how this works consider Fig. 18.45.

The voltage on the capacitor $C$ is $(1+G_V)V_{in}$, so the charge on the capacitor is $(1 + G_V)$ larger than on the identical capacitor in Fig. 18.46.

This charge must be supplied by $R_s$, and hence the frequency response of the circuit is that of a single pole whose characteristic frequency is given by $1/RC(1 + G_V)$ as opposed to $1/RC$ for the case of the simpler pole (with same values of R and C) in Fig. 18.46.

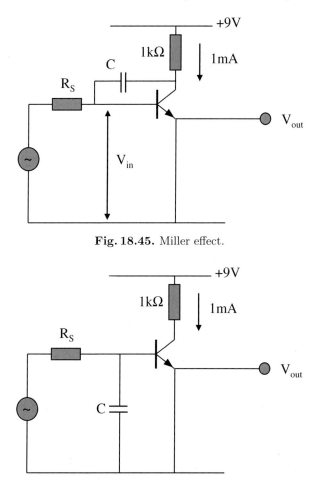

**Fig. 18.45.** Miller effect.

**Fig. 18.46.** Miller effect.

## 18.12   Nyquist Amplifier Stability Theory

One way to make an oscillator is to take an inverting amplifier with gain $-A$ and apply negative feedback to it via a network that provides 180° of phase shift at a specific and well-defined frequency. The result is that, for that frequency only, positive feedback occurs and the circuit oscillates. The only criterion to satisfy is that the gain A of the amplifier more than makes up for the attenuation of the frequency selective feedback network.

These requirements may be elegantly expressed on the Argand diagram for the transfer function of the frequency selective network, Fig. 18.47. If the path of the transfer function includes the point -1 on the real axis, then the circuit will oscillate (dotted curve). If it doesn't, the circuit is stable – and boring (solid curve)! This is called the Nyquist Stability Criterion.

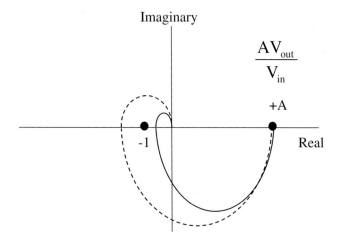

**Fig. 18.47.** Plot of the transfer function, with A the amplifier gain.

The circuit of such an oscillator is shown in Fig. 18.48 where we have used a triple pole network as an illustrative example of a suitable frequency selective network. Analysis reveals that the output of this network is antiphase with the input for a frequency $\omega = \sqrt{6}/RC$ and that the attenuation is then a factor of 29. Accordingly the gain of the amplifier must be -29 to ensure oscillation.

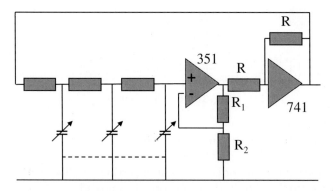

**Fig. 18.48.** An oscillator circuit.

The 351 OPA provides a high input impedance to avoid loading the frequency selective network. The gain of the 351 is $R_1 + R_2/R_2$. $R_1$ is a negative temperature coefficient thermistor which actively stabilises the 351's gain to 29 and hence keeps the oscillator amplitude constant. In practice $R_2$ is chosen such that $R_1 + R_2/R_2$ is slightly larger than 29 when the thermistor is cold. The 741 OPA provides the necessary $\pi$ phase shift to cancel the phase shift of the frequency selective network at resonance. The Argand diagram of the transfer function multiplied by the amplifier gain of 29 is shown in Fig. 18.50b and is seen (from the zoom in Fig. 18.50c) to intersect the point -1 on the real axis when oscillating stably.

Suppose now that the gain necessary to sustain oscillation is obtained from three separate amplifiers which have been cascaded and furthermore that the triple pole network is split into its three constituent sections each of which is intercalated between amplifier blocks (Fig. 18.49). The order of the circuit functional blocks is different but the functioning of the circuit is unchanged (leaving aside minor issues of different input/output impedance matching). So this circuit will still oscillate. The alarming thing is that this is exactly the circuit we con-

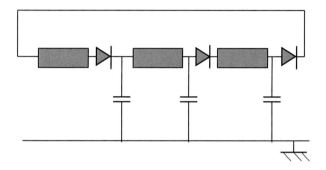

**Fig. 18.49.** Oscillator circuit with three single transistor gain stages (and no amplitude stabilisation!)

struct when we cascade three single transistor gain stages and then put feedback round the whole assembly. The three poles come ready supplied with each transistor – there is no need for additional discrete components! They are there by virtue of the Miller capacitance of each device (the capacitor of the pole) being driven by the output resistance (the resistor of the pole) of the previous stage. So in general, an amplifier which consists of three or more single transistor stages with feedback fails the Nyquist Test and is an oscillator instead! Because it has no provision for regulating the gain to ensure amplitude stability, it usually also has a very nasty waveform.

This begs the question as to how operational amplifiers could possibly work. After all they consist of many cascaded gain stages round which feedback is applied by the user.

Useful insight is gained by looking at the Nyquist plot for the transfer function of one, two and three cascaded poles as seen in Fig. 18.51 a single pole gives a maximum of 90° phase-shift (Fig. 18.51a). Two poles together are capable of giving the necessary 180° phase shift necessary for oscillation (Fig. 18.51b), but only at infinitely high frequency where their attenuation is infinite and they are thus unable to access the -1/A point, no matter how large is A. Three poles can produce 180° phase shift at a finite frequency (Fig. 18.51c). So we see that if more than 2 poles are present we cannot prevent there being some frequency

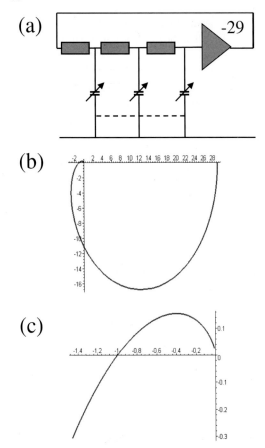

**Fig. 18.50.** (a) Circuit with an amplifier gain of -29 (b) Transfer function of circuit and (c) magnified view of transfer function.

at which 180 degrees of phase shift occurs: however we **can** make sure that the attenuation of the phase shift network is so large at this frequency that it beats the gain of the amplifier and the overall loop gain is less than unity at this frequency. This is achieved by making one pole much larger (i.e. much lower in frequency) than all the others. This is the secret of stabilising amplifiers. The technique is illustrated in Fig. 18.52 for the triple pole network and an amplifier gain of 10:

In Fig. 18.52a the three poles are equivalent and the curve includes the -1/A point (=-0.1 for a gain of 10) so the circuit oscillates. In Fig. 18.52b one pole is dominant. This first pole switches in at a frequency ten times lower than the other two and gives nearly 90 degrees of phase shift before the other two poles "wake up". By the time the two high frequency poles have got around to giving 45 degrees of phase shift each (i.e. 180 in total) the attenuation from the first

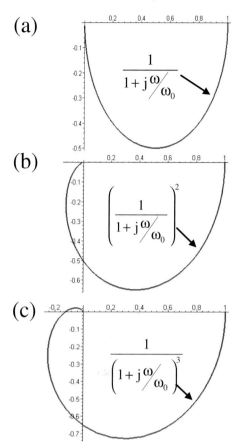

(a)

$$\frac{1}{1+j\,\omega\!\big/\!\omega_0}$$

(b)

$$\left(\frac{1}{1+j\,\omega\!\big/\!\omega_0}\right)^{2}$$

(c)

$$\frac{1}{\left(1+j\,\omega\!\big/\!\omega_0\right)^{3}}$$

**Fig. 18.51.** Transfer plot of (a) one pole (b) two cascaded poles and (c) three cascaded poles.

pole is so large that the curve fails to ensnare the -1/A (-0.1) point (where A is the amplifier gain) and so the circuit is stable.

**Problem** A 9-stage amplifier with a low frequency open loop gain of A has 9 poles. One pole turns over at frequency $\omega_0$ and the other 8 poles at $\omega_1 = 100\omega_0$. Find the maximum value of A which is compatible with unconditional stability of the amplifier.

**Solution** The open loop transfer function of the amplifier is:

$$\frac{V_{\text{out}}}{V_{\text{in}}} = A\left(\frac{1}{1+j\frac{\omega}{\omega_0}}\right)\left(\frac{1}{1+j\frac{\omega}{\omega_1}}\right)^{8}$$

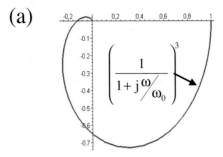

$$\omega_1 = \omega_0/10$$

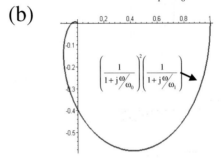

**Fig. 18.52.** Transfer functions for a triple pole network: (a) when the poles are equivalent and (b) when one pole is dominant.

At the potential oscillation frequency where the transfer function phase shift is 180°, roughly 90° of this is supplied by the low frequency pole. The remaining 90° is provided by the 8 high frequency poles, which means each provides $(90/8)° = 11°$. This means $\omega/\omega_1 = \tan(11°) = 0.2$ and hence $\omega/\omega_0 = 20$. The attenuation of the transfer function is

$$\frac{1}{\sqrt{1 + \frac{\omega^2}{\omega_0^2}}} \times \frac{1}{\left(1 + \frac{\omega^2}{\omega_1^2}\right)^4} \tag{18.36}$$

which is thus 0.04 at the danger frequency and this in turn implies that the maximum allowed amplifier gain for unconditional stability is 25. The Argand plot of this normalised transfer function is shown in Fig. 18.53a and it may be seen that the curve indeed intersects the negative real axis at about -0.04, thus illustrating the accuracy of this simple analysis. For comparison Fig. 18.53b shows the corresponding transfer function for which all 9 poles are equivalent and it is seen that the maximum allowed gain is now less than 2! In this example, for purposes of illustration, the dominant pole was chosen to be only a factor of 100 lower than the others: in a real operational amplifier, this factor would be much larger and the maximum possible gain for unconditional stability would be correspondingly larger also.

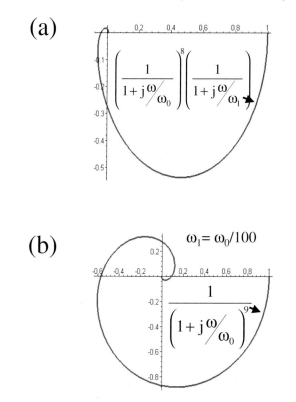

**Fig. 18.53.** Transfer plot for an amplifer with a gain of 25 (a) when one pole dominates (b) when all nine poles are equivalent.

### 18.12.1   Local and Non-local Feedback

It may be seen from the above section that negative feedback not only divides into categories of series and parallel, but it also can be "local" or "global". Local feedback is feedback applied to just one stage (i.e. round just one pole) and it is obvious from the Nyquist plot for a single pole, Fig. 18.54 that there is no way this can encircle the -1 point and so this kind of feedback is unconditionally stable.

Sensible amplifier design uses a mixture of local and global feedback. The local feedback is built in and fixed to set values in order to linearise the transfer functions of the individual circuits blocks, rather then relying on the global feedback to do this. The global feedback (at least in the case of OPAs) is left to be set by the user to determine the overall gain of the device.

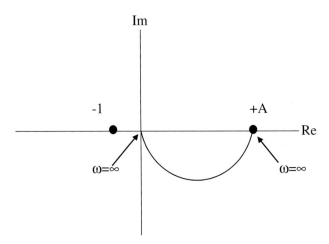

**Fig. 18.54.** Nyquist plot for a single pole.

## 18.13    Useful Circuit Tricks

### 18.13.1    Bootstrapping and the "Ring of Three"

An interesting example circuit is shown in Fig. 18.55 which illustrates several of the points which have been discussed above. This is a traditional "ring of three" transistor circuit for amplifying small voltages from low impedance sources. The 820 Ohms and 1 nF in the base of the input transistor degrade the noise performance slightly owing to the noise generated by the resistor, but are there to protect the circuit from being driven nonlinear by large amplitude radiofrequency pickup. Two kinds of AC feedback are applied. There is local series feedback applied to $Tr_1$ by the 470 $\Omega$ in its emitter. Global parallel feedback is additionally applied to the entire circuit via the 10 k$\Omega$ resistor. $Tr_1$ also derives its DC bias via the 200 k$\Omega$ feeding back from the second stage. Finally, and perhaps most interestingly, the circuit employs the technique of bootstrapping to enhance greatly the voltage gain which is provided by the second stage. The bootstrap capacitor is the 1 microfarad between $Tr_3$ emitter and the junction of the two 12 k$\Omega$ resistors. The idea is that if $Tr_2$ draws a current through its collector load, a voltage appears across the latter and its bottom end (attached to $Tr_2$ collector) drops in potential. However, because of $Tr_3$ and the bootstrap, its the resistors top end chases the bottom, forcing the bottom to drop even further. Huge gains are obtainable from this configuration (the reader now has enough technical ammunition to analyse this circuit for himself and prove that this is so). The pole associated with $Tr_2$ has been deliberately emphasized relative to the other two by introduction of the 4.7 pF capacitor in parallel to the devices own Miller capacitance. This stabilizes the amplifier. There is particular subtlety in this choice of position for the low frequency pole. As seen from the above example, the stability criterion involves the product of $A$ and $\omega_0$. If the gain of stage 2 is varied by changing the effectiveness of the bootstrapping, this

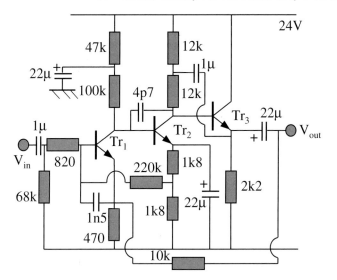

**Fig. 18.55.** "Ring of three" circuit.

affects the overall gain $A$. However it varies the value of $\omega_0$ by the same factor (since the size of the effective Miller capacitance on stage 2 is proportional to $Tr_2$ gain) so that the product $A\omega_0$ remains fixed and the amplifier stability is not compromised. Another high quality transistor circuit is shown in Fig. 18.56.

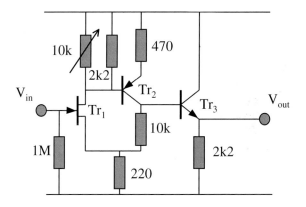

**Fig. 18.56.** High quality transistor circuit.

This has a high input impedance, a large gain and uses both series and parallel feedback via the 200 Ω and 10 kΩ resistors respectively. In this case, no deliberate pole has been introduced to stabilize the amplifier which is perfectly safe since feedback is applied round a maximum of two stages.

## 18.14     Noise

Noise, according to Ambrose Bierce, is "a stench in the ear". To the Spin Electronian, it is an unwanted voltage which gets in the way of what he is trying to measure. It comes in lots of unruly forms, most of which are frying, crackling and popping noises wholly insusceptible to mathematical analysis and generally indicative of bad circuit design, bad screening and over-cooked electronic components, all of which are avoidable with a little care. Unavoidable noise has its origins in thermodynamics and fortunately is modellable. Most of it comes in two main flavours, Johnson Noise and Shot Noise. For a more detailed discussion of noise in magnetic materials, see Bertrand Raquet's contribution to this book.

### 18.14.1     Johnson Noise

Any cavity at temperature $T$ contains blackbody radiation appropriate to that temperature and a blackbody inside that cavity continually trades energy with the cavity radiation bath. If bath and body are at the same temperature, dynamic equilibrium prevails and the body emits as much power as it receives. This situation has a 1-dimensional analogue, namely a transmission-line terminated by a resistor equal to its characteristic impedance, each at temperature $T$. In like fashion, the resistor trades electrical power with the blackbody radiation bath in the transmission line. If dynamic equilibrium is again to occur, this implies that any resistor at non-zero temperature must act as a voltage source. The spontaneous voltage produced has a "white" spectrum – that is the power per unit frequency bandwidth is uniform at all frequencies of practical interest – and its value is given by:

$$V^2 = 4k_BTRdf \qquad (18.37)$$

Where $R$ is the resistor value, $k_B$ the Boltzmann constant, $T$ the absolute temperature, $df$ the element of bandwidth considered. This noise is called Johnson noise. Note that noise generators are quantified in terms of the square of their noise voltages. This is because noise from different sources is incoherent and adds as power rather than as voltage.

### 18.14.2     Shot Noise

Diodes and transistors are characterized by a different sort of noise – shot noise, whose origins lie in the fact that the electrical charge passing through the device is quantised. With every device current I there is an associated noise current generator whose magnitude is given by:

$$I^2 = 2eIdf \qquad (18.38)$$

So for a bipolar transistor the noise performance is dictated by the components shown in Fig. 18.57. Note that only 2 noise generators are shown, one for

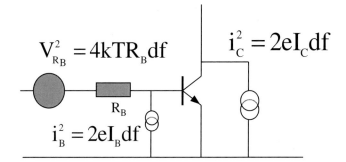

$$V_{R_B}^2 = 4kTR_B df \qquad i_C^2 = 2eI_C df$$

$$R_B$$

$$i_B^2 = 2eI_B df$$

**Fig. 18.57.** Noise in a bipolar transistor.

each of the base and collector currents. The emitter noise generator is necessarily coherent with and included in the other two.

The worked example illustrates how to use this elementary noise theory to calculate circuit parameters essential to obtaining good noise performance in a typical transistor circuit.

**Problem** A 100 ohm microphone is to be preamplified as shown in the circuit diagram overleaf. Find the transistor collector current which gives the best noise performance. The transformer has 100 turns on its primary windings and 1000 turns on the secondary.

Draw a practical circuit diagram showing all components needed to make a working preamplifier.

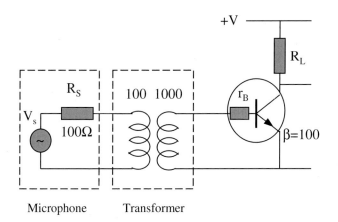

**Fig. 18.58.** Amplifing a low impedance microphone.

**Solution:**

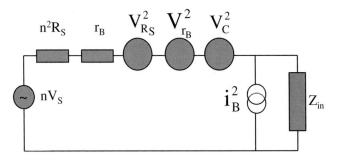

**Fig. 18.59.** Noise equivalent circuit of Fig. 18.58.

$n$ is the step-up ratio of the transformer.

$V_{R_S}^2 = 4kTn^2R_Sdf$ is the source impedance voltage noise.

$V_{R_B}^2 = 4kTr_Bdf$ is the base resistance voltage noise.

$V_C^2 = i_C^2/g_m^2 = 2eI_Cdf/g_m^2$ is the collector shot noise referred to the base as a voltage generator.

$i_B^2 = 2eI_Bdf = 2eI_Cdf/\beta$ is the base current shot noise

$Z_{in} = \beta/g_m$.

Define $R^* = n^2R_S + r_B$ in parallel with $Z_{in}$.

If the circuit is well designed, $Z_{in} \gg$ source impedance. The base current shot noise may be represented by a voltage generator so $R^* \frown n^2R_S$

$$\frac{2eI_Cdf}{\beta} \times R^{*2}$$

Then the total effective voltage noise generator in the base is:

$$V_{Tot}^2 = 4kTdf\left(R_Sn^2 + r_B\right) + \frac{2eI_Cdf}{g_m^2} + \frac{2eI_Cdf}{\beta}R^{*2}$$

The last two terms are $I_C$ dependent, so we must minimize their sums with an appropriate choice of $I_C$

$$= \frac{2eI_Cdf}{g_m^2} + \frac{2eI_C}{\beta}R^{*2}df$$

$$= \frac{2eI_Cdf}{40^2I_C^2} + \frac{2eI_C}{\beta}n^4R_S^2df$$

$$= 2edf\left(\frac{1}{40^2I_C} + \frac{I_C}{\beta}n^4R_S^2\right)$$

$$= \frac{2edfn^2R_S}{\beta^{1/2}}\left(\frac{\beta^{1/2}}{40I_CR_Sn^2} + \frac{40I_CR_Sn^2}{\beta^{1/2}}\right)$$

which has the form $x + 1/x$ which has a minimum value if $x = 1$ i.e.

$$I_C = \frac{\beta^{1/2}}{40n^2 R_S}$$

putting in values gives

$$I_C = \frac{10}{40 \times 10^2 \times 100} = 25\,\mu A$$

**Practical Circuit**

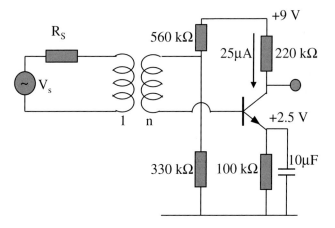

**Fig. 18.60.** Practical preamplifier circuit for Fig. 18.58.

$$Z_{\text{in}} = \frac{\beta}{g_m} = \frac{100 \times 10^{-6}}{40 \times 25} \sim 100\,\text{k}\Omega \gg n^2 R_S = 10\,\text{k}\Omega$$

so

$$\frac{1}{g_m} = 1\,\text{k}\Omega$$

$Z_c$ must be $1\,\text{k}\Omega$ at the lowest audio frequency of 20 Hz, so

$$\frac{1}{2\pi fC} \ll 1\,\text{k}\Omega \Rightarrow C \geq \frac{1}{6 \times 20 \times 10^3} \sim 10\,\mu F$$

It is apparent from this calculation that there is a correspondence between the source impedance and the collector current needed for optimum amplifier noise performance. For very low source impedances, this collector current may take impractical values. The solution here is to use an audio transformer to transform the source impedance to a higher value as described previously in the section on transformers.

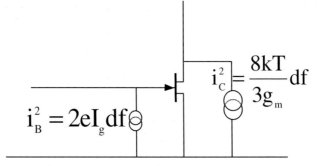

$$i_B^2 = 2eI_g \, df \qquad i_C^2 = \frac{8kT}{3g_m} \, df$$

**Fig. 18.61.** Noise in a FET.

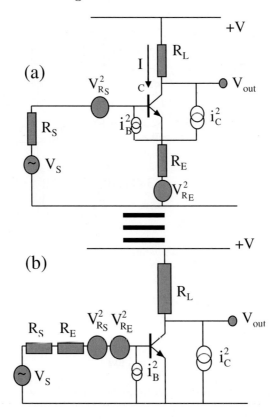

**Fig. 18.62.** The circuits offer the same noise performance but (b) is easier to analyse.

The noise analysis of FET circuits proceeds in a very similar fashion to that outlined above except that the relevant noise generators in this case are as shown in Fig. 18.61. It is worth noting in passing that applying feedback to a circuit has no effect on the noise performance (assuming that the noise generators associated with the added feedback resistors are insignificant). It is therefore advisable to remove the feedback from a circuit before proceeding with its noise analysis since the job is then considerably simpler. An example is shown in Fig. 18.62 for the very elementary case of an emitter series feedback resistor. The feedback is removed by shifting the earth point as shown. The argument which established the magnitude of Johnson noise at a particular temperature was a thermodynamic one. It is no longer valid if batteries are connected to the components involved since the Fermi level of the system is no longer a constant and thermodynamic equilibrium does not pertain. Resistors are generally noisier when they carry electrical current. By contrast, some devices are actually quieter when biased. For example, a diode with dynamic impedance R generates half the noise power of a resistor of the same impedance. Another useful example is shown in Fig. 18.63 where an active circuit is used to generate the characteristic impedance necessary to terminate a transmission line. The actual resistor involved has a value 1+A times larger than the impedance it is simulating and hence generates correspondingly less noise power than if the transmission line is terminated by a passive resistor of the same value as the characteristic impedance.

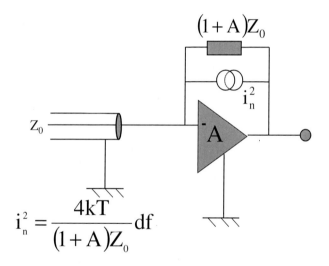

$$(1+A)Z_0$$

$$i_n^2$$

$$i_n^2 = \frac{4kT}{(1+A)Z_0}df$$

**Fig. 18.63.** Reducing noise at the terminal end of a transmission line.

## 18.15   The DC Motor

No chapter on electrics is complete without at least a rudimentary discussion of electrical machines. Here we have chosen the DC motor as an example which is sufficiently simple to afford easy analysis, yet serves to give a flavour of how all such devices behave.

A DC motor consists of a rotor of magnetic material which is wound with coils to which electrical connection is made via a commutator which reverses those connections during the cycle. The rotor rotates in a static magnetic field created either by a permanent magnet or by a set of field windings.

Passing a current through the rotor results in its experiencing a torque. When it starts to move it cuts the field lines with which it interacts and this generates a back emf which opposes the current in the rotor which causes the motion.

The equivalent circuit for a DC motor then looks as follows: The back emf

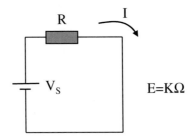

**Fig. 18.64.** Equivalent circuit for a DC motor.

$E$ is proportional to motor speed , i.e. $E = K'W$

Evidently the torque is given by :

$$\tau = B_{field}CI_{rotor} = KI \qquad (18.39)$$

But the power delivered is

$$\tau = IE \qquad (18.40)$$

Which leads to

$$KI = IK' \qquad (18.41)$$

Thus we have the interesting result that $K = K'$, i.e. that the same constant determines torque and back emf.

If the field is provided by a fixed magnet, then the factor $K$ is fixed. However if the motor is of the field-winding type, then changing the field winding current varies the value of $K$ and the motor is much more versatile. The field is linear in $I_{field}$ for small currents but limits as magnetic saturation is approached.

## 18.16   Acknowledgements

We would like to express our gratitude to Rui Borges and Eric Jouguelet for their help with the pole modelling and also to Michael Ziese for assistance with the diagrams.

## 18.17   Concluding Remarks

In this chapter we have given a very brief introduction to some of the electronic delights which all Spin Electronicians should be conversant. The interested reader is referred to Horowitz & Hill "The Art of Electronics" CUP 1989, for further edification.

# 19    Spin–Valve and Spin–Tunneling Devices: Read Heads, MRAMs, Field Sensors

P. P. Freitas

Instituto de Engenharia de Sistemas e Computadores, R. Alves Redol, 9, Lisbon, Portugal and Instituto Superior Tecnico, Departamento de Fisica, Av. Rovisco Pais, P-1000 Lisbon, Portugal

## 19.1    Read Heads and Magnetic Data Storage

Hard disk magnetic data storage is increasing at a steady state in terms of units sold, with 144 million drives sold in 1998 (107 million for desktops, 18 million for portables, and 19 million for enterprise drives), corresponding to a total business of 34 billion US\$ [1]. The growing need for storage coming from new PC operating systems, INTERNET applications, and a foreseen explosion of applications connected to consumer electronics (digital TV, video, digital cameras, GPS systems, etc.), keep the magnetics community actively looking for new solutions, concerning media, heads, tribology, and system electronics. Current state of the art disk drives (January 2000), using dual inductive-write, magnetoresistive-read (MR) integrated heads reach areal densities of 15 to 23 bit/$\mu$m$^2$, capable of putting a full 20 GB in one platter (a 2 hour film occupies 10 GB). Densities beyond 80 bit/$\mu$m$^2$ have already been demonstrated in the laboratory (Fujitsu 87 bit/$\mu$m$^2$–Intermag 2000, Hitachi 81 bit/$\mu$m$^2$, Read–Rite 78 bit/$\mu$m$^2$, Seagate 70 bit/$\mu$m$^2$ – all the last three demos done in the first 6 months of 2000, with IBM having demonstrated 56 bit/$\mu$m$^2$ already at the end of 1999). At densities near 60 bit/$\mu$m$^2$, the linear bit size is $\sim$ 43 nm, and the width of the written tracks is $\sim$ 0.23 $\mu$m. Areal density in commercial drives is increasing steadily at a rate of nearly 100% per year [1], and consumer products above 60 bit/$\mu$m$^2$ are expected by 2002. These remarkable achievements are only possible by a stream of technological innovations, in media [2], write heads [3], read heads [4], and system electronics [5]. In this chapter, recent advances on spin valve materials and spin valve sensor architectures, low resistance tunnel junctions and tunnel junction head architectures will be addressed.

Since the beginning of the nineties, and for areal densities > 1.5 to 3 bit/$\mu$m$^2$, MR read sensors have gradually replaced inductive readers. The MR sensor is shielded in order to increase linear resolution and improve high frequency response [6]. Figure 19.1 shows schematically the shielded MR sensor used for disk applications. MR heads for areal densities up to 8 bit/$\mu$m$^2$ were based on the anisotropic magnetoresistance (AMR) effect [7], where the sensor resistance is proportional to the square of the cosine of the angle between the magnetization and the sense current. For some of the higher density drives using AMR sensors, the MR element consisted of a 15 nm thick Ni$_{80}$Fe$_{20}$ sensor, with 0.7 to 1 micron trackwidth ($W$), 0.5 to 0.7 micron height ($h$), giving an MR signal close to 2%.

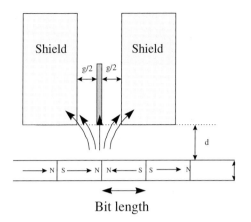

**Fig. 19.1.** Schematic of a shielded read head.

To linearize sensor output, a second Ni-Fe layer separated from the active $Ni_{80}Fe_{20}$ layer by an oxide or high resistance layer, was used to magnetostatically bias the active $Ni_{80}Fe_{20}$ magnetization to 45 degrees with respect to the sense current direction, in what is called a Soft Adjacent Layer configuration (SAL–MR) [8]. For areal densities typically above 8 bit/$\mu m^2$, it becomes difficult to maintain the desired Bit Error Rate with AMR sensors. A first generation of spin valve heads [9] with MR from 6% to 8% came into production at the end of 1997 from IBM, soon followed by Fujitsu, and Toshiba. For 15.5 bit/$\mu m^2$, written tracks are around 0.6 $\mu m$, and the bit length is 70 to 75 nm, with a read head trackwidth of 0.5 $\mu m$. With the demand for even higher areal densities (150 bit/$\mu m^2$, bit size 40 nm, written track width = 0.15 micron), read sensor trackwidth will decrease to 0.1 micron. For this trackwidth, the output of the present first generation of spin valve heads becomes too low, and a second generation of read-out heads (MR close to 20%) will come into play. At this stage and beyond, low resistance spin tunnel junction heads may become competitive.

### 19.1.1  Spin–Valve Sensors

Spin valves were introduced in 1991 [10], first sensors designed and tested in 1993–1994 [11], and first head prototypes presented in 1994 [9]. A good review of the spin valve mechanism is given in reference [12]. The spin valve consists of two ferromagnetic layers, separated by a Cu spacer (see Fig. 19.2a, where a bottom-pinned simple spin valve structure is shown). One of these layers has its magnetization pinned, while in the other it is free to rotate. The free ferromagnetic layer forms the sensing element and usually consists of Co or $Co_{90}Fe_{10}$ or a $Ni_{80}Fe_{20}/Co$ or $Ni_{80}Fe_{20}/Co_{90}Fe_{10}$ bilayer. The pinned ferromagnetic layer (Co or $Co_{90}Fe_{10}$) is coupled by exchange to an antiferromagnet (for example $Mn_{76}Ir_{24}$, or $Mn_{50}Pt_{50}$) [13] or a synthetic antiferromagnet ($NiO/Co/Ru/Co$, $Mn_{76}Ir_{24}/Co_{90}Fe_{10}/Ru/Co_{90}Fe_{10}$) [14]. Free and pinned layer easy axis can be set either parallel or orthogonal. Typical MR values for these first generations of top-pinned or bottom-pinned spin valves ranged from 6% to 10%. Here the top-pinned or bottom-pinned designations relate to spin valves where the pinned layers are above or below the Cu spacer respectively. For head applications, and apart from large MR values, the exchange field created at the pinned layer/exchange layer interface is very important. The blocking temperature (temperature where the exchange field vanishes) should exceed 300°C, to prevent accidental de-pinning of the pinned layer during head fabrication or head life. Also the exchange energy should be large (> 0.2 mJ/m$^2$) such that the exchange field prevails against demagnetizing fields at head level. Also the corrosion resistance of this exchange layer should not be worse than that of $Ni_{80}Fe_{20}$ used as reference [15]. Another factor to take into account is the coupling field $H_f$ between the free and pinned layers, which should not exceed 0.8 to 1.2 kA/m, to allow for proper biasing. This coupling arises from the competition between ferromagnetic Néel coupling (caused by interface roughness), indirect exchange coupling across the Cu spacer, and the coupled demagnetizing fields of both layers in patterned sensors.

For a shielded spin valve sensor, the head output is given by [16],

$$\Delta V_{0-p} = (\Delta R/R)\, R_\square\, I\, (W/h)\, (1/2)\, \langle \cos{(\Theta_f - \Theta_p)} \rangle \qquad (19.1)$$

$$\langle \cos{(\Theta_f - \Theta_p)} \rangle = E\, \Phi_{ABS}/(tw\mu_0 M_S)\,. \qquad (19.2)$$

Here, $\Delta R/R$ is the maximum MR signal of the spin valve sensor (6 to 8% in first generation spin valves), $R_\square$ is the sensor square resistance (16 to 20 $\Omega/\square$), $W$ the trackwidth of the read element, $h$ is the sensor height, $I$ the sense current, $\Theta_f$ is the angle between the free layer magnetization and the current direction, and $\Theta_p$ the angle between the pinned layer magnetization and the current. The average $\langle ... \rangle$ is taken over the height of the sensor. The media-flux leakage to the shields is described by the head efficiency $E = [\tanh(h/2l_c)]/(h/l_c)$, with $l_c$, the flux propagation length defined as, $l_c = \sqrt{t\mu g_R/2}$, with $\mu$ the relative free layer permeability, $g_R$ the shield to sensor separation (read half gap), and $t$ the free layer thickness, with $\Phi_{ABS}$ the media flux entering the sensor. As can be seen from (19.1) and (19.2) the head output depends critically on the head geometry, spin valve MR signal, and media parameters.

Table 19.1 shows the required read head parameters for areal densities up to 124 bit/$\mu$m$^2$, aiming at a head output in the 2 to 5 mV/$\mu$m of trackwidth [17]. It can be seen that the MR signal must be pushed to the 20% range, the free layer thickness must be decreased below 2 nm, and the read gap will decrease to 50 nm. In the following section, recent developments in spin valve materials and structures are reviewed. Also low resistance tunnel junctions are introduced, as possible candidates for read elements at areal densities near 150 bit/$\mu$m$^2$.

Consider first MR signal enhancement. This has been achieved in two ways. First, a dual symmetric spin valve can be fabricated where two spin valves, one bottom pinned, and the other top pinned, share the same free layer. MR signals can surpass 20%. Although suggested some years ago with the structure NiO/Co/Cu/Co/Cu/Co/NiO [18], only recently Read Rite and Fujitsu produced

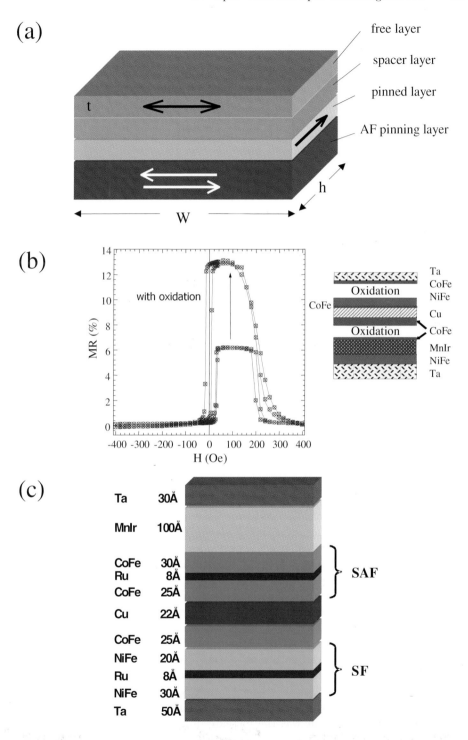

**Fig. 19.2.** (a) Schematic of a simple 4 layer spin valve structure. (b) Specular spin valve with Nano Oxide Layers. (c) Spin valve with both synthetic free and synthetic pinned layers.

head prototypes with practical exchange layers ($Mn_{76}Ir_{24}$, $Mn_{50}Pt_{50}$) working at densities greater than 30 bit/$\mu m^2$ [19]. Due to the larger thickness of this dual symmetric spin valve, it is not foreseen that this structure will prevail at much higher densities, since the sensor itself can hardly be thinner than 35 nm, for an available read gap tending to 50 nm. Another development is the control of electron scattering at the external spin valve surfaces. If the surfaces are made to be specular with respect to electron scattering, than the spin valve signal can approach that of an equivalent GMR multilayer. For NiO/Co/Cu/Co/NiO spin valves, MR signals close to 20% were obtained, but without good exchange properties [20]. More recently, Toshiba suggested the use of Nano Oxide Layers (NOL) incorporated on both sides of the Co-Fe/Cu/Co-Fe structure of a standard spin valve [21]. This allows the use of state of the art spin valve architectures in what concerns exchange fields, but increasing the MR into the 15 % to 20% range. Figure 19.2b shows one of our own specular spin valves [22], where the inclusion of NOL layers increases the MR ratio from 6% to 14%. The method of fabrication of the NOL layers, their optimum thickness and their thermal stability are at present under study.

**Table 19.1.** Read head parameters for increasing areal densities

| year | density (bit/$\mu m^2$) | $w$ ($\mu m$) | $2g_R$ (nm) | $t$ (nm) | $\Delta R/R$ (%) | sensor type |
|------|---------|-----|-------|------|------|--------|
| 1998 | 5–8 | 1.5–0.8 | 250–200 | 12–9 | 1.8–1.5 | AMR |
| 2000 | 15 | 0.5 | 140 | 5 | 6.5 | sv |
| 2003 | 62 | 0.25 | 70 | 2.5 | 12 | sv |
| 2005 | 124 | 0.18 | 50 | 1.8 | 18 | sv or TJ |

Concerning exchange fields, two approaches were followed to increase exchange and blocking temperature. The first was the use of a synthetic antiferromagnet (SAF) consisting of two Co or Co-Fe layers with similar thickness strongly AF coupled through 0.5 to 0.7 nm Ru [14] (see Fig. 19.2c). This AF coupling is of the order of 0.7 mJ/$m^2$. To avoid a spin flop transition under an external transverse field [23], a conventional exchange layer must be coupled to one of the ferromagnetic layers. Although this strong AF coupling has a weak temperature dependence [24], this is not so for the conventional AF that must still have an optimum blocking temperature, but no longer large exchange. These SAF structures have another advantage. Since the effective moment of the pinned layer is low, its contribution to the demagnetizing and coupling fields acting on the free layer is much weaker than in conventional spin valves. SAF layers have also been incorporated in dual (symmetric) spin valve structures [19]. Another recent development was the increase of exchange energies achieved in bottom pinned, $Mn_{76}Ir_{24}$ and $Mn_{50}Pt_{50}$ spin valves (> 0.3 to 0.4 mJ/$m^2$), by proper growth control, and microstructure tailoring [25]. Exchange fields in excess than 80 kA/m can now be obtained in spin valve structures with blocking temperatures above 300°C. Finite size effects were observed for the blocking temperature [26]. Higher blocking temperatures are obtained with thicker AF films, but higher exchange fields are obtained for thinner AF layer thicknesses [27]. Combined SAF structures using $Mn_{50}Pt_{50}$ as reference antiferromagnet seem to provide the best thermal stability, and the largest exchange. Table 19.2 summarizes the properties of some conventional exchange materials ($Fe_{50}Mn_{50}$, NiO) and some of the newer, more promising structures [4,13,15,25,32].

**Table 19.2.** Comparison of exchange bias materials

| materials | $T_b$ (°C) | exchange energy (mJ/m$^2$) | corrosion resistance | requires anneal? |
|---|---|---|---|---|
| $Fe_{50}Mn_{50}$ [15] | 150 | 0.13 | bad | no |
| NiO [15] | 190 | ≪ 0.1 | good | no |
| $Mn_{78}Rh_{22}$ [15] | 235 | 0.2 | fair | no |
| $Mn_{76}Ir_{24}$ [25] | 300 | 0.2–0.4* *(bottom sv, after anneal) | fair | yes (bottom spin valves) |
| $Mn_{50}Ni_{50}$ [13,15] | 375–425 | 0.27–35 | fair/good | yes |
| $Mn_{50}Pt_{50}$, MnPdPt [13,32] | 350 | 0.3–0.4 | good | yes |
| ($Mn_{76}Ir_{24}$, MnPdPt,NiO) /Co/Ru/Co [4,32] | (see text) | > 0.4 | good/fair | yes (because of MnPt and MnIr layers) |

The third avenue for spin valve sensor improvement, is concerned with the reduction of the thickness of the free layer. Here essentially two approaches are proposed, first the use of a "spin filter" spin valve, where a high conductivity layer (normally Cu) is placed under the $Ni_{80}Fe_{20}$ free layer [28], and the second the use of a synthetic free layer [29]. As seen from (19.1) and (19.2), as the magnetic thickness of the free layer is reduced, the output signal of the head increases. INESC's approach [30], has been to use a synthetic ferrimagnetic free layer consisting of two ferromagnetic layers antiferromagnetically coupled through a thin Ru layer. It is quite important to control the coupling and thickness of the two layers, since this synthetic free layer must rotate coherently, as a single unit with effective magnetic thickness $t_{eff} = (M_a t_a - M_b t_b)/\langle M \rangle$, but physical thickness $t_a + t_b$, contrary to the scissor-like motion of the magnetization in a SAF. With this architecture, the full spin valve MR signal can be maintained as the free layer magnetic thickness is reduced to less than 1 nm [30]. As with the SAF, the synthetic free layer will have a reduced demagnetizing field, and reduced coupling field. Since these structures posses Ru thin layers that hinder diffusion, and good antiferromagnets as reference layers for the synthetic free and pinned structures, they appear as good candidates for read element use. Also the thermal stability is good [30]. Figure 19.2c shows a spin valve developed at INESC using both synthetic free and synthetic pinned layers [30].

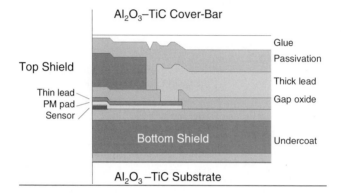

**Fig. 19.3.** Cross-section of the shielded spin valve structure.

Once good materials properties are achieved, a spin valve sensor must be fabricated [11]. The spin valve film is patterned into a stripe with dimensions $L$ (magnetic length), $W$ (trackwidth, distance between contact pads) and $h$ (height). The pinned layer easy axis is in the transverse orientation. The free layer easy axis is in the longitudinal orientation. Upon excitation by a transverse signal field, the free layer magnetization rotates out of the longitudinal direction. Due to non-uniform demagnetizing fields in the stripe, the angle of rotation depends on position along height. The sensor output is given by (19.1). For a properly biased sensor, in the quiescent state (no applied signal field), $\langle \Theta_f \rangle \simeq 0°$, and $\langle \Theta_p \rangle \simeq 90°$. Biasing is achieved with the sense current field. For small signal fields, the magnetization of the free layer rotates away from the longitudinal orientation and sensor resistance deviates linearly with field. For a recording head application, the single spin valve sensor is placed between magnetic shields to improve linear resolution. Figure 19.3 shows the cross section of the shielded spin valve structure. In the shielded configuration, all the demagnetizing fields are reduced and the free layer will require extra longitudinal stabilization [31], this in our case, is provided by permanent magnets ($Co_{66}Cr_{16}Pt_{18}$). Figure 19.4 shows the typical sensor used in a disk head (abutted permanent magnets). Table 19.3 shows head and media parameters for a 31.6 bit/$\mu$m$^2$ recording demonstrations [32].

According to Table 19.1, and for densities near 150 bit/$\mu$m$^2$, spin tunnel junctions may appear as alternative candidates to spin valves as read elements in heads.

**Table 19.3.** Head and media parameters for a 31.6 bit/µm² demonstration

| Head and media parameters | 31.6 bit/µm² SV, Fujitsu 1999 |
|---|---|
| media $H_c$ (kA/m) | 270 |
| media Mrt (mA) | 4.2 |
| magnetic spacing (nm) | 3.1 |
| write trackwidth (µm) | 0.45 |
| write head polarization (T) | 1.6 |
| read trackwidth (µm) | 0.36 |
| free layer thickness (nm) | 3.5 |
| MR height (µm) | unknown |
| read gap (µm) | 0.11 |
| output (mVpp/µm) | 2.5 |

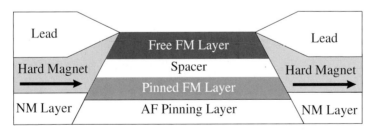

**Fig. 19.4.** Schematic cross section of a spin valve sensor with abutted magnets and thin leads.

## 19.1.2  Naturally Oxidized Spin–Tunnel Junctions

**Table 19.4.** Properties of naturally oxidized junctions

| | Area ($\mu m^2$) | TMR (%) | $R \times A$ ($\Omega \times \mu m^2$) | Structure (units: nm) | oxidation |
|---|---|---|---|---|---|
| IBM | $1.4 \times 2.8$ | 20.3 | 61 | $Ta_5/Al_{25}/NiFe_4$ $/MnFe_{10}/Co_4/Ru_{0.7}$ $/Co_3/Al_t/...$ | **natural** |
| | $2 \times 4$ | 21.4 | 61 | | |
| TDK | $1 \times 1$ | 23.2 | 49.4 | $Ta_5/NiFe_3/CoFe_2$ $/Al_t/CoFe_3$ $/RuRhMn_{10}/Ta_5$ | **natural** |
| | $1 \times 1$ | 25.8 | 36.4 | $Ta_5/NiFe_3/CoFe_2/Al_t$ $/CoFe_3/PtMn_{30}/Ta_5$ | anneal at $250°C$ |
| | $1 \times 1$ | 31.2 | 36.1 | $Ta_5/NiFe_3/CoFe_2$ $/Al_t/CoFe_3/Ru_{0.9}$ $/CoFe_2/PtMn_{30}/Ta_5$ | anneal at $250°C$ |
| | 0.237 | 31.6 | 33.5 | | |
| INESC | $3 \times 1$ | 20.6 | 63 | $Ta_7/NiFe_{20}/CoFe_3$ $/Al_{0.7}/CoFe_3/MnIr_{18}$ $/Ta_3/TiW_{15}$ | **natural** + anneal at $190°C$ |
| | $4 \times 1$ | 22.6 | 66 | | anneal at $190°C$ |
| | $4 \times 1$ | 20.7 | 60 | | anneal at $210°C$ |
| | $3 \times 1$ | 18.2 | 39 | $Ta_7/NiFe_{20}/CoFe_3$ $/Al_{0.6}/CoFe_3/MnIr_{18}$ $/Ta_3/TiW_{15}$ | anneal at $215°C$ |

In the spin-dependent tunnelling effect, electrons tunnel across an insulating barrier between two ferromagnetic electrodes. The magnetoresistance of such a junction is proportional to the product of both electrode polarizations. Although known since the mid sixties, only recently have significant room temperature MR signals, ranging from 20–40% been obtained [33,34]. This opened a realm of practical applications, among which, two of the most important are non-volatile tunnel junction random access memories (TJRAM) and MR sensors for very high-density magnetic storage (HD drives). In this section, only the naturally oxidized junctions will be described, since up to now they have provided the lowest junction resistances, needed to compete with spin valves in terms of signal to noise ratio. At the heart of tunnel junction fabrication lays the barrier fabrication. For read head applications, junction resistance must be less than $10\ \Omega \times \mu m^2$ [35,36]. Table 19.4 compares electrode structure, and junction characteristics for naturally oxidized junctions from three labs (IBM [34], TDK [36], INESC [36]). As of June 2000, resistance $\times$ area products of $20\ \Omega \times \mu m^2$ have been reported by these labs and Seagate, with TMR in the vicinity of 20%.

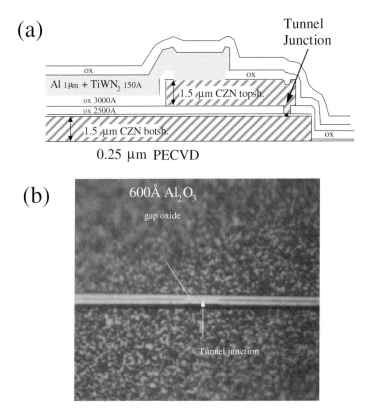

**Fig. 19.5.** (a) Schematic cross-section of the tunnel junction head (b) ABS picture of a tunnel junction head with a 60 nm read gap.

Concerning the junction electrodes, all improvements made on spin valve technology are in general applicable here. One point of concern at this moment is the thermal stability of the ultra thin natural oxidized barriers. Junctions with good resistance and TMR characteristics, ($t_{Al} = 0.6$ or $0.7$ nm, TMR $> 20\%$, $R \times A < 100\ \Omega \times \mu m^2$) show a thermal stability of at most $250°C$. This is not good enough. Figure 19.5a shows schematically the first tunnel junction head structures being fabricated at INESC, and Fig. 19.5b shows the sensor at the air bearing surface (ABS) level [36]. In our design, the TJ is at the ABS, and contacts are made through the shields. Lateral permanent magnet hard bias is used. The head read gap is 60 nm. At the moment of writing this article, first completed head prototypes show loss of TMR signal at the final lapping stages. TDK has recently announced (January 2000) working TJ read heads at 3Gbit/in$^2$ level, using a flux-guide-type structure, where the TJ read element is recessed from the air-bearing surface. At Intermag 2000, tunnel junction read head prototypes were also presented by Seagate, and Fujitsu. Apart from the required low resistance, gap smearing turns out to be an important factor to take into account in the design and fabrication of tunnel junction read heads.

## 19.2   Tunnel Junction Random Access Memories (TJMRAM)

There is a strong interest in non-volatile memory devices based on magnetic materials, Magnetic Random Access Memories (MRAMs), due to their non-volatile characteristic, radiation hardness, non-destructive read-out, low-voltage, and very large ($> 10^{15}$) read-write cycle capability [37]. MRAMs can be as fast as Dynamic Random Access Memories (DRAMs), and almost as small as Static Random Access Memories (SRAM) in cell size. To compete with CMOS embedded memories, they must be fabricated with $< 125$ nm features, bringing several technological issues regarding micromagnetics and fabrication issues for deep submicron magnetic elements. MRAMs compete also with Ferroelectric Random Access Memories (FERAMs) for non-volatile memories. Power consumption is here a major issue. They compete directly with Flash memories used where speed is not a major concern (i.e. in some storage applications). Write speed for Flash technology is in the μs range in comparison of few ns for MRAMs. Functional devices have been fabricated using the anisotropic magnetoresistance (AMR) effect, finding niche markets in satellite and military applications. With the rapid improvements in giant magnetoresistance (GMR) and spin dependent tunnel junctions, higher signal levels became available and renewed interest has arisen in MRAM fabrication [34,38,39]. Figure 19.6 shows schematically the MRAM matrix, where each cell consists of a tunnel junction. Notice the buried world line needed for write (diode or transistors needed for read selectivity are not shown). In particular for tunnel junctions, the observed tunneling magnetoresistance (TMR) at room temperature now reaches values around 40% [34]. These resistance changes can be observed in low field ranges, 1.6–2.4 kA/m, and are higher than those presently achieved with spin valve or GMR effects. As the initial high resistivity (MΩ × μm$^2$) handicap of tunneling junctions was overcome, by lowering its resistance to values of 1–10 kΩ × μm$^2$ [2,34], its integration in a memory device became realistic.

Figure 19.7 compares tunnel junctions fabricated at INESC with different oxidation technologies. Both plasma oxidation or ion beam oxidation provide good results with $R \times A$ values controllable from 200–500 Ω × μm$^2$ and up. In all cases, an Al layer, 0.7 to 1.1 nm thick is deposited, and then the oxidation

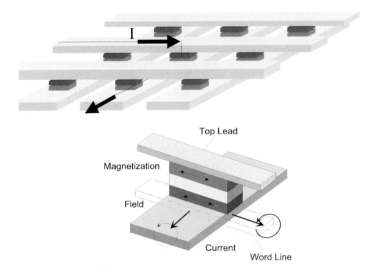

**Fig. 19.6.** Schematic layout of the MRAM matrix, where each cell consists of a tunnel junction.

time tuned to provide perfect oxidation of the $Al_2O_3$ barrier, without leaving unoxidized Al, or oxidizing the electrodes. The resistance value influences two important parameters in memory applications. Because of the insulator between the electrodes, junctions also act as capacitors. The junction's time constant, RC, sets a lower limit on the speed a data bit can be accessed. For a stand-alone junction, it is therefore desirable to have the lowest possible value for R, since the thickness required for tunnelling, 0.7–1.1 nm, and the small area, $< 1~\mu m^2$, determine the value of C. The requirement of a 1 ns time constant, would result in a resistance under 23 k$\Omega$ for a 1 $\mu m^2$ junction area. When the junction is in series with a diode, its resistance should be matched to that of the diode for maximum signal in the series device [39]. For MRAM fabrication, the TJ is deposited onto a properly planarized CMOS wafer incorporating transistors or diodes and word lines, and then a backend metallization process is realized to connect the junctions in the memory matrix [34,40]. Standard backend technology for metallization of integrated circuits requires annealing in forming gas at 400–450°C to heal transistor damage due to plasma processing. Thus for perfect CMOS compatibility, tunnel junctions should be able to cope with this final sintering step. We have previously shown that tunnel junctions can withstand thermal treatment up to 300°C [41], actually improving their performance. Figure 19.8 exemplifies the thermal stability of such devices. Interdiffusion of the exchange layer into the CoFe electrode has been pointed out as one of the causes of the loss of TMR signal above 300°C [41]. Incorporation of SAF structures as exchange layer does improve the thermal behavior, but a definite experiment where Ta interdiffusion stoppers were incorporated failed to stop the TMR decrease [42]. This means that either structural changes in the barrier itself, or polarization

changes at the interface level must be responsible for the major TMR loss above
400°C.

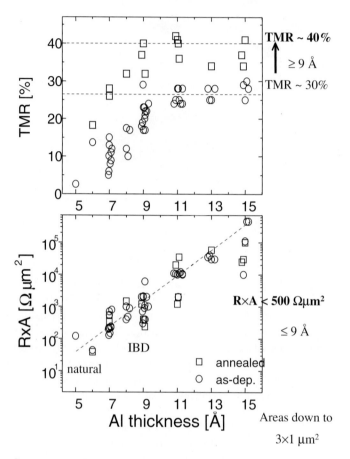

**Fig. 19.7.** Comparison of tunnel junction properties (TMR, and $R \times A$) for junctions
prepared by RF plasma oxidation, ion beam oxidation, and natural oxidation.

In a memory array matrix, to selectively read a bit, current must flow through
one single junction and alternative current paths must be blocked. This can be
achieved making use of current directionality in diodes or on/off transistor characteristics [34,38,40]. In the simplest case the basic memory cell will be a diode
connected in series with a tunnel junction. Figure 19.9 shows schematically the
vertical integration of a tunnel junction with an amorphous Si diode [38]. For
the combined device, changes in current up to 20% are obtained when the free
layer is switched (see Fig. 19.12). The problem with this device is the high resistance of the diode at the required working voltage. To lower this resistance
means increasing the diode area to tens of μm². For higher density applications,

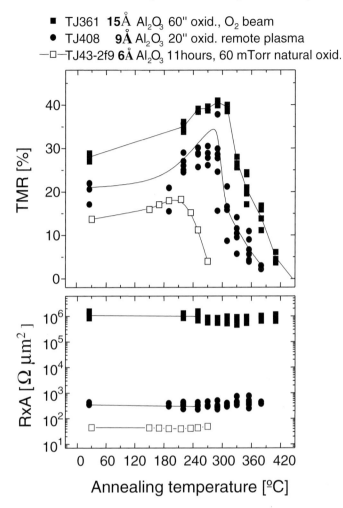

■  TJ361  **15Å** Al₂O₃ 60" oxid., O₂ beam
●  TJ408    **9Å** Al₂O₃ 20" oxid. remote plasma
—□—TJ43-2f9 **6Å** Al₂O₃ 11hours, 60 mTorr natural oxid.

**Fig. 19.8.** Thermal stability of TJ cells.

transistors seem a better solution. For write selectivity, buried world lines are used such that two orthogonal fields can be created at each junction that can switch the free layer [40]. Figure 19.10 shows schematically the integrated junction on top of 1.2 μm buried Al lines, and Fig. 19.11 shows the asteroid curve for a $3 \times 1$ μm² junction.

Under study is the optimum shape of the junction top free layer in order to minimize Barkhausen noise, and to promote single step, coherent switching [43]. At present switching times of few ns were demonstrated in real matrix cells [44], and experiments are ongoing with specially-designed chips to follow the switching mechanism with tens of picosecond resolution. Figure 19.12 shows a $3 \times 3$ bit memory, where the diodes are $200 \times 200$ μm in size, and junctions are $3 \times 1$ μm². Notice that 7 of the 9 bits have almost equal transfer curves, one

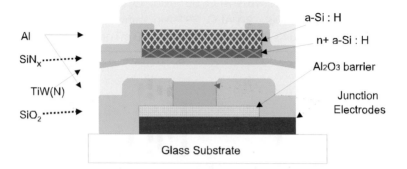

Fig. 19.9. Vertical integration of a tunnel junction with an amorphous Si diode.

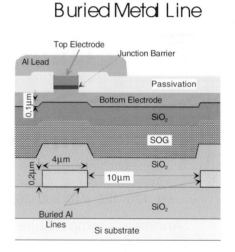

Fig. 19.10. Tunnel junction integrated on top of 1.2 μm buried Al lines.

has a damaged diode, and another a shorted junction. This example shows only some of the reliability problems that exist when moving from prototype level to a full memory device.

## 19.3    Other Sensor Applications; Current Monitoring, Position Control, Bio–Molecular Recognition.

Magnetic field sensors are widely used in a variety of less quoted applications encompassing field, current, position and speed monitoring [45]. Depending on the range of field to be detected, linearity, offset, and temperature requirements, several device categories are available. Hall effect devices [46], flux gates [47],

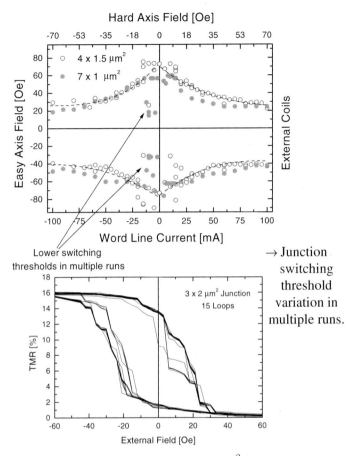

**Fig. 19.11.** Asteroid curve for a $3 \times 1$ μm$^2$ junction.

anisotropic magnetoresistance sensors [48] and giant magnetoresistance sensors (both spin valve and sensors based on AF coupled multilayers) [49–53]. Wheatstone bridge configurations with 4 sensors are usually used to reduce offset and increase output. Compared with anisotropic magnetoresistance sensors, spin-valve sensors, have almost one order of magnitude higher signals, and both cover linear ranges up to few kA/m. GMR multilayer sensors have higher linear ranges, and are promising candidates for detecting magnetic fields in a range of 3–100 kA/m, where they compete with Hall effect sensors but have order of magnitude higher outputs. For Wheatstone bridge sensor architectures, the success of the final device depends on how well the 4 GMR or spin valve sensor resistances are matched, how equal in amplitude are the opposite biasing magnetic fields applied to contiguous sensors, and how equal are their transfer curves under an applied field. Here, a new GMR sensor configuration, with 4 active GMR elements [53], biased by integrated permanent magnets is presented.

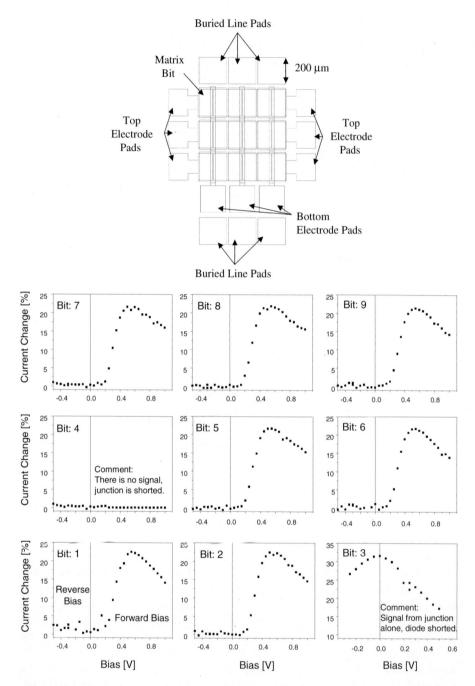

**Fig. 19.12.** Switching characteristics of the elements of a 3 × 3 bit memory, where the diodes are 200 × 200 μm large, and junctions are 3 × 1 μm². The graphs show the current change in % as a function of bias voltage.

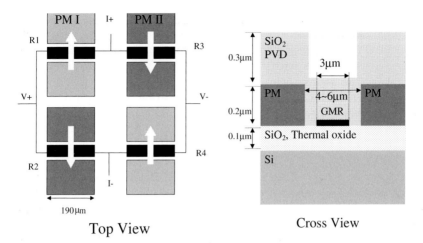

**Fig. 19.13.** GMR bridge sensor biased by integrated permanent magnets.

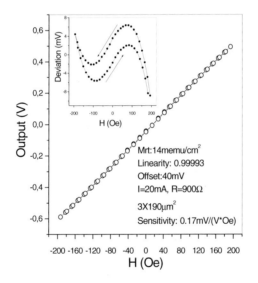

**Fig. 19.14.** GMR bridge output, as a function of an uniform external applied field.

Figure 19.13 shows the schematic diagram of the first type of GMR multilayer bridge sensor where pairs of permanent magnets (PM's) with the same $M_r t$ value (remanence × thickness product), but different coercivities ($H_c$) create the biasing fields. For our particular application, a linear range of 10 to 15 kA/m was required for the bridge. Therefore, permanent magnets were designed, in order to shift the individual GMR elements transfer curves by about ±16 kA/m. The GMR multilayer was chosen to have a linear range in excess of 40 kA/m. As shown in the top view, CoPt PM's having higher coercivity $H_{c1}$ are placed on both sides of two GMR elements (R1 and R4), while CoPt PM's with lower coercivities $H_{c2}$ are placed near GMR elements (R2 and R3). In order to create the required opposite biasing fields in contiguous GMR elements, the permanent magnets need to be magnetized after chip completion in opposite directions. The difference in coercivities must be large enough to meet this requirement. The magnitude of the biasing field is controlled by the Mrt value of the PM. Also shown is a cross section of the device, with indication of all critical dimensions. Figure 19.14 shows the type I bridge output for a device where the linear range is ±16 kA/m. Each single sensor has a dimension of $3 \times 190$ μm$^2$. Total bridge resistance is 900 Ω. The gap between the two CoPt permanent magnets is 5 μm. For a 20 mA current flowing through the bridge, an output of ±0.6 V is obtained, leading to a sensitivity of 2.12 mV/(V kA/m). Bridge offset is 40 mV. The linearity deviation is below ±1% (see inset). The hysteresis observed in the linearity deviation versus field curve comes from a 80 A/m hysteresis in the bridge transfer curve.

One application of this bridge design is given below. The precision placement of objects in moving transport trays is a common problem in automatic assembly lines. A test jig was fabricated where a robot needs to tightly place an object in a recessed box, precision machined in a tray. The uncertainty in the tray position (coming from the transport conveyor belt) is few mm from a preset value. The tray carries two small ferrite dipole magnets with dipole axis perpendicular to the tray plane, creating fields of few kA/m with cylindrical symmetry, few mm away from the tray surface. For successful placement and removal of objects from their places in the tray, the relative position of the robot grip with respect to the tray must be know to less than 100 μm. This was achieved by designing and fabricating a dual GMR bridge sensor, capable of measuring two orthogonal field components, with high spatial resolution [54]. The sensor is incorporated in the robot grip, and is scanned over the preset magnet positions that need to be known only within an area of 0.5 cm$^2$. The robot (from SONY, see schematics in the inset of Fig. 19.15b can then zero in to the "zero field" position, finding the absolute X-Y position of the reference magnets within 10 μm in each direction. The GMR elements in the two bridges are rotated by 90 degrees, and are sensitive to the field component perpendicular to the sensor length.

The integrated permanent magnets biasing the two bridges have different co-ercivities (48 kA/m and 96 kA/m respectively) allowing fringe field setting after chip fabrication. Individual bridge sensitivity in each direction is 4.37 mV/(V kA/m). Figure 19.15a shows the sensor spatial resolution for two orthogonal field directions, measured for a sensor-tray separation ($d$) of 6.5 mm and for a sensor-tray separation ($d$) of 3.5 mm (17 μV/μm). These values are given at the linear response region of the bridge, a region of about ±5 mm around the center of the magnet. In both cases a scan of ±5 cm over the magnet was realized in steps of 500 μm. The bridge drive current was 9.5 mA. Figure 19.15b shows the result of an overnight positioning test in one direction (X) ($d = 3.5$ mm), where the robot performed sequentially 1900 positioning cycles. The robot made a maximum error in positioning of 11 μm (180 μV), which is actually close to its minimum step. The insensitivity of the positioning to sensor offsets and elec-

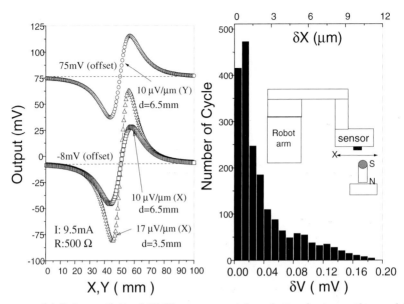

**Fig. 19.15.** (a) (left panel) Dual GMR sensor spatial resolution for two orthogonal field directions, measured for a sensor-tray separation $(d)$ of 6.5 mm and for a sensor-tray separation $(d)$ of 3.5 mm (17 μV/μm). (b) (right panel) Result of an overnight positioning test in one direction (X) $(d = 3.5$ mm), where the robot performed sequentially 1900 positioning cycles.

tronic or thermal drifts in a time scale much larger than one positioning cycle (10 to 20 s), stems from the positioning algorithm; the sensor determines its offset voltage away from the magnets, in the grip home position (earth + environment fields), and then goes over the magnet and positions itself until the measured voltage over the magnet equals the home-position offset voltage (it should coincide with the position with zero magnet fringe field). In summary, the fabricated sensor successfully allows the positioning of objects on a transport tray, with an absolute accuracy of 10 μm tested over more than 1900 cycles, insensitive to thermal or electronic offset drifts (test done along one direction only).

A second application of this new generation of magnetoresistive sensors, is the recognition of interactions at the level of bio-molecules (proteins, DNA strands, etc.), that have been previously labeled with a magnetic tag [55]. Here, the bio-molecule is tagged with a polymer particle, of radius varying from 100 nm to few μm, containing magnetite. The total particle moment is of the order of $10^7$ to $10^8$ $\mu_B$. Figure 19.16 shows results of first experiments at INESC where NANOMAG–D magnetic nanoparticles from Micromod Partikeltechnologie, Germany, are clustered near Al lines patterned on glass or Si substrates. These particles chemically bond to the required bio-molecules, and the group is then magnetically or chemically anchored to a site on a chip, which contains a 2D field-imaging array, consisting of micron size or sub-micron size magnetoresistive elements. Both tunnel junctions and spin valves are being evaluated as

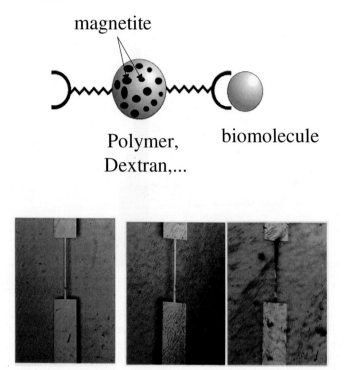

**Fig. 19.16.** Magnetic nanoparticles, with biologically active coatings, clustered on top of current carrying Al lines. A 2D magnetoresistive sensing array is being fabricated to detect particle immobilization, and at a second stage, perform nanoscale molecular recognition.

sensing elements. The goal is to be able to detect when a second bio-molecule attaches to the anchored one, involving a differential field measurement, in a somewhat noisy environment. This will allow detection of molecular recognition in nanometer size volumes, a feat up to now not possible.

So far only spin valve and GMR sensors have been discussed. Sensors based on spin tunnel junctions are also possible, profiting from the higher signal, and larger resistance of the sense element. However, S/N ratio must be addressed now. Three main noise sources exist in magnetoresistive sensors (AMR, GMR, and TMR). First, $1/f$ noise (DC to 10 kHz), that also exists in non-magnetic systems is dominant at low frequencies. Magnetic systems show an excess $1/f$ noise [56] that depend on the applied field and sense current, with large excess noise occurring in the high sensitivity points of the $R$ vs $H$ sensor transfer curves. This excess $1/f$ noise was correlated with thermal fluctuations of the magnetic moment of the free magnetic layer, and decreases once a large field is applied, reducing down to the non-magnetic metal level for saturating fields. At higher frequencies, Johnson noise (in resistors) and or shot noise (in tunnel junctions) become dominant. The thermal resistance noise (Johnson) is caused by the thermal smearing of the Fermi–Dirac electron distribution function near the Fermi level. The shot noise in metal/insulator/metal structures is caused by the discrete nature of the charge traversing the barrier. The noise limits the ultimate resolution of magnetoresistive sensors, and is basically independent of the magnetic structure. A fourth noise source of thermal origin is the magnetic domain configuration instability, leading to wall displacement, and sudden jumps in the magnetic moment of the sensor. This noise source is avoided by proper biasing schemes, insuring that no domain walls are ever present in the active region of the magnetoresistive sensor – something easier said than done! Barkhausen noise is also present in flux guides and shields, and appears then convoluted with sensor noise. Apart from these noise sources, and in particular for analog sensors, thermal drift is a major enemy (about 0.25% change in resistance per K). Wheatstone bridge configurations minimize thermal drift substantially (more than one order of magnitude). In the end, the MR sensor is integrated in a complete microsystem, with its own temperature and voltage offset drift corrections.

## 19.4   Conclusions

For areal densities in data storage to increase beyond 150 bit/$\mu$m$^2$, critical advances are needed in magnetic media, and improved read head sensitivity is required, calling for more than a "simple" scaling down of existing head technology (dual inductive thin film-and spin valve head). Achieving low noise, stable media is probably one of the most critical issues required to go beyond 60 to 80 bit/$\mu$m$^2$. As bit densities increase, bit sizes shrink, and the magnetic signal becomes weaker, to the point where the grain structure becomes a significant noise source. Grain volumes can be reduced to the point where they can no longer hold a magnetic charge, and will spontaneously demagnetize (magnetostatics and thermal activation). This thermal instability, called the superparamagnetic limit results in data loss. Both perpendicular recording and patterned media may be ways of overcoming the superparamagnetic limit that will affect presently-known longitudinal media above 150 bit/$\mu$m$^2$. Increase in media anisotropy is also a solution, limited by progress in high Bs materials for write heads, or recurring to thermally assisted writing [35]. Spin valve heads are becoming a mainstream product, but their performance, and in particular head sensitivity, must improve as densities move towards and beyond 150 bit/$\mu$m$^2$. Low resistance, spin dependent tunneling heads may compete with spin valves above 150 bit/$\mu$m$^2$.

As far as MRAM is concerned, the next few years will show whether they can become a real alternative to established non-volatile technologies (Flash, FERAM), and, if they can address part of the market taken today by conventional high capacity CMOS memories (DRAM, SRAM). In other applications, such as current, field, and position monitoring, the allowable signal to noise ratio, and ultimately the cost, will determine which sensing element to use.

# References

1. Data Storage, p.16, August 1999.
2. D. Weller, A. Moser, L. Folks, M. E. Best, W. Lee, M. F. Toney, J. U. Thiele, and M. F. Doerner, IEEE Trans. Magn. **36**, 10 (2000).
3. S. K. Khizroev, M. Kryder, Y. Ikeda, K. Rubin, P. Arnett, M. Best, and D. A. Thomson, IEEE Trans. Magn. **35**, 2544 (1999).
4. H. C. Tong, X. Shi, F. Liu, C. Qian, Z. W. Dong, X. Yan, R. Barr, L. Miloslavsky, S. Zhou, J. Perlas, P. Prabhu, M. Mao, S. Funada, M. Gibbons, Q. Leng, J. G. Zhu, and S. Dey, IEEE Trans. Magn. **35**, 2574 (1999).
5. T. D. Howell, P. E. McEwen, and A. Patapoutian, J. Appl. Phys. **87**, 5371(2000).
6. R. I. Potter, IEEE Trans. Magn. **10**, 502 (1974).
7. J. Smit, Physica **16**, 612 (1951); T. R. McGuire and R. I. Potter, IEEE Trans. Magn. **11**, 1018 (1975); O. Jaoul, I. A. Campbell, and A. Fert, J. Magn. Magn. Mater. **5**, 23 (1977); L. Berger, AIP Conf. Proc. **34**, 355 (1976); L. Berger, P. P. Freitas, J. D. Warner, and J. E. Schmidt, J. Appl. Phys. **64**, 5459 (1988).
8. N. Smith, IEEE Trans. Magn. **23**, 259 (1987).
9. C. Tsang, R. E. Fontana, T. Lin, D. E. Heim, V. S. Speriosu, B. A. Gurney, and M. L. Williams, IEEE Trans. Magn. **30**, 3801 (1994).
10. B. Dieny, V. S. Speriosu, S. S. Parkin, B. A. Gurney, D. R. Wilhoit, and D. Mauri, Phys. Rev. B **43**, 1297 (1991).
11. E. Heim, R. E. Fontana, C. Tsang, V. S. Speriosu, B. A. Gurney, and M. L. Williams, IEEE Trans. Magn. **30**, 316 (1994); P. P. Freitas, J. L. Leal, L. V .Melo, N. J. Oliveira, L. Rodrigues, and A. T. Sousa, Appl. Phys. Lett. **65**, 493 (1994); J. L. Leal, N. J. Oliveira, L. Rodrigues, A. T. Sousa, and P. P. Freitas, IEEE Trans. Magn. **30**, 3031(1994).
12. R. Coehoorn in "Giant magnetoresistance in exchange-biased spin-valve layered structures and its application in read heads", (ed.) U. Hartmann, Springer, Berlin 1999.
13. M. Lederman, IEEE Trans. Magn. **35**, 794 (1999).
14. "Spin valve magnetoresistive sensor with self pinned laminated layer and magnetic recording system using the sensor", K. R. Coffey, B. A. Gurney, D. E. Heim, H. Lefakis, D. Mauri, V. S. Speriosu and D. R. Wilhoit, U.S. Patent No. 5 583 725, 10 December 1996; H. van den Berg, W. Clemens, G. Gieres, G. Rupp, W. Schelter, and M. Vieth, IEEE Trans. Magn. **32**, 4624 (1996).
15. A. Veloso, N. J. Oliveira, and P. P. Freitas, IEEE Trans. Magn. **34**, 2343 (1998).
16. H. N. Bertram, IEEE Trans. Magn. **31**, 2573 (1995).
17. R. E. Fontana, S. A. McDonald, H. A. Santini, and C. Tsang, IEEE Trans. Magn. **35**, 806 (1999).
18. T. C. Anthony, J. A. Brug, and S. Zhang, IEEE Trans. Magn. **30**, 3819 (1994).
19. A. Tanaka, Y. Shimizu, H. Kishi, K. Nagasaka, H. Kanai, and M. Oshiki, IEEE Trans. Magn. **35**, 700 (1999); see also Ref.4.
20. W. F. Engelhoff, T. Ha, R. D. Misra, Y. Kadmon, J. Nir, C. J. Powel, M. D. Stiles, R. D. mcMichael, C. L. Lin, J. M. Sivertsen, J. H. Judy, K. Takano, A. E. Berkowitz, T. C. Anthony, and J. A. Brug, J. Appl. Phys. **78**, 273 (1995); H. J. Swagten, G. J. Strijkers, P. J. Bloemen, M. M. Willekens, and W. J. de Jonge, Phys. Rev. B **53**, 9108 (1998).
21. Y. Kamiguchi, H. Yuasa, H. Fukuzawa, K. Koui, H. Iwasaki, and M. Sahashi, Digest DB-01, presented at INTERMAG 1999, Korea; see also H. Sakakima, M. Satomi, Y. Sugita, and Y. Kawawake, J. Magn. Magn. Mater. **210**, L20 (2000).

22. A. Veloso, P. P. Freitas, P. Wei, N. P. Barradas, J. C. Soares, B. Almeida, and J. B. Sousa, Appl. Phys. Lett. **77**, 1020 (2000).
23. J. G. Zhu, IEEE Trans. Magn. **35**, 655 (1999).
24. J. L. Leal and M. Kryder, IEEE Trans. Magn. **35**, 800 (1999).
25. M. Mao, S. Funada, C. Y. Hung, T. Schneider, M. Miller, H. C. Tong, C. Qian, and L. Miloslavsky, IEEE Trans. Magn. **35**, 3913 (1999); H. Fuke, K. Saito, M. Yoshikawa, H. Iwasaki, and M. Sahashi, Appl. Phys. Lett. **75**, 3680 (1999).
26. A. Devasahayam and M. Kryder, IEEE Trans. Magn. **35**, 649 (1999).
27. V. Gehanno, P. P. Freitas, A. Veloso, J. Ferreira, B. Almeida, J. B. Sousa, A. Kling, J. C. Soares, and M. F. da Silva, IEEE Trans. Magn. **35**, 4361 (1999).
28. B. A. Gurney, V. S. Speriosu, J. P. Nozieres, H. Lefakis, D. R. Wilhoit, and O. U. Need, Phys. Rev. Lett. **71**, 4023 (1993); H. Iwasaki, H. Fukuzawa, Y. Kamiguchi, H. N. Fuke, K. Saito, K. Koi, and M. Sahashi, DIGEST BA-02, presented at INTERMAG 99, Korea.
29. V. S. Speriosu, B. A. Gurney, D. R. Wilhoit, and L. B. Brown, DIGEST, presented at INTERMAG 96.
30. A. Veloso, P. P. Freitas and L. V. Melo, IEEE Trans. Magn. **35**, 2568 (1999); A. Veloso and P. P. Freitas, J. Appl. Phys. **87**, 5744 (2000).
31. Q. Leng, M. Mao, C. Hiner, L. Miloslavsky, M. Miller, S. Tram, C. Qian, and H. C. Tong, IEEE Trans. Magn. **35**, 2553 (1999).
32. H. Kanai, M. Kanamine, K. Aoshima, K. Norma, M. Yamagishi, H. Ueno, Y. Uehara, and Y. Uematsu, IEEE Trans. Magn. **35**, 2580 (1999).
33. J. Moodera, J. Nassar, and G. Mathon, Annu. Rev. Mater. Sci. **29**, 381 (1999).
34. S. S. P. Parkin, K. P. Roche, M. G. Sammant, P. M. Rice, R. B. Beyers, R. E. Scheurlein, E. J. O' Sullivan, S. L. Brown, J. Bucchiganno, D. W. Abraham, Y. Lu, M. Rooks, P. L. Trouiloud, R. A. Wanner, and W. J. Gallagher, J. Appl. Phys. **85**, 5828 (1999); R. C. Sousa, J. J. Sun, V. Soares, P. P. Freitas, A. Kling, M. F. da Silva, and J. C. Soares, Appl. Phys. Lett. **73**, 3288 (1998); S. Cardoso, V. Gehanno, R. Ferreira, and P. P. Freitas, IEEE Trans. Magn. **35**, 2952 (1999).
35. J. J. Ruigrok, R. Coehoorn, S. R. Cumpson, and H. van Kesteren, J. Appl. Phys. **87**, 5398 (2000).
36. J. J. Sun, K. Shimazawa, N. Kasahara, K. Sato, S. Saruki, T. Kagami, O. Redon, S. Araki, H. Morita, and N. Marsuzaki, Appl. Phys. Lett. **76**, 2424 (2000); P. P. Freitas, S. Cardoso, R. Sousa, W. Ku, and R. Ferreira, *to be published* in IEEE Trans. Magn. (Proceedings of INTERMAG 2000).
37. G. B. Granley and T. Hurst, in "Sixth Biennial IEEE International Non-Volatile Memory Technology Conference Proceedings", p.138 (1996).
38. S. Tehrani, E. Chen, M. Durlam, M. de Herrera, J. M. Slaughter, J. Shi, and G. Kerszykowski, J. Appl. Phys. **85**, 5822 (1999).
39. R. C. Sousa, P. P. Freitas, V. Chu, and J. P. Conde, Appl. Phys. Lett. **74**, 3893 (1999).
40. R. C. Sousa, V. Soares, F. F. Silva, J. Bernardo, and P. P. Freitas, J. Appl. Phys. **87**, 6382 (2000).
41. S. Cardoso, P. P. Freitas, C. de Jesus, P. Wei, and J. C. Soares, Appl. Phys. Lett. **76**, 610 (2000).
42. S. Cardoso, R. Ferreira, P. P. Freitas, P. Wei, and J. C. Soares, Appl. Phys. Lett. **76**, 3792 (2000).
43. J. Shi, S. Tehrani, and M. R. Scheifein, Appl. Phys. Lett. **76**, 2588 (2000).
44. R. H. Koch, S. G. Deck, D. W. Abraham, P. L. Trouilloud, R. A. Altman, Y. Lu, W. L. Gallagher, R. E. Scheurlein, K. P. Roche, and S. S. P. Parkin, Phys.

Rev. Lett. **81**, 4512 (1998); R. H. Koch, G. Grinstein, G. A. Keefe, Y. Lu, P. L. Trouilloud, W. J. Gallagher, and S. S. P. Parkin, Phys. Rev. Lett. **84**, 5419 (2000); R. C. Sousa, A. David, and P. P. Freitas, *to be published* in IEEE Trans. Magn (Proceedings of INTERMAG 2000).

45. J. Heremans, J. Phys. D: Appl. Phys. **26**, 1149 (1993).
46. H. Blanchard, PhD thesis, EPFL, Lausanne (1999) *unpublished*; H. Banchard, C. de Raad, and R. S. Popovic, Sensors and Actuators A **60**, 10 (1997).
47. P. Ripka and S. W. Billingsley, IEEE Trans. Magn. **34**, 1303 (1998); P. Ripka, Sensors and Actuators A **33**, 129 (1992).
48. B. B. Pant and D. R. Krahn, J. Appl. Phys. **69**, 5936 (1991).
49. J. Daughton, J. Brown, E. Chen, R. Beech, A. Pohm, and W. Kude, IEEE Trans. Magn. **30**, 4608 (1994).
50. J. K. Spong, V. S. Speriosu, R. E. Fontana, Jr., M. M. Dovek, and T. L. Hylton, IEEE Trans. Magn. **32**, 366 (1996).
51. E. W. Hill, A. F. Nor, J. K. Birtwistle, and M. R. Parker, Sensors and Actuators A **59**, 30 (1997).
52. P. P. Freitas, J. L. Costa, N. Almeida, L. V. Melo, F. Silva, J. Bernardo, and C. Santos, J. Appl. Phys. **85**, 5459 (1999); P. P. Freitas, F. Silva, N. J. Oliveira, L. V. Melo, L. Costa, and N. Almeida, Sensors and Actuactors A **81**, 2 (2000).
53. W. Ku, F. Silva, J. Bernardo, and P. P. Freitas, J. Appl. Phys. **87**, 5353 (2000).
54. W. Ku, P. P. Freitas, P. Compadrinho, P. Alves, and J. Barata, to be published in IEEE Trans Magn (Proceedings of INTERMAG 2000).
55. D. Baselt, G. Lee, M. Natesan, S. Metzger, P. Sheenan, and R. Colton, Biosensors & Bioelectronics **13**, 731 (1998).
56. R. J. Veerdonk, P. J. Belien, K. M. Schep, J. C. Kools, M. C. Nooijer, M. A. Gijs, R. Coehoorn, and W. J. de Jonge, J. Appl. Phys. **82**, 6152 (1997); M. Xiao, K. Klaassen, J. Peppen, and M. Kryder, J. Appl. Phys. **85**, 5855 (1999).

# Index